Boundary Layer Flow over Elastic Surfaces and Combined Method of Drag Reduction

Boundary Layer Flow over Elastic Surfaces and Combined Method of Drag Reduction

Viktor V. Babenko
Ho Hwan Chun
Inwon Lee

AMSTERDAM • BOSTON • HEIDELBERG • LONDON
NEW YORK • OXFORD • PARIS • SAN DIEGO
SAN FRANCISCO • SINGAPORE • SYDNEY • TOKYO
Butterworth-Heinemann is an imprint of Elsevier

Butterworth-Heinemann is an imprint of Elsevier
225 Wyman Street, Waltham, MA 02451, USA

First edition 2012

Notice

No responsibility is assumed by the publisher for any injury and/or damage to persons or property as a matter of products liability, negligence or otherwise, or from any use or operation of any methods, products, instructions or ideas contained in the material herein. Because of rapid advances in the medical sciences, in particular, independent verification of diagnoses and drug dosages should be made

British Library Cataloguing-in-Publication Data
A catalogue record for this book is available from the British Library

Library of Congress Cataloging-in-Publication Data
A catalog record for this book is available from the Library of Congress

ISBN: 978-0-12-394806-9

> For information on all Elsevier publications
> visit our web site at books.elsevier.com

Typeset by MPS Limited, Chennai, India
www.adi-mps.com

Printed and bound by CPI Group (UK) Ltd, Croydon, CR0 4YY
Transferred to digital print 2012

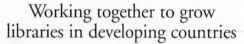

Contents

Preface

Until recently, investigators considered two types of boundary layer, the laminar boundary layer and the turbulent boundary layer, for which theoretical and experimental methods are well developed. It can been seen from the papers of Schubauer and Skramstad, Klebanoff and Lin that such separation is rigorously admissible only up to the region of instability and in the region of turbulent flow. The intermediate region—the transitional boundary layer—began to be investigated about 40 years ago.

Due to the development of visualization methods, specific coherent vortical structures were revealed and investigated in the area of the transitional boundary layer. Later it was shown that disturbance motion in this layer is continuously transforming. The motion, beginning with a two-dimensional Tollmien—Schlichting wave, passes through a number of specific stages. Each stage has a certain coherent form.

After that, a great deal of attention was devoted to revealing coherent structures in turbulent boundary layers. Many experimental investigations determined three types of specific coherent vortical structure across the boundary layer—in the outer region, at the boundary of the buffer layer and in the near-wall region of turbulent boundary layer.

We have shown that in the near-wall area of turbulent boundary layer coherent vortical structures in the longitudinal direction and their laws of variation are the same as in the transitional boundary layer under the influence of various worsening factors.

Simultaneously, with classical research on a solid smooth plate, research on different conditions of flow—surface roughness, strong outer turbulence, presence of mass- and heat-transfer and so forth, was also conducted. Methods of flow control were also widely studied; the most intensively examined was the method of polymer solution injection into a boundary layer.

Now considerable attention is being paid to experimental research on coherent vortical structures (CVSs) in turbulent boundary layers, and to various methods of controlling the revealed vortical structures. Numerical modeling of the laws determined in experiments has been successful, particularly that using the full Navier—Stokes equations.

Morkovin and Reshotko developed a new scientific direction—the receptivity of a boundary layer. The task in the beginning was to define the conditions at which unstable disturbances of the Tollmien—Schlichting-type wave exist in a boundary layer.

We have developed a new direction—the hydrobionic approach to solving hydrodynamic problems. Based on hydrobionic research, the laws of development and interaction of disturbances in a boundary layer and the methods of their active control have been elucidated.

Hydrodynamic modeling of experimental results on live high-speed sea animals has been conducted. It has made it possible to systematize the early results of modeling experiments in order to develop methods of boundary layer CVS control.

Practically all the vortical structures known in hydromechanics have been systematized. These structures occur due to the appearance of large shear stresses in a flow. For the majority of vortices the form and character of CVS development is identical. Classification of CVSs by their kinematic parameters has been performed and three kinds of CVS have been revealed. To I group the small disturbances acting on the disturbing movement of a boundary layer causes concern. Disturbances to group II are large, acting on all of the boundary layer. Such disturbances can arise, for example in angular areas of design. Group III disturbances are the largest arising at the fulfillment of maneuver and non-stationary movements on acting parts of a body and on lateral areas of a body (see Section 1.12). The analysis of the character of interaction of these three kinds of CVS is given. These three groups are systematized in 12 characteristic types of CVS.

The purpose of this monograph is, basically, to review the experimental investigations on various methods of boundary layer CVS control. The majority of examples of such methods are conducted by means of elastic coverings that model the skin coverings of high-speed sea animals.

In order to tackle this problem, it is necessary to investigate the boundary layer CVS on a rigid smooth plate, first of all, and then and on elastic plates. For this purpose, the corresponding hydrodynamic facilities and equipment have been developed. The basic results of the experimental research of hydrodynamic stability and the development of the initial stages of a transitional boundary layer, of a boundary layer on rigid and various kinds of elastic plates, are detailed in the book by Kozlov and Babenko (1978) [305]. Experimental research of the transitional boundary layer and some results of the receptivity problem are presented in [310].

In the present book, some results of laws of CVS development in the boundary layer over rigid and elastic plates are presented. Experiments have allowed a number of new laws to be discovered. The laws of development of disturbing movement in a liquid are obtained. This has made it possible to define laws and reasons for the development of disturbances at all stages of transition. In the boundary layer on a rigid plate, linear flat fluctuations are practically absent, and all disturbances concentrate near the surface in the flow over elastic plates.

The results obtained in earlier experiments have made it possible to undertake a wide program of research on the problem of receptivity. A new understanding of the problem of receptivity is presented here, based on new fundamental representations of CVS types. It is shown that the laws of interaction of various disturbances revealed in a boundary layer are the basis for the development of CVS-control methods.

Experimental results have led the to the confirmation that disturbances in the boundary layer over a rigid plate have characteristic features of a nonlinear asymmetrical wave-guide. The means of creating complex wave-guides in the form of elastic plates of a special design and a boundary layer in the flow on such plates are shown. It has made it possible to offering a hypothesis of a new method of boundary layer CVS control.

Experimental research of laminar, transitional and turbulent boundary layers on various kinds of elastic plates modeling dolphin and swordfish skin coverings were first conducted. Elastic material modeling was performed and criteria of similarity are determined. Mechanical static and dynamic characteristics of elastic composite coatings were measured. Corresponding techniques and devices were developed for these measurements. The data was analyzed according to the boundary layer CVS types. Measurements were done in air and water flows. Fluctuation spectra of a surface under the action of turbulent boundary layer pulsations have been investigated. All kinematic parameters of boundary layers in flows over rigid and elastic plates have been obtained.

Special facilities and equipment have been developed for determining the integral characteristics of boundary layers. The maximum speed of towage of a model in water was about 25 m/s.

Various combined methods of drag reduction based on the revealed physical laws of the boundary layer CVS and their interactions were experimentally investigated and then developed. Features of the structure of dolphin and swordfish skin were first explored. The findings of research on the influence of a xiphoid tip, injection of polymer solutions, elastic coatings and generated longitudinal vortical structures on hydrodynamic resistance of model are all represented.

Various projects for the application of the methods of CVS control are considered, including modern, state-of-art uses. Methodological issues relating to these experimental investigations are also discussed.

This monograph builds on existing books by the authors: [66, 305, and 310].

The supervisor for research on hydrobionics and flows on elastic surfaces was Professor L. F. Kozlov, Head of the Department of Boundary Layer Control and Hydrobionics, who has promoted the solutions of numerous practical problems for many years. His advice and ideas relating to the discussions of experimental results encouraged the wide development of the tasks considered in this book. We are sincerely grateful to Professor Kozlov and deeply regret his premature death.

The authors express profound gratitude to V. T. Grinchenko, Director of the Institute of Hydromechanics at the National Academy of Sciences of Ukraine, for his constant help and discussion of the manuscript.

Between 1993 and 1994 the research of the Professor V.V. Babenko was sponsored by the International Science Foundation, G. Soros, (Grand number of principal investigators UAW000 and UAW200).

Between 1995 and 1997 Professor Babenko conducted research relating to the contract made between the Institute of Hydromechanics NANU, Kiev, Ukraine, and the Cortana Corporation, through its President KJ Moore (Task 11 and 12 under DARPA Delivery Order 0011 of Contract MDA972-92-D-0011). The outcome of this cooperation was not only scientific reports; Babenko has received seven American patents. In 1998–1999, Babenko conducted hydrobionic research in Wissenschaftskollegs zu Berlin, Institute for Advanced Study, Berlin. In 1999 Babenko was supported for participation in the conference in Newport (Grant Number N 00014-98-1-4040 Office of Naval Research International File Office—Europe Special Programs Assistant Code 240, 223 Old Maryellen Road, Visiting Support Program, support for providing financial support for visiting the Naval Undersea Warfare Centre). In 2005–2006 Babenko worked as a visiting professor at the Advanced Ship Engineering Research Centre (ASERC, Busan, Korea) at Pusan National University, with the director, Professor H. H. Chun.

All these activities promoted research in various directions, and, as a result, it was possible to formulate the basic direction of research and to systematize the results obtained in the present book.

The authors would like to express sincere gratitude to the specific persons mentioned and to the organizations for the financial help in this research. They would especially like to thank K. J. Moore, President of the Cortana Corporation, for his creative cooperation and some very important and productive discussions. The authors also thank Dr Natalie Rozumnyuk, Institute of Hydromechanics of NASU, for her help in editing and discussing the book.

The authors believe that their findings will promote an improved knowledge in the field of turbulent wall flows.

List of Symbols

x, y, z longitudinal, normal and transversal axes of a Cartesian coordinate system

x_1 valid coordinate of a place of measurement

$\bar{y} = y/\delta$ dimensionless vertical coordinate

b width; transversal distance between elastomer layers in composite

h width of the hydrodynamic channel; height of a disposition of the lower edge of an ejector above a streamlined surface; thickness of the water layer under a membrane; thickness of a flexible surface

l, L length; l — designate length besides a symbol l in various parts designate length of a way of mixing scales of turbulent vortices in the viscous sub-layer.

S wetted surface; area of a crack

F wetted area

s thickness of a slot of an ejector

d, D diameter

R radius; resistance force $-R_{turb} = 0.62 \cdot \rho \cdot (U^2/2)\, d \cdot l$

δ thickness of a boundary layer

δ^* theoretical value of displacement thickness

δ_m^* measured value of displacement thickness

δ^{**}, δ_2 momentum thickness

$H = \delta^*/\delta^{**}$ formparameter of a velocity profile

$\bar{\delta}_{lam} = 4.64 \left(\nu \frac{x_0}{u_0} \right)^{1/2}$, $\bar{\delta}_{lam} = \alpha \frac{\nu}{\sqrt{\tau_w/\rho}}$ viscous sublayer thickness

$\alpha = \frac{\bar{\delta}_{lam} u_\tau}{\nu}$ empirical constant

k, K_s height of element of a roughness; k — spring constant; k — kinetic energy of turbulent fluctuation motion; $k = 2\pi f/u = 2\pi/\lambda$-wave number

$K_{lim} = K_S U_\infty/\nu$ parameter of roughness

c_a, C_L lift coefficient

C_f, c_f, c_w, C_{xi} coefficient of resistance

λ, $C_f = 2\tau_w/\rho U_\infty^2$ local skin friction coefficient

$C_F = \frac{2P}{S\rho U_\infty^2}$, $C_F = \frac{1}{x}\int_0^x C_f dx' = 2\frac{\Theta}{x}$ complete friction coefficient or the mean friction drag coefficient

Δy and Δz cross and transversal amplitudes of disturbing oscillation

A_{vibr}, A_v amplitude of oscillation of a vibrator tape

A_{do} amplitude of disturbing oscillation

$\varepsilon = \frac{\sqrt{\frac{1}{3}(\overline{u'^2} + \overline{v'^2} + \overline{w'^2})}}{U_\infty}$; T_u degree of turbulence

$\bar{k} = E/\varepsilon$ wave-guide amplification coefficient

$E = \sqrt{u'^2/U_\infty}$ maximal value longitudinal pulsating velocity

t time

u, v **and** w longitudinal, normal and transversal component of a velocity of the resultant flow

U, V **and** W longitudinal, normal and transversal component of a velocity of the basic flow

u', v' **and** w' longitudinal, normal and transversal component of a disturbing flow

$\sqrt{\bar{u}'^2}, \sqrt{\bar{v}'^2}, \sqrt{\bar{w}'^2}, \overline{u'v'}, \ldots$ averaged over time components of a turbulent pulsation of velocity

$\frac{2\overline{u'v'}}{U_\infty^2}$ Reynolds stress

τ_0 tangent pressure on a wall

u_H value of a longitudinal component velocity on the exterior boundary of a boundary layer

ψ flow function

$q = \rho V^2/2$ dynamic pressure

p pressure of the basic flow, $\sqrt{\overline{p'^2}} = 0.5\rho U_\infty^2 \, \mathrm{Re}^{-0.3}$

P resulting pressure

p' pressure of the disturbing flow

$\frac{2\sqrt{\overline{(p')^2}}}{\rho U_\infty^2} = \alpha_\tau \lambda$ energy of pulsations of pressure p'

α_τ coefficient of proportionality

$C_{p,\bar{P}i} = \frac{P - P_\infty}{\frac{1}{2}\rho U_\infty^2}, \bar{p} = \frac{p}{\rho u_H^2}$ dimensionless pressure; coefficient surface-pressure distributions

$\alpha = \frac{2\pi}{\lambda}$ wave number of disturbing oscillation; $k = 2\pi f/u = 2\pi/\lambda$-wave number

f **and** λ frequency and length of the wave harmonic

u mean velocity in the point of measure

λ length of the oscillation wave

λ_x disturbing wavelength of longitudinal vortices in direction of x axis

$\tilde{\lambda}_x = \lambda_x/\delta^*$ dimensionless length of the oscillation wave

λ_z three-dimensional disturbing wavelength in direction of z axis

$\beta = \beta_r + i\beta_i$ complex frequency of disturbing oscillation

$\beta_r = 2\pi n$ circular frequency

$n = 0.159 \frac{\beta_r \nu}{U_\infty^2} \frac{U^2}{\nu}$ frequency of disturbing oscillation

Δn frequency span of disturbing oscillation

$T = \frac{1}{n}$ cycle of vibration; viscous sublayer regeneration period; integral limit scale; elastomer tension

β_i increase coefficient

$c = \frac{\beta}{\alpha} = c_r + ic_i$ complex velocity of disturbing movement

c_r distribution velocity of waves of disturbing movement

c_i increase coefficient

$\alpha\delta$ dimensionless wave number of disturbing oscillation

$\frac{\beta_r v}{U_\infty^2}$; ω_r dimensionless frequency of disturbing oscillation

$\frac{c_r}{U_\infty}$ dimensionless velocity of distribution of disturbing movement

μ dynamic viscosity coefficient

$\nu = \mu/\rho$ kinematic viscosity coefficient

$\rho = \gamma/g$ density of a liquid (mass of unit of volume)

γ specific gravity (weight of unit of volume)

g gravitational acceleration

ρ_m density of a material of a flexible surface

Re Reynolds number calculated on the x

Re$_\delta$ Reynolds number calculated on thickness of a boundary layer

Re* Reynolds number calculated on the displacement thickness

Re** Reynolds number calculated on the momentum thickness

Re$_1$ $= \frac{U_\infty}{v}$ single Reynolds number

Nu Nusselt number

χ empirical constant of turbulent flow; $l = \chi y$

$\varepsilon_\tau = |\overline{u'v'}|/(|\partial U/\partial y|)$ turbulent viscosity coefficient, dissipation velocity of turbulence energy for unit mass

$k^+ = k\,u^+/\nu$ dimensionless kinetic energy of turbulent fluctuation motion

$P^+ = \frac{-\overline{vu'v'}}{u_\tau^4} \cdot \frac{dU}{dy}$ velocity of generation of turbulence energy in a turbulent boundary layer

$D^+ = \frac{v^2}{u_\tau^4} \cdot \left(\frac{dU}{dy}\right)^2$ velocity of dissipation of turbulent energy of the averaged motion of a turbulent boundary layer

y^+, $y^* = y\,u_*/\nu$ dimensionless distance from a wall

u/u_* dimensionless velocity

u_τ, $v_* = \sqrt{\tau_w/\rho}$ dynamic velocity; friction velocity; skim friction stress; dimensionless thickness of viscous sublayer

λ_z wavelength of the viscous sublayer longitudinal vortices

$\lambda_z^+ = \lambda_z u_\tau/\nu$ dimensionless wave length of the viscous sublayer longitudinal vortices

$T^+ = Tu_\tau^2/\nu$, $T^+ = \left(\frac{u_\tau \bar{\delta}_l}{4.64\nu}\right)^2$ period of bursting from a viscous sublayer of a turbulent boundary layer

$f^+ = (T^+)^{-1}$ frequency of bursting from a viscous sublayer of a turbulent boundary layer

$K_S^+ = K_S u_\tau/\nu$ dimensionless factor of roughness

$\overline{\tau} = \frac{\tau}{\rho u_H^2}$ dimensionless of shear stresses

τ shear stresses (force on unit of the area)

τ_w shear stresses on a wall, $\tau_w = \mu\left(\frac{\partial u}{\partial y}\right)_0$

Π Coel's parameter

ΔB additive constant

$R_{u'u'(0,r,0)} = \frac{u_1' u_2'}{\sqrt{\overline{u'}_1^2}\sqrt{\overline{u'}_2^2}}$, $R_{u'u'(0,r,0)}, R_{u'u'(r,0,0)}, R_{u'u'(0,0,r)}$ correlation factor, autocorrelation factor

$R_{u'u'(0,r,0,t)}$ space−time correlation

$R(r) = \int_0^\infty E(k)\cos krdr$ correlation function

$E(r) = \frac{2}{\pi}\int_0^\infty R(r)\cos krdr$ spectrum function

$\Phi(\omega); E(k) = \frac{F(\omega)u}{2\pi\Delta f}$ spectral functions

Δf infiltration band-pass filter width

$L = \int_0^\infty R_{(0,r,0)}dr, L = \int_0^1 R_{(0,r,\delta,0)}d(r/\delta)$ integral scale or macro scale of the turbulence

$R_i = \dfrac{u_1'(\xi)u_1'(\xi + x_i)}{\sqrt{u_1'2(\xi)u_1'2(\xi + x_i)}}$ coefficients of transverse correlation between the longitudinal

   components of the velocity fluctuations

$R_i(\tau) = \dfrac{\overline{u_i'(t)u_i'(t + \tau)}}{\sqrt{u_i'^2}}$ autocorrelation coefficients of longitudinal fluctuation velocity

$Ca = \rho U_\infty/E$ Cauchi parameter

θ corner of shift of phases between a stress and deformation

σ stress of visco-elasticity material

$E = \sigma/\varepsilon$ module of elasticity of an elastic material

ε elastomer strain; relative cross deformation of polymeric materials, apparent kinematic viscosity of turbulent flow

t_i, h thickness of a flexible surface

t_m thickness of a membrane

H thickness polyurethane sheet

$c = \sqrt{E/\rho}$ group velocity of the elastic wave

$\lambda = c\tau$ wavelength

c_R phase velocity of Rayleigh wave

c_t phase velocity of a flat cross-section wave

$C_m = \sqrt{(T/M)}$ phase velocity of oscillation propagation, velocity of the forced fluctuations on a dolphin's skin surface

$\omega = 2\pi f$ circular frequency, limiting frequency of fluctuation of an elastic material

$\Omega_i(\omega)$ areas of spectral functions $\Phi(\omega)$ of fluctuation of the boundary layer

$\Psi\Delta_i(\omega)$ energy dissipated in an elastic plate; areas of fluctuation energy of the boundary layer, absorbed by the plate at different modes of a flow

K_w coefficient of oscillation frequency of elastic surface

$Z = \dfrac{\sqrt{\overline{p'^2}}}{\overline{a'^2}}$ complex coefficient of rigidity

$\sqrt{\overline{a'^2}}$ amplitude of the surface oscillations

$A(f)/\sqrt{\overline{a'^2}}$ oscillation spectrums of elastic surface

$p(f)/\sqrt{\overline{p'^2}}$ spectrums of pressure fluctuations on the elastic plates

$G^* = G' + iG''; E^* = E' + iE''$ complex modulus

$tg\ \delta = G''/G = \Delta/\pi(1 - \Delta^2/4\pi^2)$ dissipation factor, loss angle tangent

$\Delta = ln(A_n/A_{n=1})$ logarithmic damping decrement

$\psi = \frac{\Delta W}{W} = \frac{2\pi}{Q} 2\pi\tau g\delta$ loss coefficient

ΔW energy, absorbed by a unit of the material volume

W energy saved by the unit of the material volume

Q parameter, characterizing the amplitude increasing with resonance

$K = 1 - h/h_0$ parameter of damping properties; damping coefficient

h_0 fall height

h recoil height

$K = -P/\varepsilon_v$ volumetric module

$\varepsilon_v = \Delta V/V$ three-dimensional strain

V volume of a elastomer specimen

$I(\tau) = 1/E(\tau); J(\tau) = 1/G(\tau)$ elastic compliance of an elastic material

η lateral of the wall; elastomer stringiness; viscosity of an elastic material; coefficient of absorption small ball beat, damping factor

$\tau = -\eta/E$ elastomer relaxation time

ξ longitudinal strain of the wall

T_F tension of an element of an elastic material

M, m oscillating mass of an element of an elastic material

$S, E' = E/t_i$ rigidity of an elastic material

P_i generalized force, size of compressing force

$\mu = \overline{\varepsilon}/\varepsilon$ Poisson coefficient

$\overline{\varepsilon}, \varepsilon$ relative cross-strain and relative longitudinal strain

$\nu = 1/\mu$ cross (lateral) number

$\tau^* = G\gamma$ shear stress

G module of rigidity

$\gamma = \frac{\Delta l}{l} = tg\beta$ relative shear of upper side of the unit cube

$\lambda = l/l_0$ strain in the direction of stress line

l_0 length of the edge of unstrained cube

$\omega_{a.res.} = \sqrt{4k/m_1}$ anti-resonance frequency

$f_P = \frac{U_\infty}{2\pi\delta}$ natural frequency of membrane vibration

ζ factor of a damping

$K_1 = E_{dynam.}/E_{static.}; k_2 = E_{long.}/E_{transv}$ anisotropy factor

$\delta_i = ln\left(\frac{Ai}{Ai+1}\right)$ logarithmic decrement

$K_a = \frac{h_i - h_{i+1}}{h_i} 100$ absorbing factor

A_i and A_{i+1} amplitudes of the preceding and the following wave

h_i and h_{i+1} fall height and recoil height of indentor

D diffusivity of the polymeric component

C concentration of polymer in a solution

c_Q; c_q; $C_q = Q/US$ discharge coefficient of polymer solution
$Q = V/t$ discharge of polymer solution
U towing velocity
V volume of an injected liquid
$C_\mu = 2Qu_{sl}/\rho U^2 \cdot S = 2\,Cq \cdot \rho_c \cdot u_{sl}/\rho U$ factor of quantity of movement
$\xi(\mathbf{Re}) = (C_{xrig.} - C_{xi})/C_{xrig.}$ coefficient of relative changing in friction
u_{sl} velocity of injection of a liquid through a slot
ΔC_f, $\% = 100 \cdot (C_f - C_{f\,inj}) \cdot C_f$ efficiency of drag reduction as a percentage

Indices

Lower

t parameter of turbulent motion
H value of a parameter on the exterior boundary of a boundary layer
∞ in approach flow
$0, w$ value of a parameter on a on a wall
Σ total value
n nasal part
los. st.; s.l. loss of stability
cr. critical
m middle
eff effective value of a parameter
i number of a node of a difference grid along a coordinate x
j number of a node of a difference grid along a coordinate y
o parameter in initial cross section of a boundary layer (at $x = x_o$)
max maximum value of a parameter
sl slot
ppm parts per million

Upper

$+$ nondimensional parameter in the correspondence with the law of a wall;
$\sim$ result of the first stage of the numerical method;
$\approx$ result of the second stage of the numerical method

Abbreviations

BL boundary layer
TBL turbulent boundary layer

T−S Tollmien−Schlichting wave
CVS coherent vortices structures
LVS longitudinal vortices systems
LDVI laser Doppler measuring instrument of velocity
LEBU large-eddy breakup device
CC continuous coating
CCF current-carrying fabric
LS longitudinal strips
RF rubber film
FP foam polyurethane
FL foam latex
PPU penopoliuretan
PU polyurethane
FE foam elastic
SAA same as above
DISA thermo anemometer
PEO polyethylene oxide (WSR-301)
OT ogive tip
SXT short xiphoid tip
LXT long xiphoid tip

CHAPTER 1

Interaction of the Free Stream with an Elastic Surface

1.1 Introductory Remarks

In the middle of the twentieth century the publication of three pioneering articles triggered a large worldwide trend in drag reduction research, both experimental and numerical. In 1948, Toms [528] was the first to report significant drag reduction at high Reynolds-number flow of a polymer solution, after whom this phenomenon is named. To date, drag reduction by polymer injection has remained an active topic from both an academic and practical viewpoint. In the 1960s, Kramer published a series of articles on achieving essential drag reduction for longitudinal streamlined cylinders with a surface elastic coating [311−318]. His designs of elastic coatings were inspired by the structure of the epidermis of dolphins. Since then, vast amounts of experimental and theoretical research on the flow over elastic surfaces has been done all around the world, either to reproduce or to understand the drag reduction effect. In the 1930s, after one decade of research, Gray [213] proposed the celebrated Gray's Paradox. He calculated that a dolphin's swimming speed is approximately 10 times the speed that could be developed by its muscular weight. This led to much biological research aimed at either substantiating or refuting Gray's assertion.

The combination of these three ideas gave rise to a novel research direction named hydrobionics. It would be nearly impossible to describe the numerous achievements in research field. Therefore, only the major research trends obtained at the key institutions will be mentioned here. Regarding polymer injection drag reduction, extensive research has been carried out in the Applied Research Laboratory (ARL) at Pennsylvania State University, USA. In the former Soviet Union research into this field was performed by the Academy of Sciences of USSR in Novosibirsk, Moscow, St Petersburg (all now in Russia) and Kiev and Donetsk (now in Ukraine). The major findings pertaining to the drag reduction mechanism are: 1) the necessity of preliminary stretch of polymer macromolecules, 2) the drag reduction effect resulting exclusively from the polymer solution in the near-wall region and 3) damping of the vertical structures by the grid-like structure of polymer macromolecules settled in the near-wall region. With regard to the development of elastic coatings for drag reduction, which is the main topic of this monograph, a review of the extensive research will be given. For the third field described above, the biological research, studies have been conducted in various countries, such as the USA, Germany, Ukraine and China.

Hydrobionics, the combined research of all three fields, was initiated in the Department of Boundary Layer Control and Hydrobionics at the Institute of Hydromechanics of the Academy of Sciences of Ukraine in 1965. Under the direction of Professor L.F. Kozlov a program of research has been conducted jointly with the Institute of Biology and Institute of Zoology of the Academy of Sciences of Ukraine and other organizations in the former Soviet Union. The results of the biological and hydrodynamic research led to the finding that the characteristics of dolphin skin are automatically regulated to reduce hydrodynamic drag. The results of this joint research between hydromechanical engineers, biologists and zoologists was recognized in 1982 as discovery N 265 [49] by the Academy of Sciences of the USSR.

Morphological studies on dolphin skin have led to the finding of specific structures and features of this skin. Various hypotheses concerning not only the epidermal structure but also other body organs have been put forward. One is that the microstructure of the external layer of the skin generates two-dimensional (2D) and three-dimensional (3D) disturbances in the boundary layer flow during the oscillatory swimming movement of body. Interaction between the disturbance and the boundary layer instability could generate a pressure fluctuation field in the near-wall region leading to a change in the boundary layer structure and causing drag reduction. Stimulated by this finding, the Institute of Hydromechanics carried out extensive experimental research on the interaction of boundary layer flow with various kinds of elastic plates simulating dolphin skin. The research topics were: the development of a natural laminar boundary layer, hydrodynamic stability, Goertler stability on curvilinear rigid and elastic plates, the turbulent boundary layer and the susceptibility of a boundary layer under various disturbances. These features were modeled as a susceptibility of various flows over elastic surfaces. The problem of susceptibility was explored in the work of Morkovin and Reshotko [366]. In the experience of Schilz [444], T−S waves were found to interact rapidly with the flat waves generated by a flexible wall.

Research on the drag reduction mechanisms of dolphins led to the development of the so-called "combined method of drag reduction." This involves the damping of the organized vortical structure of the boundary layer using an elastic coating [25, 29, 33, 34]. The body structure and organs have also been investigated for other fast-swimming hydrobionts, such as shark, tuna, swordfish, squid, amongst others. Morphological research of these animals has confirmed that they have evolved specific organ features that leads to drag reduction. Other kinds of "combined methods of drag reduction" have been proposed based on these other fast-swimming hydrobionts. For example, Babenko [34] has proposed a design with an elastic damping coating through which the polymer solutions are injected for drag reduction.

The purpose of the first chapter of this monograph is to review the major features of such combined methods of drag reduction with in-depth analyses on the flow—elastic-surface interaction; hydrobionic drag reduction mechanisms are described in the following chapters. Generally speaking, hydrobionic research has suggested a drag reduction mechanism caused by the attenuation of coherent vertical structures (CVSs). Development of flow visualization techniques and computational fluid dynamics (CFD) have revealed the existence and significance of CVSs in the generation and maintenance mechanism of turbulent boundary layer flows. The research on turbulent boundary layer flows from the latter half of the twentieth century has been mainly focused on the identification and control of CVSs. Therefore, it is natural to begin this chapter by introducing various types of CVS encountered in the flow around a body and turbulent boundary layer flows (see Sections 1.2 and 1.3). After that the fundamental findings on the flow over elastic surfaces are described in Sections 1.4 and 1.5. Sections 1.6 to 1.8 deal with the issues of laminar—turbulent transition of boundary layer flow over a rigid plate. Discussions on the interaction between flow and elastic surfaces, such as the hydrobionic principles of drag reduction, boundary layer receptivity and CVS control are given in Sections 1.9 to 1.15.

1.2 Basic Types of Coherent Vortical Structures arising in the Flow about a Body, and Methods of their Control

The formation and interaction of vortices of various scales is observed everywhere, in the atmosphere and the ocean, in airflows around buildings, in various vegetation canopies on the ground and in water, in flows over various deepenings and barriers. Significant experience from experimental and theoretical research of such problems has been collected. One of the basic problems is the lack of understanding of the reasons and laws of appearance of various sorts of vortical structures. Rockwell systematized and analyzed some types of vortical structures [437]. Below the basic types of CVS and methods of their control are listed.

1.2.1 CVS Types

CVSs in a Boundary Layer

Based on much experimental research, the results of which are presented in articles and proceedings of conferences, etc. [4, 79, 173, 174, 178, 185, 277, 337, 388, 429, 430, 452, 475, 489], the basic concepts of the characteristic kinds of CVS in transitional and turbulent boundary layers (BL) have been developed. Basic CVS types in a transitional boundary layer are analyzed in papers [66, 305, 310, 277, 475]. Nickel and Schonauer [380] were the first to visualize the initial stages of the development of disturbances in a transitional boundary layer over a flat surface. Brown [135], and Knapp and Roache [279], using smoke

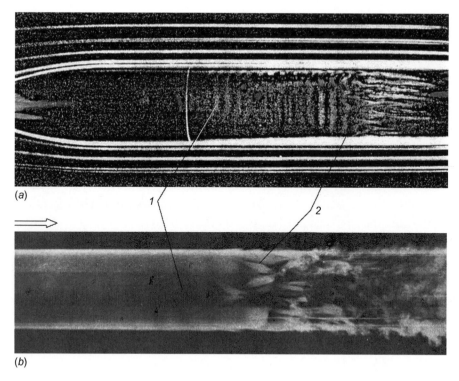

Figure 1.1

Smoke-flow visualization of the boundary layer of a body of rotation. Photograph by Brown [135]; (*a*) and Knapp and Roache [279]; (*b*): *1* — Tollmien-Schlichting waves; *2* — λ-shaped vortices.

visualization in a wind tunnel photographgraphed various stages of the transitional boundary layer over a body of revolution (see Figure 1.1).

The paper [372] provides visualization pictures of CVSs in transitional boundary layers of a flow around a motionless body of a shell-form rotating along the longitudinal axis. Similar visulization pictures have been obtained for rotating cones [496]. Stuart [475] first elucidated the features of the disturbance movement at various stages of development of a transitional boundary layer. Based on the authors' own experimental research and the results of others [66, 305, 310] the breadboard model of development of CVSs in a transitional boundary layer has been developed. For all 2D or 3D bodies, the law of CVS development in a boundary layer is identical.

CVSs in a turbulent boundary layer have also been well investigated. Original pictures of visualization of CVSs in turbulent boundary layers can be found in [4, 79, 126, 173, 174, 277, 337, 388, 429, 430, 452, 489], as well as similar research done by other authors.

A detailed analysis of CVS structure in a turbulent boundary layer was performed in [496, 105] and visualization pictures of CVSs by various authors shown. Longitudinal CVSs in the viscous sublayer of a turbulent boundary layer obtained by Klein with use of hydrogen bubbles are presented in [277]. Srinivasan (in 1987) has generalized the findings of various authors. Visualization pictures of large-scale CVSs across turbulent boundary layers are shown in [496]. Of special interest are pictures of CVSs in turbulent boundary layers obtained in the transverse direction to the flow speed. In this work hairpin-shaped large CVSs were revealed, the spacing of which spacing was sometimes greater than that of Klein's vortices in the viscous sublayer. As Reynolds number increased, the kind and size of these vortices changed. A correlation of the longitudinal CVSs in the viscous sublayer with the longitudinal CVSs on the external border of the turbulent boundary layer was found for the first time.

CVSs in the Flow over a Curvilinear Surface

Taylor and Goertler did fundamental work on the problem of flow in the presence of centrifugal forces. Taylor focused on the problem of flow in the gap between cylinders, and Goertler concentrated on flow over curvilinear surfaces. Over the past 80 years this area of research developed broadly and found wide practical use. Pictures of visualization of Goertler vortices formed on concave surfaces are presented in [496]. The authors' experiments have shown [66, 305 and 310] that the properties of liquids are such that the laws of transition are similar for any kind of flow. As can be seen from Figure 1.2a, Goertler vortices form in the same way as shown in the model of CVSs of a transitional boundary layer [310], owing to the alteration of various CVSs at corresponding stages of transition. Figure1.2b shows horseshoe-shaped vortices arising during the development of Goertler vortices. It is characteristic that these vortices are similar to vortices in a transitional boundary layer on a flat plate and to Klein's vortices in the viscous sublayer of a turbulent boundary layer. Furthermore, the reasons for and the laws determining the development of hairpin-shaped vortices are similar to those for the development of the transitional boundary layer on a flat plate [66, 305,310].

The authors' experiments [310] have shown that the Goertler vortices that formed on a concave surface still exist after the surface changes to a horizontal plate and is influenced by the flow very far from the concave section. These results have been confirmed by experiments with a profile in a wind tunnel, when the leading part of the profile had a concave section [282]. Similar measurements have been made by Chomaz and Perrier [489, p. 79] in a water flow at a velocity 10 m/s. The radius of curvature was 15 cm. The authors found that Goertler vortices began to form on the concave surface and continued develop on the convex surface.

Using this effect of stabilization of a boundary layer by Goertler vortices, basic research on designing a class of laminarizing profiles of type X63T18S with a low drag coefficient has

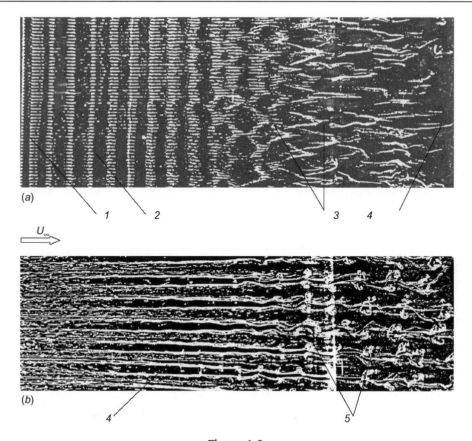

Figure 1.2

Transition on a concave wall visualized in a water towing tank. Photograph by Bippes [116]: *a* — under conditions of natural transition, R = 0.5 m, G = 9 (at the position of transition), *b* — with artificial excitation of streamwise vortices by means of regularly spaced heaters on the surface, R = 1 m, G = 9 (at the right end of the photographs). *1* — wire for generating hydrogen bubbles; *2* — Tollmien-Schlichting waves; *3* — λ-shaped vortices; *4* — Goertler vortices; *5* — horseshoe-shaped vortices at meander of Goertler vortices.

been undertaken [402]. As in [282], the profile from the pressure head side had concavity in the leading and tail parts. Goertler vortices formed over the concave leading part, stabilizing the boundary layer and shifting the zone of the turbulent boundary layer. Over the tail part, the profile concavity contributed to a reduction of separation zone of the boundary layer at zero and negative angles of attack. At positive angles of attack, Goertler vortices prevented separation on the profile trailing edge from below and stabilized the boundary layer on the upper side.

Skang investigated of the features of Goertler vortex formation on cones with cone angles between 30° and 80° [428, pp. 64–65]. Measurements were taken in a vertical set up, where an axisymmetric water jet flowed from the top. The flow was changing from a laminar to a turbulent regime. A cone was placed at the jet axis and the structure of flow on its surface was watched. Due to the rapid change of the jet's direction when it approached the cone steady Goertler vortices formed over the cone.

The influence of the leading parts of a body of revolution on its drag coefficient was investigated in [103]. Various shapes of leading parts, which had concave sections like the profile, were considered. Distributions of pressure essentially differed from those with standard forms of leading parts of bodies of revolution. Shapes with forward protrusions, like that of swordfish or dolphins, and with piecewise curvature, like some birds, for example a petrel or penguin, were considered. The pressure distributions obtained reduced body drag.

CVSs Generated in Cavities on a Body Surface

Transversal CVSs

Research on various sorts of 2D transversal cavities has been performed for around 50 years. Visualizations of such cavities are presented in [496]. Gorban investigated various kinds of cylindrical transversal cavities using the discrete vortices method [210, p. 15]. Recently much attention has been paid to investigating the influence of various shapes of leading and trailing edges of such cavities. Sine-wave channels, which represent a system of cavities that generates transversal vortices, have also been investigated. Much recent experimental research has also dealt with 3D cavities. As in [271], the authors' measurements have shown that such vortices are not axisymmetric along the z-axis. Inside cavities these vortices form closed zones in a transverse direction, between which secondary vortices appear. As a consequence, when the necessary conditions are met, a non-continuous vortex is thrown from a cavity. It bursts from the cavity in separated zones in a transverse direction; therefore, 2D cavities generate zigzag vortices, which reshape further into longitudinal vortices [174].

Three-dimensional cavities, dependent on mean flow velocity, generate disturbances of various kinds. As the velocity increases such a cavity generates a pair of longitudinal vortices with spacing equal to the cavity diameter.

Longitudinal CVSs

Extensive research has been devoted to special kinds of longitudinal structures called "riblets," which result in CVSs generated in the boundary layer. One of first researchers to look at this was Dincelacer (Germany) [166], who has drawn an analogy between the structure of shark scales and riblets. Tullis and Pollard (Canada, 1993) have published a theoretical investigation of the characteristics in the viscous zone of turbulent boundary layer of V- and U-shaped riblets.

Choi performed experimental research on this subject in the UK [489, pp. 146—160]. The characteristics of a turbulent boundary layer were studied with use of a heat-loss anemometer. Patterns of longitudinal vortices in the viscous sublayer were obtained in flows above various forms of riblets by smoke visualization.

In USA, Wallace and Balint obtained experimental data on the flow over riblets in the same way as Bechert [489, pp. 134—147]. Numerical investigations are presented in [151].

Detailed research has also been performed in Japan. In the late 1970s, Nakamura investigated the structure of turbulent boundary layer over rectangular riblets in a wind tunnel. In the 1990s, Kasagi [478] performed a large cycle of research on the characteristics of flow and patterns of streamlets, and velocity vectors above riblets were established [120, 184]. Bechert in Germany [475, pp. 278—285] has performed detailed research on various forms of riblet. A special channel filled with oil in order to increase Reynolds number was used during the modeling of shark scales. Bechert made riblets that copied shark scales, as well as various kinds of simple riblet, which modeled elements of shark scales. The optimal designs were added to elements of the wing of a model Dorn 328 and tested in a big wind tunnel. Drag reduction of 5.4% was obtained on the wing profile NASA/3M, and 10.2% on the profile DLR/HFI. Research in this direction continues.

Tornado-shaped CVSs

Nikulin [383] and Rlemp proposed a model of formation of a geophysical tornado-shaped vortex. The role of the tornado-like function of Kline vortices is discussed in [37].

Kiknadze and Krasnov [267] in their experimental research of 3D cavities revealed that the cavity generated tornado-shaped vortices. The authors' measurements have shown that generated tornado-shaped CVSs reshape into longitudinal CVSs after they leave the cavity. The authors have developed a method of generating tornado-shaped vortices by means of a cavity and obstacles of a specific structure; it was established this it possible to generate steady tornado-shaped CVSs in a boundary layer.

CVSs Generated in Front of or behind an Obstacle

Transversal CVSs

CVSs behind interceptors on wings of planes have been researched for some 60 years. The latter research has analyzed the influence of an interceptor on turbulization of the boundary layer [305, 101]. Various kinds of obstacle were investigated: rectangular protrusions or a wire in the transverse direction, and square or cylindrical pins spaced in a row in the transverse direction. The flow pattern over a vertical cylinder is presented in [445]. Such tripping devices are necessary for model tests in shipbuilding. Significant attention has been paid to the role of such obstacles for the intensification of heat exchange [522]. Bandyopadhyay also investigated the influence of small obstacles on drag reduction. For this purpose he placed transverse rectangular obstacles with the spacing such that it generated specific waves in the viscous zone of the turbulent boundary layer. The obstacles did not protrude from the viscous zone. Thin wires have also been used for the same purpose, with one placed on the wall, the next near the wall, and so on. Smalley et al. performed similar research in 2001, but in this work the micro-roughness was greater than $k = 60$. Later Bandyopadhyay repeated his investigation, but he "pulled down" the obstacles into the plate to create cavities, which generated a wave in the viscous sublayer [101]. In 1969 Babenko revealed similar structures on the surface of dolphin skin [4, pp. 3.1–3.14]. A numerical model of CVS in the flow around rectangular pins staggered on a plate has also been developed [174, p. 283].

Longitudinal CVSs

Based on hydrobionic research Babenko developed a method of controlling the CVSs in a boundary layer by means of longitudinal elements located under or on the surface of a plate [4, pp. 3.1–3.14]. The findings are presented in [4, pp. 9.1–9.24; 66; 305; 388, pp. 635; 430, pp. 113–120]. The influence of thin wires aligned on a wall in a longitudinal direction on the characteristics of a turbulent boundary layer was described in [120]. The spacing between the wires was approximately equal to that of CVSs in the viscous sublayer of a turbulent boundary layer. There are classical experiments by Shubauer and Skramstad [447] on measurements of neutral fluctuations of hydrodynamic stability. In 1960, Skramstad published an article in which he described various types of vortex generators for research in a boundary layer [144, pp. 230; 135]. Similar vortex generators were used in experiments to avoid separation on the top surface of wing profiles at large angles of attack. Now similar generators of vortices are mounted on the majority of of passenger planes in order to create pairs of large longitudinal vortices.

Monti et al. investigated the influence of longitudinal roughness on the characteristics of a turbulent boundary layer [3, pp. 286–293]. In a wind tunnel two types of plates were studied. On one, an aluminum plate, roughness of fish-scale type was etched by electrophoresis, i.e. sequential V-shaped elements 0.3 mm height. Flow velocity was 10–19 m/s. On the second aluminum plate similar roughness was obtained by means of epoxide resin. It was found that at some parameters of roughness and flow velocity longitudinal roughness can increase the drag coefficient (by 2–3 times) or reduce it (by 2 times).

CVS at Corners of a Body

During movement of real bodies large vortical structures appear at the joints of their construction, which influence both the boundary layer and the subsequent elements, for example, stabilizers or propulsors. It is known that a pair of large longitudinal vortices occurs in longitudinal flow above a rectangular corner (see Figure 1.3).

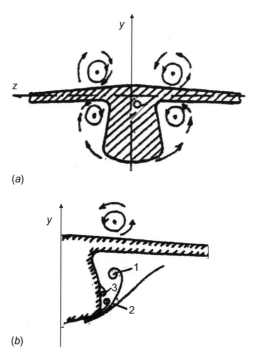

(a)

(b)

Figure 1.3
Schemes of formation of vortices at a flow on the wings of planes: a — cross-section section of the plane with vortical structures, b — origin of vortices on a sharp edge of a fuselage in area of a wing: 1 — the basic vortex; 2, 3 — secondary vortices.

Under certain conditions a pair of secondary longitudinal vortices can appear near the walls of a corner. In reality, corners consist of 2D or 3D surfaces. This leads to the formation of longitudinal vortices with a complex shape. A flow about a corner can also have a transversal direction [189]. The influence of various parameters of the shape of flow about a corner on formation of Goertler vortices has been analyzed. In all cases, a corner has a blunt point. When the corner has no curving radius at its cross-section flow in a corner steady vortices are formed [115]. Calculations were performed with the method of discrete vortices [208]. It is obvious that after such vortices are formed they distort the streamlines above and the conditions for natural formation of Goertler vortices set in [189].

CVSs in Longitudinal Flow about a Body

Recently large longitudinal vortices have been revealed in longitudinal flows over long bodies of revolution in natural conditions and in model experiments (see Figure 1.4). At a small angle of attack or longitudinal fluctuation, large vortices appear near lateral surfaces of long bodies. Research in this area continues [126, 144, 173, 174, 388, 429, 430, 445, 496].

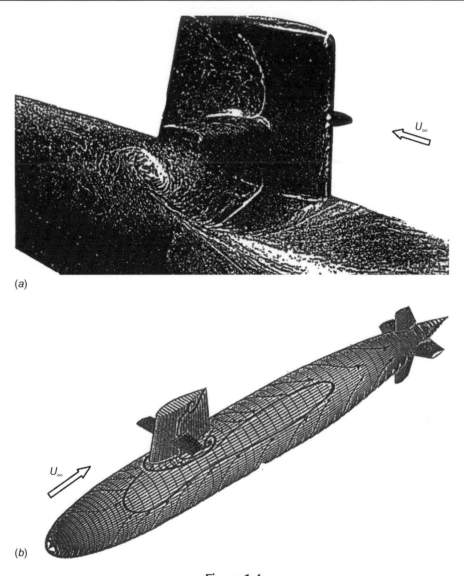

Figure 1.4
Visualization of vortices on long bodies of revolution (*a*), scheme of marked particle path (*b*),
see Farokhi, et al. [126].

It is known that ideally leading and tail parts of a body of revolution are shaped to avoid
the creation of separation vortices. However, because of design considerations it is often
necessary to shape them in another way. Shapes of separation bubbles near leading parts of
a blunt body were numerically calculated in [98, pp. 31–33]. Theoretical and experimental
data is available on the dependence of the drag coefficient on Reynolds number for two

shapes of a blunt body where the lateral surface of the cylindrical part is designed to be curvilinear to account for the separation bubble. Drag reduction occurred. Patterns of visualization of blunt leading and tail parts of a body are shown in [144, 480, 496]. At zero angle of attack separation zones appear near these parts. As the angle of attack increases the separation zones become longer and asymmetrical.

CVSs in Flows about Wings

Flow about a Profile at an Angle of Attack

Patterns of visualization of flows about a plate and a wing profile at an angle of attack can be found in [496]. It has been established that depending on the shape of a profile as the angle of attack grows separation can occur on the tail part of the profile, and on the leading part as well. When separation occurs on the leading part, closed vortices can arise. As the angle of attack increases, the separation area on the trailing edge grows and moves to the leading part of the profile. These vortices very strongly influence the lift force of a profile. At maneuvering, take-off and landing of an aircraft, it is necessary that the lift coefficient remains high at increasing angles of attack. Therefore, special arrangements are needed to avoid separation on a profile. Keeping lift at a greater angle of attack is also of great importance in shipbuilding.

CVSs at Tips of Obstacles and Wings

On a cross-section of a rectangular wing, even at zero angles, a vortex sheet separates from its rear cover and moves to the wing tips downstream. Vortical bunches form at the wing ends, which join the vortical sheet behind. As a result, a pair of tip vortices appears, which define the inductive resistance of the wing (see Figure 1.5). It is known that the magnitude

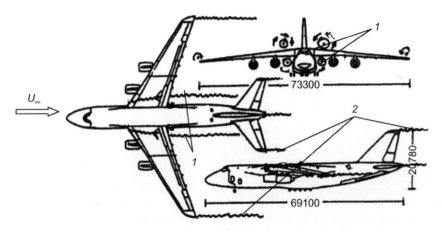

Figure 1.5
Scheme of formation of angular (*1*) and tip eddy (*2*).

of the resistance is inversely proportional to the wingspan. Angular vortices also form near rudders and the stabilizer of an airplane.

A vortex appears in the stagnant area behind an interceptor. Gorban [209] has performed calculations for CVSs behind various shape interceptors.

If an obstacle is aligned with a stream, as it is for a cabin or protruded stabilizers or rudders, longitudinal vortical bunches also form near their tips.

CVSs on a Triangular Wing

Triangular wings are widely used in engineering, especially on aircraft. In the flow about such a wing vortices typical of a rectangular wing occur near its tips. However, unlike a rectangular wing, these vortices occupy the greater part of the wing, forming the specific shape of vortices on a triangular wing [144, 496] (see Figure 1.6). In [337] results of experimental research on the flow over a triangular wings performed by Baxill and Nelson (pp. 25−94), Mario Lee and Chih-Ming Ho (pp. 365−428) are presented.

CVSs behind an Oscillating Wing

Betz investigated the behavior of gliders at fluctuation of the plane in 1912 [430, pp. 471−478]. In same work [430] Platzer et al. mentioned the research by Birnbaum (in 1925) in which he considered fluctuations of elastically fixed wing. In aerodynamics, much attention was paid to this problem in connection with the dynamics of flight during take-off, landing and maneuvering. Hydrobionic researchers [308] have widely considered the problem of fluctuations of a profile. Figure 1.7 shows the formation of vortices and a vortex trail behind an axisymmetric rectangular wing at a large amplitude of fluctuations [54]. There were different thin axisymmetric profiles with various chords b at various distances H from the channel bottom. The amplitude A of sine-wave fluctuation of profiles, frequencies of fluctuation f and flow velocity U_∞ were identical. Visualization was

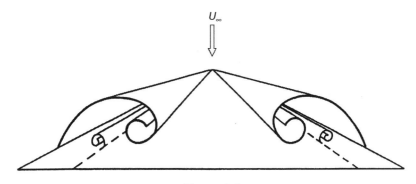

Figure 1.6
Formation of vortices in the flow about a triangular wing, Lee and Ho [337].

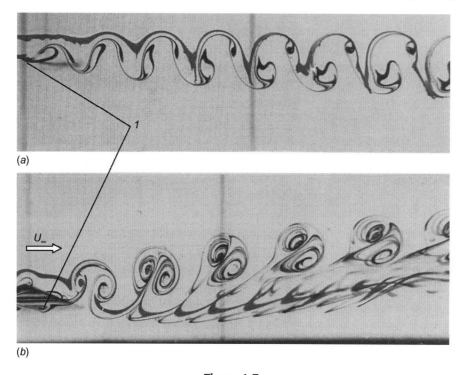

Figure 1.7
Formation of vortices for an oscillating surface (*1*) with a chord equal to 2 mm (*a*) and 25 mm (*b*) at $U_\infty = 0.3$ m/s, $A = 2$ mm, $f = 1$ Hz. For (*a*) $H > \delta$; for (*b*) $H = \delta$; Dovgiy [54].

achieved by means of a dye injected in the flow through thin tubes. Red dye was injected on the top and blue dye below. The degree of turbulence in the flow was $0.3\% U_\infty$. This made it possible to investigate the development of vortices a long distance from the fluctuating profiles. One can see the influence of the bottom and sizes of a profile chord on the form and intensity of vortices in the trail behind a profile. The greatest contribution to studying the problem of an oscillating wing has been made by Dovgij [4]. Platzer performed similar experiments [430, pp. 471–478]. A comparison of calculation and visualization of the flow over a wing when it rotates from 0° to 60° for 5.2 dimensionless time can be found in [212]. The patterns of visualization of an oscillating triangular wing with the NASA 0012 profile are shown in the work [196].

CVSs in the Flow between Lattices of Profiles

In flows in turbines, hydro transformers, compressors, pumps, multicontour metric screws and other cases when there is a lattice of profiles, zones of separation appear in both upstream and downstream parts of the profiles. Special arrangements are to be applied to avoid separation zones.

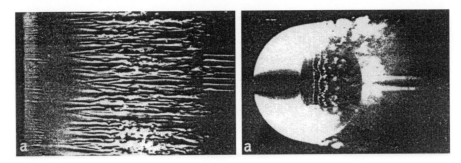

Figure 1.8
Visualization of the flow about a sphere at $Re_D \approx 10^5$, Werle [388, pp. 9–11].

CVS about High-drag Bodies

In addition to [144, 496] the paper [164] considers in detail flows and occurrence of vortices in separation zones about high-drag bodies. Patterns of visualization behind a cylinder, sphere and the end face of a cylinder, an oscillating cylinder and at its self-oscillations are shown. Integral characteristics are presented for flows about circular cylinders and various other forms, plates and other designs like a corner, flange beam, and so on.

Schmidt (in 1919) first investigated the resistance of a falling sphere. Taneda first published patterns of visualization of the flow about a sphere, cylinder, and oscillating plates and cylinders, and also behind two cylinders installed sequentially (the diameters of the cylinders were identical and changed simultaneously) [480]. At the natural development of the boundary layer on a sphere, CVSs, which are typical for other cases (see Figures 1.1 and 1.2), were observed (see Figure 1.8).

Williamson C. A. (J. Fluid MEch. 1996, V.328: 345–407) has published patterns of visualization of the flow about a very thin cylinder with a length-to-diameter ratio in the order of 1.0–4.0 and Reynolds numbers from 50 to 400. The diameter of the cylinder was 0.51 mm, 0.61 mm and 1.08 mm. It was revealed that at small Reynolds numbers Kármán vortex street appeared but was not observable, and at Re = 200 secondary vortical formations in the transverse direction with certain spacing were clearly visible. These secondary vortices arose on the crests of Kármán vortex street and had a structure similar to λ-shaped vortices in a transitional boundary layer. Similar secondary vortices were obtained in calculations by Mittal S. (Physics of Fluids. 2000, V13, No.1).

Hanchi [174, pp. 144–145] investigated the structure of vortices about a circular cylinder during start of movement. In the same reference [pp. 161] data is given for experiments for the wake of a cylinder in an oscillating flow. Visualization of vortices behind a cylinder in the vicinity of a wall are shown by Sumer et al. [388, pp. 101–103]. Brika D. Fluid Mech.

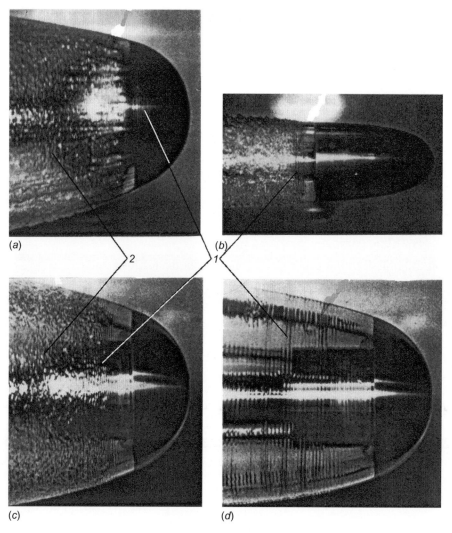

Figure 1.9
Visualization of the cavitation flow of three shapes: in sphere, $U = 35$ ft/s, natural cavity
(*a*); ogive, $U = 25$ ft/s, ventilated cavity (*b*); cut-away sphere, $U = 20$ ft/s, ventilated cavity
(*c*); cut-away sphere, $U = 12$ ft/s, ventilated cavity (*d*). 1 − Tollmien-Schlichting waves,
2 − nonlinear Tollmien-Schlichting waves, Brennen [133].

1993. V.250: 481–508), obtained visualization patterns of vortices behind a vibrating
elongated cylinder. Similar results were presented by Solnick R. L. et al. (TASME, 1957.
V.79, £% : 1043–1056).

CVSs in the flow about crosswise cylinders have also been investigated, crosswise located
steady and rotating disks.

Brennen [133] has photographgraphed the "water—air" boundary at a natural cavitation and ventilated cavitation in the flow about a sphere, truncated sphere and ogive head (see Figure 1.9). There were visible waves similar to Tollmien—Schlichting (T—S) waves, which deform as they develop into nonlinear waves and, further, into longitudinal vortical structures. At lower speeds of flow about a truncated sphere (see Figure 1.9d) T—S waves were observed on the water—gas boundary. Modeling of such flow has shown that T—S waves occur in the instability zone of the neutral curve [85]. At greater velocities of the flow about a truncated sphere T—S waves are quickly transformed into nonlinear CVS (see Figure 1.9b,c). Figure 1.8 shows that longitudinal CVSs in a combination with nonlinear forms of CVS are observed in front of and behind the separation area (see Figure 1.9a).

CVS Generated in Internal Flows

The previous sections have considered the sorts of CVSs arising in external flows. The present section describes various CVSs that can occur with internal flows. Much research has been performed on flows in curved pipelines. Large vortical formations have been revealed near the curve, the size and length of which depend on the radius of curvature [174, p. 153].

There is also considerable research on pipes with a variable section. An example is the work by Miyake Y. [ISME Intern. J. Ser.11, 1992:23—35], where vortical structures and characteristics of a turbulent flow in a channel with the sine-shaped bottom surface are presented. Other works have analyzed the flow in a channel where both walls were sine-shaped. Theoretical and experimental research supervised by Povh [424] has been performed at Donetsk University in a pipe made of consecutive convergent—divergent sections.

Last year's new direction of research on collapsible pipes has been developing [245]. CVSs in a Taylor flow between coaxial cylinders are shown in [445, 496]. CVSs visualized by Ludwig when an internal cylinder moves in the axial direction are shown in [445]. Recently experiments have been performed where the internal cylinder makes in azimuthally with sine law, or where the external cylinder is square, and so on. The paper [174, p. 300] presents findings for Taylor flow in coaxial conic vessels. It includes pictures of Taylor vortices in the gap between coaxial containers.

Batill and Nelson have performed extensive experimental research on the flow in a convergent tube. Visualizations of CVSs in the core of flow and in the boundary layer on walls of a convergent tube without and in the presence of mesh with various spacings are represented in [337, pp. 25—94]. Investigations of CVSs in diffusers have also been undertaken [496]. Of practical importance is the research on the flow structure in turbines.

The problem of swirled flow also attracts much interest. The vortical chamber is an example of such a flow. The paper [174, pp. 354—356] describes the results of Japanese research on the transportation of ceramic balls in a vortical tube. The results of research on flow in a system of "liquid—gas" inside a vortical chamber are presented in [98]. In some cases, as in [98], the investigation of characteristics of vortices on the end walls of a vortical chamber is of special interes. The shape of these vortices is similar to that observed on a rotating disk.

The characteristic shape of vortices on a rotating disk has a helicoid shape, and the structure of vortices can be laminar or turbulent, depending on Reynolds number. Patterns of such vortices are shown in [174, p. 210]. Carpenter performed similar research on the formation of vortices on rotating disks covered with elastic plates [174, p. 364].

CVS on Boundaries of Submerged or Near-wall Jets

Patterns of visualization of CVSs on boundaries of submerged jets [173, 496, 513 etc.] have now been obtained. The situation of when a submerged jet is injected (coaxially) in alignment with the direction of flow velocity or perpendicular to main flow has been investigated. CVSs in submerged two-phase jets have also been considered. For example, [174, p. 163] describes CVSs in a coaxial submerged jet filled with particles.

Two submerged jets injected perpendicularly to the main flow were investigated experimentally in [243]. The jets were spaced apart from each other either transversally or sequentially (tandem) relative to the main flow.

Patterns of visualization of CVSs in single and coaxial jets are given in [489, pp. 459–470]. Active and passive control of turbulent jets is described in [489, pp. 445–457]. The structure of the CVS shape when a nozzle is located on a wall is shown.

Investigations of submerged and near-wall jets are of great importance in various areas of engineering. Particularly important are the character and laws of CVSs of jets used in vortical chambers. Near-wall jets have great value in aircraft design. Shapes of CVSs in near-wall jets have been studied by Schober and Ferncholz [174, pp. 322; 173]. It was found that a wire placed inside a nozzle stabilizes vortices of the near-wall jet. If this wire is outside of the nozzle in the boundary layer of the plate its fluctuations leads to an essential increase in the size of CVSs in the near-wall jet.

To reduce drag of bodies moving in water, injection of polymer solutions in a boundary layer has been effectively used. In all cases, it is important to inject a near-wall jet into a boundary layer with minimal CVSs. Shapes of leading and rear edges of a nozzle were analyzed in [280] Based on the susceptibility method, the authors developed a method of injection of near-wall jets in a special form [126, 429].

A photograph of round axisymmetric submerged jets of water into atmosphere can be found in [236]. Primary and secondary instability waves in the flow from a nozzle are seen in Figure 1.10.

The approximate primary interval of the wave is 0.046 cm. Arrows indicate regions of prominent secondary waves. The development of disturbances on the boundary between

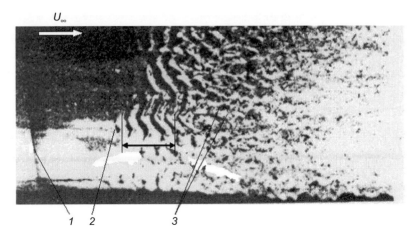

Figure 1.10

Formation of coherent vortical structures on the boundary of a submerged jet: *1* — nozzle tip, *2* — formation of a Tollmien-Schlichting wave, *3* — systems of longitudinal vortices. Arrows indicate the zone of deformation of a Tollmien-Schlichting wave [236].

two media is well pronounced. Behind the nozzle disturbances develop in the form of T–S waves, then other forms of disturbances occur, such as the development of disturbances in the transitional boundary layer on a plate or, in other cases, shift flows.

CVSs at Non-stationary (Forward, Rotary) Movement of a Body

Such vortical structures have been partly considered above, for example at fluctuation of a profile or plate. However, when fluctuations are non-stationary, the flow structure has a more complex form. This class of fluid flow still needs investigating further.

CVSs in Geophysical Problems

In most cases, geophysical problems involve similar CVSs as considered above. However, geophysical problems have a more complex character because dynamic, thermodynamic and non-stationary processes occur simultaneously. Figure 1.11 shows CVSs in the flow about a platform (experiments by A. V. Voskoboinick, 2006).

The large-scale CVSs that develop in the atmosphere or ocean are of special research interest.

1.2.2 Methods of CVS Control

Originally, methods of boundary layer characteristics control based on integral characteristics of a boundary layer were investigated [145]. In the process of studying CVSs

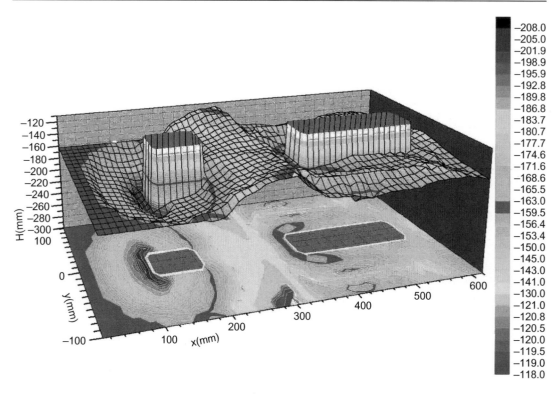

Figure 1.11

Experimental research of a washout of the obstacles base at the modeling support of the bridge.

in the boundary layer methods began to develop that specifically control the CVSs in the layer, rather than the entire layer. Nowadays methods of control of other kinds of CVS in shear flows are also well studied. Below some methods of controlling CVSs are given in relation to Section 1.2.1.

Methods of CVS Control in a Boundary Layer

The first methods of CVS control in a boundary layer were been developed to influence T–S waves. The first research in this direction was performed by Schilz [444]. Since then, plenty of similar research has been performed in different countries. In America the new scientific area for study, "susceptibility by a boundary layer to disturbances," was developed, in which the occurrence of T–S waves under influence of various external factors has been studied. Recently, active methods of T–S wave control in a system with feedback has been developed. The results of similar research are described in [3, 489]. A method of T–S wave control by means of an oscillating elastic surface was developed in [305].

Active control methods of a turbulent boundary layer were investigated in [3, 126, 204, etc.]. Such methods include:

- transverse fluctuations of a plate [150, 152] transverse fluctuations of flow above a motionless plate [146];
- longitudinal fluctuations of flow above a motionless wall [155, 251];
- normal periodic injection in the viscous sublayer through a porous strip placed across the flow [320], or suction of boundary layer through slots placed at a certain spacing along the length of the plate;
- CVSs on a moving wall, amongst others.

One of the methods of CVS control in the viscous sublayer of the turbulent boundary layer that has been extensively investigated is riblets [126, 161, 489]. The method consists of the use of riblets organized in a transversal direction; their spacing should correspond to the spacing of longitudinal vortices in the viscous sublayer. Thus, the vortices will be stabilized; the frequency of bursting from the viscous sublayer and the generation of turbulence will decrease. Hence, friction drag will also decrease. A reduction in drag of between 5–10% was achieved. In 1972 the authors designed elastic coatings in which riblets were located beneath a smooth surface, so called "reversed riblets" [33]. Such coatings were found to essentially reduce friction drag. Originally, riblets mimicked shark skin. However, an error was made; the skin of dead sharks had been studied and this differs from that of live sharks as its mucous cover is dried up. Live sharks and other high-speed fishes have a dense mucous cover over the scales, so their body surface is smooth.

Another well-investigated method of CVS control is LEBU [54, 100; 489, pp. 85–96]. For this, at a distance from a plate equal to 0.8 of a boundary layer thickness, thin profiles with a chord of 1–2 boundary layer thicknesses were installed sequentially. The spacing between profiles was $2–10\delta$. It is assumed that such profiles destroy and stabilize large CVSs on the external boundary of a turbulent boundary layer. Complex systems were developed, for example, LEBU in combination with riblets placed periodically along the x-axis.

Methods of CVS Control on Curvilinear Surfaces

Methods of CVS control on concave surfaces are based on changing the curvature of a surface, and on influencing longitudinal Goertler vortices [244]. Scientific substantiation for such approaches is given by Saric [4, pp. 8.1–8.85].

Much earlier, methods of influencing Goertler vortices had been developed using a small mechanical eddy-generator of special design on a curvilinear surface [310] and longitudinal wires with certain spacing in the transversal direction [430, pp. 113–120, pp. 319–325]. Heating the conductive wires regulated the intensity of influencing boundary layer. Longitudinal CVSs in air and water were influenced by convective fluxes. In water increased heating of wires created vertical layers of hydrogen bubbles essentially stabilizing longitudinal CVSs in the boundary layer.

In 1975 the authors developed another method of longitudinal CVS control using an elastic composite surface [100]. The external layer of elastic plates consisted of longitudinal elements with transversally varied mechanical properties installed with spacing that corresponded to the longitudinal CVSs in the boundary layer. Two longitudinal heating structures with similar spacing in transversal direction were embedded into the elastic surface. Therefore, this was a combined method of CVS control in the boundary layer: damping CVSs in the boundary layer by using elastic materials, strips of different densities in the transversal direction, and heating these strips.

Blackwelder [428, pp. 37−39] has suggested an original way of creating longitudinal "whirlpool" vortices from Goertler vortices and controlling their development with an oscillating profile.

Methods of CVS Control in Cavities

Bandyopadhyay investigated the possibility of resonant influence on a boundary layer with a system of transversal grooves in a surface [102]. Size and spacing were varied in the longitudinal direction. Thus, a wave fluctuation is generated that influences the boundary layer.

Mochizuki and Osaka [430, pp. 121−126] have researched the mechanism of drag reduction above a system of grooves similar to [163]. However, their design of grooves was different. The strips that made up the surface had longitudinal spacing equal to the cavity depth. Unlike [102], the strips were mounted not on the lateral walls of the groove but on the strips beneath, with the same longitudinal spacing. Thus, the surface was permeable, which reduced the velocity pulsations in the near-wall region of the turbulent boundary layer.

In [163], the plate had rows of round cavities covered with a film. Periodical pressure was generated inside the cavities and the surface periodically vibrated. The device in [240] had rows of open cylindrical cavities and plates like LEBU installed either above cavities or above the surface between cavities.

Methods of Controlling CVSs Generated behind a Prominence

In [522] vortices behind a prominence mounted transversally were controlled with straight or oblique slots in the top of the prominence. CVS control behind a prominence in [522, p. 336] was done in the following way. A thin slot was made in the plate at the distance $\Delta x/h = 1.75$ ahead of the prominence (where $h = 20$ mm is the obstacle height), parallel to the prominence. This slot was connected to a loudspeaker, which radiated fluctuations at various frequencies. The fluctuations were generated in the boundary layer in front of the prominence and were transferred by the flow behind the prominence.

Hupperitz and Ianke [173] influenced CVSs behind an obstacle on a plate the leading part of which was ogive-shaped with a ratio 5:1. Behind the leading edge there was a 20 mm-high obstacle, behind which a flat plate was placed. Just in front of the obstacle 105 small apertures were made and connected by a duct to a loudspeaker. At a frequency of 20 Hz the separation bubble was 63% shorter.

Such research is of great importance in aerodynamics, thermophysics, hydromechanics, etc. In shipbuilding, various forms of obstacles are widely used. Transversal redans lead to the formation of separation bubbles that essentially decrease drag. Various designs have been developed where ducts connect zones of underpressure behind redans with atmosphere. As a result, air is sucked from the atmosphere and separation bubbles grow. Juts in the form of half-cones are established on the streamline surface [378]. The apertures of the cone-faces are connected with the atmosphere. Longitudinal rows of bubbles appear as a result.

Methods of CVS Control in Corners

Various methods of control for this type of CVS are known. The most widespread is the use of a design of fairing the corners of which are rounded. For example, the rule of cross-sections is used on aircraft. It means that in the zone of junction, the area of the cross-section section must be equal to that in front of the junction and behind it. Therefore, a local depression in the body surface should be made. It is interesting that this rule is used nature—by whales and dolphins, etc., near their vertical fins.

Another method consists in "blowing-off" of the CVSs over a corner using a system of orifices for jets. These jets damp and eliminate the vortical structures.

CVS Control Methods in Longitudinal Flow about a Body

The experimental research of Farokhi, Taghavi and Barrett [126] has explored the reduction of large CVSs arising around a submarine model at maneuvering. Figure 1.12 shows such CVSs. The experiments were performed in a wind tunnel with a working section of 91.5×122 cm. Triangular inserts (vortex generators) were installed flush behind the cabin in the separation zone. The vortex generator was designed by Skramstad [144]. The vortex-generator design is such that when a separation zone occurs the triangular vortices generated are sucked out due to underpressure; rows of triangular wings were set in so that longitudinal vortical systems form behind the wings. Earlier, Bechert and Bannasch [110, 471] used rectangular plates with a similar purpose.

Another method was developed Blackwelder and Gad-el-Hak [121]. According, the surfaces of the fuselage and wings of a plane are grooved longitudinally. Such a design was used by the Russian aircraft designer Tupolev in the 1930s. This principle was also applied earlier in German constructions. The difference was that the spacing of grooves in [121] is fine

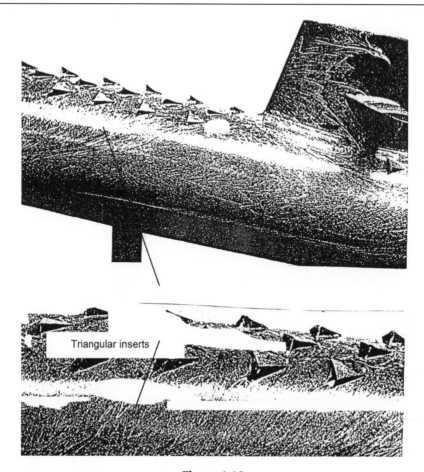

Figure 1.12
Arrangement of triangular inserts in the separation zone on a long body of rotation, Farokhi, et al. [126].

and supposedly equidistant to longitudinal CVS spacing in a boundary layer. At the tops and valleys of the riblets there are orifices connected to each other by ducts so that the process of ejection and suction is automatically controlled in corresponding places of the near-wall region of the boundary layer. In addition, heating elements can be placed in the riblets. For water applications, polymer solutions can be injected through the orifices, or the tops of the riblets can be made of an elastomer.

Separation zones occur about blunt leading and rear parts of bodies of revolution. In the early 1970s, based on theoretical and experimental research on movement of swordfish, the influence of a sword-shaped tip on pressure distributions along a body (see Chapter 6) was investigated. It was shown that it is possible to control separation areas by this method.

Apart from a sword-shaped tip, blowing-off a separation zone through a circular slot in front of the zone is widely used on bodies of revolution. Visualizations of flow where disturbances were brought in by means of a loudspeaker are presented in [489, pp. 497−514].

To avoid separation over the rear part of a rotation body, either blowing-off a boundary layer through a circular slot or spoilers increasing the sucking-in force can be utilized.

Recently, in order to avoid separation, a system of transversal grooves generating edge vortices [65] was used, as well as a system of deep longitudinal grooves [234, 235]. Figure 1.13 shows the design of the rear part of a model intended to avoid separation over such bodies. In both cases, the separation zone behind a body and base drag essentially decrease.

The original method of separation control in the rear of a body was investigated in [104]. There is a slot in front of the separation area. Inside the slot is a rotatable disk. One circular cylindrical groove is formed at location. The disk is wedge-shaped. When it rotates the groove bottom also rotates. The oblique slot in the groove bottom rotates correspondingly. As a result, a toroidal vortex is thrown out from the slot with a frequency corresponding to that of the disk rotation.

Much earlier in America the model of a body of rotation on which an external surface has been established a pluripartite propulsor has been developed and made. It was similar to the compressor in the aviation engine and was established in one or two of the lines. The engine was established on an external surface of a body, the area of which settled down before a zone separation of a boundary layer. At rotation of such an engine it was prevented from separating.

Methods of Controlling CVSs Arising at Flows on Wings

Methods of CVS control for flow about a profile at an angle of attack

For a long time in Ukraine, at the Civil Aviation University, Mhitarjan supervised theoretical and experimental research on the influence of various kinds of vortex generator on the aerodynamic characteristics of wing profiles. In the leading part of a profile, a system of transversal rectangular grooves was made to generate transversal CVS, which should influence the development of T−S waves or eliminate separation in the leading part of profile.

Slanting apertures were drilled in a transverse direction in the nasal parts of a wing section. The twirled jets were blown through apertures, with drift by a stream on the surface of a wing section. As a result, longitudinal vortices were generated in the leading part. In the tail part, a system of parallel winglets was installed on the surface, which generated longitudinal vortical pairs. Simultaneously, rows of transversal grooves were made on an aileron of the profile in the leading part.

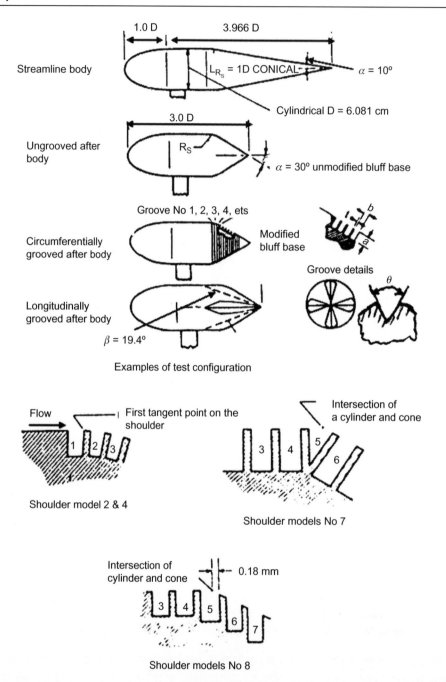

Figure 1.13
Schemes of a design of a tail part of a model for the elimination separation areas.

The interaction of longitudinal and transversal CVSs eliminated separation on a flap at greater angles of attack. Similar results were obtained much later [104, 234] but with a single influence of grooves.

Bechert, Grinblatt, Tinapp, Erk, et al. [126] undertook similar investigations in 1997. Bechert installed a vortex generator on a stabilizer, as in [109]. Grinblatt placed roughness in the leading part of a profile. Tinapp made one slot in the leading part, through which pulse disturbances were brought in. The mass and frequency of the fluctuation impulse generated through the slot varied. At 35−45° flap deviations, separation was eliminated. Erk performed similar research in the leading part of a profile. The new method of eliminating separation on a profile involved the separation bubble being blown-off by a near-wall jet injected along the leading edge in the transversal direction.

CVS control on the ends of ledges and wings

Apart from work described in [109], the methods of CVS control in the rear part of wings have been most intensively investigated in Germany. The principles of bionics are the basis of this research. At a 1997 conference in Iena, research modeling various features of the morphology of birds, fish and sea animals was considered. It was shown how various mechanisms in nature promote drag reduction, thus, increasing maneuverability and saving energy. The structure of feathers in birds has the same physical effect as the "riblets" of shark scales. When maneuvering, separation appears on a wing, as a result the top row of feathers rises and separation is eliminated automatically. To analyze the phenomenon, Bechert performed experimental research with a wing profile. On the wings of planes widely apply the rejected dashboard-flaps established in the back part of a wing. In flight, with big corners of attack, dashboard-flaps deviate under a corner. Certain preliminaries were noticed in a wind tunnel when carrying out modeling experiments. In his experiments, Bechert installed movable flaps with free hinges on the top of a wing profile. In the beginning, one row of movable flaps was used [18, p. 59; 174, p. 239], then two sequential rows [173]. Full-scale tests for a glider with an engine are described in [110]. The two kinds of plates were not continuous, and consisted of separate sections to respond to any separation. The flaps were trapezoid-shaped and permeable to eliminate both separation on the wing and also on flaps during deviation. Bechert performed the tests at small angles of attack. Numerical researches by Schatz et al. [3, pp. 385−390] defined the efficiency of such adjustable plates up to an angle of 52°.

Bionics research by Kesel [471] of dragonfly wings has revealed the specific structure of their rear edge. Bechert investigated the influence of various forms of rear edge on the aerodynamic characteristics of a wing similar to aerodynamic profile HQ17, including modeling the back edge of a dragonfly wing [471]. All the tested designs increased the coefficient of lift force of the wing at angles of attack ranging from 5o to 10o, which is within the limits of angles used by modern planes in cruiser flight.

Stanewsky [3, pp. 221−228] investigated the influence of various methods of compression shock on the upper part of a wing. The influence of various kinds of vortex generator on the aerodynamic characteristics of a profile was considered, including roles of transversal slots for sucking the boundary layer in combination with a long (passive) cavity.

Another related direction of research in aviation is the control of tip vortices on a wing, which determines additional drag. The majority of modern planes have perpendicular winglets on the ends of the wingspan, which promotes smooth shedding of induced vortices due to formation of a pair of vortices in the corner. There are plenty of designs of wing tips, such as the German patent OS3242 584, 24.05.84; the US patents US376 958, 16.11.83; US495 804, 18.05.83; US224 166, 12.01.81; and US957 049, 2.11.78; the French patents 2.521.520, 15.02.82; and 2.634.726, 26.07.88, etc.

On the basis of bionics researches, Bannasch [471] proposed a new design of wingtip (see Figure 1.14). This concept was developed in [3, pp. 304−311, pp. 312−319; 429, p. 57]. In relation to this, the long-known concept of a ring wing is interesting. The shape of wing tips were calculated using the theory of evolutionary strategy and were optimized in a wind tunnel. It is evident that the form developed is better than a straight wing for cruiser modes of flight.

CVS Control on a Triangular Wing

In [72] the method of influence on end vortices of a triangular wing was experimentally investigated. A method of control of tip vortices on a triangular wing was experimentally investigated in [8]. The measurement of vortex structure was performed with a laser knife on a wing at an angle of attack. A wire was mounted transversally at a distance of 0.25 wing chord from the rear edge. The influence of wire on vortices formed on side edges of the wing was investigated.

An active method of control of the side vortices was suggested in [119]. Slots were made along side edges in the top part of a wing. The slots were connected with a piston, which changes the resonant properties of the slot; therefore, the frequency of the vortices generated by the slot is controlled. In the slot, a loudspeaker also created disturbances. A piezoceramic strip was mounted on the upper side profile instead of the slot that generates disturbances at a given frequency.

At the Institute of Hydromechanics NASU (Kiev, Ukraine), Thiganjuk A. I. has experimentally analyzed a method of controlling such a wing by means of suction through small slots in the upper surface of a wing. Experiments were carried out in the hydrodynamic channel. The wing settled down near to the bottom of the channel. Thus movement of a wing near to the screen was modeled. Cracks settled down along a stream, across a stream and at a combination of an arrangement of cracks in longitudinal and cross-sectional directions. Such method effectively operated with the aerodynamic qualities of a wing.

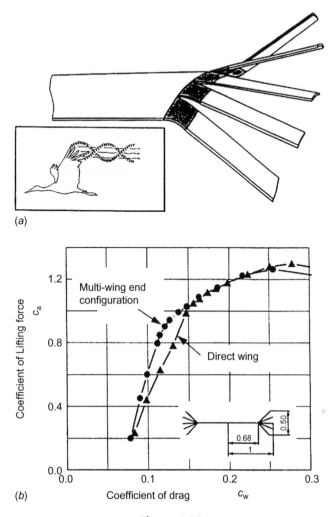

Figure 1.14
Multi-wing tip shape of a wing (*a*) and data of aerodynamic tests of this design (*b*).

CVS Control by means of a Fluctuating Wing

Results of experimental research on the influence of a fluctuating wing on the separation area for a high-drag body are described in [428, pp. 37−38]. It was shown that a fluctuating wing essentially reduces the frequency and amplitude of fluctuation of the vortical wake.

Experimental research on the influence of a fluctuating wing installed in the wake behind a plate and high-drag body was performed in [430, pp. 471−478]. It was revealed that the velocity profile behind a single fluctuating wing essentially differs from that when a flat plate is mounted in front of the wing. The mutual influence of two fluctuating profiles (a tandem type) was also investigated. Dovgiy published similar theoretical and experimental results 10 years earlier [54, 167]. In addition, his research has shown that a fluctuating wing is much more efficient near a boundary or in a domain limited by two side plates. A fluctuating wing promotes an attached flow behind a profile with a blunt rear edge and damps vortices in the stagnation zone behind an obstacle.

Hertel was first to investigate the bionics aspect of the role of a fluctuating wing as a propulsor at the end of the 1960s. He designed and patented a number of systems (propulsor for a boat, pump, etc.) based on the work of a fluctuating wing. At the beginning of the 1960s similar research had been done by many organizations in the USSR. Various designs were developed for models moving by means of fluctuating wings.

The design of such a moving model and test results are presented in [430, pp. 457−461]. This work modeled a propulsor similar to a seal mover. The data for a model with a flapping-fin propulsor, like a tuna propulsor, are described in [430, pp. 463−469].

The experimental data of traction force and vortical structure developed by the fluctuating wings of a live dragonfly are shown in [337, pp. 429−456].

The authors have developed various methods of CVS control by means of fluctuating wings. Due to the cyclic variation of a wingspan and curvature, essential reduction of CVSs on rear and side edges was achieved in [37]. CVSs were sucked inside a wing from its side surface and blown out from the rear edge in [417, 418]; and in other patent, the rear edge was made in the form of a thin rotating cylinder. This promotes a reduction in size of the corresponding vortices.

In [412] CVSs on the leading edge of a wing were controlled by a turning section in the wing. In another patent, movable mesh placed at those sites damped CVSs at lateral surfaces and the rear edge of a wing. In [413] an intensive cylindrical vortex appeared on different sides of a wing profile, or in the middle of its leading section, due to rotation of the leading section by an angle of up to 180°.

Methods of CVS Control on High-drag Bodies

Many technical designs involve transversal cylinders, behind which Kármán vortex street are formed. To control of these CVSs, various kinds of transversal or longitudinal vortex generators are used on the cylinder surface in the region of the boundary layer separation or in front of the region. Another method is oscillation of the entire cylinder, thus, changing the frequency and shape of the Kármán vortices. In [337, pp. 337−364] Williams and Amato investigated a new method of CVS control. For this purpose, a slot was made along a cylinder generatrix, on the axis of symmetry in the zone separation. A plane jet was blown through this slot at various frequencies. The design was such that liquid was blown cyclically from the slot and sucked in. The data from the measurements and visualization patterns were presented, on the basis of which an optimum frequency of jet injection was determined.

Heine et al. performed similar experiments on CVS control behind a cylinder in [126, 337]. The difference is that one or two slots were made along the cylinder at various angles in the region of the separation line. A loudspeaker was mounted outside against the slot. It generated a pressure field that forced liquid to be blown from and suck into the slot with the frequency of the loudspeaker. Tests were also performed with two slots on two sides of the cylinder under the same angle ($\approx 70°$) to its cross-section.

The simplest method consists of accommodating a plate behind the cylinder in a plane of symmetry separation zones. The plate can be rigid or elastic. The same function will be played with the flat stationary jet injection through one slot on an axis of symmetry separation zone or through two slots, placed on a lateral surface of the cylinder in the area of the lines of separation of BL.

Methods of CVS Control in Internal Flows

Some methods of CVS control for liquid flows in pipelines have been mentioned in Section 1.2.1. In [126] Weiss studied CVS control in long pipes by means of longitudinal riblets installed on the internal surface. The author stated that it was possible to reduce pressure loss by 10%. The depth of riblets was 15 μm and the pipe length 22.0 km.

Babenko investigated the influence of riblets on the internal surfaces of conic mixers and the turbine blade of hydraulic transformers.

Methods of CVS control in Taylor flows were also briefly mentioned in Section 1.2.1, and consisted of changing the shape of cylinders and creating sine oscillations of the internal cylinder in a coaxial direction.

G. J. F. Van Heijst [497] has offered two means of CVS control in Taylor flows. One of them involves a plate mounted on a motionless cylinder, which separates and fixes the gap between the cylinders. The second way consists of high-frequency vertical vibrations of the internal cylinder.

The authors have developed various methods of CVS control in vortical chambers. One of the methods was described in [390].

Methods of CVS Control in Submerged and Near-wall Jets

Various methods of CVS control in submerged jets have been developed. Some results of these methods are discussed here. A submerged jet was investigated in [489, pp. 472−484]. The top and bottom sidewalls of a plane nozzle moved forward alternately by means of a crank gear. Therefore, the border of the jet oscillated at a certain frequency. The jet direction and angle of the nozzle opening could be changed. The measurement data and visualization patterns of the jet were shown.

Another design of oscillating jet control was used in [489, pp. 485–494]. The design is based on the use of Coand effect and allows the generation of both a skewed or oscillating jet by means of a system of mechanical rods, dependent on mean flow velocity.

Ziada [173] created a mechanism of active control of a submerged jet with a feedback system. A loudspeaker was installed around a nozzle, which imposed circular pressure at a certain frequency on the jet's external border through a circular slot at the nozzle tip. A microphone behind the nozzle feeds a basic signal for controlling the frequency and amplitude of the sound radiated by the loudspeaker.

The characteristics of fluctuating jets are investigated in [179]. Slots in the top and bottom sides of a nozzle are made for the generation of a fluctuation jet. A special valve alternately feeds liquid from a reservoir to the slots. The slots are perpendicular to the jet surface and alternately blow the jet aside from the longitudinal direction.

Some methods of CVS control in near-wall jets were mentioned in Section 1.2.1. The authors have developed new methods of control for such CVSs (see Chapter 6).

CVSs in the shear layer on the surface of a cavity at cavitations flow about a body have also been investigated. Disturbances generated on the surface of a rigid body are transferred to a liquid and travel in the "water–air" shear layer. Thus, a boundary layer in air on the moving wall (liquid) environment occurs. The moving wall keeps the disturbances generated away from a rigid surface. According to this hypothesis, a method of CVS stabilization in a shear layer has been formulated in [81]. Sparck presented similar ideas in [84].

Based on the analysis of the results mentioned, it is possible to conclude that much research has been performed on the efficiency of various methods of CVS control. Some methods have an effective influence on CVS.

Babenko has developed work on receptivity [310], on which it is possible to essentially improve the efficiency of any particular method of CVS control. In [3, pp. 113–120; 430, pp. 341–350, etc.] Babenko has developed and systematized various methods of CVS control in a boundary layer (described in Section 1.13). These methods can be used for the control of any CVSs. Most of this focuses on combined methods of CVS control (see Chapters 5 and 6).

1.3 Coherent Structures in a Turbulent Boundary Layer

Experimental research on the development of disturbance structures in the boundary layer is necessary for a better understanding of flow processes and for closing the system of equation of the turbulent boundary layer. Experiments with liquid dyes conducted in 1956 showed periodic increasing and decreasing of the laminar field of the turbulent boundary layer [177]. Unrelated to this paper, a concept was proposed that gives a method of calculation of a turbulent boundary layer taking into account longitudinal periodical structures in the viscous sublayer [225, 493]. Hanratty investigated the distribution of velocity fluctuations in the near-wall field, as well as the influence of polymer solutions on the flow in the near-wall [225]. The comparison of correlation factors of transversal and longitudinal fluctuating velocity components was investigated by electrochemical and heat-loss anemometer methods with $y^+ = 13$ and makes it possible to show the dependence of stress values on the disturbance structure. It was shown that dominating vortices are strongly extended in a longitudinal direction with the ratio of longitudinal-to-transversal scale equal to 40. Spatiotemporal correlations between longitudinal shear stresses at $y^+ = 13$ have made it possible to establish that these stresses depend on the structure of disturbing motion. The vortices in this case are very bent in the *xy*-plane. Spatial correlation between the shear stress fluctuations determines the disturbance wavelength in a transversal direction: $\lambda^+ \approx 100$; the value is well agreed with the data given by Klein [274, 276, 277]. The main effect of drag-reducing polymers resulted in an increase of transversal fluctuations. In [225] simultaneously measured values of velocity fluctuations near the wall, as well as at various distances from the wall, were analyzed with the aim of discovering the near-wall vortices in the direction of the main flow.

Klein investigated the physics of the flow in near-wall areas by means of tinted jets, the method of hydrogen bubbles, and a hot-wire anemometer [274, 276, 277]. Regular motions in the viscous sublayer, and a low-velocity region were revealed. These motions are named Klein streaks. A possible mechanism of generation and collapse of the low-velocity regions was proposed.

In [277] it is shown that the streaks of decelerated flow in the viscous sublayer usually end when they are moving off from the wall. At that moment, or a little later, they start to pulse. This oscillating motion increases until the streaks collapse. This stage of the streaks' motion takes a very short time and is, therefore, called a "burst".

In 1965, Repick, independently of other authors, made the suggestion that the flow in a viscous sublayer is similar to that in a laminar boundary layer [435]. The distribution of kinematic parameters of a turbulent boundary layer near a wall depending on pressure gradients was investigated in [435, 436]. The investigation of spectral characteristics led to the evaluation of quasi-ordered vortex system dimensions, which agree with those given by Klein and other investigators.

The analysis of spatial correlation coefficients has made it possible to establish that in the viscous sublayer a pair of opposite rotating longitudinal vortices is generated. The radius of the vortices is $r = (20-50) \, \nu/u_\tau$. By means of stroboscope visualization, the flow parameters in the viscid sublayer were explored. They are average, as are the fluctuation velocity distributions, integral scale and turbulent viscosity coefficient [222, 223].

In [259] it is shown that shortly after the "burst" the large-scale motion travels to the wall and cleans the near-wall region of chaotic motion. Therefore, this stage is called "sweeping." The "sweeping" scale depends on the outer wall region parameters δ and U_∞ and forms an irregular boundary within the near-wall region. Irregularities of this border are commensurable with near-wall variables ν and u_τ. They are referred to as "pockets" and have scale $\approx 100\ \nu/u_\tau$ [493].

In 1970–1971 Kovashniy's work on the viscous turbulence was published [493, 300]. Blackwelder's investigations on the flow in the viscous sublayer was published in 1970. He studied the coherent structures in the viscous sublayer by means of a heat-loss anemometer [118, 117, 177]. This research showed that Klein's low-velocity fluid streaks are formed between two longitudinal vortices when the low-velocity fluid moves from the wall to the outer region of the viscous sublayer. Klein's low-velocity liquid streaks are formed in the region of "peaks," using Klebanov's terminology for a transitional boundary layer. These streaks are usually of $(10–20)\ \nu/u_\tau$ in width and $(100–1000)\ \nu/u_\tau$ in length. Blackwelder investigated the relationship of wavelength λ_z with external (δ and U_∞) and internal (ν and u_τ) parameters of the boundary layer, as well as with Reynolds number (Re). It was shown that over a rigid plate the value of $\lambda_z \approx 100$ and is almost independent of Re. In addition, the dynamic parameters of the "burst"—frequency and regeneration period of the viscous sublayer—are determined. A sample-averaging method was developed. The space—time correlations of the disturbing motion in the viscous sublayer were investigated. The average burst frequency was found to be in the range $600 < \mathrm{Re}^{**} < 9000$, and is comparable with the outer flow parameters δ and U_∞ as determined in [117].

In [325, 177] long vortices in the turbulent boundary layer over a smooth plate and a circular cylinder were investigated. In [181, 194, 239, 473, 489–493] the propagation of disturbing motion in the viscous sublayer over a rough plate has been investigated. As in Klein's work, the phases of ejection and burst were revealed. The instant velocity analysis showed that the ejection is defined by the low-velocity burst from the wall, and the burst by high-velocity flow toward the wall. Both the ejection and burst are independent from the rough scale and are intermittent. Therefore, the random turbulent circles (spots) generated are controlled with a kind of 3D instability mechanism.

It was revealed that ejection is located mainly near the boundary, and the burst relates to a far greater region. The liquid burst out of the viscous sublayer and of spaces between the rough clements moves from the boundary to the core of the stream and strongly influences the Reynolds stress far from the boundaries.

In [177, 232, 233, 272, 326, 327, 367, 377, 385, 516] various aspects of flow in the viscous sublayer were investigated. Ekkelman's work [170] is similar to that of Blackwelder, both conceptually and methodologically. The structure of the disturbing motion was investigated by means of an X-shaped sensor. The measurements in a plane channel filled with oil, as well as in a low-velocity wind tunnel, show that the coherent structures in the viscous sublayer are rotating. Ekkelman's sensors were smaller than those used by Blackwelder, which allowed them to make more accurate measurements in the near-wall region.

In [209] the bursting frequency was evaluated in terms of measurements of the near-wall shear stress fluctuation. In [177] two hypotheses about the scales of turbulent vortices in the viscous sublayer were considered: $l \sim y$ and $l \sim y^2$. Grass's hypothesis [177] about the 3D instability in the viscous sublayer were extended in [153, etc.], where the viscous sublayer vortices were considered in the same way as Taylor−Goertler instability. These hypotheses were verified in [118]. It was shown that the dimensionless wavelength λ_z^+ of the viscous sublayer longitudinal vortices correlates with the λ_z^+ of Goertler vortices in a wide range of Re. However, until now this agreement has not been explained by its physical causes. In [40] the following hypothesis was put forward: the flow of disturbing motion in the viscous sublayer is similar to that in the transitional boundary layer, with the same order of transition stages but in the presence of external disturbing factors [310].

The investigations [276, 300, 435, 436, 490, 492, 493] show that there are coherent vortices of two types in the turbulent boundary layer. The first type is connected with free flow and described by parameters of the outer region of boundary layer. The second one correlates with the near-wall sublayer, and is described by the viscous sublayer parameters.

The burst structure was investigated by means of hydrogen bubbles and a heat-loss anemometer [376]. The space−time correlations were measured. Figures 1.15 and 1.16 show the dynamics of formation of the longitudinal vortical system in the viscous sublayer—the scheme of formation of an inclined vortex.

Change in wavelength of longitudinal vortices on the thickness of a boundary layer has been measured by means of the hydrogen bubbles method [118]. A linear function is obtained: a change in y^+ from 5 to 30 leads to a change in λ_z^+ from 90 to 140. These

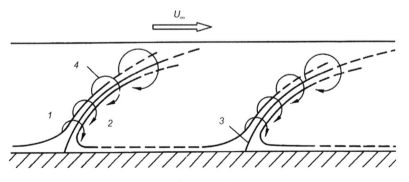

Figure 1.15
Model of low-velocity and high-velocity regions of flow in a channel: *1* − high-velocity region, *2* − low-velocity region, *3* − plane that separates the regions, *4* − inclined vortex.

results show that the vortices increase as they move away from the wall. These results connect the known data of low-scale longitudinal vortices in the viscous sublayer with the high-scale vortices in the buffer zone.

The majority of investigations of the viscous sublayer were made by means of the hydrogen bubbles method and a heat-loss anemometer. To detect the burst using a heat-loss anemometer device different algorithms and methods of statistical processing of registered pulses are used [117, 136, 153, 170, 177, 435, 436, 472]. Independent of this approach, most investigators obtained data on the shape and scale of near-wall vortices, which agree well with [177]. The main differences are in the frequency of viscous sublayer regeneration and the normalization of the burst. The one most distributed was VITA which is a method of conditional selective averaging by which the high-scale coherent structures in the turbulent boundary layer are defined [269]. Other methods of calculation of vortices of greater scales (see Figure 1.17) which have developed, for example in Russia by professor Taganov and in Ukraine by professor Shkvar are also known.

Despite the fact that much data has already been obtained many problems are still not solved. The nature and mechanism of the origin and growth of a disturbance in the viscous sublayer is not yet fully understood. Neither is the way in which disturbances in the near-wall region and the outer part of the boundary layer connect with large-scale coherent structures. A greater understanding is also needed about where a sensor must be located in the direction of z-axis, as well as what size of sensor does not corrupt the disturbing motion. We do not have a complete physical picture of the disturbing motion that develops in the viscous sublayer.

Along with the considered approach, another concept of the disturbing motion structure in a turbulent boundary layer exists—that is the wave model of turbulent shear flow. Landahl

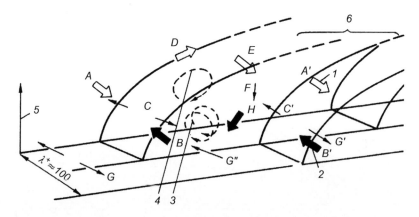

Figure 1.16
Conceptual model of bursts in the flow in an open channel: *1* — high-momentum fluid ($u > 0$), *2* — low-momentum fluid ($u < 0$), *3* — longitudinal vortex, *4* — transversal vortex, *5* — near-wall region, *6* — creating or collapsing of a vortical pair. Letters and dashes mark the flow directions.

et al. considered this concept in detail. It accounts for the turbulent flow processes by the wave theory as well as the hydrodynamic stability mechanism [177, 197, 206, 214, 265, 322, 329, 357, 360, 367, 453, 474, 490]. Two aspects exist in this approach. In paper [367] waves in the viscous sublayer are identified. In [333] the equation of fluctuations in a parallel shear flow is reduced to the Orr–Sommerfeld equation, in which the nonlinear components of fluctuating turbulent stresses are considered as force components.

Taking into account the new directions of boundary layer investigations, we can examine the large volume of papers from the point of view of the reaction of the turbulent boundary layer to various disturbances [59, 96, 125, 143, 366, 524]. Investigations of the turbulent boundary layer parameters that depend on the free-stream turbulence degree must first be noted [1, 157, 169, 176, 211, 230, 366, 381, 408, 462, 492, 493]. In papers [203, 432] the influence of acoustic disturbances on coherent structures in turbulent flows was investigated. In [136, 329] the propagation of small and finite wave disturbances in a turbulent boundary layer is described.

In [481, 511] the behavior of artificial periodical disturbances in a 2D turbulent boundary layer at a zero pressure gradient is studied. A plate with a turbulizer was installed in a wind tunnel. Periodical disturbances in the boundary layer were created by a strip with chord of $2 \cdot 10^{-2}$ m, located at 0.5 m from the turbulence promoter and at $1.8 \cdot 10^{-2}$ m from the plate surface. The strip vibration frequency was $2–20$ Hz. The measurements were made with a heat-loss anemometer and using smoke visualization to show that the disturbances with wave numbers of $0.1 < \alpha \delta^{**} < 0.6$ grew and with wave numbers of $\alpha \delta^{**} > 0.6$ were damped. In these papers, the interaction between the turbulent boundary layer and the disturbances generated from the external side of plate was investigated.

Another group of studies has investigated the disturbances generated on a plate from the lower side of a boundary layer. Similar investigations were conducted for the turbulent boundary layer parameters over a smooth curved surface [123, 181, 225]. It was shown that in the boundary layer over a concave surface, longitudinal vortices appear and Reynolds stresses change due to Goertler instability. The flows in laminar and turbulent boundary layers on a curved surface are similar, but in a turbulent flow theoretically drag can be reduced by 80%.

In [265] the turbulent boundary layer over a surface generating waves was investigated. Earlier, Schiltz [310] had used a similar method of surface waves generation. The paper [138] reviews a number of projects in which turbulent boundary layer parameters are described at different influences on the near-wall flow. In [142, 373] the growth of disturbances in the turbulent boundary layer generated by regular stripes transversally located on the surface was investigated.

The influence of a grid located near the wall on the viscous sublayer parameters is described in [128, 177, 194].

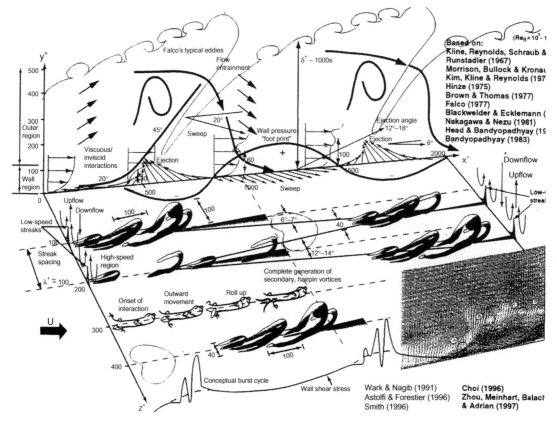

Figure 1.17
Conceptual turbulent boundary layer near-wall phenomenology by Meng (1998).

The influence of longitudinal vortices generated by various turbulizers and wings on the characteristics of the viscous sublayer has been investigated [52, 194, 272, 310].

Coherent vortical structures in a turbulent boundary layer are still intensively investigated. Meng [358, 359] has developed a generalized scheme of CVSs in a turbulent boundary layer based on results of experimental research of the flow over a flat plate undertaken by various authors (see Figure 1.17).

1.4 The Flow over Elastic Surfaces

Kramer's investigations of the flow over elastic surfaces published in 1957 are usually described as being the starting point of a new research direction in hydromechanics. In reality, however, the Russian scientist Gromeko produced a report about a theory of flow in pipes with elastic walls in May of 1883 [216].

The main stages of the development of flow over elastic surfaces are given in reviews and monographs [20, 86, 108, 112, 127, 137, 186, 192, 195, 198, 199, 245, 254, 255, 305, 360, 354, 394, 498 et al.]. Some general problems of the interaction of a flow with an elastic surface are presented in [7, 13, 201, 394 500, 504].

Kramer believed that the result of interaction of an elastic surface with a flow depended on the damping of T−S waves [315]. Because of this, the problems of hydrodynamic stability of flow over an elastic plate have been investigated since Kramer's work was published.

Benjamin and Landahl examined Kramer's results [112, 333]. They established that an elastic surface might affect the stabilized flow. However, at the same time, other unstable waves can be raised. Benjamin called these types of wave class *B* (waves on the free surface) and class *C* (waves of the Kelvin−Helmholtz type) instability, in contrast to the principle T−S instability referred to *A*-class waves. Using a simple physical model Landahl explained why disturbance dumping by the wall results in destabilization of the class *A* waves; at the same time the stability of class *B* waves is increased and that of class *C* waves almost does not change.

In [13, 15, 21, 108, 220, 252, 333, 360, 524] using strict establishment by numerical methods neutral curves have been constructed at a flow of elastic plates depending on class *A*, *B* and *C* disturbances.

Almost at the same time as Benjamin's and Landahl's work, a number of experimental [13, 108, 183, 199, 481] and theoretical [122, 199, 229, 278, 488] investigations on the effect of a elastic wall in the laminar flow were conducted. Neutral stability curves of a flow over a simple diaphragm surface were calculated, and this showed an increase in hydrodynamic stability on decreasing the diaphragm tension. In [199] the kinematic parameters of the boundary layer were experimentally investigated. In addition, calculations of shell stability were carried out based on elasticity theory. In this paper, a roughness criterion attributed to diaphragm vibration was suggested. A similar criterion for the turbulent boundary layer was put forwards in [454] and a modified roughness criterion was suggested in [285]. Galway investigated the spectral characteristics of the flow over an elastic surface. Similar investigations were carried out in [319]. The issue of hydrodynamic stability over an elastic surface was also solved by Korotkin, et al. [108] and by Amfilohiev, et al. [13, 15].

In 1965, Wehrmann's experiments involving a streamline plate with cross-section channels stuck with a thin elastic film were published [511]. In these channels pressure moved alternately, and as a result a sine wave with set parameters was generated on a wall. Simultaneously, traditional ribbon was established above the plate, causing T−S waves to be created in the boundary layer. Interaction of the sine waves generated by the wall and ribbon located near a wall was investigated. When the waves were co-phased, the longitudinal pulsating velocity reduction was fixed: with $0 < y/\delta < 0.08$ by 10%, with

$0.08 < y/\delta < 0.12$ by 50%, with $0.12 < y/\delta < 0.15$ by 20%, and in the boundary layer thickness approximately by 10%. With the phase ratio about $180°$, the longitudinal pulsating velocity was increased. Wehrmann's paper was, in reality, the first paper on boundary layer sensitivity to 2D disturbances. It showed that the interaction between 2D T−S waves and a 2D wave generated by the wall resulted in resonance interaction and in a substantial stabilization of the flow. Theoretical research on this issue is available in [500].

Georgyfalvy calculated the hydrodynamic stability over a membranous surface. The mechanical parameters of the elastic surface were varied, as in [260]. It was shown that Reynolds number of stability loss may increase by 4 times in air, and by 10 times in water.

The first experimental investigation of hydrodynamic stability over different elastic plates was made in [30, 32, 305]. It was shown that with the correct selection of material and design of the elastic coatings, and, hence, mechanical characteristics, at least one of the following effects may be achieved:

- an increase of critical Re (in the experiments the increase was by 2.1 times);
- a growth of transition region and transition Re;
- a decrease in the unstable oscillation region of the laminar boundary layer;
- a decrease in phase velocity of the disturbance movement;
- an increase of wave numbers and decrease of growth coefficients;
- a decrease in the distance from the wall of the critical layer;
- a decrease in the kinetic energy of the disturbance movement.

However, the experiments and calculations of most of the investigators showed that the increase of the critical Re obtained on the elastic surfaces cannot explain Kramer's results. Obviously, the elastic surfaces affect the later transition stages including the turbulent flow stage.

In [252, 360] the numerical analysis of nonlinear stability of the laminar boundary layer and of Poiseuille flow over the elastic surface is presented. It is shown that the elastic surface stabilizes the flow even at nonlinear stages of transition.

Research on combined factors of influence on the character of a flow in a laminar boundary layer can be related to investigations on the problem of receptivity of a boundary layer to various disturbances. For example, in [60, 61, 305] the combined influence of curvature and elasticity of the surface on boundary layer stability were studied. An increase in the rigid surface curvature destabilizes the flow and an increase in elastic surface curvature (with optimal mechanical parameters as regards to class A instability) results in stabilization of the flow. In [394] the hydrodynamic stability on an elastic surface in the oscillation flow of Newton and non-Newton fluids, and suction of the boundary layer over the elastic surface is theoretically analyzed. The application of elastic coatings combined with non-Newton fluid injection in the boundary layer was first suggested in [34].

Therefore, from an overview of the basic results of many theoretical and experimental investigations on longitudinal stability of the flow over both rigid and elastic surfaces, it can be concluded that; with respect to disturbance growth in the boundary layer at the first stage of transition, the results are rather complex [305, 310]. However, the characteristics of disturbance growth in the flow over the elastic surface at the later stages of transition have not been investigated.

The investigations of disturbance growth at the first stage of transition were followed with investigations of the elastic surface effect on the kinematic and integral parameters of the turbulent boundary layer.

Kramer's work concerning the effect of the elastic surface on the turbulent boundary layer are considered in detail in [108, 195,197]. Kinematic parameters of the flow were considered also by Laufer [195]. He tested 15 different types of elastic coating made of steel, aluminum, moire film and factory fabric with rubber fastened onto them. The tests were undertaken in an air channel. Some of tests show a slight decrease of turbulent friction.

The widest-ranging and most decisive investigations of airflow over elastic surfaces were carried out under the supervision of Blick [122]. The velocity profiles of the flow over rigid and elastic surfaces, as well as velocity fluctuation intensity and Reynolds stresses were measured. Measurements have shown that the path length of the mixture at rigid and elastic plates is identical. Reynolds stresses and the intensity of longitudinal pulsating velocities on length at some elastic plates were less than for the standard.

Turbulence intensity has also been measured in the flow over a polychlorine diaphragm of 0.057 mm thickness, which was tightened over a reservoir of 0.5 mm in depth [186]. The reservoir was filled with air, water or motor oil with a density equal to 30. Turbulent fluctuations decreased in the flow over the elastic surface. The fluctuations decreased proportionally with an increase in viscosity of the dumping fluid. Conversely, when similar tests were performed in a wind tunnel [195] the turbulent intensity, defined by values of longitudinal pulsating velocity, decreased at elastic membranous surfaces by $21-27\%$ when there was air under the membrane.

In the paper [344] Blick's results were checked. The elastic surface consisted of a film pulled taut over a frame and pasted to it. Water and a polymer solution were used as dumping fluids. Measurements were made with different thicknesses and tensions of film. The positive take was not revealed.

Aside from work on membranous surfaces, research on monolithic elastic plates has been carried out [16]. Experiments were undertaken in a specially designed flat channel in an air stream. The elastic surface was made from a sheet of foam polyurethane. The longitudinal velocity fluctuations decreased by $15-20\%$.

The dumping surface effect upon shear stress fluctuations in the turbulent boundary layer was investigated in [355]. The dumping surface was an aluminum circlular membrane of 0.00125 mm thickness. Velocity fluctuations were measured with a heat-loss anemometer at a distance of 2.4 mm from the surface. The largest deflection of the membrane from its initial position was measured with an optical sensor and was of approximately 0.4 mm. A definite effect upon the shear stress and spectrum of the turbulent boundary layer was noted. The amplitude of the surface oscillation was large, which could have created a dynamic roughness, as in [61, 344]. Such measurements were made on a rotating disk with an isotropic elastic surface [226]. Substantial friction value change was not noted in comparison with the rigid surface.

Active exposure to the turbulent boundary layer was carried out in [217]. The basic idea was to use an elastic surface to influence a turbulent boundary layer so as to change the correlation of Reynolds stresses. It was necessary that the elastic surface made only positive correlation. This means that u and v near the surface should be simultaneously either positive or negative. A decrease of momentum thickness with low Reynolds numbers was revealed.

First, as with Kramer's integrated measurements, investigations of the kinematic parameters of the turbulent boundary layer were performed on anisotropic elastic plates which have an ordered structure in the transversal direction [42, 52, 256, 258]. In these investigations, the first attempt was made to find a connection between the structures of the outer layers of elastic plates and the dynamic parameters of the turbulent boundary layer.

A similar approach was used in [197]: the evolution of the disturbance movement in the viscous sublayer near the elastic plate was photographed using of visualization methods developed by the author of the paper. The mechanical parameters of the plates were changed to achieve the largest dynamic reaction in boundary layer disturbances. Deformation of some of the elastic plates in the form of waves, such as Tollmien nonlinear waves, was revealed [355].

The elastic surface effect upon the pressure fluctuations spectrum was investigated in [165, 200, 201, 323, 339]. The measurements of surface displacement were first made using a noncontact micrometer by Jampolsky [66].

In [186, 187, 344] the integral parameters of the turbulent boundary layer depending on the elastic surface oscillation frequency were analyzed. The maximal effect was obtained when the surface oscillations' natural frequency f_c:

$$f_C = \left[\frac{\frac{T_y}{w^2} + \frac{T_x}{L^2}}{\rho_t} \right]^{1/2} \tag{1.1}$$

is approximately equal to the half frequency:

$$f_P = \frac{U_\infty}{2\pi\delta},$$

(1.2)

where f_P is the characteristic frequency of the turbulent boundary layer corresponding to the maximum power in the spectrum of the boundary layer fluctuations.

The surface oscillations of different elastic plates relating to their interaction with the turbulent boundary layer are also investigated in [16, 197, 199, 226, 250, 356]. In [356] the influence of tension of a membrane on its frequency characteristics was investigated. It was shown that deformation of a surface is defined by long-wave fluctuations of small amplitude. Similar tests have been performed on a three-layer elastic plate [250]. Its base layer had the low elasticity module and was positioned on a substrate. The interaction of this plate with subsonic and supersonic turbulent boundary layers was investigated.

Hansen's staff [226] carried out systematic investigations on elastic plates in a hydrodynamic tunnel and on rotating disks. They studied the longitudinal and transversal deformations with different parameters in water, water−glycerin and polymer solutions. It was shown that on the surface of the elastic plates small longitudinal waves are fixed in the full range of characteristics of a boundary layer and elastic properties of coverings. Large-scale transversal waves, the occurrence of which is defined by the interaction of forces of inertia and elasticity of a deformable surface, were also revealed. It was noted that the lateral waves resulted in an increase of hydrodynamic resistance. The conditions of formation of lateral waves were also investigated in [197, etc.].

Dynamic parameters, such as dynamic elasticity modules, anisotropy degree, natural frequency and the dumping factor, have been investigated on anisotropic elastic plates with an ordered structure of outer layers [56, 257, 285, 286].

The elastic surface oscillation parameters were theoretically investigated in [20, 50, 122, 220, 384, 502−504], where the fixed mechanical parameters were varied, and in [14, 40, 108, 453], where the surface parameters were specified indirectly, for example in the form of a loss angle. In [21] an elastic plate that had transverse ribbing showed the propagation of Rayleigh waves.

Investigations concerning the airflow over elastic surfaces were carried out under the supervision of Bushnell [20, 131, 137, 186, 195, 197, 226, 512]. In these papers a new hypothesis of the interaction of elastic coatings with the turbulent boundary layer interaction is proposed. According to this hypothesis the material of an elastic coating should possess such properties that in response to fluctuations of its surface prevent explosions from a viscous sublayer—the so-called "pre-burst modulation" of surface oscillations. In this model, unlike the well-known absorption hypotheses [502], a positive effect can be achieved when the surface of the coating oscillates to produce microbursts preventing the bursts from the viscous sublayer.

The combined effect of elastic surface and polymer solutions on the integral parameters of the turbulent boundary layer was investigated in [34, 455]. A summation of the effects was registered.

Eighty years after the pioneering work of Gromeko [216], investigations into the effect of elastic pipe walls on laminar and turbulent flows began [165, 199, 201, 217, 460, 484,504]. In recent years, much original research has been done in this field. Results of this research have been considered at various conferences. In 2001, the IUTAM Symposium focused on "Flow Past Highly Compliant Boundaries and in Collapsible Tubes [245]." Materials of this IUTAM Symposium are published in the form of a seperate book.

Theoretical modeling of the turbulent boundary layer over an elastic surface has been difficult. This is due to both the complexity of the problem and the lack of experimental data explaining the physical interaction of the flow with the elastic surface and the formulation of the closure hypothesis.

It should be noted that there are many theoretical papers on the subject of flow over elastic surfaces, for example [122, 260, 333, etc.]. Analysis of the above-mentioned papers shows that the investigators do not have a common view about the mechanism of interaction of the elastic surface with the flow.

Kaplan [260, 333] has made an extensive analysis of the interaction of elastic plates with a boundary layer at the stage of linear stability. His theoretical research on has focused on monolithic elastic plates.

Carpenter [10,140, 141, 159, et al.] has carried out determined and comprehensive investigations on the hydrodynamic stability of a flow over elastic plates over a long period of time. Recently, various questions of hydrodynamic stability have also been theoretically explored by Kumaran [460].

Theoretical research on the turbulent boundary layer has been carried out for a long time by Voropayev, et al. [310, 353, 502−504, 506−508], Trifonov [487] and Semenov, et al. [453−458].

1.5 Experimental Studies on the Characteristics of Elastic Plates

Many theoretical and experimental investigations of the boundary layer over elastic surfaces have been carried out [20, 108, 127, 191, 195, 305, 310, 360]. The first aspect to consider is the direct measurement of friction drag.

Most of the investigations have used membranous surfaces, which are described by the Foyght−Kelvin model and can be easily analyzed theoretically. In other papers, smooth monolithic elastic plates have been investigated. For these plates theoretical models can be constructed. Composite plates have been considered in two papers only, and it should be noted that theoretical solutions do not exist due to the complex nature of the interactions. All this experimental research gives discrepant results for four reasons:

* incorrect measuring procedures;
* lack of planning experiments;
* the efficiency of an elastic plate depends on many factors;
* there is no clear understanding of the physics of the interaction of an elastic plate with a flow.

Based on observations of dolphins swimming, Kramer first investigated the effect of a composite elastic surface on flow [108, 315]. Rotating two coaxial cylinders, he obtained a 20% increase in the speed of the cylinder with the elastic coating.

Defining the real effect of drag reduction is rather problematic. However, the results obtained have formed the basis for continuing experimental investigations, not only by Kramer but also by other researchers.

Tests at low Re were carried out in [261]. An enclosed rectangular working section with the dimensions $0.6 \times 0.14 \times 0.07$ m was installed in a hydrodynamic tunnel. A 0.406×0.086 m window was cut out in the upper side of the working section. In the working section was a 0.394×0.074 m insert in the form of a waterproof box, and this was mounted on the end faces, fixed along the longitudinal axis on flat springs with strain gauges. The depth of the insert was 0.042 m. It was filled with air or olive oil. The insert was coated with either a natural 0.15 or $3 \cdot 10^{-4}$ m-thick caoutchouc film, or with $2.5 \cdot 10^{-3}$ m-thick neoprene rubber. The mechanical parameters were determined at different elastic tensions. The flow velocity varied from 0 to 2.10^{-1} m/sec. Drag coefficient was measured depending on mechanical parameters within the Re range of $10^3 - 10^5$. Some results from the experiments are presented in Figure 1.18. The form of the curves is caused by a gap between the insert and the test work section ($6 \cdot 10^{-3}$ m). The scattering of measurement points on the solid plate was less than on the elastic ones.

The friction value was so small that it was impossible to obtain more accurate measurements with the given design of strain sensor. The conditions of the flow over the elastic plate were adverse because of the presence of greater backlashes. Due to the gap, a pressure difference was created at the leading edge of elastic plate, which grew to a force comparable with the friction value. Due to pressure distribution variations, a standing wave was created on the surface. All these conditions resulted in increased measurements and made it impossible to determine the real effectiveness of the elastic plates. For instance, the data for the most effective plate (curve 2) and the least effective plate (curve 3) are difficult to smooth and the measurement errors are comparable with the measured values.

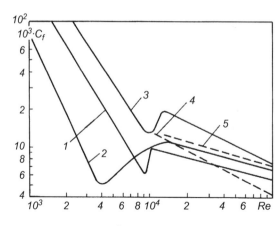

Figure 1.18

Dependence of friction coefficients over elastic plates on the Reynolds number. According to the data of work [187]: *1* — solid plate; *2* — plate with neoprene rubber, stresses 21.7 kPa, olive oil under the plate; *3* — plate of caoutchouc of a width of 0.3 mm, stresses 5.4 kPa, air under the plate; *4, 5* — solid plate in laminar and turbulent flow [144].

Although there was an indication that the effectiveness of an elastic plate bore a certain relationship to its mechanical parameters, conclusive results were not obtained. Therefore, to obtain reliable results it would be necessary to: 1) change the design of the elastic plate, 2) change the manner of fixing it to the insert, 3) decrease the gap between the insert and the test section, and 4) extend the range at Re.

These disadvantages were partly avoided in an aerodynamic tunnel [122]. The gap between the insert and work section was decreased to $1.6 \cdot 10^{-3}$ and the Re range was extended. However, the elastic surface design and the method of fixing it to the insert were not changed. Furthermore, the fixing of the insert on the strain suspension was poorer and the gap remained wide. Despite the balancing device for adjusting the insert surface to a zero angle of incidence, the fixing of the insert on the strain support resulted in the influence of the insert momentum parameters on the measurements of friction. The plate remained too sensitive to the pressure distribution. The work section of the wind tunnel was $1.2 \times 0.5 \times 0.35$ m, the insert was 0.66×0.2 m and the 0.13 m-deep box was filled with different damping liquids. The box could also be filled with foam polyurethane. A 0.064 mm-thick polychloridvinyl film coated the box. The transverse tension of the film was constant ($T_z = 54$ N/m) and the longitudinal tension was varied. Figure 1.19 shows the friction drag of plates obtained in [122].

The maximum friction drag reduction was about 25–33% in comparison with the solid plate. However, because the measurements were done in airflow the drag value was low, despite increased Re. Therefore, the measurement errors were of the order of the measured values. The obtained results were qualitative.

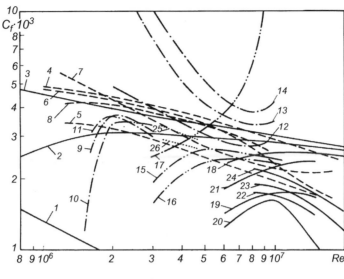

Figure 1.19

Drag over elastic surfaces at different Reynolds numbers: *1, 2, 3* — drag of a solid plate in laminar, transition and turbulent flows after Schfichting [144]. Blick's measurements [159]: *4* — solid plate; *5* — foam polyurethane, porosity of 16 pores/cm, soaked up with water and coated with polyhlorvinyl (PHV) film of $h = 6.4 \cdot 10^{-5}$ m thickness with longitudinal tension $T_x = 35$ N/m; *6* — same as the *5*, and $T_x = 23$ N/m; *7* — foam polyurethane, porosity of 10 pores/cm, soaked up with water, $T_x = 31$ N/m; *8* — same as *7*, but dry foam polyurethane (everywhere $T_z = 54$ N/m. Galway's measurements [179]: *9* — standard, *10* — foam polyurethane of $h = 6.4 \cdot 10^{-3}$ m coated with melynex film of $h = 127 \cdot 10^{-4}$ m; *11* — same as *9* and covered over with melynex film. Measurements of Taneda and Honji [210]: *12* — standard; *13* — porous rubber ($h = 5 \cdot 10^{-3}$ m), soaked up with water and coated with PHV film of $h = 2 \cdot 10^{-4}$ m; *14* — same as *13*, h = $10 \cdot 10^{-2}$ m. Kramer's measurements [194]: *15* — standard, *16* — elastic surface with elasticity of $E = 1.03$ MPa and coefficient of absorption small ball beat $\eta = 0.7$; *17* — same as *16*, but $E = 0.103$ MPa, $\eta = 0.88$; *18* — standard, *19–21* — columellar coating; *19* — $E = 11$ MPa; *20* — $E = 5.5$ MPa; *21* — $E = 4{,}1$ MPa; optimal coating [27]; *22* — columellar coating; *23* — ribbed, which fitted with liquid viscosity 7500cST; *24* — same, viscosity = 10000cST. Measurements of Lissamen and Harris [27]: *25* — standard; *26* — membrane surface.

The disadvantages mentioned in [122, 261] do not make it possible to verify whether the stabilizing properties of an elastic surface depend on its mechanical parameters as was shown in [220]. Furthermore, underpressure arises for flows over a membranous surfaces depending on the velocity above the elastic surface; this deforms and tears the membrane from its framework. To eliminate this disadvantage it is necessary either to balance the pressure under the membrane with the flow pressure or to test a monolithic surface.

In [199] an attempt was made to eliminate the disadvantages discussed. Two plates and an airfoil with an elastic coating were investigated in a wind tunnel. Figure 1.19 presents some results of the drag measurements for the 1.22×0.84 m plate. It was made of two aluminum sheets $6.4 \cdot 10^{-3}$ m thick. The leading edge of the plate was shaped as an ellipsoid with the semi-axis equal to $50.8 \cdot 10^{-3}$ m and $6.4 \cdot 10^{-4}$ m. The rear edge was blunt. A turbulence promoter of diameter $3.56 \cdot 10^{-4}$ m was installed in the leading section of the plate. In one of the aluminum sheets was a 1×0.3 m window in which a foam polyurethane sheet was installed. The curves show that the diameter of the turbulence promoter was too small to form a turbulent flow over the plate. The thin film, made of melynex, does not change the boundary layer parameters. The foam polyurethane sheet coated with the same film, gave a drag reduction of 20 percent, although the elastic surface area was only 1/3 of one side of the surface.

Despite the positive results, the manner and location of fixing the elastic surface onto the plate was unsatisfactory, as well as the measurements. Similar experiments in water flow are described in [481]. Two plates were tested in a towing tank of dimensions $80 \times 8 \times 3$ m. The maximal towing speed was 3 m/s, and Re $= 6 \cdot 10^5 - 1 \cdot 10^7$. The elastic surface effect upon both transition from laminar to turbulent boundary layer (the first measurement series) and the turbulent flow over the plate (the second measurement series) was studied. During the first series, the plate was towed vertically and partly submerged, and during the second series the plate was towed horizontally and submerged entirely. The design of plates in the two series differed. In the first series a duralumin plate $6 \cdot 10^{-3}$ m thick was coated on both sides with porous rubber $5 \cdot 10^{-3}$, $6 \cdot 10^{-3}$ or $10 \cdot 10^{-3}$ m thick. On the end faces of the elastic sheets, on the rigid aluminum base, a rigid edging of the same thickness as the elastic sheet was fixed. The width of the edging was $5.0 \cdot 10^{-2}$ m, and its forward part had an ogive shape. The entire plate was coated with $2 \cdot 10^{-4}$ m-thick vinyl film. The plate was 1×2 m in size. The basic standard was a solid aluminum sheet $6 \cdot 10^{-3}$ m thick. The second plate was different in shape and made of acryl sheet $3 \cdot 10^{-3}$ m thick. On this sheet at a distance of $5 \cdot 10^{-2}$ m from the edge of both sides, porous rubber sheets of $5 \cdot 10^{-3}$ m or $10 \cdot 10^{-3}$ m thickness were glued. The forward end of the rubber was cut off at $3 \cdot 10^{-2}$ m from the plate edge. The same film also coated the entire plate. The rubber had been impregnated with air or water.

Figure 1.19 shows some results of the second series. All the results obtained were negative. In the first series only the elastic plate of $6 \cdot 10^{-3}$ m thickness showed a slight drag reduction.

Despite the correctness of the elastic plate fixing, a number of methodical mistakes were made. The plates were structured differently. The leading edge of the second plate created a high-pressure gradient. The elastic plates were different from the solid base standards in shape and thickness. The forces were measured on the basis of dynamometry, which gives errors comparable with the forces themselves. The mechanical parameters of the elastic plates were not optimal for the considered range of towing speeds.

Kramer [108, 315] carried out the most methodologically accurate towing tests of elastic plates. He tested composite plates that had optimal mechanical parameters for the towing speed. It is due to errors in experimental methods that most of the investigators could not reproduce Kramer's results. The elastic surface effect upon friction reduction was investigated by towing of a large aspect ratio model by a motorboat. Detailed data concerning the elastic surface structures and the experimental procedure are shown given in Chapter 2 and in [108, 127, 195, 315]. In the first series of experiments (curves *16* and *17*, Figure 1.19) both an decrease and increase in friction (the latter due to low elastomer durability and its detachment from the base) were registered. In the second series the complex composite columellar coating was tested with varying mechanical parameters of composite. Very promising results were obtained—the friction drag was reduced by 57%. However, the strength of these coatings at high velocities was insufficient. Because of this, a third set of coatings was constructed—ribbed coatings. The effectiveness of these coatings was less than those in the second series but they remained sufficiently high. The behavior of Kramer's coatings is of one pattern and show, as is the transition retardation. The shapes of the curves are similar to curves *9–11* and curve *3*. Curve *20*, for which the value of C_f after a maximum essentially decreases, raises doubts.

So far, the results of Kramer's experiments remain the best. To date, theoretical and physical explanations of the data obtained have not been established. Although Kramer eliminated most of the disadvantages of the above-mentioned methods, doubts about the obtained results remain. However, the experiments were full-scale, and the high level and quality of the experiments is undoubted. A catamaran towed the model with large elongation. The model was submerged between fins to stabilize the flow and was well fixed. The data obtained shows good damping of vibrations by the relevant coatings, because the motor vibrations were transferred to the mounting and then to the model.

In a review paper [354] Hoyt presents the results obtained by Nisevanger. The measurements were made on a 0.915 m-long floating model with the bow part coated with an elastic coating imitating Kramer's coating. Drag increase was registered. Ritter obtained similar results in a hydrodynamic tunnel [197]. The conditions and methods of these experiments are unknown.

Many investigations have been carried out Blick's team in the USA. The results of these studies are shown in Section 1.3. Luney [122] reported drag reduction as high as 60%. For the same plates, Smith and Blick registered drag reduction as high as 37%. However, due to the problems relating to the technique used to make the measurements, there is doubt about the reliability of the specified values, as the elastomer material used (foam polyurethane), and its structure and mechanical properties, do not make it possible to achieve significant drag reduction. It is likely that this drag reduction effect can only be achieved with composite elastic plates. In [344] Blick's results were thoroughly checked: the drag reduction was 10%.

In [137] Boggs and Frey investigated coatings like Kramer's on a hydroplaning boat with zero deadrise. On the basis of these investigations the patent was received. Perier also performed an experiment on influence of an elastic covering on drag reduction for movement in the water environment [137].

Ritter and Massum investigated the friction drag of six types elastic surface stuck to a flat plate [137]. Two of these coatings show drag reduction. One of them, made of rubber and resembling Kramer's coating, gave a 7−14% drag reduction. The other showed a 7−15% reduction in drag, except for in a high-velocity regime, when turbulent fluctuations can lead to large surface waves creating a roughness effect.

Ritter and Portos [137] investigated a coating like Kramer's elastic material, and found friction drag increased in comparison with a solid plate. This negative result was obtained because of the manner in which the coating was fixed. Underpressure arising in the flow over the elastic plate caused a heave of the surface, which increased resistance. The negative result also indicated the unsatisfactory joining of the elastic covering with the cowl.

Experiments by Nisewanger [137, 481] also show drag increasing in comparison with a solid plate, while drag reduction was obtained in [261].

The experimental work gives discrepant results due to following:

- incorrect measuring techniques, because experimenters did not consider the high deformability and absorption of elastic surfaces, nor the dependence of the elastomers' mechanical properties on temperature, etc.;
- there is no clear understanding of the physics of the interaction of an elastic plate with a flow;
- passive elastic coatings are efficient only over a narrow range of flow velocities.

In [42, 285−290, 502, 503] the hypotheses of the damping of fluctuation energy with elastic surfaces has been described, as has the frequency and structure of the interaction between the elastic surface and flow. In addition, the methodology of experimental investigations using elastic surfaces of different types has been developed. As a result, a reliable drag reduction was obtained in many investigations. The increased drag in other experiments was also explained.

Following this methodical basis, Hansen [226], McMichael [356] and Chapter [137] registered drag reduction. Bushnell's team made the most active study in the field under consideration. A detailed review of their investigation with different elastic surface plates is presented in [20, 137]. The most substantial results are:

- a hypothesis concerning the elastic surface effect on pre-burst modulation of the viscous sublayer [137];

- data about the control of elastic plates by means of heating [186];
- the effect is obtained as: $C_{felast}/C_{frigid} = 0.4$[186];
- maximal drag reduction is registered when the natural frequency of membrane vibration is approximately equal to half of the frequency $f_P = \frac{U_\infty}{2\pi\delta}$, corresponding to maximal spectral density of boundary layer fluctuations.

In the latest reviews [137, 195, 354] the conclusion was reached that elastic surfaces have good prospects in the application of drag reduction. However, it was noted that the main disadvantage is a narrow range of velocities. From an analysis of both the authors' papers and those of other investigators, the authors' have demonstrated [289, 305, 310] two mechanisms of elastic surface effect upon friction drag: the first mechanism consists of delaying the transition and the second mechanism consists of the combined action of delaying transition and turbulent friction reduction. The shapes of curves *9–11* are conditioned with the first mechanism and Kramer's curves (curves *15–24*, Figure 1.19) with the second. The other curves in Figure 1.19 show the elastic surface effect upon the turbulent boundary layer (for this a turbulence promoter was installed on the models).

Over a long period of time Semenov, Kulik and others carried out theoretical and experimental research on the influence of elastic coatings on drag reduction [106, 149, 323, 324, 456]. In their research they used monolithic coatings on a silicone rubber base. As is underlined in [106, 149] they applied vulcanized polydimenthylsiloxane, manufactured by Kazan Synthetic Rubber Plant in Russia, which has the structural formula $[-Si(CH_3)_2-O-]_n$. Research on drag reduction with this material has been carried out in Russia on buoyant models, and in England and America on models in a cavitation tunnel at speed up to 6 m/s [106]. Unfortunately, the design of the models and the joining of the elastic insert with cowl are not explained. The most interesting aspect is the technique of measuring the dynamic characteristics of this material [323, 324]. However, neither in the theoretical nor the experimental parts of this research is there any physical representation or model of the interaction of a coating with a flow, in the light of modern understanding of the turbulent boundary layer structure. Interesting research on the influence of ageing on this material has been done [106]. In accordance with the tests of this material in Russia, the characteristics were stabilized and were kept for a long time; they were in the order of 10%. Parallel tests of same material in England and the USA have shown that the material has practically no positive effect, and with ageing its resistance increases.

The greatest mistake in almost all the experimental research until now has been the stereotypical thinking. Theoretical and experimental research on elastic surfaces has been done based on the experience gained using rigid surfaces for various tasks. Elastic surfaces have a completely different molecular structure and set of mechanical characteristics. Only elastic materials and water–polymer solutions possess the property of highly elastic deformation and other specific characteristics (see Section 2.1). Therefore, when investigating elastic materials it is first necessary to study the basic material properties. A lack of this knowledge has led to the wrong choice of fastenings of elastic coverings in almost all previous experiments. Only Kramer in his later designs understood these issues and correctly designed and fixed elastic coatings onto models. Some positive results have been obtained through accidental coincidence of the necessary conditions for carrying out the experiments at several values of flow velocity.

The following conclusions can be drawn from the results of the above-mentioned research on the structure of a boundary layer and, in particular, in the flow over an elastic surface. Considerable progress in this field has been achieved; of special importance are the numerical methods of direct solutions involving Navier–Stokes equations. The experiments have been essentially improved by the development of laser equipment, making it possible to study various kinds of CVSs using contactless means. Tiny sensors of pressure pulsations have also been sufficiently developed. However, a number of important problems have not yet been solved. The fine structure of movement at various stages of development of a transitional boundary layer and the viscous sublayer of a turbulent boundary layer has not been studied in the flow over a flat rigid plate. A theoretical or phenomenological model of the formation and development of various kinds of CVSs has not been created. The problem of susceptibility of the boundary layer over a rigid plate also has not received further development.

In spite of great efforts and some successes in the analysis of flow over an elastic surface, many questions have remained. In particular, there is no clear understanding of the movement of the elastic boundary and its interaction with generated pulsations in turbulent boundary layer. Equations that are still used for movement of an elastic plate were derived based on the experience gained from the theory of elasticity of monolithic elastic plates. However, in view of modern understanding of CVSs in the boundary layer over a rigid plate, the load acting on the elastic boundary from a boundary layer is completely different. Therefore, it is necessary to perform special experiments on elasticity theory problems for the flow over an elastic boundary. Such tasks of hydroelasticity have not been solved. For long time, various kind of closing conditions for a movable wall were used in problems on the flow over an elastic boundary. However, until now there has been no experimental check and physical substantiation of such conditions of closing.

Integral characteristics of a boundary layer have primarily been investigated in the experimental research of flows over elastic plates. Until now research on the fine CVS structure near the elastic boundary has not been done. There is no clear understanding as to what should be an elastic boundary for successful interaction with a boundary layer. There is no physical picture of fluid flow structure over an elastic boundary. The theories created in this area reflect only typified conceptions about the physical character of an interaction with an elastic border. These theories are based on out-of-date assumptions about the flow near an elastic border.

Experimental research on the flow on various kinds of elastic plate has shown that there is a narrow class of designs of elastic plates that will find practical application in the future. However, neither theoretical nor experimental research has been performed in this area. Kramer and Hertel were the first to try to apply the so-called hydrobionics approach to solving this problem. This is based on the modeling of the elastic coatings of live high-speed hydrobionts, which are optimized through evolution. Studying the structure of the skin of hydrobionts can provide an understanding of the physics of the flow above an elastic border, as well as other technological issues. Thus, the overwhelmingly negative experimental results indicate an absence of knowledge about the hydrodynamics of live high-speed hydrobionts. However, these results are also a reflection of other problems: the incorrect fastening of artificial coatings to a base; the wrong design of elastic coatings; errors in experimental technique, etc. All these things predetermined the negative results in the majority of experiments.

On the basis of above-mentioned issues, in the authors' opinion, it is now necessary to do the following:

- develop a method for the experimental investigation of CVSs in a boundary layer;
- experimentally investigate the fine structure of flow at various stages of development of a boundary layer in the flow over a rigid plate;
- on the basis of the obtained results, develop a breadboard model of CVS structure in the transitional boundary layer and specify a similar breadboard model for CVSs in the turbulent boundary layer;
- consider the hydrobionics approach in relation to the investigation of the hydrodynamics of dolphin swimming;
- experimentally investigate the fine structure of flow at the development of a boundary layer over an elastic surface;
- on the basis of hydrobionics approach, develop the problem of susceptibility;
- consider a boundary layer as a wave-guide;
- develop methods of CVS control in a boundary layer;
- give a physical substantiation of the mechanism of interaction of a flow with an elastic border;
- experimentally investigate the characteristics of the turbulent boundary layer and the elastic boundary.

1.6 Experimental Investigations of Coherent Vortical Structures in a Transitional Boundary Layer on the Flow over a Rigid Plate

A low-turbulence hydrodynamic test set up consisting of special equipment and devices was designed and constructed at the Institute of Hydromechanics of the Ukrainian Academy of Sciences (see Figure 1.20). The set up included a rig that was an updated version of that

Figure 1.20

A low-turbulence hydrodynamic test bench: *1* — hose, *2* — nozzles, *3* — dashpot, *4* — diffuser, *5* — pump, *6* — working site, *7* — rubber lining, *8* — adjusting basic plate, *9* — frame of the case, *10* — hoses to return current of water to the pump, *11* — lodgement, *12* — confuser.

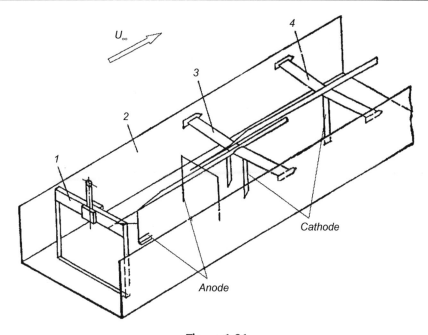

Figure 1.21
Scheme of the arrangement of supports and the vibrator in a working site: *1* − vibrator,
2 − working site, *3* − support for reception of tellurium jets, *4* − support for reception of
velocity profiles.

proposed in [519]. The basic technical data of the test set up was as follows: length of the set up 7 m, length of test section 3 m, cross-section area of the test section 0.09 × 0.25 m, range of operating speeds 0.05−1.5 m/sec, confuser *4* has a contraction factor equal to 10. The following main devices were mounted on the set up *6*: a duplex bottom of the test section and a removable cover of the test section. The set up was equipped with a system for pouring off a boundary layer in corners of the test section. The cover of test section could be tilted to various angles. The photographing cart moves along rails laid near the test section.

The tellurium method [520] was employed and special equipment was constructed for measuring the velocity field and recording the neutral oscillations. Tellurium wires were attached to special supports (see Figure 1.21). To indicate neutral oscillations, small oscillations of various types were introduced into the boundary layer by means of specially designed oscillators. Supports were installed downstream, making it possible to obtain small tellurium jets. The oscillator frequency changed in the course of the experiment; the amplitude of the tellurium jet oscillations was recorded. The voltage applied to the tellurium wires during photography of the velocity profile was approximately 500−600 V, whereas 10−20 V were required for recording the tellurium jets.

To generate tellurium jets, tellurium wires were mounted on supports *3* in the working section *2* at a distance of 10–15 cm from a vibrating tape *1*. Tellurium wires were placed on supports *4* 15–20 cm downstream in order to identify instantaneous velocity profiles. Supports *3* and *4* were fabricated in various forms to mount small tellurium wires. This made it possible to take photographs of velocity profiles in different parallel planes at each time interval. Thus, velocity profiles in three coordinate directions were obtained. Simultaneously, small wires were mounted on other support in order to generate tellurium jets. This indicated the development of fluctuations of sine waves in parallel planes and in all three coordinate axes. Such a technique allowed the simultaneous registration of both average and pulsating movement in a fluid rectangular parallelepiped. Sizes of parallelepiped were determined by the height and width of the working section, and by a length up to 0.5 m, which was the camera image size. However, quantitative measurements were made in a smaller volume in order to omit optical distortions of the camera. Wortmann's tellurium method still remains the unique method for obtaining an instantaneous spatial picture of disturbance development. Two photograph panels were mounted on the cart, making it possible to take pictures through the transparent walls of the test section simultaneously from above and from one side.

Small oscillations were introduced into the boundary layer mainly by using two oscillators. The mechanical oscillator consisted of a frame on which two motors were mounted, making it possible to obtain the required frequency range of the oscillating strip. Amplitudes of oscillations varied due to the eccentricity of the various bearings through which the oscillations were transferred from the axle to the rod. Two pushers were attached to the rod which was fastened to the oscillating strip. The second oscillator consisted of an electric relay that made the above-mentioned pushers vibrate with the strip. This oscillator produced non-sinusoidal oscillations.

The experimental investigation of hydrodynamic stability was conducted on the bottom of the test section of the hydrodynamic set up. The method of obtaining the points on the neutral curve involved determining the maximum amplitude of the tellurium jet oscillations at a fixed point in the test section with the help of photography. The velocity profiles were photographed simultaneously. In this way, the points of the second branch of the neutral curve were determined. The first branch was plotted by comparing the minimum amplitudes of oscillations at the same frequency.

Details of the hydrodynamic set-up design, the device with adaptations and the measurement technique are described in [30, 31, 305, 310, 519, 520]. To obtain reliable results, it was important to use high-quality tellurium wires. The technique of producing the tellurium wires is stated in [305, 519]. A special device for vacuum sputtering the tellurium on the constantan wire was developed. The basic feature at vacuum sputtering is that a constantan wire is reeled up on a cross piece executed from quartz glass. Inside of the flask a little metal tellurium is placed. Sputtering is made in a vacuum chamber by heating the quartz flask to 600°.

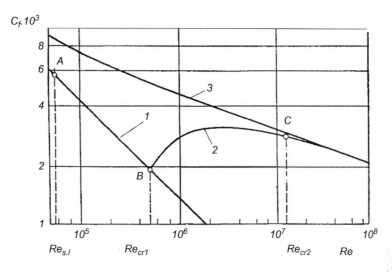

Figure 1.22

Laws of drag for smooth longitudinal flows around a flat plate in laminar (*1*), transitional (*2*) and turbulent (*3*) modes of flow in the boundary layer.

Some photographs of CVS features at various stages of transition are show below. The following classification of transition stages was proposed in [475]: I — instability of small wavy disturbances; II — amplification of 3D wave; III — zigzag-like development along with a longitudinal vortical system; IV — concentration of vortex intensity and growth of shear layer; V — destruction of the vortex; VI — development of the turbulent spot.

The dependence of a smooth flat plate drag on Reynolds number is plotted in Figure 1.22. It shows theoretical curves for laminar flow in a boundary layer (*1*), for a transitional regime of flow (*2*), and for turbulent flow (*3*). In these experiments, Reynolds number of stability loss $Re_{s.l}$, based on x was equal to $5.4 \cdot 10^4$ [305]. Point *A* in Figure 1.22 characterizes Reynolds number of stability loss and corresponds to the beginning of area I of turbulence rise. From point *A* to point *B*, transition areas I–V sequentially originated and developed. From point *B* to point *C*, transition area VI developed. Thus, the process of turbulence appearance should be considered as a continuous process of flow development from point *A* to point *C*, and can be characterized by three Reynolds numbers: $Re_{s.l}$, Re_{cr1}, Re_{cr2}.

Under the most favorable conditions of flow, transition of a laminar boundary layer to a turbulent one would occur from point *A* to point *C*. However, factors such as: ambient turbulence; curvature, waviness, roughness and surface vibration, pressure distribution on the outer boundary of the boundary layer and others can essentially accelerate the origin of turbulence; so that points *B* and *C* begin moving toward the practically immovable point

A. Under unfavorable conditions, the distance between points *A* and *B*, as well as distance between points *B* and *C*, can decrease to very small values. Conversely, in some cases [31, 32] point *A* can be displaced to the right simultaneously with points *B* and *C*, so that distance between them increases dramatically. Velocity profiles in a boundary layer without introducing disturbances have been investigated. In Figure 1.23 velocity profiles photographed using the tellurium method are presented. The nondimensional velocity profile completely coincided with Blasius's theoretical velocity profile [445]. Thus, it confirms that the tellurium method makes it possible to receive instantaneous velocity profiles in a laminar boundary layer. In the domain of the boundary layer, tellurium visualization is precise and, generally, is not washed away. Near the external border of the boundary layer, the tellurium starts to wash away, similar to a nondisturbed flow. However, the external border of boundary layer differs very clearly.

Similar velocity profiles were photographed when two tellurium wires were mounted at various *x* coordinates. In all cases, the velocity profiles were identical, which confirms the low-turbulence degree of the flow and plain-parallel flow in the hydrodynamic set up. At the same time, when sine wave disturbances were introduced in the field of the maximal

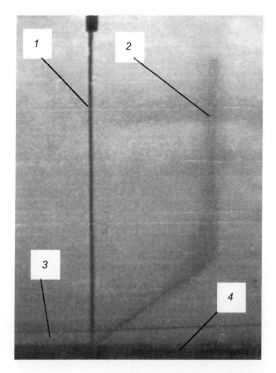

Figure 1.23
Velocity profile of a laminar boundary layer over a flat rigid plate at $x = 74$ cm from the start of a plate: *1* — tellurium wire, *2* — velocity profile, *3* — tellurium jet, *4* — plate.

amplitudes of disturbances the velocity profiles had excesses according to phases of fluctuations. Such velocity profiles are presented in [520] and will be partially presented below at visualization of the flow by means of tellurium jets.

Velocity profiles have also been photographed for the situation when tellurium wires were placed parallel to the plate. A vibrator was not mounted. The tellurium wire was situated along the OZ axis, parallel to the bottom at a height of 5.5 mm. The flow velocity was 18 cm/s. Figure 1.24 presents the velocity profiles simultaneously photographed from above

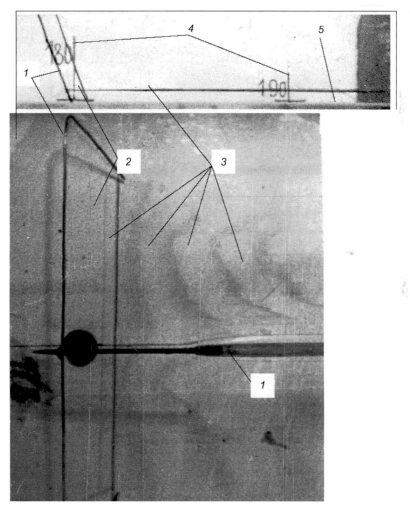

Figure 1.24
Simultaneous photographs of velocity profile from one side (*a*) and top (*b*) at $U_\infty = 18$ cm/s and $y = 5.5$ mm: *1* — support for visualization of velocity profiles, *2* — tellurium wire for visualization of velocity profile, *3* — velocity profiles, *4* — distance mark from the start of the working section (cm), *5* — plate.

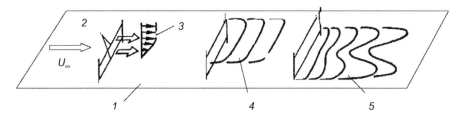

Figure 1.25

Scheme of deformation of a velocity profile in a boundary layer: *1* — plate, *2* — tellurium wires, *3* — velocity profile in plane xoy, *4* — velocity profile in plane xoz in a non-disturbing flow, *5* — velocity profile in plane xoz in a critical layer of a boundary layer.

and one side. It is possible to see the deformation of velocity in the XOZ plane. When the tellurium wire was located in the main flow, a deformation of the velocity profile was not observed.

Figure 1.25 shows velocity profile deformation in the transversal direction along the plate. It was revealed that the velocity profile in the transversal direction is better at illustrating the development of the disturbances along *x*. Unlike the velocity profile in plane XZ, it is practically impossible to measure similar deformations of the velocity profile with the velocity profile in plane XY. Even at low degrees of turbulence of flow and in the absence of introduced disturbances, 3D deformation of plane natural disturbance is observed. This process was registered by photographing the velocity profile presented in Figures 1.24 and 1.25. When the tellurium wire was placed at the coordinate of a critical layer (approximately displacement thickness) the deformation of a velocity profile becomes essential. Depending on the coordinate, the development of such deformation varies, which finally leads to a transformation of the initial small plane disturbances into small 3D disturbances. Disturbance movement along a plate at the natural development of boundary layer has also been investigated by means of tellurium wires, which generate tellurium jets into the boundary layer. The support was in the form of a thin horizontal strip located across the flow. Eight tellurium wires were soldered to the strip. The edges of the strip were up-curved by 90°. Three tellurium wires were soldered to these vertical strips. This allowed for the generation of tellurium jets simultaneously in the horizontal and vertical planes and made it possible to analyze the development of disturbance movement in the flow volume. Visualization of the flow has shown that at low levels of turbulence in the basic flow, jets develop plane-parallel without disturbances on the entire length. Figure 1.26 presents simultaneous photographs of sine-wave and non-sine-wave disturbances taken from one side and above. Visualization has shown that the disturbing movement is quickly transformed from plane into 3D. Double jet thickness is obtained at the folding of a sine-wave disturbance, and irregular steady spatial loop at a non-sine-wave disturbance.

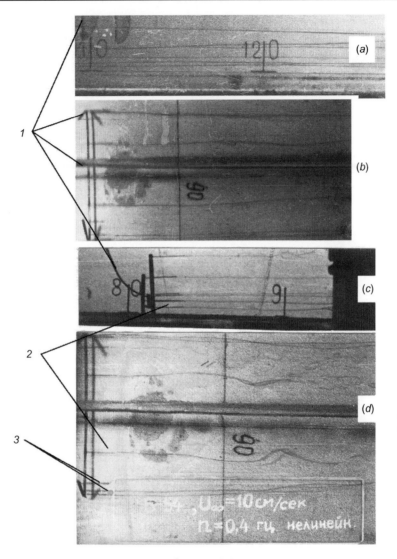

Figure 1.26
Simultaneous photographs of disturbance movement from one side (*a, c*) and top (*b, d*) at sine wave (*a, b*) and non-sine wave (*c, d*) fluctuations at $U_\infty = 10$ cm/s: *a, b* − *n* = 0.3 Hz; *c, d* − *n* = 0.4 Hz; *1* − support, *2* − bottom wire and jets, *3* − top wire and jets.

Development of the amplitude of normal velocity v' along the test section of the hydrodynamic set up is presented in Figure 1.27. In each photograph, the oscillation frequency was close to that of second neutral oscillation, according to the neutral curve [32]. It was determined that at the beginning of the plate, in the region of the point of stability loss, amplitudes of the velocity oscillations v' were the most sharp and the wave-generating process was expressed more clearly. From photos it is possible to see that

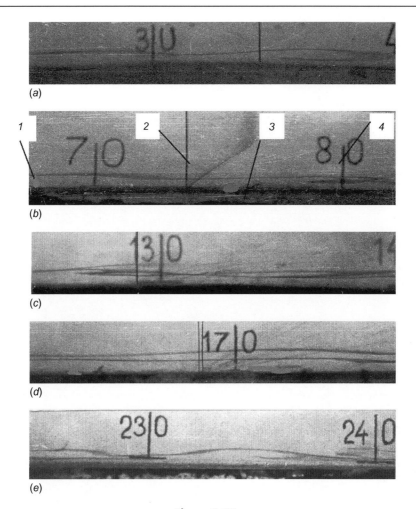

Figure 1.27
Photographs of disturbing motion at different places along the plate: $a - U_\infty = 10.5$ cm/s,
$n = 0.55$ Hz; $b -10.5$ cm/s, 0.5 Hz; $c - 0.5$ cm/s, 0.4 Hz; $d - 16$ cm/s, 0.37 Hz; $e - 16$ cm/s,
0.3 Hz; 1 – tellurium jet, 2 – wires from visualization velocity profile, 3 – plate, 4 – distance
mark from the beginning of the working section (cm).

in the beginning of a plate, in the filed of a point of loss of stability, amplitude of
fluctuations of speed v was the greatest. The process of formation and development of a
wave has been expressed more clearly. During the development of a wave the crest is
extended forward and upward. At the greater coordinate x, the wave becomes flatter and
folds and turns up becoming more extended.

Figure 1.27 shows that fluctuations develop to nonlinear very rapidly. In one wavelength
the crest of the wave is turned and extended forward. However, if this leads to forming by

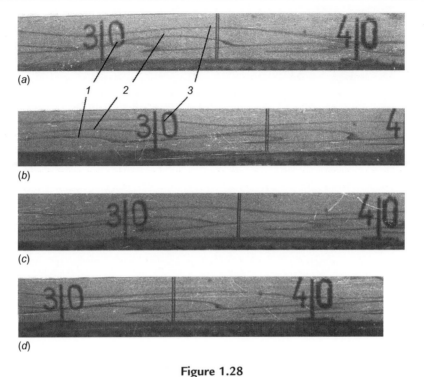

Figure 1.28

Photographs of consecutive phases of development of disturbing movement: *a–d* — consecutive still ober 0.5 s; *1, 2, 3* — tellurium jets are injected at various distances *y* from plate.

only the top part of the wave a sine-wave curve will result. After the collapse of the waves crest development of a fluctuation in the boundary layer is not observed. It seems as though the wave moves to a field of photographing as if it were frozen without deformation and development. It is possible to assume that such a shape of a wave is determined by the reflection of the wave from the rigid wall. With growth Re the length of a wave while turning is not fixed but increases. However, development of a wave has similar character, as in the zone of stability loss.

The process of wave generation and the propagation of disturbances along the plate is represented in Figure 1.28. Three tellurium jets were photographed simultaneously. The interval between frames was 0.5 s; the frequency of the applied oscillation *n* was 0.83 Hz, and U_∞ of 10.5 cm/s. The numbers on the photographs denote *x*-coordinate, taken from the beginning of the working section. Figure 1.28 shows how the generation and folding of the wave crest occurs, as well as the propagation of disturbances across the boundary layers. It is evident that the characteristics of a wave (amplitude, length and form) are different and depend on the distance from the rigid plate.

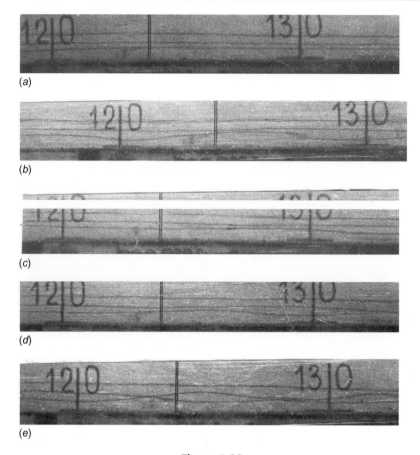

Figure 1.29
Photographs of velocity amplitudes V depending on oscillation frequencies at $U_\infty = 13$ cm/s:
$a - n = 0$, $b - n = 0.68$ Hz, $c - n = 0.79$ Hz, $d - n = 0.87$ Hz, $e - n = 1.3$ Hz.

Figure 1.29 shows visualization of disturbances movement in the boundary layer depending on frequency of the oscillation strip. It is clear that the characteristics of disturbing movement essentially vary at different frequencies. Estimating the fluctuation amplitude and other parameters of disturbing movement, one can define the frequency of neutral fluctuation for the given place of measurement.

Similar measurements were done at a positive pressure gradient. All measurements were carried out at the bottom of the hydrodynamic set up, on which lateral marks on the glass indicated the distance from the beginning of working section, i.e. from the beginning of the bottom. It is necessary to note that real beginning of the coordinates of the plate was in the confuser area, because there was the beginning of the boundary layer thickness. Therefore, the real coordinate x was defined by the velocity profile measured in each station of measurements. The design of working section was such that it had two bottoms,

Figure 1.30
Photographs of velocity profiles at a positive pressure gradient.

the technological and second one, above which measurements were made. A positive gradient of pressure was created as the end of the upper bottom descended to touch the technological bottom. Thus, a long diffuser was created.

Figure 1.30 presents the photograph of a velocity. Each 0.5 s electric contacts were sealed in and a tellurium cloud was injected in the flow, which visualized a velocity profile. It was possible to obtain successive velocity profiles, to investigate the dynamics of deformation of the velocity profile under certain conditions of experiments. In that photograph, two velocity profiles and tellurium jets are presented. The vibrating tape was absent. Despite that, the tellurium jet has fluctuations because natural disturbances began to develop in the boundary layer at a positive pressure gradient. The velocity profile differs from those presented in Figures 1.23 and 1.27b. At a positive gradient, the velocity profile is fuzzier and has inflection on practically the entire thickness of the boundary layer. Markers indicate the horizontal position of the second bottom. The markers drawn (with numerals) on the distant glass of the working section are sharp. Nearby markers are fuzzy. These markers make it possible to define distortions in photographs as the photographing was done with the optical lenses, which had changed the focal length of the camera.

Figure 1.31 shows velocity profiles photographed from above. At the start of the working section where the pressure gradient influenced the flow insignificantly, the velocity profile was rather uniform. Its deformation was regular. In the absence of a pressure gradient distortions of velocity profiles were practically not observed in the same place. Downstream, 3D deformation of the velocity profile essentially increased. In comparison with similar measurements in zero-gradient flow (see Figure 1.24) the positive pressure gradient essentially strengthened 3D deformation of the boundary layer disturbances at a lower velocity and smaller coordinate. The obtained results add to and agree well with results by Wortmann [520].

Figure 1.32 shows the development sine-wave disturbances introduced in a boundary layer at a positive gradient of pressure. The characteristics of the disturbing movement have

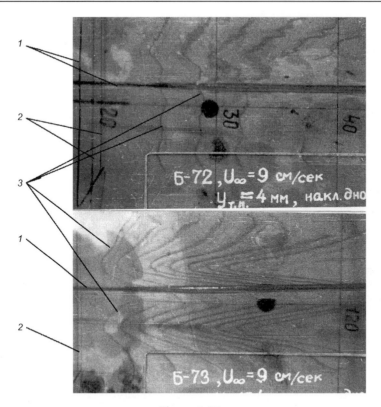

Figure 1.31
Photographs of velocity profiles at a positive pressure gradient: *a* — at the start of the working site, *b* — in the middle of the working site, *1* — support, *2* — tellurium wires, *3* — velocity profiles.

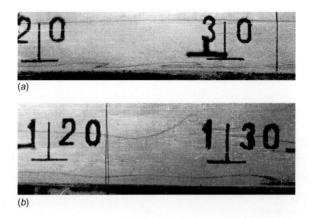

Figure 1.32
Photographs of velocity amplitudes v' at different places along the plate: $a - U_\infty = 9.5$ cm/s, $n = 0.77$ Hz; $b - U_\infty = 11.3$ cm/s, $n = 0.7$ Hz.

essentially changed in comparison with zero-gradient flow. At the start of the working site the disturbances became essentially nonlinear. The wave crest changed in another way: there was slow advancing of a wave onto its decelerated part, then a folding of the wave and further distribution of tellurium jets occurred in the form of a frozen loop. As the *x*-coordinate increased, disturbances propagated across the entire thickness of the boundary layer.

Measurements have shown that in each place along the working section, at $U_\infty = const$ there is a narrow range of frequencies with maximal amplitude of fluctuation. Moving downstream along a working section it was revealed that the maximum of amplitudes moves to the area of lower frequencies. Increasing the speed of flow over the plate, for the same coordinate *x*, leads to a shift of maximum amplitudes of oscillations in opposite direction, into the area of higher frequencies.

When values of speeds v' and w' were fixed, maximal amplitudes in both cases coincided. Depending on flow speed, at the specified distance between the source of disturbances and measurement station, the amplitude of disturbances was 5−10 times greater and then, as a rule, faded less intensively.

The influence of location of the oscillating strip $\bar{y} = y/\delta$ over the boundary layer thickness on the amplitude of fluctuation of transversal velocity v' of disturbances was also investigated. The obtained dependences are presented in Figure 1.33a for $\frac{\beta_r v}{U_\infty^2} = 430 \cdot 10^{-6}$, Re* = 570 (continuous lines) and for $\frac{\beta_r v}{U_\infty^2} = 255 \cdot 10^{-6}$, Re* = 810 (broken lines).

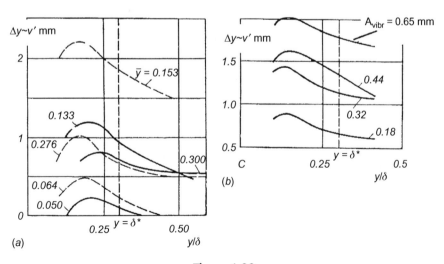

Figure 1.33

Distribution of amplitudes of velocity v' across the boundary layer thickness depending on sizes $\bar{y}(a)$ and A_{vibr} (b).

Dependences for velocities u' and w' have a similar character. Amplitudes of disturbance movement are maximal when the source of disturbances lies within the limits $(0.1-0.3)\bar{y}$.

Results of research of the distribution of transversal velocity across the boundary layer thickness at various values of A_{vibr} are presented in Figure 1.33b for $\frac{\beta_r v}{U_\infty^2} = 430 \cdot 10^{-6}$, Re* = 570. The velocity v' is in direct proportion to size A_{vibr}. At $A_{vibr} > 0.7$ mm, the velocity v' becomes so large (accordingly u' and w' increased) that a vortex forms immediately behind the vibrator. Thus, maximal values of velocity components of disturbances are within the limits $(0.15-0.3)\bar{y}$.

1.7 Distribution of Disturbing Movement across the Thickness of a Laminar Boundary Layer over a Rigid Surface

The dimensionless wave number $\alpha\delta^*$ and phase velocity c_r/u of disturbance motion are practically independent of the disturbance source location across the boundary layer thickness [305, 310]. At the same time, the measurements have shown that values $\alpha\delta^*$ and c_r/u depend on the amplitude of the oscillator (see Figure 1.34). When the disturbance amplitude rises to 0.7 mm, $\alpha\delta^*$ and c_r/u tend to attain ultimate values throughout the entire

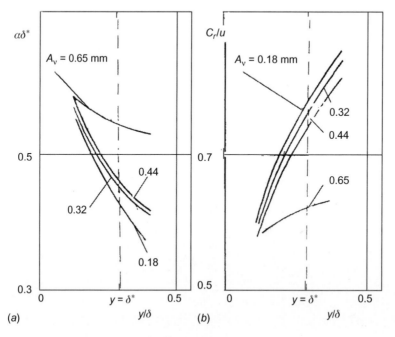

Figure 1.34

Influence of fluctuation amplitude of a vibrator tape on distribution of wave number and phase velocity of a disturbing movement on a boundary layer thickness.

thickness of the boundary layer under the given experimental conditions, and at $\beta_r \cdot v/U^2 = 525 \cdot 10^{-6}$, $\mathrm{Re}_{\delta^*} = 653$ these being 0.5 and 0.65 respectively.

It has been found that, for investigations of hydrodynamic stability, the following conditions have to be present: the degree of turbulence in water should not exceed 0.04%; the amplitude of strip oscillations should be of the order 0.2−0.4 mm at a velocity in the range 10−20 cm/s. The transversal velocities v' generated by the oscillator should not exceed 2% mean flow velocity.

The distribution of wave numbers and phase velocities of the disturbing motion across the boundary layer thickness was investigated at optimum positioning of the oscillating strip and its optimum amplitude (see Figure 1.35). It was revealed that at increasing Reynolds number these values change across the boundary layer thickness more drastically. For each frequency of fluctuation there is a certain character of the dependences, shown in Figure 1.35. At lesser $\bar{y}$ magnitude of $a\delta^*$ increases, i.e. wavelength decreases.

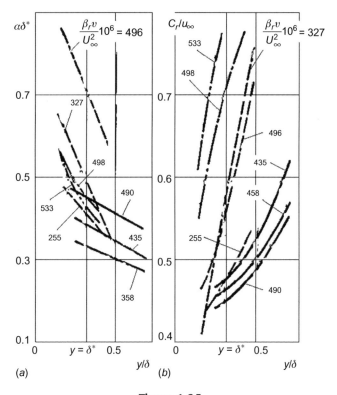

Figure 1.35

Distribution of wave number (*a*) in phase speed (*b*) of a disturbing movement along thickness of a boundary layer depending on frequency of fluctuations: Re* = 404 (continuous curves); 654 (dot-dashed), 810 (shaped).

The character of the dependence for phase velocity is opposite: approaching the surface the magnitude of phase speed decreases.

In the area of $Re_{loss\ st.}$ ($Re^* = 404$) the described laws have another characteristic in comparison with other Reynolds numbers. It is only in the area of stability loss that wave numbers change moderately across the boundary layer thickness. At growing x, the dependences essentially change and depend on the frequency of fluctuation. Phase velocity is maximal and essentially changes across the boundary layer thickness. Moving away from the wall, the wavelength and phase speed increase.

Such complex patterns of disturbance development occur as early as at the initial stages of transition and make it possible to draw a conclusion. Under real conditions disturbances that are introduced into a boundary layer from a wall or from an undisturbed flow will cause fluctuations that have various wavelengths and phase speeds across the boundary layer thickness. This will lead to a complex character of disturbance development in the boundary layer, and to the development of a spectrum of fluctuations. Only in the critical layer, in the area of stability loss, will fluctuation develop according to laws of the linear theory of stability.

In order to solve the Orr–Sommerfeld equation the following basic assumption was made: in the region of critical layer $c_r \approx u$ at any frequency of neutral fluctuation. Figure 1.36 presents measurements of distribution $c_r \approx u$ versus δ (u is local velocity at the point of measuring, see Table 1.1). General behaviors are opposed to that in Figure 1.35, where it can be seen that the value of c_r increases with growth of value y/δ. The growth rate of averaged velocity u at a point of measurement with the increase of y/δ forestalls the growth of velocity c_r, which stipulates the behaviour of curves in Figure 1.36. From above data it follows that $c_r \approx u$ at $y = \delta^*$ only when the frequency of fluctuation of the disturbing motion is equal to the frequency of neutral oscillation, and the disturbance is introduced and located in the region of the critical layer. In all other cases, when the frequency of the disturbing motion is more or less than that of the second neutral oscillation, $c_r \approx u$ only at $y/\delta = 0.4-0.6$. More at the ratio n/n_{II} than at the higher values of y/δ the mentioned equality is valid, and the curves are steeper. The amplification of oscillation amplitude (curve 8, Figure 1.36) does not change this law, although $c_r \approx u$ at smaller values of y/δ (curves 6, 7, Figure 1.36). The general distinctive features of the laws are as follows: near the wall c_r considerably exceeds u, and the larger x the greater the difference; above $y/\delta = 0.4-0.6$ value c_r becomes less than u. However, near the wall the fluctuations damp due to viscosity in spite of large values of c_r. Nearby $Re_{loss\ st.}$ the curves, as functions of y/δ, change less strongly in comparison with that at large x.

Figure 1.37 shows the dependences of c_r/u on oscillation frequency. It was found that if $y_{vib.} \approx y_{tel.wire} \approx \delta^*$ then, under all test conditions for (x, U_∞), $c_r \approx u$ only at $n = n_{II}$, which is the frequency of the second neutral oscillation, but at $n \neq n_{II}$ this relation becomes

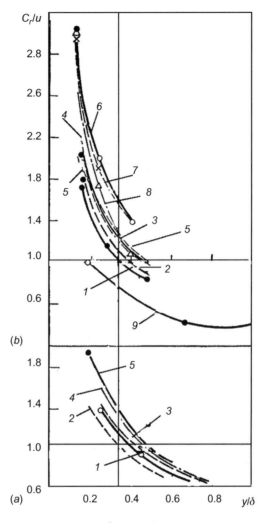

Figure 1.36

Distribution of dimensionless speed c_r/u on a boundary layer thickness in region $\text{Re}_{\text{st.loss}}$ (*a*) and at $\text{Re} > \text{Re}_{\text{loss st.}}$ (*b*): $a - x_1 = 0.4$ m, $\bar{y} = 0.248$ ($1 - n / n_{II} \approx 1$, $n_{II} = \beta_r \cdot \nu/U_\infty^2 = 4.4 \cdot 10^{-4}$; $2-1.02$; $3-0.82$; $4-1.12$; $5-1.29$); $b - x_1 = 1.3$ m ($1 - \bar{y} = 0.256$ and $n / n_{II} = 1.05$; $2-0.45$ and 1.12; $3-0.256$ and 1.8; $4-0.256$ and 1.43; $5-0.256$ and 2.18; $6 - n / n_{II} = 2.34$ and $A_v = 0.32$ mm; $7-2.34$ and 0.18 mm; $8-2.34$ and 0.65 mm; $9 - n / n_{II} = 1$ (membrane surface, experiment B-37 [305]).

$c_r/u > 1$. At frequency $n < n_{II}$, which is in the region of the unstable oscillation of neutral curve, c_r is closer to u than when crossing neutral curve into the region of ... stable oscillations. At $n > n_{II}$ the value of c_r increases drastically at the beginning, until it becomes $1.2-1.5$ times u, and then as n increases, the value of c_r decreases smoothly. It is typical

Table 1.1: Conditions of carrying out of experiments.

Number of Curve	Number of Test	x_1 (m)	U_∞ (m/s)	u (m/s)	$y_{tei.\,wire} \cdot 10^3$ (m)	$y_{vibr} \cdot 10^3$ (m)	$\delta^* \cdot 10^3$ (m)	n_{II}	n_I	$\bar{y} = \dfrac{y_{t.w}}{\delta}$	y_w/δ	δ^*/δ
1	XXXI	0.86	0.18	0.059	3.5	2.0	3.79	1.25	0.53	0.31	0.176	0.3
2	XXXII	1.16	0.18	0.064	4.0	2.0	4.42	0.93	0.40	0.272	0.14	0.3
3	XXXIV	1.4	0.18	0.056	4.0	2.0	4.85	0.79	0.36	0.25	0.125	0.3
4	XXXV	0.9	0.18	0.050	3.0	2.0	3.88	0.98	0.51	0.23	0.16	0.3
5	XXXVIII	1.78	0.17	0.090	6.0	5.6	5.6	0.91	—	0.33	0.20	0.3
6	XLI	2.39	0.105	0.052	4.0	—	3.56	0.69	—	0.29	0.18	0.3
7	B4,B9	1.04	0.105	0.040	3.2	2.5	5.0	0.5	0.3	0.19	0.15	0.3
8	B10,B11	1.3	0.13	0.0572	3.2	2.9	6.28	0.5	—	0.155	0.145	0.3
9	Membr.	—	—	—	5.9	5.15	—	—	—	0.285	0.256	—
10	Membr.	—	—	—	1.0	—	—	—	—	0.482	—	—
11	B48	1.55	0.1	0.034	3.5	3.3	6.84	0.3	—	0.3	0.19	0.3
12	B59	0.51	0.117	0.042	3.5	2.9	4.96	1.0	—	0.29	0.22	0.3
13	B65	0.4	0.095	0.0333	3.5	3.0	4.22	0.7	—	0.31	0.17	0.3
14	B66	1.2	0.113	0.0396	4.0	—	5.7	0.45	—	0.25	0.2	0.3

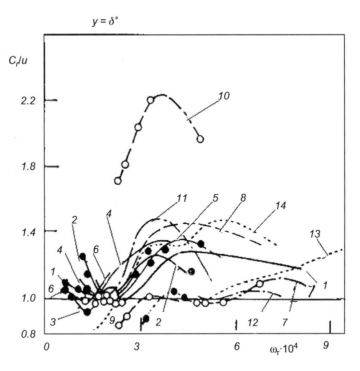

Figure 1.37
Dependence of phase speed on frequency of fluctuation: designation of curves 1, 2, 4 are presented in Table 1.1, the curves 3, 5, 6 and 7 correspond to numbers of curves 5, 11, 13 and 8 in Table 1.1; the curves 8, 9, 10 correspond to a membrane surface, and 11 — generalized dependence for a visco-elastic surface.

that as x decreases or U_∞ grows value of n increases, that is $\beta_r/U_\infty \neq \omega_{rII}$, which is accompanied by curves smoothing (curves 7, 12, 13, Figure 1.37), and not far from $\mathrm{Re_{l.s.}}$ at any n practically the equality $c_r/u \approx 1$ is valid.

Thus, depending on the frequency and location through the boundary layer thickness of introduced disturbances, the velocities of their propagation are significantly different. Only disturbances that are introduced in the region of critical layer at $\mathrm{Re_{l.s.}}$, are propagated with velocity $c_r = u$ in a wide region of frequencies. A disturbance generated by a wall (at $y/\delta < 0.33$), for example, propagates faster, while one that comes into boundary layer from outside ($y/\delta > 0.33$) propagates more slowly than u. Generally, when these disturbances come to the region of the critical layer they propagate with velocity $c_r = u$ at $n = n_{II}$. The following rules also remain in force: oscillations with a frequency greater than n_{II} propagate with a velocity $c_r > u$. It is this that underlies the occurrence of additional harmonics. It also follows from the data in Figures 1.36 and 1.37. Even if, in an ideal case,

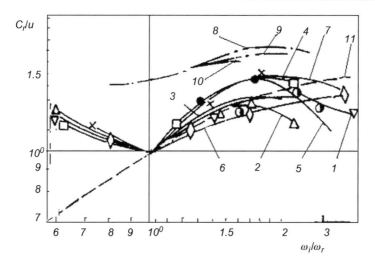

Figure 1.38

Dependence of phase speed on dimensionless frequency of fluctuation. The designation of curves is given in Figure 1.37.

the disturbances are generated in a boundary layer in the area of the critical layer only at frequency of the second neutral fluctuation, when they reach neighboring areas on the thickness of a boundary layer, phase speed automatically changes in this or that direction. New fluctuations will appear in the neighboring layers of the boundary layer thickness, which will begin to interact with the basic introduced fluctuation. This is the reason for the form of the fluctuations, as shown in the photographs in Section 1.5. The nature of shift flows automatically and leads to the occurrence of nonlinearity of the introduced harmonious fluctuation and the formation of the subsequent nonlinear stages of development of disturbing movement in the transitional boundary layer.

Figure 1.38 contains the data from Figure 1.37, which were made dimensionless by the frequency of second neutral oscillation ω_{rII}. Taking into account the accuracy of the experiments, this gives:

$$c_r/u = (n/n_{II} - 1)^2 \quad \text{at} \quad n < n_{II}, \tag{1.3}$$

$$c_r/u = 1{,}6 \ \lg(n/n_{II}) + 1 \quad \text{at} \quad n > n_{II}. \tag{1.4}$$

The investigations have shown that at each point along the working section oscillations were observed in a strictly definite frequency range. Figure 1.39 represents the dependences of wave numbers and phase velocities on disturbing motion oscillation frequency. With the maximum values of these points a region has been plotted, limited by a dot-dashed curve, in which no oscillations at any frequency were observed in the boundary layer under the given conditions of the experiments. The curve has been named the limiting neutral

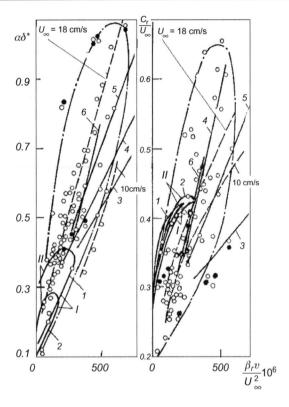

Figure 1.39

Dependence of wave number (*a*) and phase speed (*b*) on the frequency of disturbing fluctuations. Calculated data: *1* — Tollmien [446] for dimensionless frequency and Shen [443] for phase speed; *2* — Shlichting [446]; present measurement: I and II-branches of a neutral curve; *3* — $U_\infty = 9.5$ cm/s and *4*—11.3 cm/s (measurement at a positive pressure gradient); *5* — $U_\infty = 9.1$ cm/s and *6*—11.5 cm/s (measurement at non-sinusoidal disturbances).

curve [305, 310]. As measurements were done at $A_v = 0.32$ mm and 0.05%, at an increase in these parameters the limiting neutral curve attains a certain value. It limits the area of instability connected with nonlinear effects and final disturbance ranges. Dashed curves show dependences of wave numbers and phase velocities on the oscillation frequency along which the cluster of points settle down at a definite velocity of the mean flow. Neutral curves were not plotted experimentally within these coordinates. Despite the scatter of the experimental points, the obtained measurement results have shown that certain regularities are observed between the wavelength and phase velocity on the one hand, and the oscillation frequency of the exciting motion on the other. These regularities are, however, not single-valued and depend on the mean flow velocity.

The dark points characterize the parameters received from measurements of speed.

Neutral curves plotted in non-traditional coordinates, "wave number—phase velocity," have made it possible to obtain the empirical dependences for disturbances that develop with frequency of the second neutral oscillation:

$$\alpha \delta^* = 1.4 \cdot 10^3 \frac{\beta_r \upsilon}{U_\infty^2} + 0.15$$

$$\frac{c_r}{U_\infty} = 0.7 \cdot 10^3 \frac{\beta_r \upsilon}{U_\infty^2} + 0.2$$

(1.5)

Here $\alpha = 2\pi/\lambda$ is wave number of disturbing fluctuation; λ length of T−S wave; δ^* replacement thickness, $\beta_r = 2\pi n$ circular frequency; υ coefficient of kinematic viscosity.

1.8 Physical Process of Laminar—Turbulent Transition of a Boundary Layer over a Rigid Plate

The results mentioned above make it possible to answer the questions necessary for an understanding of the physical picture of flow in a boundary layer:

*   What influences the occurrence of the initial stage of transition?
*   How do the subsequent stages of transition appear and what defines the law of their alternation?
*   What factors influence the process of transition of a boundary layer and how?
*   What are features of flow at different stages of transition?
*   How does uniformity of flow pattern in a boundary layer display itself, etc.?

Even in perfect conditions of flow there are preconditions to the occurrence of transition which relate to the nature of a boundary layer. The gradient of increase of a boundary layer thickness is very great at the beginning of a plate, while the absolute value of the boundary layer thickness is small. Therefore the critical layer in which there is an exchange of energy between the mean and disturbing movements shifts rather close to the external border of the boundary layer [305].

Following from photographs of Blasius's velocity profile, the thickness of a boundary layer and its external border are defined rather precisely. At the beginning of a plate the thickness of a boundary layer changes very drastically; therefore, the approaching flow is directed at large angle relative to the externally border of the boundary layer. So, at the beginning of plate, on the external border of the boundary layer, there are additional stresses caused by the distortion of streamlines of undisturbed flow. At insignificant irregularities of flow caused, for example, by its non-parallelism, those stresses lead to the appearance of sine-wave fluctuations of the flow and local pressure on the external border of the boundary layer. In addition, as the critical layer in this place is located very close to the external border of the boundary layer these fluctuations easily get into the critical layer. Therefore, fluctuations will start developing in the critical layer, and unstable fluctuations can appear.

While flow velocity is insignificant, amplitudes of such random fluctuations are small and damped by the flow. At increasing speed, the energy of the mean stream and energy exchange in the critical layer increases. Amplitudes of fluctuations increase, which leads to the development of T–S waves. In relation to this, the stability of flow at the beginning of a plate is weak. This explains the location of the stability loss point, its characteristic attributes and the shape of the neutral curve.

It is known that a body with elastic properties is characterized by a range of self-frequencies caused by a ratio of viscosity and inertial forces. In addition, while the viscosity forces in a liquid basically depend on temperature, the inertial forces depend on flow velocity. Therefore, the ratio of viscosity and inertial forces changes at different flow velocities. Any flow velocities getting into a boundary layer will cause an oscillatory process, which will develop according to the parameters of disturbance movement and self-frequencies of a liquid. Therefore, in all cases, during the initial stage of interaction of disturbances and mean flow, the process of their interaction will occur in the form of a sine wave. This is the physical basis of the shape of the neutral curve, which describes the area of self-frequencies of fluctuations in a laminar boundary layer.

The sine-wave process will have various frequency–amplitude characteristics and time of existence, will fade, intensify or result in turbulence, depending on various factors—type of disturbing movement, energy and frequency range, place of appearance of disturbing movement on x and y, velocity of the mean flow and its degree of turbulence, sharpness or smoothness of the front edge of the plate, its roughness, and so on.

There are some special cases of occurrence of turbulence, and these can be considered on the basis of a neutral curve [305, 310, 444, 446 etc.]. Assuming that initial turbulence of a flow is not the source of disturbance movement, then at a low degree of turbulence ($\varepsilon < 0.05\%$) and small amplitude of disturbing movement ($v'/U_\infty < 1.5\%$) the spectrum of frequencies of disturbance movement with a range of the order $\frac{\beta_r v}{U_\infty^2} 10^6 = 40 - 429$ (and above), passing the area of instability, will not have time to increase the amplitude of fluctuation up to the critical magnitude that can lead to the appearance of turbulence. Such disturbing movement will always be stabilized. The low-frequency spectrum of disturbance movement with a range $\frac{\beta_r v}{U_\infty^2} 10^6 = 0 - 40$ will pass the area of instability that has the greatest extent on x in the diagram of a neutral curve. It means that the development of turbulence in this case depends on that coordinate x where this disturbance is introduced into a boundary layer. If a disturbance is introduced before the points of stability loss, there is enough time for amplitude of fluctuation of disturbance movement to increase above a critical value.

With increase in initial turbulence or amplitude of disturbances, or under other adverse conditions, the whole spectrum of frequencies of unstable fluctuations can lead to the appearance of turbulence, because increasing fluctuation amplitudes in the instability region of ordinary and limiting neutral curves occurs very rapidly. The maximal amplitude of disturbance movement according to the measurements done was reached when the source of disturbances was at the distance $(0.1-0.3)\bar{y}$. Hence, the magnitude of increase of fluctuation amplitudes also depends on where on the boundary layer thickness the disturbances were introduced, and at which frequency.

Thus, it is possible to note three features:

- there are only sine-wave disturbances in the flow over a plate that is stabilized in time;
- sine-wave fluctuations also increase disturbances sequentially, passing through all stages of transition;
- under worsening conditions, the rate of increase is so great that stages of turbulence follow one after another very fast.

Figure 1.40 presents different variants of disturbing motion behavior, which were obtained from the analysis of the results presented in Sections 1.5 and 1.6. The behavior of a tellurium jet at $x =$ const at increasing oscillation frequency is shown in Figure 1.40a. When frequency increases, the shape of the wave and crest change. At first wave, the amplitude increased and wavelength decreased; then, when the frequency was increased further, the amplitude decreased too. With an increase in the frequency the wave crest began deforming, stretching forward and upward relative to the flow, sharpening, gaining a saw-like shape. After reaching certain frequency it folded fast and in that "formed" state continued moving downstream, not going to ruin. When frequency increased further, the effect was as a reverse process of wave crest deformation, but without the generation of the saw-like shape stage. Wave amplitude became smaller and smaller, wavelength decreased further, and wave shape became plain and purely sinusoidal. At last, starting from a certain "limiting" frequency at its furthest growth, the tellurium stream became smooth, as it had been at the start of the tests before the disturbances had been introduced into the boundary layer.

Such evolution is similar to the development of a tellurium jet along x at $n =$ const, and to the evolution of streamlines at subsequent stages of transition.

It was determined that for each frequency of oscillation a certain distance between the visible oscillations and the folding of the wave existed. That distance was conditionally named the zone or stripe of stabilization. When frequency of the oscillation increased, the stabilization zone decreased at first, and then increased. Near the second neutral oscillation, the zone of stabilization was at its minimum. With the growth in frequency, the stabilization zone simultaneously moved upstream to the vibrator strip. Figure 1.40a

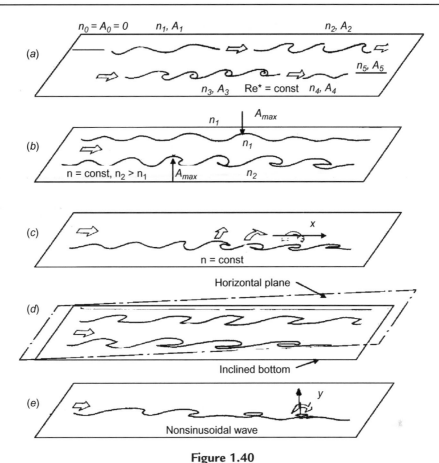

Figure 1.40

Schemes of development of disturbing motion along the plate. Arrows show the direction of disturbance movement: a — Re* = const; b — n = const, $n_2 > n_1$; c — n = const; d — inclined bottom; e — non-sinusoidal wave fluctuations.

presents a process of development of disturbing motion that would correspond to the motion in the plane of neutral curve upwise and in parallel with y-axis along the curve Re* = const.

A scheme of the behavior of the tellurium jets propagating along the plate at n = const is shown in Figure 1.40b. The propagation of the tellurium stream, which characterized the disturbing motion, revealed that the amplitude of oscillation downstream gradually increased in accordance with the shape of the neutral curve. After passing the location on a plate that corresponded, for a given frequency, to the second neutral oscillation, the amplitude began to decrease. Observations of the disturbing constant, but different, frequency oscillations ($n_2 > n_1$) propagating along the plate showed that with the growth of frequency the amplitude of the tellurium jet oscillations first grew sharply, then decreased slowly after passing the location corresponding to the second neutral oscillation. With that, depending on the position of measurement and the frequency oscillation, the shape of wave and crest could become like that shown in Figure 1.40a at n_2 or n_3. Thus, the character of development of streamlines at linear and nonlinear stages of transition is shown.

Figure 1.40b presents the process of development of disturbing motion that corresponds to the case as if it were moving in the plane of a neutral curve parallel to the x axis, where the coordinates of Reynolds number values were located—along a straight line:

$$\frac{\beta_r v}{U_\infty^2} = \text{const} \tag{1.6}$$

While simultaneously photographing the tellurium jets from the top and side, it was determined that behind the vibrator strip the purely plain sinusoidal oscillation transformed very quickly to being sinusoidal simultaneously in two mutually perpendicular planes, *YOX* and *ZOX* (screw movement, the subsequent nonlinear stages of transition, Figure 1.40c). The wave crest, after the phase of its transition from a sharp to a rounded shape, simultaneously rotated along a spiral into a longitudinal vortex (along axis *OX*). If two tellurium jets moved in parallel with each other at the same plane—*XOZ*, then they "screwed" in opposite directions inside, to meet each other outside. After the crest-bending phase the wave-folding phase occurred and, at once, the oscillations in planes *YOX* and *ZOX* stopped. Fluctuations in the plane *YOX* also disappeared; only a jet was observed, moving ahead along a plate without shift (fluctuations were completely stabilized).

These experimental investigations also made it possible to identify the physical characteristics of the point of stability loss. The "point of stability loss" is defined as the point with coordinate x that corresponds to Reynolds number of stability loss at a given flow velocity. The results of this investigation confirm that the point of stability loss should be recognized as a small region that extends along the *OX* axis and is characterized by the following:

- The oscillation amplitudes of velocity components of disturbing motion in this region are maximal.
- The wave of transverse oscillations is sharper, the wave-generating process is expressed more clearly, and wave folding occurs faster.
- The frequency of unstable oscillations and of oscillations with maximum amplitudes are higher and the range of unstable oscillations is wider.
- The wavelengths of unstable oscillations are shorter.
- The phase velocity is large and the coefficients of growth are especially large.
- The limiting frequencies in this region are maximal.
- All fluctuations propagate with a speed $c = u$.
- The fluctuations concentrate in narrow conical layer close to $y = \delta^*$.

Figure 1.40d presents the scheme of oscillations of tellurium jets for flow over an inclined plate when an adverse pressure gradient is generated on the plate. A specific shape of wave crest has been identified. When the frequency was increased the turning up of the crest occurred more slowly, and the wave was not folded but moved in a plane loop that was oriented in the *YOX* plane downstream as though in the frozen kind. The form of the wave at once became nonlinear in character.

Non-sinusoidal oscillations as depicted in Figure 1.40e produced a less clearly expressed sinusoidal wave, the shape of which was somewhat similar to that in Figure 1.40d. With a growth in frequency, an even sharper twisting of the tellurium jet and faster generation of longitudinal vortices were observed. When the wave folded, the turn of the longitudinal vortex occurred in such a manner that its axis became parallel to axis *OY*. In contrast to the sinusoidal oscillations, rather than the double thickness of the stream, an irregular stable 3D loop resulted.

Plane disturbances do not develop across the entire thickness of a boundary layer, but in a conical layer. The 3D deformation arising on a crest under the action of Goertler instability of streamlines of a plane wave in the process of its development leads to amplification of this deformation and the further alternation of stages of transition.

Klebanoff [272, 273] described measurements in a wind tunnel for the distribution of transversal velocity component $W + w'$ across the thickness of a plane boundary layer formed after the introduction of plane wave disturbances with a frequency corresponding to the second branch of a neutral curve. Measurements were registered in places on a z axis, in the middle between the peaks and hollows (z_1 и z_2). The phase of a given wave fluctuation was simultaneously registered. As a result, on a crest of a plane wave the 3D deformation was revealed to be similar to that described in [305].

According to research by Tani [482] $W + w'$ is minimal on the concave section of streamlines and maximal on the convex section. Furthermore, it changes its sign at $y/\delta = 0.2$, and profiles at z_1 and z_2 are mirror-reflected. This also agrees with the data in [305] obtained with the tellurium method and indicates that on a crest of a plane wave the standing wave was generated in the z direction with a symmetry plane of height $y/\delta = 0.2$ and at $\lambda_z/2$, equal to the distance between z_1 and z_2. Such deformation of a plane wave is the initial phase of the formation of longitudinal vortical structures.

Under Helmholtz's theorem [445] the flux of the velocity vector rotor through any section of a vortical tube is identical along the whole tube at the present time, i.e. $\omega_1\sigma_1 = \omega_2\sigma_2$. Hence, the plane wave vortical movement arising owing to Goertler instability of streamlines should remain at all phases of the wave. Near a crest, the vortical movement covers a small area on δ, i.e. σ_1 is small, hence, ω_1 is great. It follows that 3D deformation is clearly observed. Near concave streamlines the vortical movement can propagate on the greater part of a boundary layer thickness: σ_2 is great, ω_1 is small. It should be noted that a prominent feature of the ordered vortical movement is that u' compensates w': where w' is small, u' increases.

Photographs of streamlines from above and one side (see Section 1.5) confirm the appearance of vortical movement as early as at the initial stage of transition. The view from above shows some asymmetry: at the stage of a plane wave, 3D deformation arises. So, in the early stage of this deformation the preconditions appear for sine-wave fluctuations of longitudinal vortical systems in a z direction.

The regularities of development of the plane T–S wave and Benny–Lin vortices along the length of the working section of the hydrodynamic set up have been investigated (see Figure 1.41). The obtained regularities in logarithmic scale are represented as straight lines. The scatter is caused by the fact that values λ_x and λ_z were considered at different velocities of flow and near the second neutral oscillation. If wavelengths are considered for the second neutral oscillation only, then scatter will decrease essentially. This is permitted to have the empirical relations:

$$\text{for plane wave} \quad \lambda_x = 2.95 \cdot 10^{-4} \ \text{Re}^{0.46} \ m, \tag{1.7}$$

$$\text{for longitudinal vortices} \quad \lambda_z = \text{Re}^{-0.24} \ m, \tag{1.8}$$

dimensionless relation of wavelength of longitudinal vortexes $\lambda_z/\lambda_x = 3.4 \cdot 10^3 \cdot \text{Re}^{-0.7}$

$$\tag{1.9}$$

Expressions (1.4) and (1.6) can be also obtained in dimensionless form using, for example, some characteristic linear scale of the boundary layer. It follows from Figure 1.41 that at first stages of transition, when the plane wave is subjected to weak space deformations, the wave has a principal role. Observed at the beginning of the working section, the small-scale

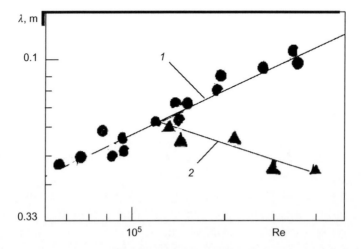

Figure 1.41
Variation of plane wavelength λ_x (1) and wave length λ_z (2) of longitudinal vortices along the plate.

wave along z has the length λ_z, which is very close to λ_x. While developing along x the wavelength λ_z increases as does λ_x; however, beginning from Re $= (1.1-1.35) \cdot 10^5$ the value of λ_z decreases whereas λ_x goes on to grow.

These regularities are characteristic at separate stages of transition. Divergence of curves 1 and 3 according to the data of Figure 1.41 occurs somewhat earlier than the fifth stage of transition. However, on the whole these regularities are in good agreement with the measurements of coefficients of intermittence and friction [305, 310]. Also as previously found, the regularities of flow at the next stage begin to determine themselves at a previous stage of transition. For example, before the sixth stage of transition has arrived, at the fifth stage, more small-scale vortical structures have been beginning to form, and have been beginning to decrease.

The results presented are illustrated well on the model (see Figure 1.42), which depicts the flow structure at different stages of transition. The whole and unified process can conditionally be divided into the following stages (see Figure 1.41 and Section 1.5):

0. Pure laminar flow in a boundary layer up to the region of Reynolds number of stability loss.
I. Amplification of plane T−waves.
II. Deformation of this wave in plane xz from the point of view of the observer who moves with the speed of the wave propagation.
III. Deformation of the wave in plane yz, wave amplitude becomes a periodical function of z with "peaks" and "valleys" in this direction.
IV. Amplification of the wave having been deformed, which causes the stretching of "peaks" forward and up and generating longitudinally-oriented "hairpin"(Λ)-like vortexes; appearance of turbulent spots.
V. Separation and "folding" of heads of those vortices and the merging of the longitudinal parts of sequential vortices, i.e. the generation of a vortical system with counter-rotating neighboring longitudinal vortices.
VI. Change of the shape and intensity of vortexes; transition from straight line to zigzag-like trajectory of vortices in the xz plane.
VII. Separation of peripheral parts of bending along z vortices; intensive development of turbulent spots; formation of a turbulent boundary layer.

Depending on external conditions, the alternation of these stages can occur with different velocities. The merging of the stages or the development of the turbulent boundary layer may occur just after the stall of heads of Λ-vortices (i.e. after the fifth stage).

As the plane wave strengthens at the first stage of transition (see Figure 1.42a) the wave amplitude grows and streamlines curvature increases too. Decelerated near a wall, particles of fluid on the wave crest are taken away into the higher-speed flow region and because of

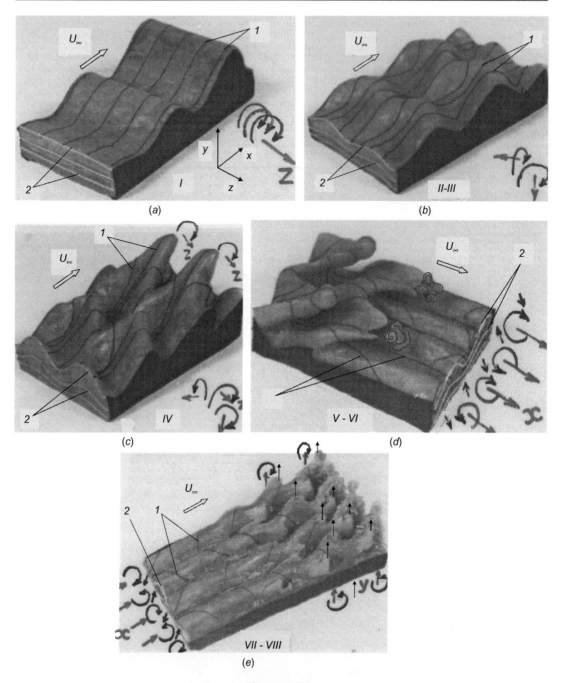

Figure 1.42

Model of transition stages of a laminar boundary layer to turbulent one: *a* — stage I, *b* — II—III, *c* — IV, *d* — V—VI, *e* — VI—VII stages of transition; *1* — streamlines; *2* — lines of equal velocities.

their very small non-uniformity along z they are subjected to different pressure. This may be imagined as a wave crest that is blown off by the wind. The result is that on the crest the front of the plane wave begins to be bent in plane xz (second stage of transition). The second reason of deformation of the wave front is micro-roughness of the plate (roughness, waviness, scratches etc.). Because stage *II* takes place at the start of the plate then the value of δ here is small. Therefore, the front of the plane wave when moving in trough and near the wall is subjected to the influence of wave non-uniformity and, coming on crest, the front is modified (heredity). It can bee seen in Figure 1.42b that the deformation of a wave front happens both near a wall, in trough, and on the wave crest.

Visualization showed that the boundary layer is very sensitive to the smallest unevenness on the bottom. This also explains the third reason of deformation of the front of plane wave: micro-unevenness of the leading edge with some delay (relaxation) contributes to modifying the wave shape.

At stage *II* of transition Goertler instability of streamlines of the plane wave causes the simultaneous deformation of the wave front not only in plane xz but in plane yz too. From here it becomes clear why in experiments the space deformation of disturbances was observed as early as the first stage of transition.

At the second stage streamlines have different curvature depending on z (see Figure 1.42b). This results in dependence of Goertler stability on z and accelerates the development of spatial deformation of plane waves.

In [232] it was shown how the initial vorticity changes. Since the streamlines are directed normally to the wave front this increases vorticity much more, as well as the tendency to bend the lateral sides of the wave front forward, from having been deformed in plane xz, and down along spiral in places between extremums of the wave front.

As soon as disturbing motion becomes 3D and component w' appears, further development of disturbances can be considered by analogy with the behavior of vortex line: its bending in any plane causes rotation of the plane in the opposite direction to that of the vortex line rotation [281, 348].

The result is that the areas in the region of the wave front minimum go down, and in the region of maximum go up. This is the coming of stage *III* of transition (yet at the second stage a standing wave is seen in vertical section yz).

At the fourth stage this process increases in accordance with Helmholtz's theorem. As it can be seen from Figure 1.42c, the "peaks" are stretched forward and up, and streamlines coming into high-speed layers of fluid accelerate themselves and, moving at a large angle to the wave front, enlarge the vorticity. Thus, on the "peaks" of the crest the vorticity increases drastically (the peripheral sides of hairpin-vortices bend down along spiral). Under the influence of velocity of the above layers of the boundary layer the upper part of hairpin-vortices stretches up, and crests incline toward the wall. As a result of the large twist of streamlines near peak (see Figure 1.42d) the head part of the vortex is "unscrewed," and free flow ("wind," which is more high-speed upper layers of fluid) stalls it and takes it away. Vortex peak, under the influence of vorticity of peripheral parts and "wind," having been set free from the vortex base separates into small vortices ("burst" of the turbulent spot).

It should be noted that turbulent fluctuations appear as early as stage *II*, while spots appear at stage *III* of transition. At the same time, as a consequence of the increase in amplitude of the disturbing motion and the non-uniformity of boundary layer thickness along z, the high-speed layers of fluid in the hollow approach the wall. Any small unevenness on the wall becomes a source of turbulent spots because, due to decreasing δ, their roughness becomes larger than allowable. Thus, the staggered order of turbulent spots is understandable.

The fifth stage of transition will now be considered. Under unfavorable factors and large Reynolds stresses the stall of the vortex head serves to accelerate the appearance of the turbulent boundary layer. If ε and U_∞ are small then "headless" hairpin vortices catch up with each other and, because of their large angle of inclination to wall and strong vorticity of lateral sides, they merge forming the classical system of Benny–Lin longitudinal vortices. The features of the plane wave have been kept until this time. If plane xy is drawn on Figure 1.42a,b,c, at any z a steady plane linear wave or non-linear wave is obtained. After stage *V* has begun, the plane wave disappears totally.

Owing to the energy exchange between mean and disturbing motions, voricity increases and longitudinal vortices begin to grow in transversal cross-section. Since they are constrained in the z direction, at first growth takes place along y. As Re increases, Reynolds stresses and size and energy of vortices grow. They begin to enlarge, merging with each other, and that causes their non-steady oscillations.

Along-front wave non-uniformity, which has already caused the front deformation in the first stage and beginning of second stage, remains on all stages of transition (heredity); in particular, it was observed at the third stage in the form of zigzag-like behavior in two neighboring layers of fluid. At the fourth stage it manifests itself in the rows (see Figure 1.42c) of hairpin vortices when they develop downstream and move with shift, alternately outstripping one another. At the sixth stage of transition, such motion increases due to the growth of vortices energy and results in magnifying the zigzag-like development of the system of longitudinal vortices. And since vortex dimensions and energy are large (occupying all δ) then the process of oscillation of longitudinal vortices in the z direction happens with large growth.

Due to the outstripping character of development of longitudinal vortices (and, thus, generation of shear) vortices with a vertical axis of rotation on their lateral surfaces appear (see Figures 1.2, 1.40 and 1.42e). Amplitude of oscillation in the z direction of longitudinal vortices and dimensions of peripheral vortices increase until peripheral vortices separate. To some extent this process is suggestive of the oscillation of a flag vertically located in wind, with the shedding of a vortical sheet alternatively from both sides. After the peripheral vortices shedding, the stage of the turbulent boundary layer begins.

The above-mentioned results and the breadboard model of alternation CVS at stages of transition constructed on their basis were obtained at the specified perfect conditions of flow. Much research on the characteristics of a transitional boundary layer has been performed for the flow over a rigid plate in the presence of various worsening factors. The following conclusion can be drawn. Worsening factors—increase of turbulence level in mean flow, positive pressure gradient, wall vibration etc.—do not change the laws of alternation of structures of disturbance movement in a transitional boundary layer. They lead only to the acceleration of alternation of transition stages so that turbulence occurs after stage V of transition.

Numerous experiments from various authors have shown that the viscous sublayer of a turbulent boundary layer has common features with a transitional boundary layer. The authors have offered a hypothesis [40], that CVSs in the viscous sublayer of a turbulent boundary layer have similar appearance to those presented in Figure 1.42, but at various worsening factors. Research has shown [305, 310 etc.] that stage I of transition is practically absent in the transitional boundary layer and there is rapid alternation of CVS beginning from the second stage, so that turbulence sets in after stage V of transition. The shape and type of CVS with worsening factors are shown, for example, in photographs in Section 1.5.

Experimental research on CVSs in the transitional boundary layer performed in [305, 310, etc.] models the flow in the viscous sublayer of a turbulent boundary layer. Until now, the correlation between CVSs in the external area of turbulent boundary layer and in its viscous sublayer has not been understood. In the opinion of the authors, large vortices are periodically separated from the external border of the turbulent boundary layer under the influence of mean flow, as can be seen in Figure 1.42d. These vortices go to the wall and impact upon the external border of the viscous area. Thus, they collapse, "flatten" and deform the viscous area. Such a picture is presented by Shlanchauskas [464]. The sharp jump of disturbances on the external border of the viscous sublayer initiates periodic bursting from the viscous sublayer. Then a new viscous sublayer forms in the viscous sublayer in the area of bursting. Such an event occurs in the area of the plate periodically in various places. There is also another view: according to this burstings from the viscous sublayer occur for the same reasons described above at the development CVSs of a transitional boundary layer. Possibly both of these mechanisms of formation of ejections (burstings) from the viscous sublayer take place, both under the

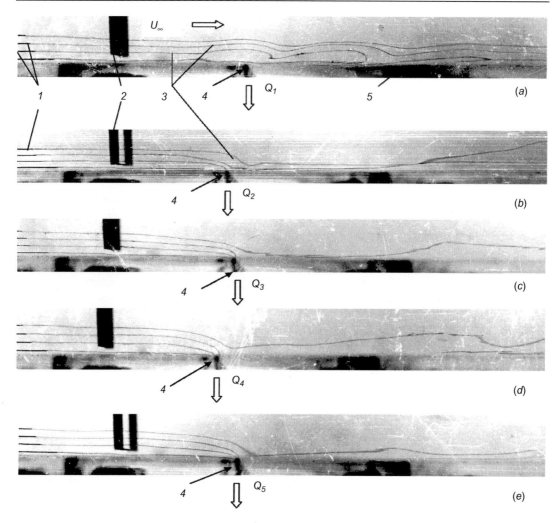

Figure 1.43

Modeling flow in a viscous sublayer of a turbulent boundary layer by means of pour out of a boundary layer through a crack: 1 — tellurium wires, 2 — vibrator, 3 — tellurium current streamlets, 4 — crack, 5 — rigid plate, $Q_1 = 0$, $Q_2 < Q_3$, $Q_4 < Q_5$, Q_3 — maximal.

action of large external vortices and owing to the development of CVSs in the viscous sublayer.

This process of formation of a new viscous sublayer has been investigated experimentally (see Figure 1.43). To do this a slot was made in the bottom of a working section located across flow. It made it possible to "pour out" the boundary layer. Because the flow velocity was small, it was enough to regulate the flow rate through the slot by the removal of a certain amount of liquid equal to the volume of the boundary layer thickness. Figure 1.43a shows the development of sine-wave disturbances along a plate without removing fluid. By regulating the amount of removed liquid the boundary layer near the slot was either partially or completely removed. It can be seen that in all cases fluctuations do not propagate behind the slot. However, the new boundary layer formed behind the slot becomes unstable under action of great gradients of thickness δ. Essentially greater disturbances develops there, in comparison to those in front of the slot.

Thus, it is possible to state that a new viscous sublayer with greater gradients of external loadings is formed in the viscous sublayer after bursting. These worsening factors cause the formation of nonlinear stages of CVSs in new viscous sublayer, similar to those in the transitional boundary layer. This explains why photographs of CVSs in the viscous sublayer of a turbulent boundary layer and in a transitional boundary layer have many common external features and characteristics.

Figure 1.43 shows the formation of a new boundary layer after removing liquid through the slot, modeling flow in the viscous sublayer of a turbulent boundary layer; this agrees well with the pattern of flow after ejection (bursting) from the viscous sublayer (see Figure 1.17).

Figure 1.44 presents a photograph of CVSs in a turbulent boundary layer throughout the thickness along a plate, obtained by smoke visualization [496]. The turbulent boundary layer consists as separate "clouds" of turbulence moving along the plate. Between clouds the dark color designates the so-called 'pockets' of the decelerated liquid (2) directed to a plate at an approximate angle of 45°. The pressure in pockets is higher than in the clouds of turbulence filled with vortices. Therefore, the turbulent boundary layer does not merge into one structure, and moves in the form of alternating areas. The basic flow "blows off" large vortices from the external border of the boundary layer, which go to the plate along the area of turbulence. These vortices maintain a velocity of the order $(0.8-0.9)$ U_∞ alongside the gained impulse of movement to the plate. Considering the pressure in the pockets above the turbulent zone, these vortices do not cross pockets, but move along a pocket to a wall. Figure 1.45 shows a photograph of a turbulent boundary layer obtained by means of smoke visualization in transversal cross-section relative the longitudinal velocity [105]. It is evident that there are pockets of decelerated liquid in the transversal direction as well. The distance between pockets remains constant at various Reynolds numbers. As velocity grows, the thickness of pockets decreases both in longitudinal and transversal directions.

According to Figures 1.44 and 1.45, Bandyopadhyay named the large vortical structures on the external border of the turbulent boundary layer "hairpin vortical structures." Indeed, these vortical structures are reminiscent of the hairpin structures in the transitional boundary layer and Klein's vortices in the viscous sublayer. Figure 1.46 shows a photograph of CVSs in the viscous sublayer of a turbulent boundary layer achieved using the method of hydrogen bubbles [274, 276, 277]. It is apparent that Klein's vortices are formed from the

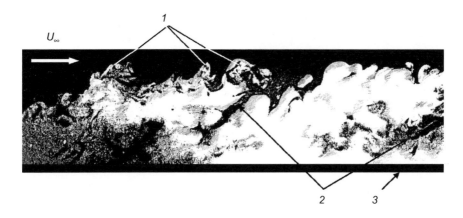

Figure 1.44

Photograph of CVSs of a turbulent boundary layer on entire thickness along a plate: *1* — large vortices on external border of a boundary layer, *2* — pockets of the braked liquid, *3* — plate.

preceding zigzag vortical structures. Figure 1.47 presents the scheme of development of CVSs in a boundary layer. The thickness of the transitional boundary layer changes in the transversal direction according to the pattern of development of CVSs (see Figure 1.42). Such non-uniformity of thickness δ develops also in a turbulent boundary layer that is well seen from visualizations by Bandyopadhyay (Figure 1.45) [105]. At high Reynolds numbers, such non-uniformity disappears owing to high velocity of mixing of CVSs on the external border of the boundary layer. The Roman numerals designate the stages of transition in accordance with the scheme in Figure 1.42.

For an understanding of the scheme of CVS development in a boundary layer at various stages, a body of research from various authors has been used, including the authors', and is partially listed in Section 1.5. It is necessary to consider the figures presented in Sections 1.1 (Figures 1.1 and 1.2), 1.2 (Figure 1.17) 1.5 (all figures) and 1.7 (Figures 1.42, 1.44—1.46). In the flow over a flat plate *1* in the laminar boundary layer *2*, plane T−S waves appear *8*, which transform into a 3D wave *9*. This wave has a sine-wave shape in planes *xoz* and *yoz*. During its development, hairpin vortices *10* form (stage *V* of transition, see Figure 1.42). The head parts of these vortices go to the external border of the boundary layer; thus, a valley in the longitudinal direction appears in a vortex. Under the action of more accelerated shear layers of the boundary layer, the head part *14* of the vortex *11* separates, and the root part of the vortex *11* folds in two longitudinal vortices *12* (stage *VI* of transition). At meandering of longitudinal vortices *12*, vertical vortices *13* appear on their lateral surfaces. As vortices *12* develop, vortices *13* separate (stage *VII* of transition). A similar picture was presented in Figure 1.2. Owing to the destruction of vortices *13*, the destruction of longitudinal vortices occurs and a turbulent boundary layer forms.

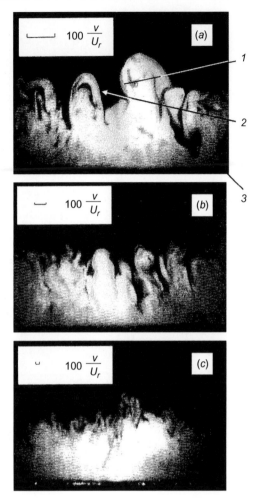

Figure 1.45
Photographs of large CVSs of a turbulent boundary layer in the transversal direction at Reynolds numbers $Re_\theta = 600$ (*a*), 1700 (*b*), 9400 (*c*): *1* — large vortices on the external border of a boundary layer, *2* — pockets of decelerated liquid, *3* — plate.

Longitudinal vortices can collapse as well without meandering during the increasing of their size downstream.

Development of disturbances movement occurs in narrow plane cone layer located in a boundary layer with a symmetry plane in a horizontal plane, containing a critical layer. Measurements of kinematical characteristics have allowed indicating existence of two areas in a boundary layer in which there is a jump of energy. The first jump of energy occurs after separation and destructions of head parts of hairpin vortices. After destruction of head

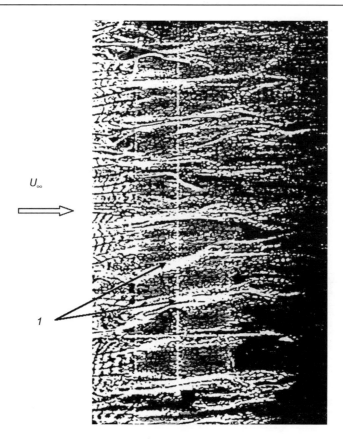

Figure 1.46
Photographs of longitudinal CVSs of a viscous sublayer (1) of the turbulent boundary layer, obtained by Klein (1967).

parts, energy of a boundary layer drastically decreases. The second jump of energy occurs after the destruction of longitudinal vortices. Thus, the energy of boundary layer also decreases and the viscous sublayer of turbulent boundary layer forms.

The turbulent boundary layer comes after stage *VII* of transition if perfect conditions of flow remain. With various deteriorative factors, turbulence can come after the fifth stage of transition. In both cases, the previous history in the form of longitudinal vortical systems in the transitional boundary layer continues developing at the formation of the turbulent boundary layer. Thus, the turbulent boundary layer is not a continuous body, and maintains typical features of CVSs of the transitive boundary layer. An essential thickening of the turbulent boundary layer occurs, which at a certain stage of development consists of large hairpin structures *15*, separated in the longitudinal direction by pockets of decelerated liquid *17* and in the transversal direction by pockets *16*. As in the transitional boundary layer, the head parts of these hairpin structures are "blown off" by undisturbed flow. Large vortices *18*

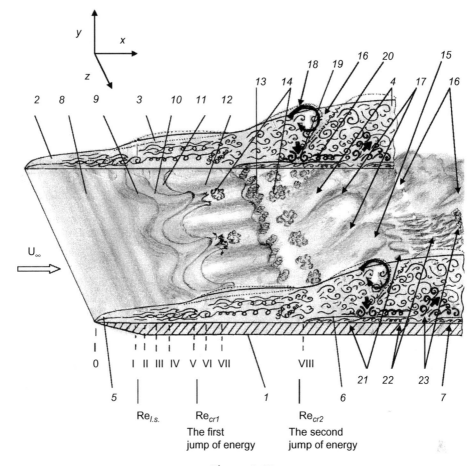

Figure 1.47

Scheme of development of CVSs in a boundary layer: *1* — plate, *2* — laminar, *3* — transitive, *4* — turbulent boundary layer (TBL), *5* — critical layer of a laminar boundary layer, *6* — viscous sublayer of TBL, *7* — renovated viscous sublayer, *8* — flat T−S waves, *9*−3D T−S waves, *10* — hairpin vortices, *11* — destruction of hairpin vortices, *12* — longitudinal vortices, *13* — vertical vortices, *14* — torn off vortices, *15* — hairpin vortices in TBL, *16* — cross-section pockets of TBL, *17* — longitudinal pockets of TBL, *18* — "blown off" large vortices, *19* — flattening T−S vortices, *20* — formation CVSs in buffer areas, *21*−3D waves in a viscous sublayer, *22* — Klein vortices, *23* — emissions (bursting) from a viscous sublayer.

go downward into less accelerated areas of the boundary layer. Reaching the external border of the viscous area, these vortices "flatten out" *19* and cause a deflection of the external border and deformation of flow in the viscous area. In the buffer area the destruction of these vortexes causes CVSs appear in the form of longitudinal vortices *20*. The generated deformations in the viscous area initiate burstings *23* from the viscous sublayer, owing to the accelerated development of CVSs in the viscous sublayer.

As shown in Figure 1.43, 3D waves *21* appear in the viscous sublayer, similar to those of the transitional boundary layer. Because the flow in the viscous sublayer is subjected to intensive disturbances from turbulent boundary layer core, disturbance movement in the viscous sublayer is practically immediately transformed from the stage of linear waves to 3D waves, which further develop as Klein's vortices *22* (Figure 1.46). These Klein vortices develop in two ways. They can increase in size and, similarly to hairpin vortices of the transitional boundary layer, collapse beginning at the head part. In the second way, this process is accelerated due to the action of large vortexes *18* getting to the boundary. In both cases, the destruction of Klein vortices leads to the destruction of the viscous sublayer and bursting of decelerated liquid from the viscous sublayer. Because of this, constant statistically uncertain bursts from the viscous sublayer occur in the area of its external border. Using various methods of averaging, it has been possible to define the probability characteristics of such bursts and the sizes of CVSs in the viscous sublayer (see Figure 1.17).

Thus, there are three areas across the thickness of the turbulent boundary layer containing characteristic CVSs:

- the area of the external border of the turbulent boundary layer in which large vortices with a primary axis of symmetry along axis *oz* form;
- the area of the buffer layer with a periodic occurrence of vortices with their primary axis of symmetry along axis *ox*;
- the area of the viscous sublayer with Klein vortices with a primary axis of symmetry along axis *ox*, and with periodic bursts upwards at an angle to axis *oy*.

The model of CVS development in a transitional boundary layer (see Figure 1.42) is uniform. Similar to this model, CVSs develop at the initial stages of the turbulent boundary layer, developing across the entire thickness and at all regimes of flow in the viscous sublayer of transitional boundary layer. With growing Reynolds number, sizes of pockets *16* and *17* decrease and it is practically impossible to not consider them.

Figure 1.48 presents numerical modeling results for prescribed shear in a flow. It has shown that CVSs appear similar to the patterns of visualization of various shear flows (see Section 1.1, Figures 1.8−1.10, and Section 1.7, Figures 1.44−1.47). Thus, the CVS types considered above for the transitional boundary layer (see Figure 1.42) are a reflection of a shear flow, and are typical for various cases of shift flows.

1.9 Hydrobionic Principles of Drag Reduction

The results mentioned in the previous sections lead to the conclusion that the important results for understanding the character and development of CVSs in a boundary layer have

E. Balaras, U. Piomelli and J. M. Wallace

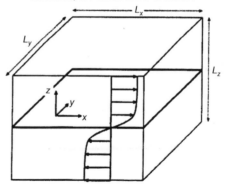

1. Sketch of the computational domain.

Self-similar states in turbulent mixing layers

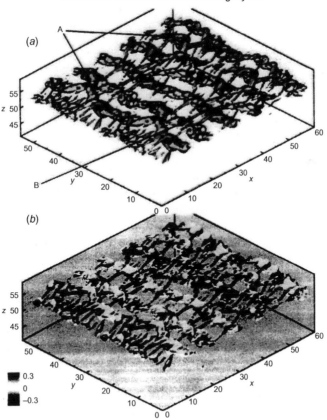

Figure 1.48

Coherent eddies during the early stage of the evolution (θ/θ_0). Case 3 (small random noise): a — iso-surfaces of $Q = 0.01 \ \Delta U^2/\theta_0^2$; b — iso-surfaces of Q, colored according to streamwise vorticity.

been obtained for the flow over a flat rigid plate. Considering its great practical value, research in this direction has continued to develop.

It is known, that the external cover of high-speed water animals and birds has a specific structure and is elastic. It is interesting to investigate the hydrodynamic value of these external coverings. It is particularly useful from the point of view of energy consumption, as movement in nature is optimal.

Increased attention to bionics research followed the publication of work by Gray [213] and Kramer [311]. The English physiologist Gray [213] undertook a cycle of research on the bio-energetic aspect of swimming dolphins. His evaluations showed that the energetic expenditure of dolphins does not correspond with the velocities of swimming they can reach. At a velocity of 10 m/s a dolphin should require power that is 7 times more than what its muscles can produce. Later, in 1949, this was termed "Gray's paradox." The facts that could explain this paradox had been sought in different ways and in different countries. A number of peculiarities of the physiological structure and hydrodynamic nature of dolphins and their skin have now been discovered.

In 1938 Kramer received his patent for a model that mimics the structure of the hair-covering of water animals and birds. In 1957 he undetook a series of experiments with a cylinder onto the surface of which various types of elastic coating were applied [311]. The design of these coatings imitated the structure of upper layers of dolphin skin. Some coatings achieved drag reduction of 57%. Kramer received three patents for the designs of coatings. His results have partially explained Gray's paradox, and they stimulated interest in both hydrobionics and new methods of a drag reduction all over the world.

In 1948 Woodcock studied the features of dolphins swimming. Bainbridge in 1958 [95] and Sickmann in 1962 [465] investigated the kinematic parameters of dolphins swimming. Same problems are also considered in [479]. In 1963, Hertel published a monograph in which he considered various topics of bionics [231]. Hertel investigated swimming dolphins and developed a series of engineering devices imitating a dolphin propulsor.

Shulejkin (1895−1979) published the monograph *Physics of the Sea*, in which he systematized the findings of hydrodynamic research of fishes and dolphins. Lavrent'ev (1900−1980) developed a theory explaining the swimming of fishes and other water animals. He initially used a mathematical means—a method of flat cuts.

Logvinovich [346, 347], and later Wu [521] and Lighthill [340], applied the general provisions of the theory of a slender body and theorems of conservation of momentum and energy, and because of this it has been possible to develop the theory of floatation of fishes. Later, Kozlov [308, 309] and Romanenko [438, 439] modified this theory for water animals.

Lang [334], Kayan [263] and Babenko [83, 86, 291, 389] have performed hydrodynamic analysis of dolphin fins [334], and Shpet has done this for whale fins [463].

Through experimental research with live dolphins, Romanenko [439] and Shakalo [303, 304, 307, 409] showed a reduction in the pulsating characteristics of the boundary layer under conditions of acceleration at non-stationary motion.

Pershin, Tomilin and Sokolov have formulated their understanding about the work of cetacean fins in their paper [399]. They stated that the cetacean fin is an active self-adapting design, which automatically changes its hydro-elasticity with dependence on swimming velocity.

Babenko [5, 23–25] revealed that the entire body of a dolphin is an adaptive system, and has shown the mechanisms of modification of its shape and size during swimming, in dependence on its speed. In 1969 he made the discovery that skin of dolphins is an active self-adapting system, with dependence on conditions of swimming [5, 23–25, 46, 49].

In 1978 Protasov and Staroselskaja [431] published a monograph in which the delineations of various fishes were given, and this was done applying the rules used in shipbuilding.

In 1979 Pershin published a monograph, in which he considered the bionics problems of swimming and flight in nature [399]. In 1988 he published new book, *Fundamentals of Hydrobionics,* in which the basic findings published by various authors in the journal *Bionics* were systematized.

Kozlov organized bionics investigations in Ukraine, at the Department of Hydrobionics and Boundary Layer Control at the Institute of Hydromechanics, part of the Academy of Sciences of Ukraine. For a number of years he was the coordinator of the hydrobionics research undertaken in various USSR organizations. The findings of this research were published in the journal *Bionics.* He has written many papers and monographs on various problems of bionics. In 1983 he published the monograph *Theoretical Bio-Hydrodynamics* [308] in which the theories on various aspects of swimming of hydrobionts are set out.

The results obtained at the Institute of Hydromechanics and other USSR organizations over the last 30 years have shown that high-speed marine animals have a number of differences compared with rigid bodies. Live organisms have specific features; therefore, they cannot be investigated using the usual engineering approach to the flow over a rigid body. The hydrobionic research of hydrobionts requires a new technique. The bionic approach can refine the search for more effective mechanical systems. For the development of original techniques for the realization of such research obtaining good results, a knowledge and understanding of basic the principles of construction of organisms is necessary.

The purpose of the research discussed here is the study of the adaptation mechanisms of hydrobionts for economical consumption of energy at high swimming velocities.

The finding of research has been that all systems of an organism are interdependent and function for reaching peak efficiency, including drag reduction and increase of propulsor efficiency with minimum energy consumption. The problem consists of defining the correlation between morphology of a hydrobiont body and the hydrodynamics of flow about it. The approach involves the exploration of the morphological properties of hydrobiont systems, including the hydrodynamic correspondence and interaction of systems of an organism.

Many authors have carried out hydrobionics research, for example Bechert, Dinkelaher, Bannash, Carpenter, Padley and others. This book presents the basic findings published in the journal *Bionics*.

The problem of interaction of high-speed hydrobionts with a flow is of interest from the point of view of defining the laws of the physical interactions with the environment, the search for mechanisms relating to the economical expenditure of energy and for the development of new technologies. Therefore, attention is given to study of features of organism systems and their interaction with a flow during the movement of water animals, and how it relates to reduction of the energy consumption. As the animals live in a water environment, the power influence of the biosphere on the organism is taken into account. Some peculiarities of the hydrodynamic influence on a body are considered at movement in the water environment. According to these peculiarities, certain structures of the specified systems of the body of water animals have been analysed. Detailed descriptions of geometric parameters of the body, speeds of swimming, energy usage, non-steadiness of motion and the morphology of the systems of structures have been studied. Morphologically, of particular interest has been the dolphin skeleton, the arrangement of the nervous system and the level-by-level arrangement through thickness and along its length of motor muscular system, the structure of the skin covering and circulatory system [5].

From the results of measuring the temperature on the surface of the skin, the elasticity, shock-absorbing properties and other mechanical parameters of the skin, the turbulence of a boundary layer are indicated at various regimes of movement of dolphins [25−29, 39, 41, et al.]. The work of the organism's systems and mechanisms of regulation of the mechanical properties of skin is considered. The ways of influencing drag reduction of movement of a body are given [307, 399].

In 1983, on the basis of their research, Babenko together with Sokolov, Kozlov, Pershin et al. made discovery No. 265 about the unique properties of the outside coverings of dolphins for drag reduction [49].

The study and ordering of the principles of hydrobionics is essential to the understanding of the peculiarities of the structure and functioning of hydrobionts and their systems, and also for modelling of these peculiarities in engineering. More than 30 principles of hydrobionics have now been revealed. Some principles are considered in [77, 86, 87]. The most important are:

- the principle of interconnectedness;
- the principle of multifunctionality;
- the principle of combined adaptive systems;
- the principle of automatic control;
- the principle of heat regulation;
- the principle of receptor regulation.

The principle of receptor regulation is as follows: when a rigid body moves various hydrodynamic forces influence its surface. In engineering, a pre-established safety factor is that the body must not be locally deformed. All hydrobionts and, in particular, cetaceans have nerve-endings located in their skin, at a distance of about 10 μm from the body surface. Evaluations have shown that pulsations of pressure and speed in the boundary layer must be registered very effectively by such receptors. Because the surface is innervated, a live organism perceives the power influence of the hydrodynamic field of the external environment, and especially gradients of hydrodynamic loading. Receptors automatically react to any vortical disturbance. This principle means that through evolution hydrobionts have developed adaptations to avoid the painful influence of the environment.

The structure of skin of hydrobionts has developed under the action of hydrodynamic and physical fields of the environment, in particular, under influence of the boundary layer structure typical for the range of speeds at which each particular group of hydrobionts swim. Simultaneously, specific structures in the skin alter the structure of the boundary layer and form a boundary layer that reduces hydrodynamic resistance and, thus, the painful action of the environment on skin receptors.

The hydrodynamic characteristics of the majority of types of hydrobionts have been investigated in detail. Models of hydrobionts and those of some of the organisms' systems have been developed. In the 1970s, in Leningrad, Krolenko first made a model of a dolphin. At the Institute of Hydromechanics NANU, Kiev, Ukraine, under the leadership of Professor Kozlov, the dermal coverings and fins of hydrobionts were modelled [44, et al.]. Research on the movement of models of a seal, shark and tuna has also been performed in various organizations in the USA [430].

The fact that the biosphere for all fast-swimming hydrobionts is identical means that the laws of interaction with a flow should also be identical according to characteristic dimensionless parameters of movement (Reynolds number, Struhal number and others). There are various adaptations in the structure of the body and dermal coverings of hydrobionts.

Plenty of experimental research on boundary layer characteristics has been done in flows over various kinds of artificial elastic surfaces. However, until now only a few researchers have attempted to model the dermal coverings of hydrobionts. The research of Kramer and that of Babenko was the first in this field. However, Kramer tried to model only the external layer of dolphin skin.

Hydrobionic research has shown that the structure of dermal layer of the dolphin is complex and controlled: all systems of an organism are interconnected and function optimally, including the greatest possible friction drag reduction.

It is important to find out which feature of dolphin skin provides drag reduction. Therefore, it is necessary to investigate the structure of the boundary layer and CVS development in the boundary layer over the dermal layer of live dolphins, and their analogues. In order to resolve these issues the hydrobionic approach has been developed.

The methodology of the hydrobionic approach is shown in Figure 1.49. The hydrobionic approach was developed in [43]. First of all, the physiological and functional features of life of the particular hydrobiont are studied. Based on this research, the principles of hydrobionics are developed, which are inherent either to the particular hydrobiont, to a species, or to all biological organisms. These principles specify an essential difference from the methods of research used for various solid and non-deformable bodies in engineering. It is, therefore, necessary to consider these principles in all types of further research in order to obtain optimum results.

In parallel, but in view of the principles known by this time, a morphological investigation of the hydrobiont is made. During morphological research, it is necessary to use all knowledge from engineering and physics relating to the problem under consideration. Some results of morphological research on the dolphin's cutaneous covering are summarized

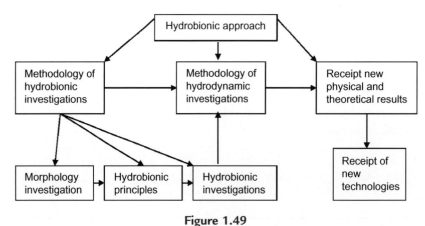

Figure 1.49
Components of the hydrobionic approach.

below. During this research, it is necessary to investigate the features of structure of the external layers of the skin that directly interact with the flow, the systems of the organism that function in the external layers of skin and in adjoining zones of its body. It is also necessary to pay attention to the structure of other layers of the skin in considering the elastic characteristics of the cutaneous covering. Finally, it is important to consider other influences of the environment on the organism. In particular, it is necessary to consider some features of CVSs described in Section 1.1.

To define the functional features of the morphological structures of a body according to the cumulated knowledge about the functioning of hydrobionts, a method of hydrobionic research is being developed, and based on this method some hydrobionic investigations have been performed. An example of such research is on the pulsating characteristics of the boundary layer during swimming of a dolphin under various conditions of movement [303, 304, 307, 409, 438, 439].

Based on the results obtained and in parallel with hydrobionic investigations, a method of hydrodynamic research on the revealed features is being developed. The methodology of this hydrodynamic research also differs from traditional engineering methods, in that it is also necessary to consider the principles of hydrobionics. It is necessary to develop π-parameters, which should underlie the technique of experimental hydrodynamic research. For example, on the basis of analysis of the similarity parameters obtained [27] experimental installations were designed [30–32, 305, 310 et al.] on which various elastic surfaces modeling the structure of dolphin skin (obtained from hydrobionic research) were tested. The results of hydrodynamic research on a range of elastic coverings will depend on how well the criteria of similarity are met.

Based on results of hydrobionic and hydrodynamic research, various approximations and theoretical models for numerical calculations are being developed. Based on the complex research performed using the hydrobionic approach, new technologies are being developed, examples of which will be discussed later.

In Figure 1.50, two schemes of the structure of dolphin skin are presented. The first scheme is constructed based on traditional morphological research by Sokolov. The second scheme is constructed based on morphological research by Surkina, who was guided by Babenko's hydrobionic approach in the analysis of skin slices. Due to such a complex approach, it was possible to reveal a number of features and to define the complex interdependence of all layers of the skin based on the hydrodynamic functioning of dolphin skin.

At the same time, Agarkov was researching the system of nerves and circulatory systems of blood and lymph in the skin. Using the hydrobionic approach it was possible to define the interrelation of functioning of all systems of the organism directed at drag reduction and economical energy expenditure [73].

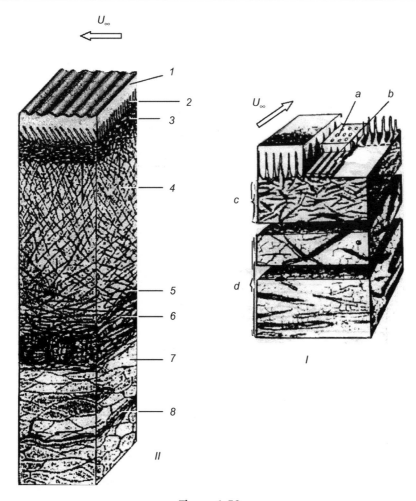

Figure 1.50

Scheme of skin structure of dolphin (*I, II*): *I* — according to Sokolov (1955): *a* — dermal nipples; *b* — longitudinal epidermal partitions; *c* — under-nipple layer of derma; *d* — under-skin fat cellulose; *II* — according to Babenko and Surkina (1971): *1* — epidermis; *2* — nipple layer of derma; *3* — under-nipple layer of derma; *4* — net-like layer of derma; *5* — connective layer; *6* — skin musculature; *7* — under-skin fat cellulose; *8* — membrane.

Some results of this research are presented below. The surface of dolphin skin when placed in a bath of seawater looks hydraulically smooth. However, if the water is blown off, or the skin is wiped then longitudinal folds relative to the longitudinal axis of the body are seen in slanting light on the concave section of the body (see Figure 1.51). Different species of dolphin have different sizes of microfolds, which correlates with the size and speed of swimming of the particular species.

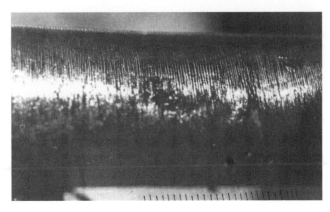

Figure 1.51
Microfolds on the surface of dolphin skin.

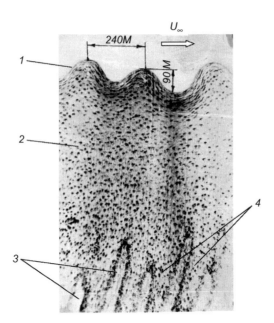

Figure 1.52
Cross-section of microfolds on the surface of dolphin skin: *1* — surface of skin, *2* — epidermis, *3* — dermal papillae, *4* — epidermal papillae.

Figure 1.52 shows sizes of microfolds on fixed material. Microfolds on the skin of a live dolphin are smaller and are 'Π'-shaped instead of sine-shape. The presence of microfolds was also mentioned in [470].

Calculations have shown that the specified sizes are the maximum permissible for rough plates until speeds of 8 m/s. As live dolphins have smaller microfolds, they do not influence roughness, even at greater speeds of movement. The hydrodynamic value of these microfolds is described in [25]. It is by means of these microfolds that the disturbances of the T−S wave type are generated in a boundary layer, interconnected with the work of skin muscles and the oscillatory movement of the body. The presence of microfolds correlate with the location of dermal papillae in skin that are clearly visible from Figure 1.52.

Dermal papillae are located along dermal rollers (see Figure 1.50) under a certain angle of slope. Measurements [477] have shown that angles of slope of dermal papillae change depending on the location of measurement along a body (see Figure 1.53). Angles of slope were measured on histological preparations counter-clockwise in the plane of a correct slice between longitudinal axis of a papilla and the tangent to the epidermis bottom surface. Such a slice was considered correct when the contours of the dermal papilla had a regular shape, and when the plane of the slice was along a dermal roller. The least inclination of dermal papillae was found in the leading part of a body ($\alpha = 10° - 25°$) where the dynamic pressure is greatest, in the middle part, which is the most movable and, presumably, where there is a transition to a turbulent boundary layer; it was also seen in the dorsal (spinal) and ventral (abdominal) parts of the caudal peduncle, where the gradients of pressure are great at oscillatory movement of the caudal fin. The greatest corners of an inclination of dermal papilla ($a = 55° - 80°$) was found in the area of a neb, in front of the eyes and also in lateral parts of a back part of a body (in places of small gradients of pressure) from below and behind lateral fins and behind a fin located on a back (in the area of hydrodynamic shadow). The hydrodynamic value of the inclination of dermal papillae is also connected with CVSs in a boundary layer and amplitude−frequency characteristics of disturbance movement of a boundary layer, which cause corresponding fluctuations in a dolphin's cutaneous covering.

At change of swimming speed, the parameters of movement and physical parameters of the environment change accordingly, and the tuning of dermal papillae occurs in response. Because each papilla has a nerve ending, and lymphatic and blood capillaries, it is reasonable to assume that the geometrical and mechanical characteristics of papillae change in response to changes to the circulation and temperature of the blood. Figure 1.54 shows a photograph of the changed shape of a dermal papilla, which is often observed on histological slices. Alterations in the degree of blood filling can explain the change in papilla shape. Dolphin skin can also respond to changes by altering skin tension by means of the cutaneous muscle.

Figure 1.55 presents the arrangement of dermal rollers in dolphin skin [477]. The scheme was established based on numerous slices of skin along the entire body. Dermal rollers basically settle down along the longitudinal axis of the body and correlate with the oscillatory movement of the tail part. Dermal rollers and the epidermal protrusions located between them have different densities. Dermal papillae and the epidermal papillae located between them also have different tissue structures and densities. Thus, these parts of the

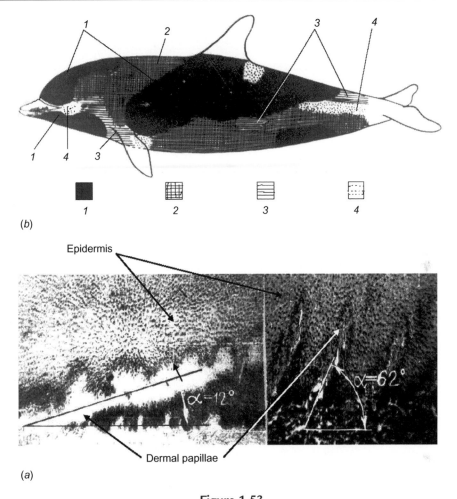

Figure 1.53

Angles of inclination of dermal papillae in dolphin skin: a — angles of inclination in various location on skin, b — distribution of angles of inclination along a body, $1 - \alpha = 10°-25°$, $2-25°-40°$, $3-40°-55°$, $4-55°-80°$.

skin constitute a system of longitudinal and transversal inhomogeneities — a specific lattice. Such a structure suggests that oscillations caused by pressure pulsations in the boundary layer will propagate in dolphin skin in a way that is different to that in a monolithic elastic material. This specific structure of the skin means that it can be considered as a specific wave-guide that leads to the attenuation of disturbances with certain amplitude—frequency parameters, while disturbances with other parameters freely propagate in the skin.

Surkina [476] also revealed that the cutaneous muscle has a specific structure and complex configuration: the muscular fibers in it are sloped relative to the longitudinal axis of the body (see Figure 1.56). Considering great thickness of the skin and its specific structure,

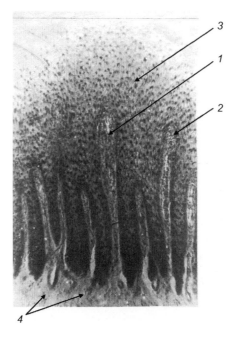

Figure 1.54
Change in the shape of dermal papilla caused by various degree of blood filling: *1* — the increased shape of dermal papilla, *2* — normal shape of dermal papilla, *3* — epidermis, *4* — location of dermal rollers.

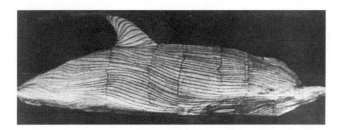

Figure 1.55
Direction of dermal rollers in dolphin skin [477].

particularly the adipose layer, the functional value of the skin muscle is, foremost, the need to preserve smoothness of the external layer and shape of the body.

With an increase in the speed of swimming, shear stress increases on the surface of a body. These forces tend to create raw skin on parts of the body. In addition, underpressure along a body increases at increasing speeds, as do separation loadings on the skin. Thus, a dolphin's muscles serve various functions: the basic muscles create propulsion force by means of the

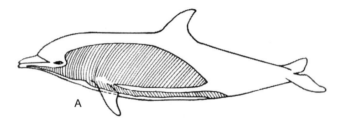

Figure 1.56
Distribution of the skin muscle along a body of a dolphin [476].

tail mover and ensure a high level of maneuvering during swimming. The function of the skin muscle is effective control of drag reduction at various regimes of swimming. To do this the skin muscle regulates tension in the skin, providing smoothness of skin and body. Another important function of the skin muscle is the reflex creation of such tension, which leads to resonant interaction with disturbances in a boundary layer and to the generation of high-frequency fluctuations on the skin surface. Kidun [266] demonstrated these oscillations on the surface of the skin of dolphins swimming at various accelerations.

It has been shown [5, 36, 46] that the arrangement along a body of the locomotor muscles, the skin muscle (see Figure 1.56) and angles of inclination of dermal papillae (see Figure 1.53) correlates and is defined by the hydrodynamic functions of these structures.

One further assumption has been made about the role of the skin muscle. The external layers of the skin have two crossed structures: surface microfolds and a longitudinal systems of dermal rollers and epidermal projections located deeper in the skin. The first structure promotes the generation of flat 2D disturbing movements in a boundary layer. The second structure promotes the generation of longitudinal 3D disturbing movements. The next task is to investigate how and what the function is of generating such disturbances in a boundary layer. However, presence of these structures makes it possible to state that various structures of disturbance movements are generated in the boundary layer over a dolphin. Thus, it is necessary to investigate the susceptibility of the boundary layer to various vortical structures. It is obvious that the skin muscle allows a regulating process of interaction of various disturbances in a boundary layer. It is also evident that the morphological structures found in the skin allow dolphins to form a boundary layer with minimal drag for economical energy consumption.

It is not only the structure of surface layers of the skin but also that of the deeper parts of the skin that defines the optimum functioning of its controllable elastic-damping properties. The following structural aspects contribute to this effect: the direction, thickness and structure of connective-tissue fibers, the quantity and size of fat cells, the thickness of the dermal layer, the skin muscle, the structure, thickness and properties of hypodermic fatty tissue, the two parallel blood-filled layers in the skin, the two membranes and tension bars connecting the skin muscle, motor muscles and membranes in the skin, etc. During swimming at various speeds, various loadings, for example, distribution of pressure and shear stress, form along a body. The functional value of the layers of skin is the preservation of a well-streamlined shape of body and optimum mechanical characteristics of dermal coverings. It is known that the mechanical characteristics of elastomers essentially depend on the thickness, temperature and rigidity of the material of the base. The same dependence is true for dolphin skin.

On the basis of above-mentioned research (as well as other work not mentioned) it is possible to draw some conclusions. The hydrobionic approach has made it possible to define new functional values of dolphin dermal coverings, in particular:

- Dolphin dermal coverings have an optimum structure and mechanical characteristics for drag reduction at a certain range of swimming speeds.
- The systems of an organism work in an interrelated way and allow for the automatic control of the characteristics of external layers of skin.
- The interaction of a boundary layer with the dermal coverings are such that it generates characteristics of a boundary layer that promote essential drag reduction.
- In the boundary layer of dolphins interactions occur between the various disturbances generated in the boundary layer from its various borders.
- The boundary layer on the body of a dolphin together with the external layers of its skin can be considered as a complex wave-guide.
- The structure of flow near the skin essentially differs from the structure of a boundary layer over a rigid surface.

1.10 Experimental Investigation of Coherent Vortical Structures in a Transitional Boundary Layer over an Elastic Plate

As a result of the hydrobionic approach, new tasks of hydromechanics were discovered. Experimental research on kinematic characteristics of the flow near a dolphin's surface were undertaken by Shakalo and Romanenko [303, 304, 307]. The results have revealed essential differences in the characteristics of this boundary layer in comparison with that of a rigid surface.

Plenty of experimental research has been carried out on the integrated characteristics of the boundary layer over elastic surfaces. Attempts have been made to design an elastic surface capable of reducing drag. However, research on the physical picture of the flow over an elastic surface has not been done. There are no representations of dependence of the flow structure on design features of elastic surfaces.

According to hydrobionic approach it is necessary, first of all, to carry out experimental research of the kinematic characteristics of the boundary layer over analogues of dermal coverings and develop parameters of similarity. In [220, 305] some criteria of similarity are presented. According to the findings of morphological and hydrobionic research, three kinds of elastic plates have been developed (see Figure 2.22) that reflect some features of the structure of dermal coverings. In Chapter 2 the techniques of experiments on elastic surfaces are described. Below the results of experimental research on the physical features of the flow over those elastic plates are presented.

According to the method described in Section 1.5, research was done by means of Wortmann's tellurium method [519].

The above results have made it possible to determine the difference between the development of the disturbing motion in the boundary layer over elastic plates and that over rigid plates.

The basic laws of fluid flow over elastic plates are the same as those over rigid plates; consequently, regularities of boundary layer thickness development are analogous, and the basic preconditions for a disturbing motion in form of a plane wave arising in boundary layer are also maintained [305, 310]. Therefore, the shape of the neutral curve for an elastic plate is similar to that for the standard [305, 310]. However, the regularities of evolution of disturbing motion over elastic plates differ essentially from those over the rigid standard.

The results of investigations of hydrodynamic stability of the flow over elastic plates presented above have led to the conclusion that with the right selection of mechanical properties, an elastic plate can efficiently stabilize the laminar boundary layer at sinusoidal and non-sinusoidal type of disturbing motion.

In experiments it was found that with a favorable selection of mechanical properties at least one of the following effects was observed, independently of form of disturbing motion: Reynolds number of stability loss, Reynolds number of transition and wave numbers increased; frequency ranges of unstable oscillations of the laminar boundary layer, phase velocity of disturbing motion, coefficients of growth and ordinate of critical layer and kinetic energy of disturbing motion decreased.

Both the character of development and principal characteristics of the disturbing motion changed at not only at the first stage of transition but at all following ones too.

Development of the amplitude of normal velocity v' along the test section of the hydrodynamic stand is presented in Figure 1.57. Similar to experiments on a rigid plate, visualization of the disturbing movement across the thickness of a boundary layer was carried out by means of three tellurium jets. The mechanical characteristics and

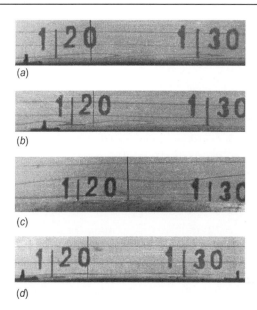

Figure 1.57

Photograph of amplitudes of velocity v' over membrane surfaces. Frequencies of oscillations are near the second neutral fluctuation: a, b, d — water under membrane, c — air; a — experience B3, $n = 1.1$ Hz, $U_\infty = 11$ cm/s, $E = 1.92 \cdot 10^4$ N/m^2, $T = 79.0$ N/m; — B8, 1.1 Hz, 12.4 cm/s, $1.92 \cdot 10^4$ N/m^2, 79.0 N/m; c — B18, 0.5 Hz, 11.7 cm/s, $1.55 \cdot 10^4$ N/m^2, 36.0 N/m; d — B38, 0.4 Hz, 13.0 cm/s, $1.35 \cdot 10^4$ N/m^2, 25.0 N/m.

corresponding dimensionless coefficients of an elastic plate are presented in [305]. The disturbing movement was visualized in the flow over the same simple membrane surface, which consisted of polyvinyl film mounted on a metal skeleton (see Chapter 2). The character of the disturbing movements in Figures 1.56a and 1.56b differed due to the change of speed of the basic flow. On this membrane surface a zone of stability loss was at the given coordinate x. In this location on x all typical attributes of a point of stability loss were indicated [305]. However, in comparison with a rigid plate (see Figures 1.27−1.29) the shape of the wave and size of its amplitude essentially differed: the shape of wave at once became nonlinear, peaked, but without collapse of the wave, and amplitude of oscillation essentially decreased by 3−4 times (compare with Figure 1.29). Furthermore, in the flow over the membrane surface the disturbance was observed only near to the plate. It is possible to assume that only the coordinate of critical layer was changed. The only exception was the membrane with air beneath it (see Figure 1.56c). Although the tension of the film was reduced, the air beneath the film stretched the membrane. As a result, disturbing movement rebounded from the surface as from a drum, and it transformed to the external border of a boundary layer.

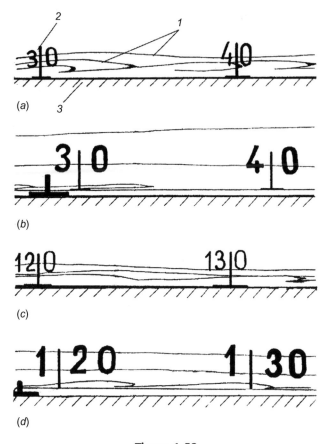

Figure 1.58

Development of disturbing motion through thickness of a boundary layer during flow over a rigid (*a, c*) and simple membrane surfaces (*b, d*): 1 — tellurium jets, 2 — distance marks along working section (cm), 3 — surface along which measurements are made.

With a change in the membrane tension, the amplitude of the disturbing movement essentially decreased (see Figure 1.56d). The surface damped oscillations of the vibrator practically at any coordinate of x.

Figure 1.58 gives a graphic representation of photographs of three tellurium streams, which characterize the evolution of the disturbing motion along a simple membrane and rigid plates. The disturbing motion in the region of the point of stability loss for rigid plate is shown in Figure 1.57a,b: the amplitude of vibrator strip oscillation for rigid plate was 0.32 mm, $n = 0.74$ Hz; and for membrane one 0.24 mm, $n = 0.98$ Hz.

The disturbing motion over the rigid plate spread along the whole thickness of the boundary layer and was characterized by all of the signs characteristic for point of stability loss [305].

In the flow over the membrane plate, in spite of its neutral curve form [305], which differs essentially from that for the rigid plate, the disturbing motion was generated near the surface. The shape, amplitude and type of the wave differ from the analogous ones over the rigid plate and show that oscillations become nonlinear.

The disturbing motion appears near the surface in the flow over a damping surface. The shape and type of wave differ from similar waves in the flow over a rigid surface. Moreover, the difference in the tendency toward twisting the wave crest is obvious. The same differences were found when the characteristics of the disturbing motion were compared to a frequency close to the first neutral frequency for the flow over a rigid surface, only the amplitude of oscillation over a flexible surface was still less.

Comparison of the disturbing motion in flows over both surfaces in the region of $x = 120$ cm showed that the wavelengths were greater on a rigid surface (see Figure 1.58c), the shape of wave was sloping, twisting of the wave crest occurred in a manner typical for a rigid surface and the disturbing motion was applied to the entire thickness of the boundary layer. In the flow over a membrane surface (see Figure 1.58d) the wavelength λ increased compared to that at $x = 30$ cm, but decreased compared to λ on a rigid surface at $x = 120$ cm. The shape of wave also varied but remained the same as at $x = 30$ cm. The disturbing motion was observed approximately in the same section of the boundary layer as in the flow over a rigid surface (see Figures 1.58c,d). The oscillation frequency of the wave was 0.68 Hz on a rigid surface and 1.1 Hz on a membrane surface.

As can be seen from Figure 1.58, the disturbing motion in the flow over a simple membrane surface becomes clearer but is more similar in shape to the disturbing motion with a positive pressure gradient (see Figures 1.31 and 1.32), the results of investigation of which will be outlined separately.

During study of hydrodynamic stability on a rigid surface it was discovered that there is a specific distance between the appearance of visible oscillations and the "folding" of the wave for each oscillation frequency. This distance was arbitrarily named the "stabilization zone." The stabilization zone initially decreases and then increases as the oscillation frequency increases. The stabilization zone is minimum near the second neutral oscillation and shifts upstream to the vibrator strip as frequency increases. The principles of wave "folding" in the flow over membrane surfaces were essentially the same. The stabilization band seemingly characterized the rate of increase of the disturbing motion. Thus, the stabilization band on practically all tested damping coatings was less than in the flow over a rigid surface. For example, the stabilization band was 20 cm over a rigid surface at $x = 70$ cm and oscillation frequency of 0.74 Hz (frequency close to the second neutral oscillation—oscillations appeared at $x_{begin} = 60$ cm and $x_{end} = 80$ cm, respectively). The stabilization band was 18 cm ($x_{begin} = 67$ cm and $x_{end} = 85$ cm, respectively) in the flow over a simple membrane surface at the same location (experiment B2) at oscillation

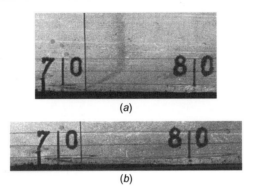

(a)

(b)

Figure 1.59

Photograph of profiles of average velocity U (*a*) and amplitudes of velocity v' (*b*) at a flow on membrane surfaces that is heated from below: *a* — experience BT44, $n = 0.0$ Hz, $U_\infty = 13.9$ cm/s, $E = 1.04 \cdot 10^4$ N/m^2, $T = 15.0$ N/m; *b* — experiment BT44, $n = 0.56$ Hz, $U_\infty = 13.9$ cm/s, $E = 1.04 \cdot 10^4$ N/m^2, $T = 15.0$ N/m; *n* — second neutral fluctuation.

frequency of 0.76 Hz, it was 10 cm ($x_{begin} = 70$ cm and $x_{end} = 80$ cm) at 0.96 Hz, and it was 6 cm ($x_{begin} = 72$ cm and $x_{end} = 78$ cm) at 1.09 Hz close to the frequency of the second neutral oscillation. The stabilization band increased with a subsequent increase of frequency.

Figure 1.59 shows photographs of visualization of disturbance movement in the flow over simple membrane surfaces. In this case, hot water (temperature 32°C) was poured under the membrane, and then the working section was filled with water with a temperature of 7.3°C. The temperature of water under the membrane and that of the water above gradually equalizes. Owing to the heating of the membrane its mechanical characteristics changed: they became minimal for all membrane surfaces. As a result, the disturbing fluctuations were stabilized and concentrated in a very narrow area close to the membrane surfaces. The shape of the wave indicates nonlinearity of disturbance movement. Moreover, a stabilization band was not found. After a peak wave was formed, the disturbances with this shape of wave moved downward along the flow without being deformed, as in a solidified type. The profile of average velocity is typical for the boundary layer on a heated plate.

It is important to note that such a temperature difference exists for swimming dolphins. But dermal coverings have special structures that essentially reduce heat irradiation and the temperature difference on the surface of the skin.

The nature of the disturbing motion varied in the flow over complex membrane surfaces, combining the characteristics of disturbing motion in flows over rigid and simple damping surfaces. The design of membrane and visco-elastic surfaces is described in Section 2.5 and in Figures 2.22 and 2.23. A graphical representation of photographs of disturbing motion at frequency of $n = 0.8$ Hz and flow velocity of 13.6 cm/s (experiment P11) is presented in

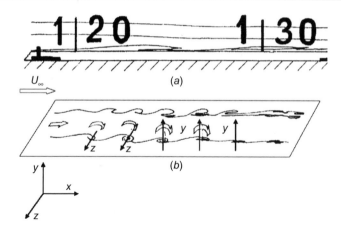

Figure 1.60
Graphical representation of photograph of disturbing motion (*a*) and scheme of its development (*b*) during flow over a complex membrane surface.

Figure 1.60a, and a scheme of the behavior of disturbing motion over those complex surfaces is in given in Figure 1.60b.

Three characteristics of disturbing motion were found: the stabilization band behaved in the same way as that for the flow over a rigid surface. The stabilization band approaching the vibrator became constricted and then expanded as frequency increased. At the same time, unlike a rigid surface, "double folding" of the wave was observed. Part of the wave crest was "folded" while the upper part, not breaking down, continued to move above it. The upper part of the crest was then "folded" and two stabilized crests moved parallel to each other downstream. The second characteristic included the fact that the disturbing motion was not immediately stabilized with a further increase of oscillation frequency, but was rotated inside, so that if the axis around which the crest was twisted were previously parallel to the flow surface, this axis became perpendicular to the surface lower downstream. The crest was twisted so rapidly at these frequencies and the wavelength decreased so much that seemingly small vortices parallel to the surface were initially formed, which became stabilized after their rotational axis was turned by 90°. Moreover, the impression was created by the movement of disturbing motion downwards along the flow that the lower layer of the fluid lying under the oscillating layer moved opposite the flow so much that it was strongly retarded, while waves moved above this layer without increasing and without being damped immediately, but were stabilized considerably downstream compared to a rigid surface.

Detailed results of research on the development of disturbing movement in flows over various kinds of elastic surface are presented in [305]. Membrane surfaces have no practical value. However, for such designs it is convenient to write down the equation of movement

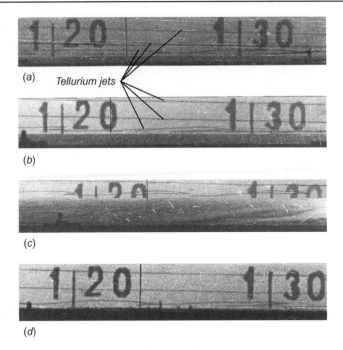

Figure 1.61

Photograph of profiles of amplitudes of velocity v' at a flow on visco-elastic surfaces (PPU):
a — experience P23, $n = 0.83$ Hz, $U_\infty = 14.9$ cm/s, $E = 0.31 \cdot 10^4$ N/m^2, $T = 0$, thickness $t = 5$ cm;
b — P19, $n = 0.67$ Hz, $U_\infty = 12.4$ cm/s, $E = 0.31 \cdot 10^4$ N/m^2, $T = 0$, $t = 5$ cm (non-sinusoidal
disturbances); c — P27, $n = 0.83$ Hz, $U_\infty = 11.7$ cm/s, $E = 3.5 \cdot 10^4$ N/m^2, $T = 0$, $t = 5$ cm (with
membrane); d — P35, $n = 0.83$ Hz, $U_\infty = 10.0$ cm/s, $E = 0.5 \cdot 10^4$ N/m^2, $T = 0$, $t = 3.1$ cm. The
second neutral fluctuation makes: $a - n = 0.83$ Hz, $b - 0.67$ Hz, $c - 0.67$ Hz, $d - 0.46$ Hz.

and experimentally check up the meaning of certain mechanical characteristics. So, the thickness of liquid under a membrane, its density, tension of the membrane and elastic characteristics of membrane surfaces, influence of nonharmonic disturbances, and temperatures of liquid under membrane were varied.

The greatest practical interest relates to the visco-elastic surfaces made of foamed polyurethane (FPU). Thus, these coverings were investigated, onto the surface of which the same film made of polyvinyl was placed. The experiments were also done without a film, and when dry or FPU saturated with water was under a film. In Figure 1.61, photographs of disturbing movement are shown for various visco-elastic coatings.

The nature of the disturbing motion in the flow over a visco-elastic surface also differed from the nature of flow over a rigid surface. Graphical representation of photographs of the behavior of the disturbing motion in the flow over a thick plate of foamed polyurethane are shown in Figure 1.62. Everywhere the oscillation frequencies were similar to the frequency

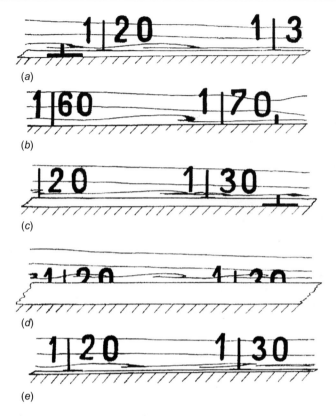

Figure 1.62
Graphical representations of photographs of disturbing motion at a flow on visco-elastic surfaces (PPU): a — experience P23, $n = 0.83$ Hz, $U_\infty = 14.9$ cm/s, $E = 0.31 \cdot 10^4$ N/m^2, $T = 0$, thickness $t = 5$ cm; b — P24, 0.77 Hz, 14.9 cm/s, $0.31 \cdot 10^4$ N/m^2, 5 cm; c — P19, 0.67 Hz, 12.4 cm/s, $0.31 \cdot 10^4$ N/m^2, 5 cm (non-sinusoidal disturbances); d — P27, 1.4 Hz, 11.7 cm/s, $3.5 \cdot 10^4$ N/m^2, 5 cm (with membrane); e — P35, $n = 1.0$ Hz, $U_\infty = 10.0$ cm/s, $0.5 \cdot 10^4$ N/m^2, 3.1 cm.

of the second neutral oscillation. The oscillation amplitude decreased strongly compared to that on a rigid surface, the shape of the wave was similar to that in flow over a simple membrane surface, but the lower layer adjacent to this surface was strongly retarded as in the case of flow over complex membrane surfaces (see Figure 1.60). The disturbing motion propagates in the immediate vicinity of the damping surface.

Figure 1.62a demonstrates the development of the disturbing movement in the area of a point of stability loss over the FPU surface. It is interesting to compare the behavior of the disturbing motion in a similar situation for a rigid surface (see Figures 1.58a and 1.28). The behavior of the non-sinusoidal disturbing motion in the flow over a sheet of foamed polyurethane (see Figures 1.61b and 1.62c) was the same as that over a rigid surface (see Figures 1.26 and 1.40), with the exception for that the wave crest folds almost immediately behind the vibrator without the development of oscillation.

The process of wave folding is the same as on a rigid surface and is similar to the process of double folding of the wave on a complex damping surface. The difference from double folding includes the fact that the beginning oscillation in the *yox* plane breaks down very rapidly, almost immediately behind the vibrator with rotation to the *zox* plane. All the oscillations in the *yox* plane then stop while oscillation in *zox* plane continues for some time until complete stabilization is achieved. The behavior of the tellurium jet after double folding is easily visible in the region of $x = 130$ cm, while separation of the jet in the *zox* plane is visible in the region of $x = 130$ cm (see Figure 1.62c).

The oscillation amplitude of the vibrator strip was 0.24 mm during sine-wave disturbances on visco-elastic surfaces and 0.3−0.4 mm during non-sine oscillations. The oscillation was damped immediately behind the vibrator (experiment P17) in the first series of measurements in the region of $x = 30$ cm. The amplitude increased to 0.7 mm (experiments P18) in the second series at $x = 80$ cm. However, the oscillations were damped rather well, even in this case. It was found that the fluid near the surface was severely retarded at frequencies close to the frequency of the second neutral oscillation, but it seemingly slid over the lower layer of fluid so that an impression was created that the boundary layer of the fluid moved opposite to the flow. A similar pattern was observed in the flow over complex membrane surfaces.

When the oscillation amplitude of the vibrator strip decreased to 0.4−0.5 mm the oscillation amplitude of the fluid was less than that in the flow over a rigid surface and the perturbing motion itself was concentrated immediately near the surface.

The structure of the disturbing motion during flow over a foamed polyurethane surface coated on the outside with film is shown in Figure 1.62d. Although the nature of the disturbing motion was similar to the nature of the motion on a surface not covered with film, the frequency and oscillation amplitude increased. The same was observed in experiments on a thin sheet of foamed polyurethane, which had worse stabilizing properties than a thick sheet (see Figure 1.62e).

Measurements were made in region III of the natural transition in the flow over a simple membrane surface (see Figure 1.63) similar to measurements in the flow over a rigid surface (see Figures 1.24 and 1.31). The speed of the mean stream was 9 cm/s, coordinates of tellurium wire $x = 96$ cm, $y = 3$ mm. It appeared that the deformation of profile $U(z)$ in the flow over a membrane surface was about same as in the flow over a rigid plate, at the coordinate of tellurium wire $x = 180$ cm (see Figure 1.24). The distance of tellurium wire from the surface was such that in both cases it was in the area of the critical layer. From here, as well as from the shape of velocity profile $U(z)$, it is possible to conclude that in

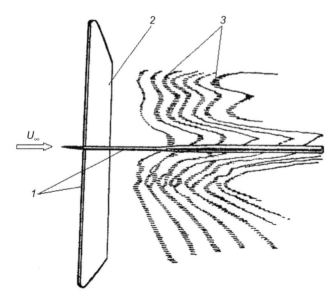

Figure 1.63
Graphical representations of photographs of velocity profiles during flow over a simple membrane surface: *1* — support for tellurium wire, *2* — tellurium wire, *3* — velocity profile *U(z)*.

the flow over surfaces vibrations arise of a membrane that lead to the acceleration of nonlinear deformation of a flat wave in comparison with a rigid plate. The vibrations arise owing to fluctuation of the vibrating strip. At the same time, in comparison with the flow over a rigid plate at a positive gradient of pressure (see Figure 1.31) deformation of profile *U(z)* in the flow over a membrane surface is essentially less.

The wavelength of Goertler vortices decreased considerably in the flow over an elastic surface. Thus, considering the shape of the tellurium jets, which characterize the disturbance development on various elastic plates, it is possible to draw the following conclusions:

1. On all the plates the disturbing motion became nonlinear at once: the wave took the indicated but non-sinusoidal shape.
2. The disturbing motion, as against the rigid plate, propagated in a very narrow conical layer with a small angle of slope near the surface. At that, the disturbances concentrated along the axis of this liquid layer, which is located considerably closer to the wall than the critical layer on the standard, and lower than the layer the disturbances became stabilized very fast. Lower than that layer the phase velocity, as against standard, decreased: $C_r/u < 1$.
3. If the mechanical parameters of elastic plates differed by one or two orders from the optimal ones then Reynolds number $\text{Re}^*_{\text{l.s.}}$ increased as compared to the standard. The

disturbing motion developed very slowly at nonlinear stages of transition. If, moreover, parameters differed by 4 orders then $Re^*_{1.s.}$ became either less or comparable to that on the standard, and the disturbing motion developed very fast to the seventh stage of transition (a fast interchange of stages), after which the disturbances came to stabilization.

1.11 Distribution of Disturbing Movement on the Thickness of a Laminar Boundary Layer on an Elastic Surface

The measurements were conducted in the same hydrodynamic set up as described previously. Various types of elastic damping surfaces replaced the rigid bottom surface in the test section. The design of the elastic damping surfaces, the technique of measuring their mechanical properties and the specific values of these properties are presented in Chapter 2.

The distribution of transversal pulsating velocity v' of the disturbing motion throughout the boundary layer thickness was investigated at the optimum oscillator strip arrangement and its optimum amplitude. Figure 1.64a illustrates the results for a simple diaphragm surface with air underneath, at $Re^* = 770$, $x_1 = 168$ cm, and $U_\infty = 11.7$ cm/s (experiment B18, $E = 1.55 \cdot 10^4$ N/m^2, $T = 36.0$ N/m). Figure 1.64b illustrates results for a simple membrane surface with water underneath it, at $Re^* = 690$, $x_1 = 122$ cm, and $U_\infty = 13$ cm/s (experiment B38, $E = 1.35 \cdot 10^4$ N/m^2, $T = 25.0$ N/m). Figure 1.63c presents results obtained under the same simple membrane surface as in Figure 1.63b but the mechanical properties were different. In this case, $Re^* = 710$, $x_1 = 163$ cm, $U_\infty = 10.4$ cm/s (experiment B13, $E = 1.92 \cdot 10^4$ N/m^2, $T = 79.0$ N/m). Figure 1.63d presents results of a complex membrane surface at $U_\infty = 13.4$ cm/s. Curve 1 corresponds to $Re^* = 700$, $x_1 = 123$ cm (experiment P6, $E = 2.5 \cdot 10^4$ N/m^2, $T = 16.0$ N/m); curve 2 — $Re^* = 826$, $x_1 = 175$ cm (experiment P7, $E = 1.8 \cdot 10^4$ N/m^2, $T = 16.0$ N/m); and curve 3 — $Re^* = 1000$, $x_1 = 225$ (experiment P8, $E = 1.8 \cdot 10^4$ N/m^2, $T = 16.0$ N/m). The dot-dashed line in Figure 1.63d, represents a situation when a foamed polyurethane plate is at $U_\infty = 14.9$ cm/s, $Re^* = 900$, $x_1 = 183$ cm (experiment P23, $E = 0.31 \cdot 10^4$ N/m^2, $T = 0$). In Figure 1.64 the vertical dotted lines characterize the boundary layer thickness $y = \delta^*$. The coordinate x_1 defines the virtual beginning of a boundary layer defined by results of measurement of a velocity profile. As compared with the flow over a rigid plate a series of specific features were observed. One of them was that with an increase of oscillation frequency the exciting motion tended to shift toward the outer boundary of the boundary layer with a simultaneous increase in oscillation amplitude. Such a pattern was observed only with simple membrane surfaces with poorly chosen mechanical properties.

In the case of a flow over a complex diaphragm surface, oscillations were observed only in the region not exceeding values of $0.3-0.4$ for y/δ^*; in the flow over a foamed

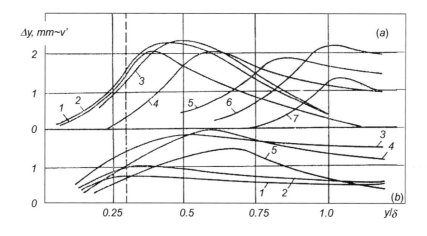

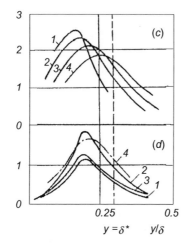

Figure 1.64

Distribution of velocity v' amplitudes throughout the boundary layer thickness in a flow around simple membrane surface (*a, b, c*), intricate membrane and visco-elastic surfaces (*d*): *a*: 1 — $(\beta_r v/U^2_\infty) \cdot 10^6 = 175$, 2—259, 3—332, 4—387, 5—526, 6—680, 7—920; *b*: 1—134, 2—170, 3—188, 4—227, 5—262; *c*: 1—475, 2—680, 3—875, 4—1170; *d* : 1—4—340.

polyurethane plate oscillations were in the vicinity of the plate surface, in the layer of liquid not exceeding $y/\delta^* = 0.2-0.3$ (see Figure 1.64d). Deterioration of boundary layer stabilization from damping surfaces resulted in an increase in relative thickness of the boundary layer in which the disturbing motion was observed, which was independent of the damping plate design. The plate design affected only the oscillation range of value y, at which fluctuations were observed.

The same conclusions can be drawn based on laws of distribution of values $\alpha\delta^*$ and c_r/U_∞ throughout the boundary layer thickness (see Figure 1.65). The flow above simple

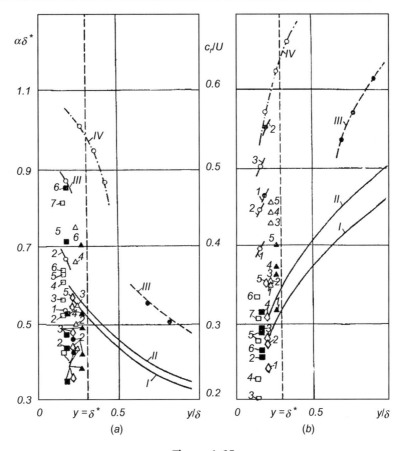

Figure 1.65

Distribution of wave number (*a*) and phase speed (*b*) on thickness of a boundary layer at a flow of elastic surfaces: simple membrane surfaces — clean points (experiment B13), black points (B18); complex membrane surfaces — clean triangles (P6), clean rhombuses (P7), black triangles (P8); PPU — clean squares (P23), black squares (P24); simple membrane surfaces: *I, II* — (B37), *III* — (B18), *IV* — (B13).

membrane surfaces is plotted by solid curves (*I, II*) (experiment B37, $E = 1.35 \cdot 10^4$ N/m², $T = 25.0$ N/m), by hatch (*III*) (experiment B18, $E = 1.55 \cdot 10^4$ N/m², $T = 36.0$ N/m), and dot-dashes (*IV*) (experiment B13, $E = 1.92 \cdot 10^4$ N/m², $T = 79.0$ N/m). The figures designate frequencies of fluctuation according to Table 1.2.

In Figure 1.65 it follows that in the flow over a complex membrane and polyurethane surfaces, values $\alpha\delta^*$ and c_r/U_∞ in contrast to a rigid wall (see Section 1.6, Figure 1.35), are practically independent from the boundary layer thickness. In the case of a flow over simple membrane surfaces, when their mechanical properties were unfavorable for boundary layer stabilization, values $\alpha\delta^*$ and c_r/U_∞ were distributed in the boundary layer

Table 1.2: Parameters of experimental points of Figure 1.65.

Number of Experiment	Figures in Figure 1.65	$\frac{\beta_r v}{U_\infty^2} 10^6$
B37	1	227
	2	262
B13	1	300
	2	420
	3	615
	4	876
B18	1	259
	2	332
	3	387
P6	1	220
	2	254
	3	340
	4	442
	5	492
P7	1	110
	2	152
	3	168
	4	210
	5	248
P8	1	124
	2	150
	3	168
	4	199
	5	220
	6	286
P23	1	80
	2	117
	3	127
	4	158
	5	196
	6	238
	7	287
P24	1	68
	2	114
	3	140
	4	168
	5	204
	6	227

thickness about the same as for a rigid plate flow. The difference was only in the character of laws and absolute values $\alpha\delta^*$ and c_r/U_∞.

In Figure 1.66, the dot-dashed line shows an ultimate neutral curve, and the hatch show the dependences of $\alpha\delta^*$ and c_r/U_∞ on oscillation frequency at various velocities of the flow over rigid surfaces (see Figure 1.39). The usual and ultimate neutral curves for damping

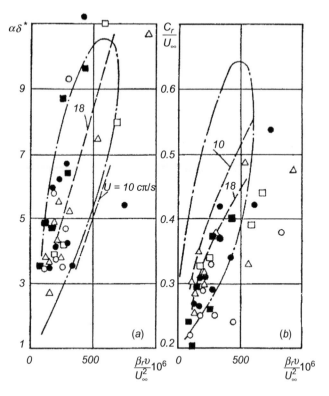

Figure 1.66

Dependence of wave number (*a*) and phase speed (*b*) from frequency of disturbances fluctuations at a flow of elastic surfaces. The same designations, as for Figure 1.65.

surfaces in these coordinates were not plotted. However, dots, triangles and squares denote the same cases of flow over damping surfaces as shown in Figure 1.65.

A comparison of data for rigid and damping surfaces illustrates that in both the cases, for the wavelength and phase velocity, on the one hand, and the oscillation frequency of the disturbing motion, on the other hand, ambiguous features are observed that depend on the mean flow velocity. On the whole, for the flow over damping surfaces, the values $\alpha\delta^*$ increase and the values c_r/U_∞ decrease.

The increase in the values of $\alpha\delta^*$ signifies a reduction for the value of λ in comparison with a rigid plate. This was indeed photographed in the experiments. Since the increase coefficients of the exciting motion, C_i are proportional to a value of λ^2, so the increase rate drops, and the hydrodynamic stability of the flow rises. As it turns out, the reduction of value λ is not a sufficient condition for an increase in stability. The salient point is that a reduction of λ for the same value of velocity v' results in the growth of the slope of the oscillation wave. This can lead to preconditions for a premature development of further stages of the transition region of the boundary layer, or to the destruction of the wave ridge and to the appearance of secondary oscillations that have a high-frequency nature. Therefore, it is important to recognize that, simultaneous to the growth of $\alpha\delta^*$, a reduction in the amplitude of velocity v' is required. This relationship was observed on the best damping surfaces tested.

A reduction of the value λ and with it a reduction of value C_i additionally signifies the possibility of an increase in the extent of the transition zone from the laminar boundary layer into the turbulent one. The same follows for Figure 1.66, from which it is evident that even during unstable oscillations the values of c_r are decreasing.

Direct measurements of values of C_i were also conducted and dependences of values of C_i on the wave number were plotted in the same way as described in papers [305, 310]. The corresponding graphs are given in papers [31, 32, 305]. The same physical features were revealed as in the case of the flow over a rigid plate. The difference was in that the values C_i were lower for the flow over all the effective damping surfaces. For the best damping surface tested, the values C_i were decreased by half.

Using the value C_i, the Reynolds number of transition at low free-stream turbulence can be determined easily and dependably for the case of a flow over a rigid surface [305]. Employing this technique and knowing the characteristic values of quantities C_i for any damping surface, the corresponding Reynolds number of transition can be determined.

Thus, according to measurements on the elastic plate's critical layer, it coincided practically with value δ^* too. In another words it was found that in the process of complicated interactions the relations of characteristic thicknesses of the boundary layer changed.

Just as on a rigid plate, the wavelength λ_z of plane oscillations in dependence on x_1 (see Figure 1.67) were measured at small ε over all types of simple and complex membranes as well as on visco-elastic plates.

On simple and complex membrane plates (curves 2, 4) the regularities $\lambda_x(x_1)$ differ from standard ones. The large scatter of points of curve 2 at small values x_1 is caused by membrane oscillations and peculiarities of development of disturbing motion through the boundary layer thickness. The regularities $\lambda_x(x_1)$ on visco-elastic plates, made of PPU-1 and PPU-2 and covered from outside with a membrane (curves 6, 7), are different in character. The membrane with heating (curve 3) has dependence equidistant to them, and there is another dependence for plate PPU-1 without film (curve 5). So, each type of elastic plate is characterized by its own regularity of evolution of disturbing motion along x_1. In another words, it was shown that due to a change in conditions on the lower boundary of wave-guide (boundary layer) the characteristics of oscillations in the wave-guide vary.

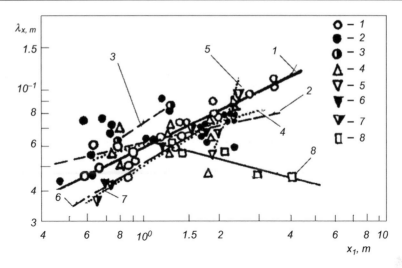

Figure 1.67

Variation of wavelength of plane wave λ_x and longitudinal vortices λ_z along plates: *1* — rigid horizontal plate, *2* — simple membrane plate, *3* — the same as *2* with heating, *4* — complex membrane plate, *5* — visco-elastic PPU-1, open, *6* — the same as *5* with outer film, *7* — the same as *6* with different mechanical properties of surface — PPU-2, *8* — regularity for λ_z on rigid plate.

The conclusions from Figure 1.67 become more obvious if the results are represented in dimensionless form: $\tilde{\lambda}_x = \lambda_x/\delta^*$.

Curves in Figure 1.68 have been plotted for second neutral oscillations. Although all the curves in Figure 1.68a are smooth, it can be seen that every elastic plate has its own type of regularity. The same curves, as those in Figure 1.68a, represented in dependence on Reynolds number Re*, have a certain minimum. Analogous relations but in another dimensionless form [59] do not have such information capacity. Values of minima are noticeable especially in Figure 1.68c built in normal coordinates. It is characteristic that on the rigid plate extremum occurs at Re $\approx 1.35 \cdot 10^5$; for simple membrane plate it corresponds to lower Reynolds number Re (10^5) in case of unfavorable mechanical characteristics, and otherwise to the same Reynolds number for a simple membrane plate or to a larger Reynolds number (Re $\approx 2.35 \cdot 10^5$) for a complex membrane plate.

It can be seen from Figure 1.68 regularities of development of value $\tilde{\lambda}_x$ from the left and right from extremum are different. Since λ_x grows monotonically along x_1 (see Figure 1.67) then such a character of the curves in Figure 1.68 is explained only by the fact that in the process of boundary layer evolution the low of growth of value δ^* changes. In addition, this, in its turn, is connected with the flow character at different stages of transition. For instance, Reynolds number Re $\approx 1.35 \cdot 10^5$ for a rigid plate agrees well with Reynolds number Re at which character of development of value λ_z changes (see Figure 1.67). Thus,

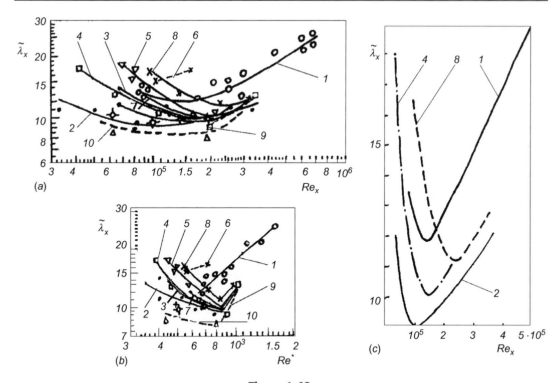

Figure 1.68
Dependence for wavelength $\tilde{\lambda}_x$ on Reynolds number calculated using length (a, c) and displacement thickness (b): *1* — rigid plate. Simple membrane surface: *2* — tests B1—B4, *3* — B8—B10, *4* — B11—B14, *5* — B20—B24, *6* — BT43 (with heating). Complex membrane surface: *7* — P9—P12, *8* — P5—P8. Visco-elastic surface: *9* — P23—P24, *10* — P25—P35.

being based on form of curves in Figure 1.68c, it is possible to judge the areas of development, in terms of Re$_x$, of nonlinear stages of transition.

From data of Figure 1.68c according to [310], the following empirical relations have been obtained:

$$\text{for rigid plate (curve 1) } \tilde{\lambda}_x = 7.19\cdot 10^5\ Re_x^{-1} + 3.56\cdot 10^{-5}Re_x + 1.7 \qquad (1.10)$$

$$\text{for simple membrane (curve 2) } \tilde{\lambda}_x = 2.28\cdot 10^5\ Re_x^{-1} + 1.76\cdot 10^{-5}\ Re_x + 4.96 \qquad (1.11)$$

$$\text{(curve 4) } \tilde{\lambda}_x = 6.3\cdot 10^5\ Re_x^{-1} + 2.2\cdot 10^{-5}\ Re_x + 2.5 \qquad (1.12)$$

$$\text{for complex membrane plate (curve 8) } \tilde{\lambda}_x 17.17\cdot 10^5 Re_x^{-1} + 3.13\cdot 10^{-5}Re_x - 3.15 \qquad (1.13)$$

For flow over a rigid plate, the point of intersection of curves *1* and *8* (see Figure 1.67) corresponds to the second stage of transition [66, 305], when the evolution of disturbing motion makes it possible to identify a deformation of velocity front in plane *xz*. The point of intersection of curves *2* and *4−7* shifts downstream at a distance two times greater than that for a rigid plate (delay of development extent of the first stage of transition). From that point there will be the start, in the boundary layer, of development of 3D deformations, which will proceed slowly too. Therefore, the curve for an elastic plate, analogous to curve *8*, will take its own beginning from this point of intersection and dispose in parallel to axis of abscissas.

It was shown in Section 1.6 that the character of development of disturbing motion when flowing over rigid plate is conservative as to any outer conditions. Investigations carried out on elastic plates leads to the idea that the character of development of disturbing motion depends substantially on many factors, which are defined by the properties of the lower wall of wave-guide (boundary layer). All of them may be grouped into two types: the first type, there is essential stretch, along *x* (and along Re), of separate stages of transition [66, 305]; the second type, conversely, initial stages supersede one another very rapidly and followed by a stage of developed system of longitudinal vortices, the extent (of stage) of which is much stretched along *x* too.

On a sheet made of PPU-1 without film (see Figure 1.69) fluctuations *u'* are less than those with film. On plate 10*a* (see Section 2.5) the difference between curves *5* and *6* testifies to the presence of smaller-scale longitudinal vortical structures in the boundary layer, and transition occurs more smoothly. Both the presence of two maxima and the dependence of the curves on *z*, as for plates 11 and 11*a* (see Section 2.5), is the sign of a nonstationary system of longitudinal vortices. When reducing elasticity of plate 11*a* as compared to 11, vortical systems become a smaller scale on *z* (see Figure 1.69b). It is such flow structure that explains the shape of similar curves in [44].

All above data make it possible to conclude that for the first type of transition the model of disturbing motion in the boundary layer over elastic plates looks different. At the first stage of transition under influence of wall oscillations, the shape of the plane T−S wave becomes pointed and strongly stretched (compare Figure 1.67 and [66]). This stage is strongly extended on *x*. The second, third and fifth stages of transition are absent. The nonlinear plane wave is transformed very fast into the fourth stage, which, without shedding of heads of hairpin vortices, turns into the sixth stage, much-extended on *x* as well as in the first stage. Transition to turbulence follows evolution and shedding of peripheral vortices.

The second type of transition differs from the first one, as on the rigid wall at increased ε too there is a very rapid interchange of the first five stages of transition. Then, without shedding of heads of hairpin vortices, the sixth stage comes, which, as at the first type of transition, is very stretched on *x*. However, during its evolution, owing to surface oscillations the system of longitudinal vortices becomes nonstationary. The transversal

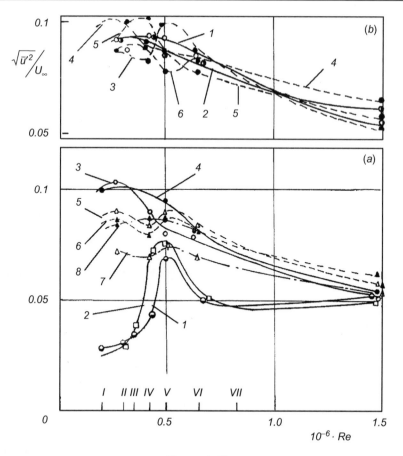

Figure 1.69

Variation of longitudinal fluctuation velocity at different stages of transition when flowing rigid and elastic plates: *a*: *1* — rigid plate, *2* — rigid plate with vibrator, *3* — plate PPU-1, *4* — PPU-1 with film, *5* — plate *10a* between built-in stiffeners, *6* — plate *10a* over built-in stiffeners, *7* — the same as *5* with vibrator, *8* — the same as *6* with vibrator; *b*: *1* — plate *11*, *z* = 0; *2*—*11*, *z* = 0.01 m; *3*—*11a*, *z* = 0; *4*—*11a*, *z* = 0.01 m; *5*—*11a*, *z* = 0.02 m; *6*—*11a*, *z* = 0.04 m.

shape of vortices is deformed permanently: the form of cross-section of vortices enlarged in scale distorts in such a way that a it leads to the creation of a system of longitudinal (in several layers) small-scale vortices, which, in their turn during evolution, merge with each other and form large-scale vortices with an irregular shape of cross section. The evolution of these systems is accompanied by the appearance of peripheral vortices, the destruction of which just leads to turbulence.

Thus, due to oscillations of elastic plates and their surface waves, the disturbing motion has definite specific features.

Experimental studies conducted previously and the above-mentioned data lead us to the conclusion that adequate selection of mechanical properties of damping surfaces makes it possible to obtain at least one of the following effects:

- increase of stability loss in Reynolds number;
- increase in the extent of the transition zone and in the value of the transition Reynolds number;
- reduction in the frequency range of unstable oscillations in the laminar boundary layer;
- reduction in the phase velocity of the exciting motion;
- increase in the wave number value and reduction of increased coefficient values;
- reduction in the vertical coordinate of the critical layer;
- reduction of in the kinetic energy of the exciting motion.

1.12 Receptivity of the Boundary Layer to Different Disturbances

The problem of the receptivity of the boundary layer to different disturbances arose in connection with an investigation on the effect of different agents on stability [445]. First, investigations on the effect of pressure gradient on the laminar boundary layer transition in the turbulent boundary layer and on flow stability were needed. This made it possible to design and investigate a laminar aerofoil, which are widely used in aviation. Furthermore, the effect of roughness elements on transition, separation and heat transfer of profiles was investigated [445]. Researches of the influence on hydrodynamical stability of effect of the waviness of a streamline surface, its fluctuations and as dependencies of viscosity of superficial temperature [96, 125, 206, 305, 442, 445, 524].

Both the factors active at the flow stability "from inside," i.e. acting at the lower boundary of the boundary layer, and the factors acting "from outside," i.e. acting at the upper boundary of the boundary layer have been investigated. They include free-stream turbulence, sound fields, flow compressibility and fluctuations.

The accumulated experimental data concerning various aspects of transition from the laminar boundary layer to the turbulent boundary layer posed the problem of generalizing this obtained data. The generalization consists of developing both a common theory of the boundary layer and a common method of boundary layer investigation. Morkovin [366] made the first step in that direction. He formulated the receptivity problem, involving the determination of the mechanism of normal modes excitation from exposure to the flow disturbances, free-stream turbulence, anisotropic disturbances, wind tunnel sound fields, roughness and so on.

The receptivity problem had been stated earlier, in fact, in papers by Schiltz [444] and Wehrmann [511] and also in other areas of hydromechanics. Schiltz was the first to investigate the interaction of two 2D sinus waves, one of which was being generating by the wall, and the other was a natural T−S wave. This paper marked the beginning of boundary layer experiments concerning the interaction of two or more 2D forward and oblique waves, as well as the interaction of the main normal wave and subharmonics. The main conclusion was that when the phase shift of two 2D waves is equal 180° the main Tollmien wave becomes substantially smaller in amplitude or is fully stabilized.

In the further problem of susceptibility investigated by Reshotko [366]. The overall objective of her work consisted in defining the influence of a degree of turbulence at the basic flow for carrying out of the investigation in low-turbulent wind tunnels on characteristics of a boundary layer and models or at a flow of plates. It was important to find the mechanism of comparison of the results received at this with the results of experiments on natural devices at their flight in air. It is known that in flight by inertia in air the degree of turbulence is very small. In this connection results of natural researchers will not always agree with modeling experiments. The problem of susceptibility has grown out of the need for a more exact definition of Re. The calculations received on the basis of the linear theory of stability did not allow for correctly estimating Re transition. Reshotko has formulated the main problems:

> *Susceptibility by means of which the concrete forced disturbances are included into a boundary layer, and character of its influence on the disturbing flow. If initial disturbances are great enough, they can grow due to forcing mechanisms up to nonlinear level and directly to lead to turbulent flow. If they are small, they will aspire to excite free disturbances in a boundary layer. These free disturbances are normal modes of a boundary layer and they often name Tollmien–Schlichting waves.*
>
> *The phenomenon of susceptibility differs from stability both from a physical and mathematical viewpoint. From a physical viewpoint, it is character of reaction of a boundary layer on some external disturbances. Mathematically it already any more a task of the decision of the homogeneous equations with homogeneous boundary conditions, but a task where the equations and/or boundary conditions are not homogeneous.*

Despite the formulated substantive provisions of the problem, there are a number of essential elements lacking. Morkovin and Reshotko considered the problem of receptivity to be a reduction in the reaction of a boundary layer to external disturbances, the consequence of which "is the neutral decision of the equations of the disturbed movement having the same frequency and phase speed, as particularly considered forcing disturbance." With this approach, the complex form of the structure of disturbing movement at various stages of transition and the structure of the concrete forcing disturbance is not considered, and the reaction of the boundary layer is considered only from the viewpoint of the development of 2D disturbances. Thus, nonlinear effects and the spatial 3D character of development of disturbances in the boundary layer are not considered.

Lamford [59] formulated the receptivity problem in connection with the turbulence problem with more generalization:

> *... it is clear that a hydrodynamic system cannot be fully isolated from an outer medium. Therefore, a whole theory must allow the opportunity to introduce a light outer*

noise in the much-generalized form. However, we consider this noise not as a fundamental cause of the large-scale motion of flow, but as a certain complication, which can be analyzed after the isolated system, were investigated. What philosophical reasons here would not arise, one of the basic achievements of the described approach is the strict explanation of the chaotic behavior, which is not demanding any external source of accident.

In fact, the receptivity problem was formulated in electrodynamics [59]. A linear converter of a sound signal was used, a sinus curve $A \cdot e^{-i\omega t}$ was fed at an input and a signal $\chi\ (\omega)$ $A \cdot e^{-i\omega t}$ was raised at an output, where $\chi\ (\omega)$ is a transfer function.

Section 1.1 contains the classification of various types of CVS in various hydrodynamic problems.

In real conditions of a flow about a body disturbances always appear in a boundary layer owing to nonstationary motion. In accordance with size of these disturbances they can conditionally be divided into three groups. Small disturbances that influence the disturbing motion of a boundary layer belong to group I (see Section 1.1); these include external disturbances, i.e. that act from the outer boundary of a boundary layer—turbulence of external flow, vortical and thermodynamic disturbances, acoustic and electrodynamics fields in the environment and so on. The internal disturbances of group I influence a boundary layer from its internal boundary, i.e. from a surface. Surface irregularities, roughness, various obstacles and cavities, vibration of surface, its thermodynamic characteristics, etc., are examples of internal disturbances of group I.

The research of these authors has shown that under ideal conditions of flow certain CVSs form in a boundary layer. These structures can be conditionally divided into groups, so that the transition of a laminar boundary layer into a turbulent one consists of eight stages (see Section 1.7, Figure 1.42). A typical type of CVS corresponds to each stage. The physical nature of a boundary layer is such that each consequent kind of CVS follows from development of the previous type of CVS. An important phase of investigation is exploring the features and characteristic types of CVS and their development at the influence of external and internal disturbances of group I onto a boundary layer. Similar CVSs exist also in the near-wall area of a turbulent boundary layer (see Section 1.7, Figure 1.47). Some research has been done towards this and still develops the problem of receptivity of a boundary layer to various disturbances [310]. The interaction of various outer disturbances of group I with characteristic CVSs of a boundary layer is considered in this case. It should be noted that the outer disturbances cannot be eliminated. These disturbances can be damped or transformed into favorable ones for drag reduction. The internal disturbances can essentially be reduced.

Group II contains disturbances of large size, which are operational on the entire boundary layer. Such disturbances can arise in angular areas of designs; for example, in the region of connection of a body and wing, two pairs of large longitudinal vortices (LLVs) appear (see Section 1.1, Figures 1.3 and 1.5). These vortices influence the entire thickness of a boundary layer.

The disturbances of large size arising at maneuvering or at oscillatory motion of a vehicle pertain to group III. LLV form on protruding parts of a body and on its leeward shaded areas, which affects both the neighboring and downstream boundary layers (see Section 1.1, Figures 1.4 and 1.6).

Research on the influence of such LLVs on a flow over a body is limited, and when such research has been conducted only the structure and kind of LLVs arising at nonstationary motion of a body were considered. However, some research on the influence LLVs on the characteristics of a boundary layer has been carried out, including the development of research of the influence of LLVs on the characteristics of a boundary layer and its CVSs. Thus, it is necessary to develop the problem of receptivity for the case of interaction of CVSs with LLVs, with disturbances of group II. Here LLV refers to "large longitudinal or transversal vortices."

A third aspect of the problem of receptivity has virtually not been investigated. This is research into the influence of CVSs on LLV formation, interaction of disturbances of groups I and II, and interaction of disturbances of groups I and II with CVSs.

The current state of the problem of flow about a body in an environment is summarized as follows: the existing calculation methods for potential flow about a body and the characteristics of a boundary layer are idealized and do not take into account an actual pattern of fluid flow near a body. The results of such calculations can be used as a reference and need to be corrected due to various disturbances influencing a boundary layer and the character of potential flow (for example, the influence of LLVs).

One of the methods for defining the size and character of the corrections can be the method of receptivity of a boundary layer to various disturbances. Theoretical methods of calculation of receptivity of a boundary layer have been developing but only for the initial stages of transition. The authors have developed a methodology and experimental methods of researching receptivity of a boundary layer in a generalized representation.

In relation to the above discussion, the problem of receptivity of a boundary layer has three parts:

*   receptivity of a boundary layer to disturbances of group I (small disturbances of CVS type);
*   receptivity of a boundary layer to disturbances of group II (large disturbances of LLV type);
*   receptivity of a boundary layer to disturbances of groups I and II in various combinations.

The text below provides a review of the research relating to the first point, namely the receptivity of a boundary layer to group I disturbances.

The results of the experimental research described in Section 1.5 (see Figure 1.28) and in Section 1.6, indicate that disturbing movement in a boundary layer develops in conical layer throughout the thickness of a boundary layer. Thus, due to a difference of speed in shear layers of the boundary layer, monoharmonic disturbances develop at once across a boundary layer thickness in the form of subharmonics. Oscillations appeared in different layers of the boundary layer and begin to interact with each other moving along x. All the worsening factors specified above influence the character of such fluctuations developing a boundary layer and the features of their interactions with each other. This is the physical nature of a boundary—the nature of shear flows (see Section 1.7, Figure 1.48). Thus, the problem of receptivity *a priori* took place in the structure of a boundary layer itself. The same is true for the 3D disturbances, which in shear layers of a boundary layer begin to form subharmonics and then interact with each other.

Thus, the problem of receptivity can be considered as the interaction of various disturbances in a boundary layer. It is possible to imagine that typical kinds of CVSs exist in a boundary at each stage of development. Then, outside disturbances introduced at each stage of transition will interact with typical natural CVSs of the boundary layer. However, the real picture is more complex. Disturbances introduced into the boundary layer will cause interaction of their own generated subharmonics, which have arisen under the action of external disturbances. In addition, simultaneously introduced disturbances will interact with these subharmonics.

For simplicity, the problem of receptivity will be considered using the hydrobionics approach. In Figure 1.70 the generation 3D disturbances in a boundary layer by means of the functioning of rows of dermal papillae is illustrated (see Section 1.8, Figures 1.49 and 1.53). Depending on the degree of blood filling, longitudinal CVSs of various sizes and shape are formed in the boundary layer. In Figure 1.71, the scheme of interaction of boundary layer CVSs with CVSs generated by the specific structures of a dolphin's skin is shown.

The method of investigation of receptivity of the boundary layer is as follows. The interaction of natural disturbances of the boundary layer with disturbances initiated by asperities, roughness, wall vibration, turbulence, other 2D or vortex disturbance of ongoing flow [59] is examined. It is convenient to represent the real disturbance as a combination of a simpler wave or vortex motions of different form, frequency and amplitude.

For the first step, the structure of flow at different stages of natural transition from the laminar boundary layer to the turbulent boundary layer is investigated to obtain standard and metrology measurements.

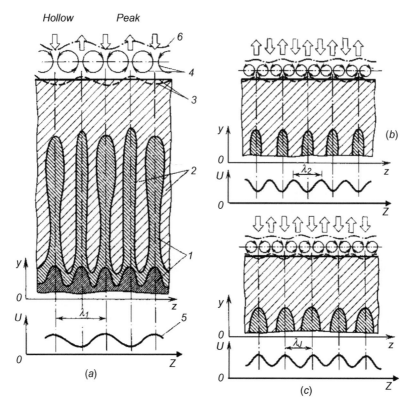

Figure 1.70

Scheme of location of longitudinal vortices in boundary layer of dolphin when disposing one (*a*) and two (*b, c*) vortices between the rows of dermal nipples: *1* — longitudinal lines dermal rollers with dermal nipples, *2* — longitudinal lines epidermal nipples, *3* — surface of a skin, *4* — longitudinal vortices, *5* — distribution longitudinal velocities component *U(z)*, *6* — change of boundary layer thickness δ (*z*).

Various deteriorative factors (increase of turbulence level of free stream, positive pressure gradient, wall vibration etc.) do not affect the disturbance form, but promote generation of the stages transition to turbulence and may lead to shortened transition — turbulence arising after stage *V*. Then, depending on the form and type of outer disturbance the disturbance parameters are determined—amplitude, intensity, type—2D or 3D. With this, the coordinate of the disturbance source is determined as well as the location of disturbance motion generators in the undisturbed flow, boundary layer or on the wall. For this purpose, visualization of the disturbing motion is made at the stages of transition [30, 305].

Such a visualization is presented in Figure 1.72 where a transversal velocity profile *U(z)* at the nonlinear stages of natural transition is pictured. It reveals a 3D deformation of the flat

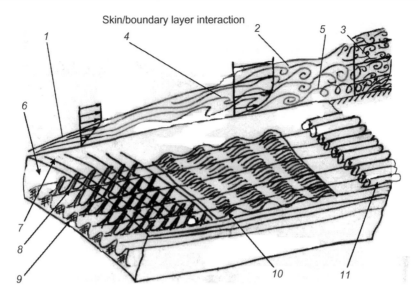

Figure 1.71

Scheme of forming of two aspects of coherent vortical structures: sinusoidal wave and longitudinal vortices in boundary layer, with the help of skin: *1* — laminar boundary layer (BL), *2* — transaction BL, *3* — turbulent BL, *4*—2D disturbances, *5* — longitudinal vortex in BL (3D disturbances), *6* — dolphin skin covers, *7* — transverse microfoliation, *8* — dermal nipples, *9* — longitudinal lines dermal rollers, *10* — sine waves in BL generated by cross-section microfolds, *11* — longitudinal vortex in BL generated by longitudinal lines of dermal rollers and nipples.

part of the e velocity profile, it is nonstationary and dependent on the y $U(z)$ at the coordinate. The revealed vortex tubes, indicated by heavy lines, are also registered. The generators were developed by Babenko [305, 310]. The visualization made it possible to discover the location of "peaks" and "valleys" [190], which are advantageous for taking the measurements. In [310] photographs visualization of boundary layer receptivity of 3D disturbances, which were initiated by a grid, are given. In Figure 1.72b you can see the investigation results of the disturbance motion in the boundary layer over an elastic plate made of the porous elastomer penoelast (PE) with a smooth external surface. Natural indignations are stabilized by an elastic surface on all thickness of a boundary layer.

For elucidating the character of interaction of natural disturbances with introduced ones in a boundary layer the receptivity of a boundary layer at all stages of transition was investigated. Thus, the field of kinematic parameters in planes xy and xz is investigated in detail at the certain values of coordinates x, y and z.

First, visualization of every transition stage was carried out by means of the tellurium method. The detailed measurements were taken in certain locations. The measurements were made using a DISA heat-loss anemometer or with a laser Doppler velocity indicator

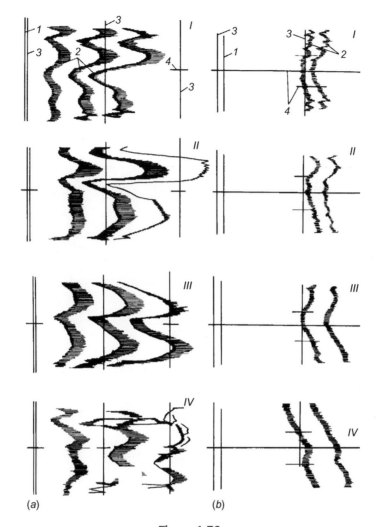

Figure 1.72

Development of disturbance movement at natural transition of the boundary layer over the solid plate — *a*: $U_\infty = 6.7$ cm/s, $x_{t.w} = 1.1$ m, $y_{t.w} = 3$ mm (*I*); $y_{t.w} = 4$ mm (*II*); $y_{t.w} = 5$ mm (*III*); $y_{t.w} = 6$ mm (*IV*); and elastic plate — *b*: $U_\infty = 5.0$ cm/s, $x_{t.w} = 2.2$ m, $y_{t.w} = 2$ mm (*I*); $y_{t.w} = 4$ mm (*II*); $y_{t.w} = 6$ mm (*III*); $y_{t.w} = 8$ mm (*IV*); 1 — tellurium wire, 2 — tellurium chords, 3 — transverse marks every 0.1 m, 4 — longitudinal marks (central mark along the axis of test section — Figure 1.20).

(LDVI). Then, at certain points of the boundary layer, the 2D or 3D coherent disturbances were inserted and the interaction of the disturbance with natural disturbance was investigated. The interaction was investigated both by visualization and using quantitative methods [59, 310]. In [47, 48, 51, 55, 57, 59–61, 310] boundary layer receptivity was investigated at every transition stage to 2D sinus disturbance and to 3D longitudinal disturbance of Benny–Lin type vortices initiated by the disturbance generators developed

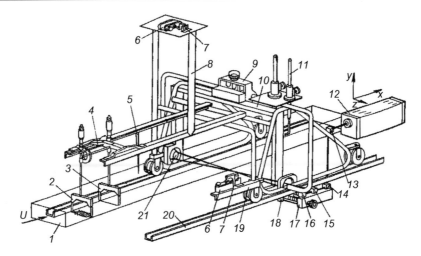

Figure 1.73

Accommodation on the trolley of the equipment for research of a boundary layer: *1* — working site, *2* — support with vortex generators, *3* — support with tellurium wire, *4* — micro coordinate device, *5* — anode, *6* — camera, *7* — relay of automatic release, *8* — frame, *9* — coordinate device, *10* — framework, *11* — gauges heat-loss anemometer, *12* — laser, *13* — prism of full reflection, *14, 16* — rotary prisms, *15* — beam splitter, *17* — adjusting device, *18* — focusing lens, *19* — trolley, *20* — rails, *21* — photographic detector.

by Babenko [66, 305, 310]. With this, the interaction of structures and their modification at every stage of natural transition was investigated.

The dependence of character of receptivity on the level of disturbances defined at a various degrees of turbulence of a stream and intensity introduced disturbances was explored. It is obvious that linear and Goertler stability is a special case of the general problem of receptivity when the intensity of own, natural and introduced indignations are small.

Receptivity has been investigated with a low-turbulence hydrodynamic test set up (see Section 1.5, Figure 1.20) [30, 305, 310]. Methods, measurement techniques, devices and apparatus are described in Sections 2.5 and 3.1 [310]. In Figure 1.73 the scheme for mounting of the equipment on a trolley and moving along the working section is shown.

Figure 1.74 indicates coordinates along the working section where visualization and quantitative measurements were made. The insert is located in the end of the working section. The Roman numerals designate numbers of stations of measurements, the results of which are shown in the following chapters.

The boundary layer over the bottom of the working section is a conservative system. This means that certain stages of transition correspond to certain values of U_∞, and coordinate x. Under these conditions finite flat or 3D disturbances were introduced and the interaction of

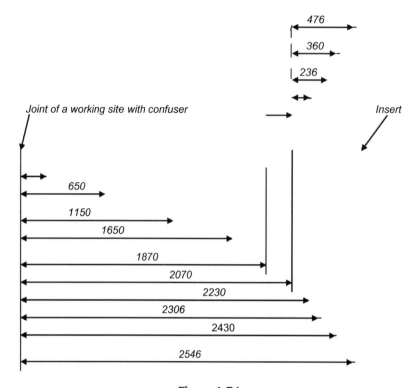

Figure 1.74
Scheme of an arrangement of sections along a working site for measurement of kinematic characteristics of a boundary layer.

the introduced disturbances with natural ones were analyzed, i.e. the response of a boundary layer in the presence in the boundary layer of CVSs typical for the given stage of transition before the external disturbances were introduced (see Section 1.7, Figure 1.42) [59, 462].

The receptivity to 2D disturbances was investigated at a high turbulence level ($\varepsilon > 1\%$) by means of LDVI. The vibrator strip located at the area of maximum fluctuating velocity across $\bar{y} = y/\delta^*$ introduced in the boundary layer finite disturbance with amplitude of $A_{vibr} = 0.8-0.9$ mm (compare with when stability is investigated, $A_{vibr} = 0.24-0.32$ mm). The profiles of average and fluctuating longitudinal velocities at different stages of transition with natural turbulence were used as standard profiles (see Figure 1.75) [47]. The velocity profiles were measured in characteristic locations determined according to Figure 1.42.

The numbers above each profile correspond to the number of the stage of transition in Figure 1.42. An average velocity was chosen such that would receive CVSs in accordance with the scheme in Figure 1.42. Horizontal lines near each profile indicate the location of the vibrator strip relative the boundary layer thickness.

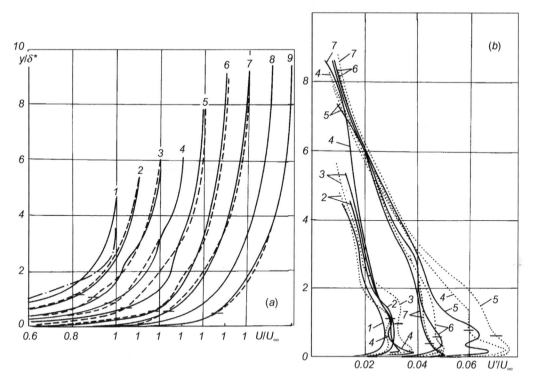

Figure 1.75

Profiles of average (*a*) and fluctuating (*b*) longitudinal velocity components at $U_\infty = 0.09$ (*1*); 0.12(*2*); 0.15(*3*); 0.18(*4*); 0.21(*5*); 0.27(*6*); 0.35(*7*); 0.4(*8*); 0.6(*9*) m/s, at natural transition (solid lines) and with 2D disturbances introduced in the boundary layer (dashed lines). The dot-dahsed line is Blasius profile, the vibrator tape positions are marked with horizontal lines [310].

Velocities of flow typical for various stages of transition were determined at the bottom of working section in nine stations (or IV section on the insert) (see Figure 1.74) on the basis of visualization of a boundary layer by the tellurium method (see Figure 1.72). Profiles of longitudinal average and pulsating velocity (see Figure 1.75) were measured with a laser anemometer. Then the vibrator strip was placed in that station, which was used for research of hydrodynamic stability of the boundary layer [305]. At the same velocities, the strip was placed at a height corresponding to maximum values of the longitudinal pulsating velocity, at each flow speed. The amplitude of fluctuation was about 1.5−2 that for research of hydrodynamic stability. The frequency of fluctuation was chosen to correspond to the second branch of the neutral curve for each flow speed [305].

Introducing 2D disturbances into a boundary layer, as with natural transition development, leads to the distortion of average velocity profiles at the same values of y/δ^*. During the first five stages of transition, at the introduction of 2D waves, this distortion decreases and then becomes constant. However, the fluctuating velocity profiles and their maximal values differ from the standard ones and indicate accelerated transition at the initial stages [310].

Research on the development of a boundary layer along the working section has been carried out at higher levels of turbulence than the sections specified in Figure 1.74 at various values of δ by means of the tellurium method. Pictures of visualization were obtained. In Figure 1.72, an example of such visualization is shown. Based on the analysis of these results, characteristic places along coordinate z corresponding to peaks and hollows of longitudinal vortical structures (see Figure 1.70) and to places between them were determined. In these characteristic places, numerous measurements of average and longitudinal speeds of the boundary layer were made at natural transition. In the same places, measurements of receptivity of the boundary layer to 2D disturbances were executed simultaneously. Numerous structures of velocity were determined. In Figure 1.74, profiles of average and fluctuating longitudinal velocity components measured along a longitudinal axis of a working site result and are shown.

The receptivity of the boundary layer to 3D disturbances was also investigated in [301, 57, 60, 61]. The measurements were made under same conditions. To generate uniform 3D disturbances of a predetermined scale, vortex generators were located in the boundary layer (see Figure 1.76) [310]. At certain locations (at peaks, i.e. along the line behind vortex generators (at $z_1 = 0$), at valleys, or along the middle line between the neighboring vortex generators (at $z_2 = \lambda_z/2$), and between them (at $z_3 = \lambda_z/4$)), the average velocity profiles $U(y)$ and the fluctuating velocity profiles $u'(y)$ were measured by means of LDVI (λ_z is wavelength of the disturbance in transversal direction (as in Figure 1.70) and also the distance between vortex generators [305]).

The investigations were made at stages *I*, *V* and *VIII* of transition. The disturbances were generated by vortex generators of three sizes. The λ_z of disturbance were of 3−6 times less than that of natural transition. The sizes of vortex generators were selected taking into account the above-mentioned recommendations and δ^*/δ ratio, which is 0.3 for the laminar boundary layer, 0.125 for the turbulent boundary layer and between these values for the stage of vortical structures.

The wavy profiles $U(z)$ were revealed across the boundary layer [47, 310] as well as the corresponding inflections in $U(y)$ profiles. The fluctuating profiles have the following features:

- maxima of velocity fluctuations are distanced from the wall;
- a characteristic form, depending on z_i, which is connected with generation of longitudinal CVSs;
- fluctuation levels decrease at optimal interaction between natural and introduced 3D disturbances.

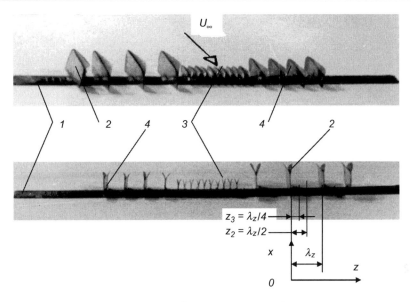

Figure 1.76

Generation of the 3D disturbances: *1* — plate having grooves for installation of wings (vortex generators), distance between grooves = 2 mm, *2* — large wings, *3* — small wings, *4* — average wings. Vortex generators in a working site (see Figure 1.73).

Investigations have also been made of the complex interactions of disturbing movements when disturbances introduced into a boundary layer simultaneously from above and below a surface interact with natural disturbances characteristic for a corresponding stage of transition. Disturbances introduced into a boundary layer from below a surface were created by means of an elastic plate having a regular structure of the external layer. Both 2D and 3D disturbances were generated in a boundary layer from below. Such interactions of disturbing movements are called complex disturbances [47, 310].

At introduction of 3D disturbances in a boundary layer from below, and 2D disturbances from above from the surface of elastic plates, the resonant effect was found. It involves the greatest interaction of identical structures of disturbing movement at natural transition and introduced disturbances, and fast transformation of introduced disturbances by means of elastic plates.

Regulating the intensity of the 3D disturbances generated from below of the surface it is possible to increase or weaken the development of the characteristic stages of transition over an elastic plate. According to Sections 1.9 and 1.10 natural transition on elastic plates is characterized by rapid organization of 3D disturbances (longitudinal CVSs) [305]. So, it is possible to regulate the character of complex interactions at the corresponding organization of generating 3D disturbances from below.

Introduction of flat disturbances on a rigid plate leads to accelerated transition. At complex interactions on elastic plates the boundary layer is stabilized more intensively then at interaction only with 3D disturbances. When complex 3D disturbances are generated from above and below the plate, stabilization of the boundary layer occurs more intensively and strongly as higher energy and sizes of 3D disturbances arising at natural transition with generated disturbances coincide.

According to this, the boundary layer can be presented as a dynamic system possessing a spectrum of eigenfrequencies, which respond to introduced disturbances depending on resonance and the structures of introduced and natural disturbances [47, 310].

Experimental research on complex interactions have made it possible to draw two basic conclusions [310]. First, the process of interaction of disturbances in a boundary layer can be presented as superposition of "frozen," so-called natural CVSs in a boundary layer at each stage of transition and the CVSs introduced into boundary layer from outside. Second, the interaction of these types of CVS has resonant character depending on their type and characteristics (amplitude, frequency, etc.).

As an example of such interaction, it is convenient to assume that disturbance develops in a boundary layer according to the CVS model presented in Figure 1.42. So, when a disturbance is introduced in a boundary layer, for example, a flat wave will pass through all stages of transition. Interaction of the introduced disturbance will occur only at that coordinate x (at those stages of transition) that is characterized by flat or quasi-flat structures of CVSs in the non-perturbed boundary layer. However, interaction in this case will occur only involving similar characteristics of CVS. If parameters of waves essentially differ, interaction of CVSs will not occur.

If longitudinal vortical disturbances are introduced into a boundary layer, for example, they also pass CVSs to the non-perturbed boundary layer at all stages of transition. Interaction will occur in the place where CVSs exist of the hairpin vortex type at the development of the natural boundary layer. However, the condition of interaction will be the same: it is necessary that the shape and parameters of disturbances are similar.

Therefore, the longitudinal vortical disturbances brought into a boundary layer amplify only in the area of maximal amplifications of the neutral stability curve for longitudinal vortical disturbances. This will occur in the place where the boundary layer is unstable to longitudinal vortical disturbances and where at natural transition CVSs develop in the form of hairpin vortices.

Thus, flat disturbances arising due to vibrations of a wall excite the boundary layer mostly at the initial stages of transition. Conversely, longitudinal vortical disturbances caused, for example, by elements such as non-uniform roughness or curvature of a surface excite a boundary layer downstream at nonlinear stages of transition. The disturbances of the complex form going from the outer area of flow can be presented as consisting of wave and vortical components. Then wave modes will excite initial stages, and vortical modes the subsequent stages of transition. However, interaction of introduced and natural disturbances will occur only based on the interaction of resonant character [25]. Therefore, the boundary layer will not respond to just any external disturbance.

These events are illustrated in the experiments in [305], in which at each location along x flat disturbances in a wide range of frequencies were introduced. Fluctuations in the boundary layer were observed only in a certain range of frequencies, about which a so-called limiting neutral curve has been constructed. Outside of this range of frequencies fluctuation faded very fast. The boundary layer responded to a certain spectrum of flat fluctuations in the area of first stages of transition at resonant interaction of introduced disturbances and eigenfrequencies of the liquid. The resonant character of receptivity of a boundary layer is such that structures of introduced and natural CVSs interact when their amplitude−frequency−wave characteristics are similar, resulting in damping or amplification of the disturbing movement.

The same research experiments were performed in the flow over membrane surfaces. Fluctuations of the vibrator strip caused fluctuations in the elastic membrane in the location of the strip. When phases of these fluctuations were same in some places along x, the area of unstable fluctuations limited by the neutral curve essentially increased. Furthermore, visualization (see Sections 1.9 and 1.10) has shown that the membrane surface appears to deflect the disturbances to the outer border of a boundary layer, destabilizing the boundary layer and accelerating the alternation of stages of transition.

A more complex character of receptivity is observed at the interaction of several disturbances simultaneously introduced into a boundary layer from the external border and from the surface. Such a picture exists in a real flow over a plate; for example, when natural CVSs in a boundary layer interact with the disturbances caused by vibration, sinuosity, roughness of the wall or turbulence in the mean flow, etc., CVSs (LLVs) of groups II and III interact. In this case, it is possible to use the approach described above for complex disturbances. If in one disturbance any mode of wave propagation (flat or vortical) prevails over others, the character of interaction of natural and introduced disturbances is same as in the simplest cases.

All the above statements are valid not only for the transitional boundary layer but also for the turbulent boundary layer. Experiments [40] have shown that after bursting in the viscous sublayer, stages of disturbances development are formed in a similar way as in a transitional boundary layer at various worsening factors [310] (see Page 145, on dispersion). Therefore, the considered mechanisms of receptivity are also typical for near-wall areas of a turbulent boundary layer.

According to experiments on the structure CVSs in the near-wall by many researchers and to the generalized schemes by Meng (see Section 1.2, Figure 1.17) and Babenko (see Section 1.7, Figure 1.47), the structures of CVSs in a turbulent boundary layer, the structure of CVSs in the near-wall areas and periodicity of their formation make it possible to distinguish precisely the flat-wave character of CVS formation, as in the form of longitudinal CVS. Such patterns of CVSs in the near-wall region resemble the shapes of CVSs generated into a boundary layer by structure of dermal coatings of dolphins (see Figure 1.71).

1.13 *The Boundary Layer as a Heterogeneous, Asymmetric Wave-Guide*

The laws of development of disturbing movement throughout the thickness of a transitional boundary layer mentioned in Sections 1.5−1.7 and 1.9−1.11 have made it possible to discover a number of features of disturbances development in flows over a rigid and elastic surface. Under certain conditions, the energy of pressure pulsations in a boundary layer can cause propagation of fluctuations inside an elastic plate. In this case, it is necessary to solve the dual problem of development of disturbances in a boundary layer that will interact with the caused fluctuations in the elastic surface. In the previous section some results of experimental research on such complex interactions were presented.

The analysis of the characteristics of the disturbing movement has revealed that these characteristics are similar to the characteristics of the fluctuations propagating in wave-guides. For the analysis of the revealed features, the basic formulations of a wave-guide [495] are discussed, which will be applied to both a boundary layer and an elastic surface.

A wave-guide is a segment of environment restricted in one or two directions and involved in the process of transferring waves. A wave-guide can be a layer or a pipe filled with a liquid or gas, or it can be a bar or plate (solid wave-guides). Waves can propagate in a wave-guide in the form of flat waves, as in infinite environments (a layer or pipe with rigid walls), or as normal waves at sufficient thickness of the layer. Such normal waves are formed at sequential reflections from wall. Simultaneous propagation of longitudinal and shear waves is also possible in wave-guides.

In liquids and gases, only longitudinal waves can propagate, which are elastic waves, the direction of propagation of which coincides with the direction of displacement and speeds of particles of the environment. The basic parameters of a wave are wavelength, the period of fluctuation or frequency, a wave vector.

One of characteristics of a wave is polarization. The plane in which the fluctuations of a transverse wave occur is perpendicular to the direction of propagation. It is this feature of transverse waves that makes polarization of the wave possible, which involves asymmetry of propagation of disturbances, for example, displacements and speeds, relative to the direction of propagation. Polarization can arise at distribution in anisotropic environment, or at refraction and reflection of a wave on the border of two environments. In Figure 1.77 examples of polarizing waves are shown.

The dispersion of a wave is a dependent of phase speed of a monochromatic wave from frequency. Dispersion is caused by physical properties of the environment, presence of borders of a body, and so on.

Nonlinear effects are changes of the shape of a wave during its propagation. To define nonlinear effects it is necessary to consider nonlinear terms of equations of hydrodynamics and the equation of state. A characteristic feature of nonlinear effects is their dependence on the amplitude of a wave. From the spectral point of view, this process corresponds to the transfer of energy to the higher and more strongly absorbed harmonic components of

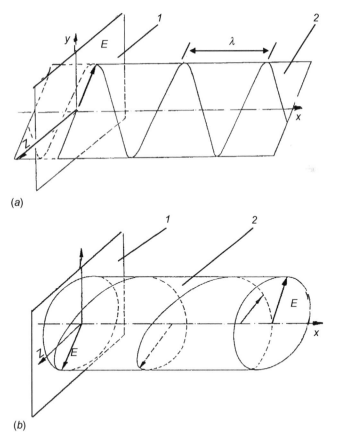

(a)

(b)

Figure 1.77
Types of waves polarization: *a* — plane polarization wave, *b* — wave, polarized on a circle;
1 — plane of oscillation, *2* — surface (plane) of polarization, *x* — direction of a wave propagation,
E — vector representing the propagating disturbance.

waves. Because the shape of a wave varies in during its propagation, absorption of a wave also depends on distance.

Generation of harmonics occurs when at collinear interaction and equality of frequencies $\omega_1 = \omega_2$ resultant wave is the harmonic $\omega_3 = 2\,\omega_1$. At absence of absorption for harmonics, linear growth of a harmonic in space is observed. Absorption of sound limits the linear growth of an acoustic harmonic in space. Therefore, in the beginning the amplitude of second harmonic grows linearly, then processes of dissipation slow down its growth, stabilization of amplitude occurs, and then its recession caused by fading of the harmonic is observed. The distance of stabilization is determined by the ratio $l_{st} = \ln 2/2\alpha$, where α is coefficient of absorption of the basic frequency. At distance $x > l_{st}$ the processes of dissipation leads to the fact that the amplitude of the second harmonic exponentially drops with increasing x.

Normal waves are harmonic waves propagating in a wave-guide without changing form. In all wave-guides for each normal wave there is a so-called critical frequency below which it does not propagate and turns into an oscillation with amplitude varying along the wave-guide by exponential law. Normal waves in plates and bars are harmonic elastic disturbances. Unlike elastic waves in infinite solid environments, normal waves in plates and bars satisfy not only equations of the theory of elasticity but also boundary conditions on the surface of the plate or bar. In most cases, these conditions are reduced by absence of mechanical stresses on surfaces. Normal waves in plates and bars are the same elementary waves as longitudinal and shear waves in an infinite environment, in the sense that any complex wave movement breaks up into a sum of normal waves and flux of elastic energy is equal to the sum of fluxes in all normal waves. Normal waves in plates are subdivided into two classes: Lamb's waves, which have an oscillatory displacement of particles in the direction of wave propagation, which is parallel to the plane surface and perpendicular to the plane. The second class comprises transverse normal waves possessing only one component of displacement (which is absent in Lamb's waves), that is parallel to the plate plane and perpendicular to the direction of wave propagation (see Figure 1.78).

Thus, deformation in a transverse normal wave is pure shear. By the character of deformation, transverse normal waves are divided into symmetric and antisymmetric ones. In symmetric waves, movement occurs symmetrically relative to the median plane $z = 0$, and in different waves they move in each half-plane of a plate in antisymmetric waves. Normal waves, as well as Lamb's waves, are characterized by the fact that at specified ω (circular frequency) and h (half-thickness of a plate) only a certain number of waves can extend. All properties of Lamb's and normal waves are defined by parameters of elasticity and density of the material, frequency ω and the transversal size of the wave-guide.

At distribution of a normal wave in a bar, as well as in a plate, only one normal wave of each type can propagate at low frequencies. Thus a zero wave of longitudinal type has the

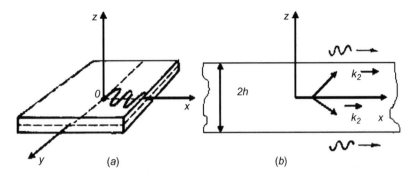

Figure 1.78
Transverse normal symmetric wave in a plate 2 *h* thick (*a*), and the same wave presented in the form of sum of two shear modes, propagating at an angle to the direction of its propagation (*b*); k_t — wave vectors of the shear mode forming a normal wave; *x* — direction of wave propagation; *y* — direction of oscillating displacement of particles.

phase and group speed equal $\sqrt{E/\rho}$, where E is Young's modulus and ρ is density of the material. Thus, a zero wave is usually bending mode.

Surface waves are elastic waves propagating along a free surface of a solid body, or on a border of a solid half-space with vacuum, gas, liquid or another solid half-space [386]. Elastic surface waves are a combination of non-uniform longitudinal and shear waves, the amplitudes of which exponentially decrease moving away from a boundary. Surface waves are of two classes: 1) with vertical polarization, at which the vector of oscillatory displacement of particles of environment in the wave is located in a plane perpendicular to the boundary surface (a vertical plane); 2) with horizontal polarization, at which the vector of oscillatory displacement of particles of environment is parallel to the boundary and perpendicular to the direction of wave propagation.

The elementary and frequently observed waves with vertical polarization are Rayleigh waves, propagating along a boundary of a solid body with a vacuum or rather rarefied gaseous environment. Rayleigh waves are the elastic waves propagating in a solid body along its free boundary and fading with depth. In flat Rayleigh waves, in homogeneous isotropic elastic half-space there are two components of displacement, one of which is directed along the direction of wave propagation (axis *x*), and the other component is perpendicular to the free border into the half-space.

Their energy is localized in the surface layer of thickness from λ to 2λ, where λ is wavelength. On depth λ density of energy in a wave is ≈ 0.05 the density near a surface. Particles in a wave move on ellipses, the greater axes of which are perpendicular to the boundary, and the small ones parallel to the direction of wave propagation. The phase speed of a Rayleigh wave is $c_R \approx 0.9\, c_t$, where c_t is phase speed of a flat transverse wave (see Figure 1.79a).

In an anisotropic environment, the structure and properties of a Rayleigh wave depend on type of anisotropy and direction of wave propagation. These waves can propagate not only on flat but also on curvilinear free surfaces of a solid body. Thus, their speed, distribution of displacement and stresses throughout depth, and spectrum of allowable frequencies, which can become discrete instead of continuous, vary.

If a solid body borders with a liquid, a fading wave of Rayleigh type can propagate on the boundary. When propagating, this wave continuously radiates energy into the liquid, generating there a non-uniform wave moving away from the boundary (see Figure 1.79b). Phase speed of such wave is equal within percent to c_R, and coefficient of attenuation on wavelength is ~ 0.1. This means that at distance 10λ the wave fades approximately by e times. The distributions of displacement and stresses throughout depth in such a wave in a solid body are similar to distributions in a Rayleigh wave.

An addition to a fading surface wave, a non-fading surface wave always exists on the boundary between a liquid and solid body, traveling along the boundary border with phase speed less than speed c_w of a wave in the liquid and speeds of longitudinal and transverse waves in the solid body. This surface wave is a wave with vertical polarization; it has a completely different structure and speed to a Rayleigh wave (see Figure 1.79c). It consists

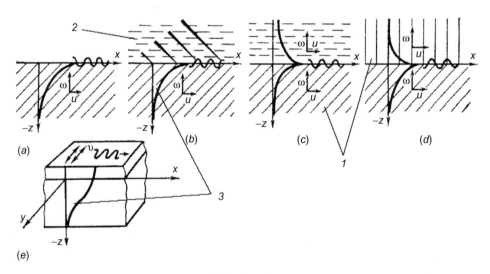

Figure 1.79
The scheme of surface waves of various type: *1* — solid environments, *2* — liquid, *3* — character of change of displacement amplitude moving away from the boundary between two environments, *x* — direction of wave propagation, *u, v, w* — components of displacement of particles in the given environment; *a* — Rayleigh wave on free border of solid bodies, *b* — fading wave of Rayleigh type on the boundary of a solid body and liquid (inclined lines in liquid indicate wave fronts of a departing wave, their thickness is proportional to amplitude of displacement), *c* — non-fading surface wave on the boundary of a solid body and liquid, *d* — Stoneley wave on the boundary between two solid bodies, *e* — Love wave on the border between solid half-space and solid layer.

of a weak inhomogeneous wave in the liquid, the amplitude of which slowly decreases moving away from the border, and two strongly inhomogeneous waves in the solid body (longitudinal and transverse). Owing to this, the energy of the wave and movement of particles are localized in the liquid, instead of the solid body.

If solid environments have a common boundary and their densities and modules of elasticity do not strongly differ, than a Stoneley wave can propagate along the boundary (see Figure 1.79d). This wave appears as if of two Rayleigh waves—one in each environment. Vertical and horizontal components of displacement in each environment decrease moving away from the boundary, so that energy of the wave is concentrated in two boundary layers of thickness $\sim \lambda$. Phase speed of the Stoneley wave is less than the magnitudes of longitudinal and transverse speeds in both environments.

Waves with vertical polarization can extend on the border of solid half-space and liquid, or a solid layer, or even a system of such layers. If the thickness of the layers is much less than the wavelength, then movement in the half-space is similar to that of Rayleigh waves, and phase speed of the surface wave is close to c_R. Generally, movement can be such that the energy of the wave will be redistributed between solid half-space and layers, and phase speed will depend on frequency and thickness of layers.

Apart from surface waves with vertical polarization (waves of Rayleigh type), there are also waves with horizontal polarization, called Love waves. These waves can propagate on the border of solid half-space with solid layer (see Figure 1.79e). Such waves are transverse; they have only one component of displacement v, and elastic deformation in the wave represents pure shear. Displacement in a layer is distributed by cosine, and exponentially decreases with depth in half-space. Depth of penetration of the wave into half-space varies from fractions of λ up to many λ, depending on thickness of the layer h, frequency ω and parameters of the environment. The condition of existence of a Love wave is defined by presence of a layer on half-space. At $h \to 0$, depth of penetration of the wave in half-space tends to infinity, and the wave becomes a spatial one. Phase speed c of Love waves are limited by phase speeds of transverse waves in the layer and half-space $c_{t1} < c < c_{t2}$.

Acoustic wave-guides serve for propagation of waves of a zero mode (with uniform distribution of amplitude across cross-section) and other modes of fluctuations. Acoustic wave-guides on spatial waves are strips, tapes or wires in which certain normal waves are excited (see Figure 1.80):

$$c = c_t[1 + (a/l)^2 tg^2(kh)^{-1/2}] \tag{1.14}$$

In his book, Pier provides basic equations of distribution of fluctuations in a wave-guide. The intensity S of a sine wave changes with time and with distance by the law:

$$S = S_0 \cos(\omega t - kx + \phi), \tag{1.15}$$

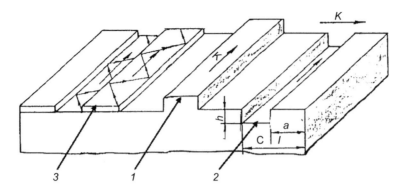

Figure 1.80
Types of acoustic wave-guides for surface acoustic waves: *1 —* ledge, *2 —* groove, *3 —* metal film.

where S_0 is amplitude of wave, ω circular frequency, $k = 2\pi/2$ wave vector, φ initial phase of the wave.

$$S = S_0 \cos \theta, \tag{1.16}$$

where $\theta = \omega t - kx + \varphi$, $S = $ const at $\omega t - kx = $ const, i.e. $x = Vt$, $V = \omega/k$ is phase speed of the wave, frequency $f = V/\lambda$, where λ is wavelength, $\omega = 2\pi f$, period $T = 1/f$, $V = \lambda/T = \lambda f$.

The power of a wave is equal to a square of intensity:

$$P = 2S^2 = 2S_0^2 \ \cos^2(\omega t - kr + \omega) = S_0^2[1 + \cos^2(\omega t - kr + \omega)] \tag{1.17}$$

where average power $P = 2S_0^2$.

By focusing attention on the environment in which the wave processes occur while studying a wave, it is possible to explore very complex interactions. However, another approach exists. The behavior of many waves does not depend on the specific nature and character of disturbances related to the wave. It is possible to learn much about waves by treating them as certain kind, or mode, of disturbance of the environment in which wave moves. So, the equation of a sine wave describes one particular type of wave or mode, one form among many which can propagate in an environment. (For example, two modes, depending on the plane of polarization, can be distributed on a string; yet two modes is a wave ahead and wave back.)

When considering any disturbance of an environment it is necessary to take into account the probability of excitation of many modes at all possible frequencies. However, it is possible to investigate many interesting phenomena by analyzing only one or two modes. The term "mode" is used for waves propagating with a constant speed in an environment. This will, in part, concern non-uniform environments.

Group speed is an inclination of a tangent to curve $\omega = f(k)$. The group speed is a speed of energy moving in direction x. If E is linear energy (energy per unit of length), power P is speed of a flow of energy:

$$P = V_{gr}E \tag{1.18}$$

V_{gr} can be either greater or less than V_{middle}.

The electromagnetic waves can propagate in wave-guides only when their frequency is above so-called 'critical frequency', which will be designated ω_{kp}. At power equal to one ($P = 1$) the amplitude S_0 is great at small group speed and is small at large group speed:

$$P = V_{gr}E = A \cdot V_{gr}S_0^2. \tag{1.19}$$

If the properties of the environment gradually change with distance (for example, the diameter of the wave-guide gradually decreases) then the critical frequency ω_{cr1} of the chosen mode grows with distance. If the diameter of a wave-guide changes more slowly, the wave of given frequency is distributed in such way it is as if the properties of environment along its way do not change.

A wave of single power, which is distributed at some mode with frequency ω exceeding the critical frequency, will now be considered. At the narrowing of a wave-guide ω_{cr1} is increased, and, hence, ω/ω_{cr1} decreases. Thus, at distribution of a wave along a narrowed wave-guide the group speed decreases and the amplitude S_0 is increased, i.e. the electrical field intensity is increased.

At sufficient narrowing, the frequency ω_{cr1} can be equal to ω, and then the chosen mode cannot distribute further on the pipe of a wave-guide; it will be reflected so that a wave with negative speed will run back along the wave-guide. Based on formula (1.19) it is possible to conclude that the amplitude S_0 should become infinitely large in the section of a wave-guide where $\omega = \omega_{cr1}$. Actually, this is not so. All these ratios are not correct in the case when $V_{gr} \to 0$. However, in narrowed wave-guides an electrical field of a wave does grow in accordance with the reduction of section of the wave-guide and the critical frequency grows.

The dependence k from ω can be presented by a power series:

$$k = k_0 + \left(\frac{\partial k}{\partial \omega}\right)_0 (\omega - \omega_0) + \frac{1}{2}\left(\frac{\partial^2 k}{\partial \omega^2}\right)_0 (\omega - \omega_0)^2 + \frac{1}{6}\left(\frac{\partial^3 k}{\partial \omega^3}\right)_0 (\omega - \omega_0)^3 + \cdots \tag{1.20}$$

index '0' at individual derivatives shows that they depend on ω_0 and k_0.

Figure 1.81 shows the often-used element of centimetric waves—such a contour is known as directed branches of communication. A wave-guide is a metal pipe on which the electromagnetic wave can be distributed. Two identical wave-guides W_1 and W_2 have a

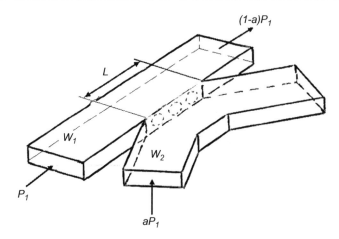

Figure 1.81
Circuit of two wave-guides with one common site of length L.

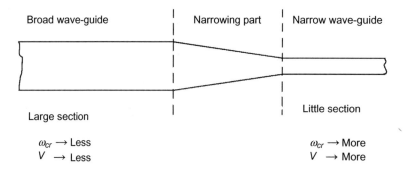

Figure 1.82
Diagram of wave parameter dependence on wave-guide geometry.

common wall of some length L, in which apertures are done so that the wave of one wave-guide interacts or contacts with a wave of other wave-guide. The result is that some share of power P_1 in the first wave-guide is transferred to wave-guide W_2, and part $(1-a)P_1$ remains in wave-guide W_1. The share of power that is transmitted from one wave-guide to the other depends on length of L on which the waves are connected, on the force of connection, i.e. on the size of apertures, and on frequencies. At some value of connection $a = 1$, i.e. all power passes from one wave-guide to the other.

In his book, Pier G. ("Almost all about waves," Moscow, Mir, 1976, p. 176) provides methods of calculation of various wave-guides and characteristics of the fluctuations extending in wave-guides are described. In Figure 1.81, the scheme of connection of two wave-guides is shown. For such a scheme corresponding methods of calculations of the characteristics of a wave-guide are mentioned. Figure 1.82 shows the scheme of how wave parameters change in a narrowed wave-guide. Using such schemes it is possible to change the parameters of waves propagating in wave-guides.

The results presented in Section 1.7 and [310] lead to some conclusions. Both types of transition, disturbance amplification at transition stages, including stage *V*, can be described by means of wave (motion) theory [35].

Visualization of disturbance amplification [12, 310] showed that a 2D wave could be specified by variance (dependence of harmonic wave phase velocity on frequency), which is followed by oscillation form changing with disturbance propagation. The variance is influenced specifically with velocity gradient and with the boundaries of the boundary layer.

Only in the critical layer the given 2D sinusoidal wave propagates with the second neutral oscillation frequency according to the Orr–Sommerfeld equation as a T–S wave. Away from the theoretical layer and with other disturbance frequencies a set of interacting harmonic oscillations are generated in the boundary layer. This can be presented, for example, as follows [132]:

$$f(x,t) = F(x,t) \ \exp[i(k_0 x - \omega_0 t)], \tag{1.21}$$

where $\omega_0 = \omega(k_0)$ is dependence of wave number on wave frequency (dispersion); and $F(x, t)$ the envelope of harmonic oscillation set. Equation (1.21) specified a modulated harmonic wave. Having spread out function $\omega(k)$ in Taylor's line, the following is obtained:

$$f(x,t) = \int_{\infty}^{-\infty} \tilde{f}(k) \ \exp\left[i\left(k_0 x + \aleph x - \omega_0 t - \frac{d\omega}{dk}\left(\frac{\aleph t}{k_0} - \cdots\right)\right)\right] dx, \tag{1.22}$$

where $f(k)$ is spatial Fourier spectrum of initial disturbance, $\aleph$ small parameter $k = k_o + \aleph$. Quantity $d\omega/dk = c_{gr}$ is wave group velocity. Thus, equation (1.21) can be stated as follows:

$$f(x,t) = F_0(x - c_{gr} t) \ \exp[i(k_0 x - \omega_0 t)], \tag{1.23}$$

This means that the envelope $F_0(x - c_{gr}t)$ is being propagated without changing its form with small *t*. These conclusions agreed with results of the experiments.

However, for a description of a valid picture of development of wave disturbances in a boundary layer more complex mathematical apparatus as at greater *t* and enveloping curve $F(x, t)$ it is characterized by a dispersion is required. This leads to the subsequent changes of structure of flow (alternation of stages of transition).

The disturbing movement in transitional boundary layer, except for dispersion, is characterized by one more property—nonlinear interaction. Measurements have shown that even at very small ε, generation of 2D disturbances in a boundary layer is accompanied by nonlinear effects, and is characteristic for all stages of transition. As follows from the figures in Section 1.6 and [64, 305, 310], dispersion and $c_r(y/\delta)$ excite additional harmonics, which interact with each other in a nonlinear fashion. This can be noted from photographs in Section 1.5 and [30, 32, 305]. Dependence of disturbing movement on amplitude is also an indication of nonlinearity. It is known that if nonlinear effects predominate over dissipated ones [495] the sinus waves go over into saw-shaped ones while growing. This is noted in Section 1.5 and [305, 310].

It is known [132, 495] that a nonharmonic wave, consisting of a package of harmonic waves of different frequencies, during its development changes form because of dispersion. The relationship between phases of the harmonics of a wave changes. The wave form also modifies, owing to nonlinear properties of the medium.

With colinearity, interaction and equality of frequencies $f_1 = f_2$ the second harmonic $f_3 = 2f_1$ is generated. The harmonic at first linearly rises and then stabilizes and attenuates. The length of the stabilization is determined by the equation:

$$l_{st} = \ln 2 / 2\alpha_i, \qquad (1.24)$$

where α_i is basic frequency attenuation coefficient. Investigations of hydrodynamic stability [32, 305] can be considered from the receptivity viewpoint (see Section 1.11). Generated 2D waves introduced into a boundary layer interact with natural 2D disturbances in the region where the measurements are made. It is seen that the introduced wave, which has the frequency of the second neutral oscillation in the measurement region, grows higher than others do. At other frequencies interaction between introduced and natural disturbances occurs less intensively. For every introduced disturbance frequency, a stabilization region was registered (see Sections 1.5 and 1.7 and [305]).

The disturbance motion amplification in transition boundary layer is characterized by another property of wave motion—wave polarization. The polarization phenomenon [495] consists of violation of symmetry of the disturbance distribution about the direction of propagation (see Figure 1.77). There is linear, elliptical and circular polarization. It can become known due to the absence of symmetry in the wave generator, wave propagation in an anisotropic medium, wave refraction, or reflection at the boundary of two media. This phenomenon is caused by dispersion and nonlinear effects and it is well illustrated in Sections 1.5 and 1.7 and [56, 310]. Although in the beginning the 2D wave propagates symmetrically the following stages of transition appear very rapidly. At these stages a linear polarization occurs first [310] (transition stages *II* and *III*), and then a circular polarization occurs [495] (transition stages *IV–VI*).

Receptivity of a boundary layer to 2D and 3D disturbances can be considered from the point of view of wave motion theory. From that point of view the conclusions about the mechanism of interaction of various disturbances in a boundary layer [310] can be considered as follows: all above noted properties of wave motion remain at interaction of

2D and circular polarized waves introduced into the boundary layer with wave motion of natural boundary layer at different stages of transition. This accounts for the nonlinear interaction mechanism and its resonance character, because oscillations with the same polarization interact with each other strongly.

Taking this into account and resting upon the well-known electromechanical and electroacoustic analogy [242, 451] it is possible to conclude that the transitional and turbulent (in its viscous sublayer) boundary layer can be considered as a non-symmetric wave-guide [469].

The heterogeneous wave-guide is characterized by the fact that its permittivity can change under the square law [469]:

$$k(r) = k_0 - k_r r^2, \tag{1.25}$$

where k_0 is permittivity of outside surface; k_r permittivity in the direction of wave-guide axis; r radius of wave-guide. The dialectic penetrability of a heterogeneous wave-guide can also be presented with another law, for example, as an exponential curve.

The asymmetrical wave-guide, for example, when flat is characterized by that permittivity of its external surfaces and is various (metal, glass, air, etc.).

Parameters of a heterogeneous non-symmetric wave-guide cannot be theoretically analyzed at this time. In Figure 1.83, schemes of homogeneous and heterogeneous wave-guide are presented. A symmetric heterogeneous wave-guide differs from a homogenous one by its continuous changing of refraction factor across the section. Because of this, a ray does reflect from the outer surfaces but moves inside the wave-guide following a sinus trajectory. Consequently, the heterogeneous wave-guide can be specific with considerably less losses compared to the homogeneous one.

The heterogeneous wave-guide differs from a homogeneous wave-guide, which consists of layers of materials having various dielectric properties. Thus, production materials with such properties that their permittivity changes on layers so that the wave passing on layers was gradually reflected from each layer. This is how harmonicity forms with disturbing movement in a wave-guide that does not reach a final surface of a wave-guide. In addition, properties of layers in a wave-guide are picked up in such a manner that losses in the passage of a disturbing movement through a wave-guide are minimal.

In [64, 305] it is shown that a boundary layer can be presented as a wave-guide with the upper boundary moving asymptotically away from the lower one and being penetrable for outer energy. There is doubt about the legitimacy of this hypothesis of considering a boundary layer as a nonlinear asymmetrical wave-guide. Standard wave-guides are characterized by a large throughput of waves moving inside it. According to offered hypotheses, a rigid surface from below and steady flow from above form such a wave-guide

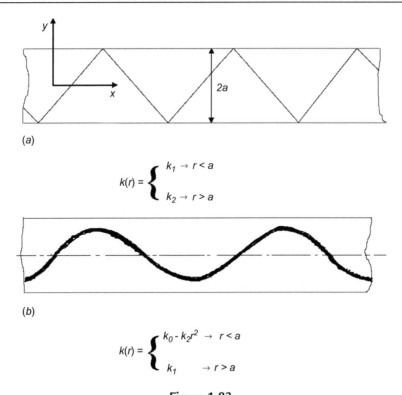

$$k(r) = \begin{cases} k_1 \to r < a \\ k_2 \to r > a \end{cases}$$

$$k(r) = \begin{cases} k_0 - k_2 r^2 \to r < a \\ k_1 \qquad \to r > a \end{cases}$$

Figure 1.83

Homogenous (*a*) and heterogeneous (*b*) types of wave-guides; k = dielectric constant.

(boundary layer), which is characterized by viscous dissipative properties, decelerating propagation of a wave in a boundary layer. However, specific properties of a boundary layer as wave-guide are that a disturbing movement moves in it with a constant change of form and polarization of fluctuation. The losses of energy due to dissipation are compensated by picking up energy from the mean flow.

Disturbances in a free flow propagate almost without modification. However, when they are introduced into a boundary layer and interact with it they are amplified and lead to boundary layer reaction, which clearly becomes known and can be registered. With this, certain structures of disturbance motion are created [305]. For example, with free-stream turbulence intensity $\varepsilon < 0.05\%$ the maximum is $E = \sqrt{u'^2}/U_\infty$ [305]. In this case, the wave-guide amplification coefficient is:

$$\overline{k} = E/\varepsilon = 48 \tag{1.26}$$

Although the disturbance motion in the wave-guide propagates in a narrow cone layer, the main part of its energy is concentrated in the critical layer of the transitional boundary layer or in the region of outer boundary of the viscous sublayer of a turbulent boundary layer.

Such features of a boundary layer as wave-guide can be explained by two factors. First, physical characteristics of flow in a boundary layer model the typical asymmetrical homogeneous wave-guide, described by high throughput. Secondly, additional charging of energy from the basic movement compensates the dissipative properties of a boundary layer [305]. All this leads to the fact that wave-guide disturbing movement extends in a narrow cone layer, but its basic energy concentrates in an area of the critical layer of the transitional boundary layer or in an area of the external border of the viscous sublayer of the turbulent boundary layer.

One further property of this wave-guide should be noted—self-pumping of energy. As soon as the disturbing movement begins dissipating (losing energy), due to this dissipation and development throughout the thickness of a boundary layer, a critical jump of energy appears (see Figure 1.47): heads of hairpin vortices collapse and disturbing movement again concentrates in a narrower layer on y/δ, which is accompanied by the reduction of pulsating velocities (see Figure 1.47). At further "swelling" of the disturbing movement second critical jump of energy occurs and the turbulent boundary layer is formed. Transmission of energy in the wave-guide again concentrates in a narrower layer on δ—in the viscous sublayer. The same picture is also observed in the viscous sublayer. Burstings from the viscous sublayer occur periodically and chaotically in the plane of the viscous sublayer, after which new viscous sublayers are formed.

The amplification coefficient of the wave-guide (i.e. of a boundary layer) depends on the transition stage and the outer disturbance magnitude. For example, at $\varepsilon < 1\%$ $\bar{k} = 7$. The amplification coefficient can be changed by changing the wall properties.

It has been shown [132], both in water or atmosphere, that a necessary condition of appearance of a wave-guide is the occurrence of a layer in which the velocity of disturbances is maximal relative to the outer layers. The critical layer of a boundary layer, in which phase speed is always higher in comparison with the other layers of a boundary layer, corresponds to this condition (see Section 1.6).

One more validation of the mentioned hypothesis is the jumps of energy of the disturbing movement during the development of a boundary layer by analogy to the condition of development in diverging or converging wave-guides [310, 495].

Sections 1.6, 1.9 and 1.10 present the results from experimental research on the physical picture of a flow near three kinds of elastic surfaces. The main results, which essentially differ from similar results for a flow over a rigid plate, are noted below:

- The disturbing movement and vertical coordinate of pulsating velocities maxima are located essentially closer to the elastic surfaces in comparison with the standard.
- The rate of increase of disturbances and amplitude of disturbing movement is $2-4$ times less than that of the standard.

- The ratio c_r/u is equal to unit at the second neutral fluctuation, in other cases it is less than unit and reaches the value 0.5, while for the standard $c_r/u > 1$ at all values of oscillation frequency not equal to the second neutral fluctuation.
- The disturbing movement propagates in the region of smaller y/δ relative to the coordinate of the critical layer for the standard, and that for elastic plates values of λ, c_r and c_i decrease.
- With growing Reynolds number, λ increases essentially more slowly than at the standard, and it can be assumed constant.

These and other features of structure of the flow over an elastic surface can be explained only from the position of receptivity (see Section 1.11). It is obvious that disturbing movement in a boundary layer causes disturbances in the elastic surface, which interact with the imposed disturbances. Much research has been done on the problem of receptivity, including the flow over elastic plates, and has shown that such representation is valid [310]. However, the mechanism of such interaction remains still uncertain.

According to basic categories of the theory of wave-guides and laws for the flow over the standard mentioned, it is possible to state that the results of flow over elastic plates can be analyzed from the point of view of representation of the boundary layer as a wave-guide. It is known that by changing the mechanical characteristics of the material of a wave-guide, or the mechanical properties of layers of a wave-guide or its borders, as well as wave-guide design, it is possible to control wave-guide characteristics. According to these data, in the flow over elastic plates, the characteristics of a wave-guide, in this case, the characteristics of a boundary layer change. It makes it possible to assume that in the flow over elastic plates the boundary layer should be considered as a complex wave-guide, the bottom border of which essentially influences its characteristics.

This statement is substantiated both by the general statement of the theory of wave-guides and by the hydrobionics approach.

Section 1.8 describes some features of the structure of dolphin skin. Looking at the structure of dolphin skin (see Figures 1.50, 1.54 and 1.55) and the schemes of wave-guides (see Figure 1.80), it becomes obvious that during evolution the structure of dolphin skin gained all the attributes of wave-guides. Some of these will be mentioned briefly.
Figure 1.79 illustrates schemes of surface waves. It is obvious, that Rayleigh waves appear at flow over the surface of a dolphin's skin. Disturbances that get into dolphin skin propagate inside the skin in the form of the waves of Stoneley and Love. Near the surface of skin there are longitudinal rows of dermal papillae, which are located in longitudinal rows of epidermal emergences. Thus, there are specific longitudinal layers, the densities of which differ slightly. These are the necessary preconditions for existence of these waves. Change of spacing of dermal rollers along a body length (see Figure 1.54) makes it possible to control the characteristics of these waves. Changing the angle of inclination of the

dermal papillae (see Figure 1.53) influences the character of fading of these waves. External transversal folds (see Figures 1.51 and 1.52) influence the characteristics of Rayleigh waves. Figures 1.81 and 1.82 present schemes of engineering wave-guides that allow for changing the parameters of a wave. Similar structures are observed in the outer structure of dolphin skin. This skin has a complex interconnected multilayered structure. The thickness and mechanical characteristics of the skin change along the body. The outer layers change their mechanical and geometrical parameters by means of the skin muscle (see Figure 1.56). Thus, it is possible to automatically control the wave-guide properties of skin. Standard engineering wave-guides are intended to transfer oscillations with the least losses over a given distance. The length of wave-guides in dolphin's skin is limited and ends at the fin mover. The purpose of wave-guides in dermal coverings is interaction with the disturbing movement acting on the skin from the boundary layer. Such interaction is necessary for drag reduction of a body. Therefore, the wave-guide character of dermal coverings should be considered in interrelation with wave-guide properties of a boundary layer. It was revealed on the simplest models of artificial elastic coverings that a disturbing movement from a boundary layer is transformed to the surface. It allows a disturbing movement of a boundary layer to interact with three types of surface waves.

The structure of dolphin skin has features that allow it to change the structures of a boundary layer, transforming all parameters in a boundary layer for effective drag reduction. The structure of the skin is such that it transforms a part of the energy to a boundary layer, and part of the energy to transform super surface waves, due to change of skin structure in the end of the body. This part of energy in the end of the body is transformed into a boundary layer, which is accelerated in the tail part of the body due to the work of tail fin mover and dumped in the wake behind the body.

Full-scale measurements of the characteristics of a boundary layer and parameters of fluctuations of a dolphin's dermal coverings were undertaken by Shakalo [409], Romanenko [438, 439] and Kidun [266]. Their findings have shown essential differences compare with same characteristics in the flow over a rigid body and they validate the proposed hypothesis about a complex wave-guide.

1.14 Control Methods of the CVSs of a Boundary Layer

An important problem for modern hydrodynamics is the search for new ways of saving energy saving. Until now the energy losses at movement of a body in a liquid has been basically defined by an estimation (theoretical or experimental) of the momentum thickness, i.e. on the basis of integrated characteristics of a flow. Now more attention is being paid to understanding the laws of formation and development of CVSs in flows of a liquid or gas. As has been shown in Section 1.1, practically in all types of flow different kinds of CVS have been identified. Experimental research has made it possible to systematize CVSs in the transitional boundary layer and offers a model of CVS development at various stages of transition.

Kline has formulated eight types of CVS in a turbulent boundary layer [276, 277, 441]:

1. low-speed streaks in the field of $0 < y^+ < 10$;
2. ejection of a low-speed fluid from a wall;
3. movement of high-speed fluid to a wall;
4. vortical structures of several assumed types;
5. forced inner shear layers in the near-wall area $y^+ < 80$;
6. near-wall "pockets" precisely observed as areas of marked particles used in some methods of visualization;
7. large (delta-shaped) discontinuities of longitudinal velocity;
8. large-scale movements in outer layers (for boundary layers: with convexities, layers of high quality and deep hollows, completed fluid of free stream).

Kline has analyzed all these types of CVS. It is known that on a flat plate the majority of turbulence ($\sim 70\%$ or more) in the inner layer is generated via CVS mechanisms, and the minority ($\sim 30\%$ or less) is generated by fluctuations in the outer layer.

In Sections 1.3 and 1.8 schemes of CVS in a turbulent boundary layer by Meng and Babenko were presented, which consider modern results of CVS research and refine Klein's classification. Based on results of much experimental research it is possible to offer a hypothesis: the majority of energy exchanges in a boundary layer occur in the near-wall areas and are accompanied by different characteristic kinds of CVS. Therefore, effective control of the boundary layer characteristics is possible at smaller energy expenses. For this purpose it is necessary to influence not the entire boundary layer, but certain kinds of CVS in the boundary layer. Hence, the energy consumptions necessary for the control of coherent structures is approximately 1 order less than that needed for traditional methods of boundary layer control. Section 1.1 describes some results of CVS control methods.

The basic results of research on boundary layer receptivity (see Section 1.13) and representations of the boundary layer as a wave-guide (see Section 1.13) are used to systematize methods of CVS control. In addition to the idea of structural character of disturbance development at different stages of transition, the idea of influencing CVS with frequency characteristics forms the basis of control methods development.

A fluid particle has a dual peculiarity of its motion: it moves along a certain trajectory and has a spectrum of oscillations during motion where it is possible to segregate certain energy-carrying frequencies for each stage of transition or for every type of CVS.

The principal factor (third approach or principle of influence) is that energy of influence must be of the same order as that of disturbing motion at the corresponding stage of transition.

Figure 1.84 summarizes the developed methods of influencing the coherent structures of a boundary layer, which consists of seven groups: mechanical (I), dynamic (II), kinematic (III), electric (IV), acoustic (V), combined (VI) and active ones with a feedback system (VII) [70].

Mechanical methods will first be considered. The method of changing the shape of a surface is a technique that has been used in engineering for a long time. Application of this method for CVS control in a boundary layer relates to minor alterations of shape to influence CVSs. In one example [421] the rear part of the plate was put forward, if necessary, and riblets added (see Figure 6.68). The geometrical parameters of the riblets were regulated.

The method of stationary plates, for example, involves mounting stationary plates intended to influence CVSs in a boundary layer. A vortex generator established on the upper surface of an airplane wing on its forward part can be referred to.

Stationary plates can be placed in a boundary layer to destroy large-scale vortices on its outer boundary. An example of such plates is the large-eddy-breakup device (LEBU), i.e. rings established equidistant to a cylindrical surface [100].

The method of oscillating plates, for example, involves vortex generators on a wing that can change their size or oscillate at a specified frequency. Oscillatory plates are fluctuations of LEBU.

The method of oscillating surface is illustrated by the design proposed in [413]. It involves pulling out the leading part of a wing or body of revolution, which can oscillate in the longitudinal direction at a specified frequency, if necessary (see Figure 6.70). Thus CVSs having certain frequency characteristics are generated in a boundary layer.

Transversal obstacles or grooves are considered in Section 1.1 as are longitudinal obstacles or grooves. The use of riblets has been investigated for a long time. The authors experimentally investigated transversal and longitudinal non-uniformities, which if necessary, can change their geometric parameters (see Figure 6.69). Longitudinal stationary vortex generators are illustrated in Section 1.2 and in Figure 1.76.

The authors experimentally investigated variants of longitudinal oscillating vortex generators. Regular roughness serves the same purposes: the organization of longitudinal and transversal vortices. Dimples are new and are a poorly studied method of influencing boundary-layer CVS. Depending on the flow under consideration, dimples can be located in a longitudinal or transversal direction. This method is described in detail below.

Dynamic methods are based on the principle of mass transfer and differ from known methods in that the influence on boundary-layer CVSs is made from the surface. Such

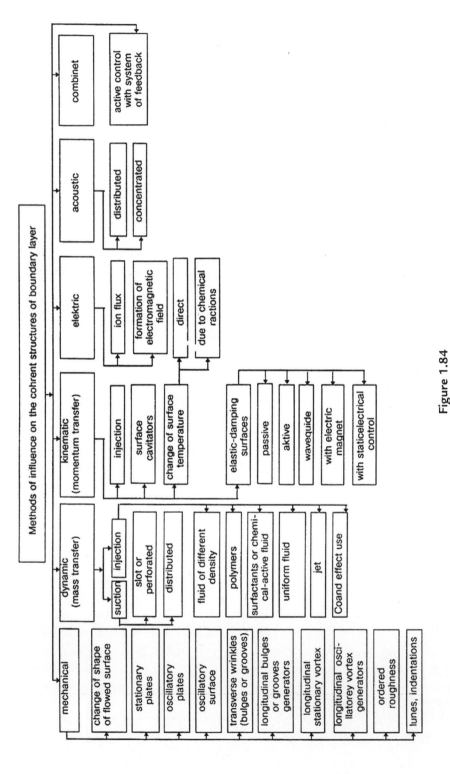

Figure 1.84

Methods of influence on the coherent structures of the boundary layer.

methods include specially profiled slots, perforated surface or distributed suction of a boundary layer. An example of such method can be found in [80] (see Figure 6.56). Examples of the application of dynamic methods are presented in Chapter 6.

Kinematic methods are based on the principle of momentum transport (impulse). They consist of injection (as in the previous case), the installing of separate surface cavitators, regulating the temperature of the surface by heating, or by chemical reactions with a flow. A separate class is elastic surfaces: single-layered, multilayered, isotropic or anisotropic, passive or active ones. Examples of this method will be presented below.

Electric methods involve the interrelationship of CVSs in a boundary layer with a stream of ions or with an electromagnetic field around of a body. Recently methods of control of boundary layer CVSs by means of electromagnetic elements placed in a surface have been developed (Proc. of the 2nd Intern. Symp. on Seawater Drag reduction. ISSDR 2005. May 23−26, 2005. Busan, Korea).

Acoustic methods can be distributed or concentrated. These methods are based on the principle of generating an additional pulsating directional field of pressure.

The most promising are the combined methods, which provide simultaneous use of a combination of considered methods and/or combinations with active methods of control.

Some of the considered methods have been studied experimentally.

1.14.1 Experimental Investigation of a Flow of the Localized Hollows

In the engineering, various kinds of surface deepening are often formed. The role of deepenings for increasing heat- and mass-exchange has been experimentally investigated. However, the problem of flows about small deepenings is poorly studied. Since 1990, experimental research on various 3D deepenings [62, 65, 68−70, 74, 75 et al.] has been done. Experiments were carried out in a hydrodynamic tunnel. The technique of the experiments consisted of initial visualization of flow in the deepening and subsequent quantitative measurements in characteristic locations, which had been discovered during visualization of the flow over the deepening.

An insert in the bottom of the working section provided the ability to mount two replaceable dimples with various shapes. It was also possible to turn the insert about on its vertical axis of symmetry. This made it possible to establish the dimples at various angles, and establish dimples side-by-side or one after another relative to the flow. By means of four dyed jets, different colors visualized the flow over one or two dimples simultaneously. Passive and active dimples were investigated. By means of a micromotor it was possible to change the depth of a dimple or to make its bottom oscillate at a specified frequency.

In addition to visualized patterns of flow by means of laser Doppler anemometry, longitudinal profiles were measured. At a low velocity a vortex with a transversal rotation axis in the direction OZ arose in the dimple. The location and shape of the vortex depended on the shape of dimple. Its ends were attached to lateral walls of the 3D dimple.

In the dimple of regular spherical shape, the axis of rotation was about the center, and its ends were attached to lateral walls. The maximum velocity in the dimple was on the axes and minimum velocity on the sides because of constant ratio $\omega \cdot G$, where ω is angular velocity and G is area of the vortex cross-section. The movement remains vortical also in outer side of the dimple, but the axis of its rotation is turned and directed streamwise along OX.

In the flow over the tandem of two deepenings (see Figure 1.85b), a steady pair of longitudinal vortices forms behind the ellipsoidal cavity, and meandering arises after the

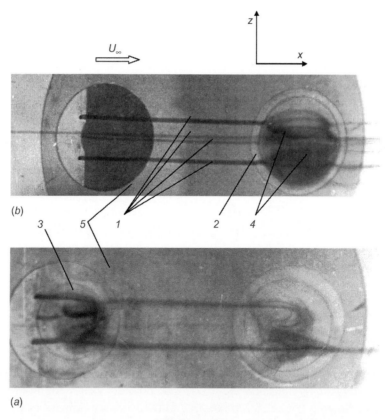

Figure 1.85
Visualization of flow in semispherical and ellipsoidal cavities: *1* − color jets; *2* − semispherical cavity; *3* − ellipsoidal cavity, *4* − single-band hyperboloid vortices; *5* − flat plate. View from above: *a* − single cavity, *b* − tandem of the ellipsoidal cavity and spherical cavity. $U_\infty = 10.3$ cm/s.

second circular cavity: a longitudinal vortex is formed periodically on the right and left of the longitudinal axis of the cavity.

The basic result is that 3D dimples allow the generation in a boundary layer of both flat disturbances and two longitudinal vortices. The dependence of dynamic characteristics of generating of disturbances from dimples, which was correlated with CVSs in the boundary layer, has been determined. Research on receptivity of the boundary layer to disturbances generated by dimples have been carried out.

Experimental research on the flow over 3D dimples of round and ellipsoidal shapes in water have made it possible to plan and carry out research on the flow over 3D dimples of various shapes. The influence of dimples on each other has also been investigated. For this purpose, dimples of various forms and with various obstacles in front of and behind them were sequentially installed.

In Figure 1.86a, generators of longitudinal disturbances [310] are established in front of the elliptic deepening. The longitudinal vortical systems generated alter the flow inside the deepening. Therefore, the shape of vortices in the deepening becomes similar to a double-strip hyperboloid. In other words, two separate vortices in the transverse direction are formed in the deepening. When the generators of longitudinal disturbances are established just behind the elliptic deepening, the pair longitudinal vortices flowing from the deepening interact with the system of longitudinal vortices generated by the generators of longitudinal disturbances. The interaction of these vortices leads to their stabilization (see Figure 1.86b).

The necessary conditions were first obtained for generating vertical steady tornado-like vortices in a boundary layer by means of dimples, one end of which is attached to the plate, and other lies on the outer border of the boundary layer. Pictures of visualization were taken with a video camera.

1.14.2 Concentrated and Distributed Methods of Formation of Longitudinal Vortical Systems in a Boundary Layer

Section 1.11 and [310] mention the device for the concentrated formation of longitudinal vortical systems (LVS) in a boundary layer by means of wings of various sizes mounted on and near rigid and elastic surfaces. A photograph of transversal cross-section of outer layers of a dolphin's skin is also presented, as is an explanation of how LVS are formed in the boundary layer by means of the distributed method due to the structure.

The experimental distributed method of introducing LVS consists in gluing longitudinal strips or thin wires onto the plate in flow. According to Section 1.12 it is obvious that introduction of LVS into the boundary layer will be effective when the basic conclusions

U_∞

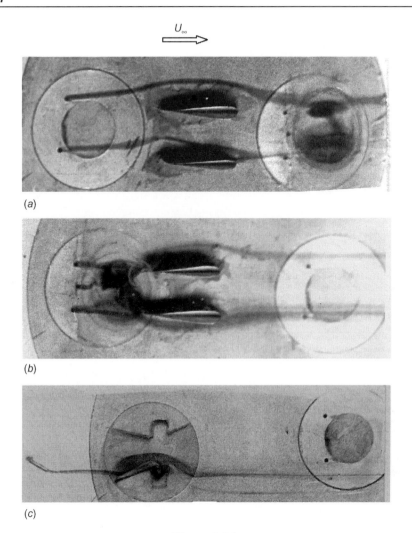

(a)

(b)

(c)

Figure 1.86
Visualization of the flow over deepenings of various shape and various combinations of
deepenings and ledges: *a* — vortex generators in front of the elliptic deepening; *b* — vortex
generators behind the elliptic deepening; *c* — the cone-shaped deepening with pockets for
formation of vertical vortices; *d* — the square deepening behind the elliptic deepening; *e* — the
transversal ledge in front of the elliptic deepening.

from research on the receptivity problem are satisfied. It is necessary that the sizes and
intensity of the introduced LVSs are comparable with similar CVSs in the undisturbed
boundary layer. Nevertheless, longitudinal wires and strips glued onto the plate created an
additional roughness and, to some extent, their influence on the boundary layer was similar
to the action of riblets.

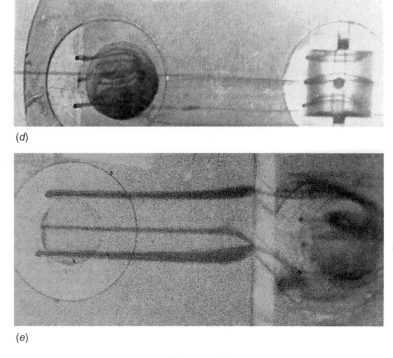

(d)

(e)

Figure 1.86
(Continued)

It is expedient to increase the intensity of influence of these strips on the boundary later, but not increasing their size at the same time and allowable hydraulic roughness. The results of numerous experiments with riblets of various shapes are widely known. Another two methods, which are distinct from riblets, are offered below. The intensity of the influence upon the boundary layer increases.

One of these ways involves the use of slightly heating longitudinal wires. Another uses a complex elastic surface, the upper layer of which is made of longitudinal elastic strips with various modules of elasticity (see Chapter 2).

According to the developed experimental technique, visualization of the structure of flow in a boundary layer was carried out before quantitative measurements were taken. As an example, photographs of the flow over the elastic plate *7a* (classification of plates is given in Chapter 2), onto the surface of which longitudinal thin metal wires $1.2 \cdot 10^{-4}$ m high were glued (see Figure 1.87). Visualization was achieved with the tellurium method. Velocity profiles in the transversal direction were photographed. Three tellurium wires were located at various distances from the surface. This revealed the simultaneous development of disturbing movement at various distances from the plate. The elastic plate was placed on the insert (see Figure 1.74).

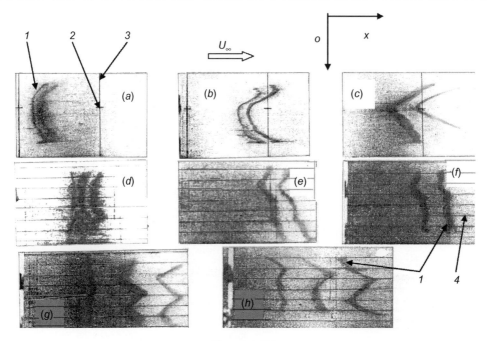

Figure 1.87
Visualization of $U(z)$ at $U_\infty = 0.05$ m/s before elastic plate $7a$: $x = 1.6$ m, $y = 0.002$ m (a),
0.004 m (b), 0.006 m (c); and over it: $x = 2.18$ m, $y = 0.002$ m (d), 0.004 m (e), 0.006 m (f),
0.008 m (g), 0.01 m (h); 1 − velocity profiles $U(z)$, 2 − marker of a longitudinal axis of symmetry
of a working site, 3 − marker of distances lengthways x (0.1 m), 4 − longitudinal wires.

In Figure 1.87 the results of measurements are shown. The first three photographs (*a, b* and
c) characterize the development of the disturbance movement in the boundary layer above a
rigid surface at $x = 1.6$ m (Re $\approx 1.1 \cdot 10^5$). Other photographs were taken above an elastic
plate, which is placed on the insert, at $x = 2.18$ m (Re $\approx 1.4 \cdot 10^5$). It is evident that, at
greater Reynolds numbers, disturbing movement in the flow over the elastic plate with
longitudinal wires located on its surface were stabilized. The elastic plate was made of
longitudinal elastic strips. All this has led to the stabilization of the disturbing movement
near the surface. At $y = 0.01$ m $\approx \delta/2$, the velocity profiles becomes typical for LVS of a
boundary layer. In addition, LVSs above the elastic plate have different parameters from
the rigid plate (see Figure 1.72).

Figure 1.87 illustrates two sequential velocity profiles in each series. The tellurium cloud
visualizing velocity profiles in the transverse direction were released each 0.5 s. Three
consecutive velocity profiles were photographed in series *g* and *h*. It is evident that with
increasing coordinate y local longitudinal velocity increases, and distances between velocity
profiles also increases.

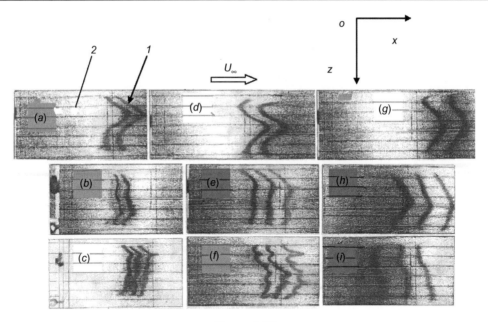

Figure 1.88
Visualization of $U(z)$ at $U_\infty = 0.06$ m/s over elastic plate 7a: $x = 2.18$ m, $y = 0.002$ m (a, b, c), 0.003 m (d, e, f), 0.005 m (g, h, i); without heating (a, d, g), with heating of longitudinal wires at consumed electrical capacity 7.8 V (d, e, h) and 12.2 V (c, f, i); 1— velocity profiles $U(z)$, 2 — longitudinal wires.

At low flow velocity the elastic plate effectively stabilized the disturbing movement. To increase the intensity of the introduced LVSs wires were heated by passing a current through them. Because the elastic material had good thermo-insulating properties heating the wires led to a change in the mechanical characteristics of the elastomer in the area of wires. However, insufficient heating led to small convective fluxes in the near-wall layer of the boundary layer. Results of the experiment are illustrated in Figure 1.88. Photographs are organized in such a way that columns (a, b, c), (d, e, f), (g, h, i) contain photographs at identical coordinate y. The horizontal rows of the photographs correspond to the following: first row—without heating; second row—7.8 V current; third row—12.2 V current.

Heating wires led to stabilization of a boundary layer at all specified distances from the plate. More intensive heating leads to the occurrence of small-scale LVSs near the surface, with λ_z that is twice the distance between wires. Moving away from the wall, LVSs were not revealed: $\lambda_z = 0.012$ m.

The measurements performed validate the efficiency of the method of interaction of disturbances in a boundary layer—receptivity. However, it is obvious that such a method can change the wave-guide characteristics of a complex wave-guide (see Section 1.13). These experiments are only one of the variants of the combined method.

1.14.3 Combined Method of Drag Reduction

The combined method of CVS control was first put forward in [34], in which the simultaneous use of elastic coatings and distributed polymer solution injection were substantiated for CVS control in a boundary layer.

Simple combination of various methods of CVS control according to Figure 1.84 does not guarantee the achievement of positive result for a particular problem. In order that a combined method of CVS control provides maximum efficiency it is first necessary to develop, for each particular case, a physical substantiation for the simultaneous use of two or more methods. A basis for such physical substantiations can be the method of receptivity or, in other words, the revealed laws of interaction of various disturbances, and the offered hypothesis about a complex wave-guide. The method of concentrated or distributed formation of LVSs in a boundary layer considered above in combination with various elastic surfaces is an example of a combined method of CVS control.

Distributed formation of longitudinal vortical disturbances in a boundary layer is also carried out at the manufacturing of elastic plates, the external layer of which has a specific regular structure in the longitudinal direction [66, 73]. At increased speed U_∞ the dynamic loadings of flow grow, which causes a reaction in the elastic material leading to longitudinal disturbances being generated in the boundary layer from below, as well as by means of heated wires.

The second type of the combined method consists of the simultaneous use of a xiphoid tip and injection of polymeric solutions [45, 93, 305]. A model for the introduction of a homogeneous liquid through one or two slots (or two or more kinds of liquid with various densities) into a boundary layer has been designed. An electric field is created along the model; a current arises when the nasal and tail parts of the model are connected.

The third type of combined method uses the generation of gas bubbles in a boundary layer through electrolysis [335]. Various methods of injection of gas bubbles into a boundary layer by means of thin slots, current-carrying wires, strips or diaphragms placed streamwise or transversally are shown. Figure 1.89 illustrates a scheme of introduction of longitudinal CVSs, stabilized by means of gaseous microbubbles.

It is known that microbubbles can be effectively injected into a boundary layer through a crack located across the basic stream or through a porous surface. Efficiency is reached when the size of the microbubbles do not exceed 50 μm. In Figure 1.89 systems of vertical layers of microbubbles are injected into the boundary layer. The layers are located at a

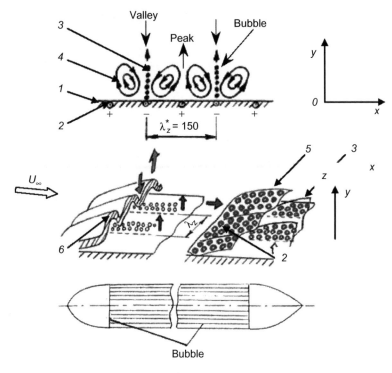

Figure 1.89

Schemes generation of microbubbles in a boundary layer: 1 — streamline surface, 2 — electric wires, 3 — vertical layers of microbubbles, 4 — longitudinal vortical systems, 5 — horizontal layers of microbubbles.

distance $\lambda_z = 150$, which approximately corresponds to a step of Klein vortices. In this case such vertical layers stabilize Klein vortices existing in a boundary layer or form vortical systems of the set scale. Thus, the size of the microbubbles has no effect in this case, and has no limit in terms of efficiency. The combination of injecting a layer of microbubbles through a cross-section crack and through longitudinal current-carrying elements or periodically located cracks is also viable.

In [81, 84 et al.] a method of simultaneous injection into a boundary layer of several types of liquid, including at a flow of elastic surfaces, has been demonstrated. Thus, various methods of organization in a boundary layer of various CVSs are applied. Examples of other combined methods are mentioned above [84, 85]. In the Proceedings of the Second International Symposium on Seawater Drag Reduction (ISSDR) in 2005 further research was presented in which other variants of the combined method of CVS control in a boundary layer were investigated.

1.15 Physical Substantiation of the Interaction Mechanism of the Flow with an Elastic Surface

The investigation results presented in this chapter and in [30, 32, 40, 42, 305] leads to the following conclusions concerning the interaction between flow and elastic plates, given the correct selection of elastic plate parameters:

- Reynolds number $Re_{l.st}$ of stability loss increases;
- the transitional zone and Reynolds number $Re_{l.st}$ of transition increases;
- the area of frequencies of unstable fluctuations of laminar and transitional boundary layers decreases;
- phase speed of disturbing movement decreases;
- the wave number increases and growth rate of disturbances at first stages of transition decrease;
- the vertical coordinate of distribution of the basic energy-carrying fluctuations of disturbing movement decreases; kinetic energy of disturbing movement decreases;
- elastic plates are favorable for forming longitudinal vortices in boundary layer;
- flow structure and disturbance motion behavior in boundary layer are modified if the wall is elastic;
- the characteristics of a boundary layer, considered as a wave-guide, can be controlled by modifying the structure and properties of the elastic surface.

An elastic surface made in the form of longitudinal strips, having various mechanical characteristics, promotes the generation of longitudinal vortical structures in a boundary layer. In Figures 1.87−1.89, other methods of organization of LVS in boundary layer are shown.

Several above-mentioned effects are usually observed simultaneously in the flow over elastic surfaces because they usually become apparently interconnected. Investigations have shown that the criteria of similarity used correctly model the features of the flow over the considered elastic surfaces. For more complex designs of elastic plates, as follows from Section 1.14, it is necessary to develop more complex equations of movement of such plates and correspondingly refine the criteria of similarity.

Taking the above-mentioned aspects into account, the integral result of flow interaction with damping surfaces can be presented in the form of the dependence of the plate drag on Reynolds number (see Figure 1.90). The solid lines and numbers *1−3* denote the curves for laminar, transitional and turbulent modes of flow, respectively, along the plate. Points *A*, *B* and *C* denote the typical Reynolds numbers during flow over a rigid plate and the dashed lines denote the regularity of drag of damping surfaces.

In the flow over rigid surfaces, Reynolds number of stability loss remains practically constant. Experimental research on hydrodynamic stability over various kinds of elastic

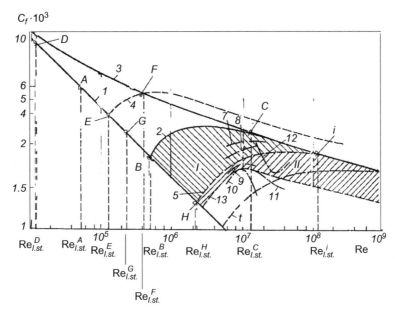

Figure 1.90

Dependence of drag of longitudinal flow over rigid and elastic plates on Reynolds number: *1* — laminar, *2* — transition, *3* — turbulent BL, *4* — resistance of the elastic surface destabilizing a boundary layer, *5* — resistance of an elastic surface at retardation of transition, *6* — optimum resistance of an elastic surface at retardation of transition; experiments of Kramer [315]: *7* — rigid standard, *8* — elastic surface at $E' = 1.63 \cdot 10^8 \text{H/M}^3$, $d = 57\%$, $\eta' = 10-400\text{cSt}$, *9* — $E' = 4.34 \cdot 10^8 \text{H/M}^3$, $d = 44\%$, $\eta' = 1200\text{cSt}$, *10* — $E' = 2.17 \cdot 10^8 \text{ H/M}^3$, $d = 47\%$, $\eta' = 300\text{cSt}$, *11*, *12* — resistance of emerging model, *13* — resistance of the elastic surface reducing turbulent friction; *I* — efficiency from application of an elastic coating at retardation of transition, *II* — efficiency from application of an elastic coating at turbulent drag reduction, *I* and *II* — peak efficiency from application of an elastic covering.

plates have shown that $Re_{l.st.}$ of elastic plates varies over a wide range depending on the mechanical characteristics of the material of elastic plates and their design.

Investigations on simple membrane surfaces showed that the hydrodynamic stability of the flow along the surfaces deteriorates if they have non-optimal mechanical properties [305]. If the results found in experiments B1−B5 are interpolated, the Reynolds number calculated with respect to *x* comprises $Re_{l.st.}^D \approx 1.35 \cdot 10^4$. The point of loss of stability under these conditions is designated by letter *D*. The $Re_{l.st.}^D$ has decreased four times in comparison with $Re_{l.st.}^A$, corresponding to a point *A* at a flow of a rigid plate. It will be assumed that deterioration of hydrodynamic stability leads to a decrease of the transition zone and to Re_{cr1} approaching $Re_{l.st}$ (similar changes occur during flow over a rigid surface under unfavorable conditions). In fact, oscillations, which do not coincide in phase with oscillations of the boundary layer, occur in a destabilizing boundary layer of an elastic surface due to the effect of flow pressure fluctuations, while the oscillation amplitude of the elastic surface may exceed even the permissible height of the roughness.

Thus, the effect on the flow of the destabilizing surface can be similar to that of a rough or wavy rigid surface. It is assumed that the transition steps $I-V$ denoted according to the classification suggested in Section 1.7, Figure 1.42, are reduced during flow over a destabilizing surface and point B, corresponding to $Re_{l.st.}^B$ of a rigid surface, is shifted upstream to point E by a value proportional to the ratio $Re_{l.st.}^A/Re_{l.st.}^D$. The critical Reynolds number during flow over a destabilizing elastic surface then comprises $Re_{l.st.}^E = 1.25 \cdot 10^5$. The drag curve from point F is shifted above the solid curve with strongly destabilizing properties of the elastic surface because the pressure and velocity fluctuations in the laminar boundary layer will be greater than those during flow over a rigid surface.

Step VI of transition to turbulence occurs from point E to point F during flow over an elastic surface. The drag of this surface, due to the factors indicated above, will be greater than that of a rigid smooth plate and corresponds to curve 4 in Figure 1.90. The nature of curve 4 may differ from that indicated in this figure, as a function of the degree of destabilization of the boundary layer by the elastic surface.

From research on which elastic surfaces have the best mechanical characteristics, the Reynolds number was $Re^* = 850$ or $Re_{l.st} = 2.4 \cdot 10^5$. The letter G in Figure 1.90 denotes the point corresponding to this value of Reynolds number. Due to the main effects enumerated above and found during investigations on an elastic surface, it may assumed that not only the point of loss stability is shifted downstream but the entire transition zone increases as well.

It will be assumed that at elastic surface stages of transition $I-V$ increase the extent proportionally to the ratio $Re_{l.st.}^G/Re_{l.st.}^A = 2.4 \cdot 10^5/5.4 \cdot 10^4 = 4.45$ in comparison with a rigid surface. Then $Re_{l.st.}^H$ accordingly increases during flow over an elastic surface and equal to $2.22 \cdot 10^6$.

In Figure 1.90, the origin of transition stages to turbulence VI will be denoted by the letter H during flow over this damping surface, the end of this step will be denoted by the letter i, and the drag curve on this step by the number 5. The advantage in drag reduction during flow over these damping surfaces compared to a rigid surface can be evaluated by using the cross-hatched zone I in Figure 1.90. The method of hydrodynamic drag reduction by using elastic surfaces based on the retardation of the transition of a laminar boundary layer to a turbulent boundary layer is called the distributed damping method.

According to the results of the investigations, the coefficient of increase of Re_{cr1} number or the laminarization coefficient is equal to 4.45. The use of elastic surfaces may increase this coefficient to 10 [220]. The number 6 with this coefficient denotes the drag curve of a damping plate, constructed in Figure 1.90. From the results of these calculations, curve *6* characterizes the maximum effect of laminarization of the boundary layer by using elastic surfaces. The results of tests by Kramer [311, 315, 317], who investigated the drag of a large aspect body of revolution with different damping surfaces attached to the cylindrical part, towed by a launch, are denoted by curves *7–10* in Figure 1.90.

The best results obtained for for tests of the surface designated by the curve *10*. Drag was reduced in this case by 59% compared to the rigid reference. This means that 83% of the length of the model had a quasi-laminar flow over it.

Curves *11* and *12* present the results of testing of floating rigid model of a body of revolution, having a laminarized shape. This shape permitted the preservation of the laminar flow in the boundary layer over a long length of the body. Curve *12* characterizes the tests of the model at different speeds of motion. It is obvious that the given shape of body permits to have different length of the laminar boundary layer at different velocities. Curve *1* is constructed from results of testing of a model with regard to the fact that 60% of its surface has a laminar boundary layer over it at any speed.

Comparing curves *7–10* and *11–12*, one can conclude that damping surfaces reduce hydrodynamic drag not only by extending the transition, but also by laminarization of the boundary layer. If the envelope of the curve can be drawn by curves *10* and *11* and if it can be interpolated in both directions, then curve *13* is found. This curve characterizes zone *II* of reduction of hydrodynamic drag and includes zone *I*.

The distributed damping method actually eliminates a mixed flow over the body if it is completely covered by a damping surface with correctly selected mechanical characteristics. In this case there is significant extension of the transition on the initial section of the body and the flow structure in steps *I–VI* differs from the similar structure along a rigid surface. The boundary layer is then formed along the damping surface and differs in its structure from a turbulent boundary layer, which would occur at the same values of x along a rigid surface.

It is known that damping surfaces have optimum stabilizing properties in a limited range of flow velocities. Therefore, if the mechanical properties of a damping surface are constant, the value of C_f begins to increase with an increase of Reynolds number and may not only reach but also exceed the parameters of curve *3*. In order for the right branch of curve *13* to be parallel to curve *3* one must regulate the mechanical properties of the damping surface with an increase of Reynolds number so that its stabilizing properties are optimum all the time.

Efficiency of elastic surfaces is defined not only by the correct selection of their mechanical characteristics but also by the choice of optimum design considering the physical features of flow over an elastic surface.

In most papers the interaction of flow and an elastic surface are considered as a static or quasi-static process. The field of external loadings is represented in the form of bending mode of pressure. With this simplified approach, many features of the interaction are not taken into consideration.

To develop and describe a complete rheological model of a material that is exposed to a flow it is necessary to consider the real structure of an elastic surface and distribution of outer loading.

Foundations of structural and kinematic–dynamic interaction of a boundary layer with a surface are put used as a basis for the development of schemes of elastic surfaces. These foundations are presented in [52, 55, 59, 305, 310].

The structural principle involves satisfying the correspondence between the structures of the boundary layer disturbance motion and of the elastic plate. This means that if there is a 2D intensive wave in the critical layer at the initial transition stages it is desirable that the outer layer of the elastomer generates a 2D disturbance motion under the action of the boundary layer stress field.

The kinematics–dynamic principle demands that the mechanical properties of the composite elastic surface layers maintain kinematic and dynamic correspondence to the boundary layer parameters. So, a 2D wave in the critical layer at the initial transition stages must have energy high enough to generate deformation of the outer layer of the elastomer. With this, the stimulated boundary layer force frequency must correspond to the natural frequency of the elastomer layer for layer. In such a case the outer load leads to oscillation generation. These oscillations, with certain correspondence with stimulating force phase, dump the outer load. In the case of the wave-guide structure, the outer load results in surface waves in the elastomer and at the liquid-wall boundary. The suctioning of these surface waves will take up energy from the 2D wave and so the boundary layer energy will be dumped.

The coordination principle means that along the length of a surface in each place the two first principles should hold [5]. In other words, modifying the structure and properties of disturbance motion must lead to modification of the structure and properties of the elastomer (or its outer layer).

With sufficient length of the plate and small ε, all stages of boundary layer transition are formed. These stages have certain lengths and structures. In this case, at the zero stage (see Sections 1.1, 1.9 and 1.10, Figures 1.1, 1.8–1.10, 1.42, etc.) up to the region of $Re_{l.st}$ all oscillations must attenuate, so there is no necessity to use an elastic plate. But the experiments showed oscillation before $Re_{l.st}$, so the monolithic elastic plate has to be placed. After this region, the surface has to be divided into sections in accordance with the stages of transition. At stage I, the elastic surface structure must be conformed to 2D sinus waves. Because there is a spectrum of flat fluctuations in a boundary layer, ordering in an elastomer should be determined by the condition of conformity to the second neutral fluctuation.

The elastic surface section in the region of formation of longitudinal vortices in a boundary layer must satisfy the structure, dimensions and directions of these vortices. At the turbulent boundary layer stage, the elastic surface structure must satisfy the viscous sublayer flow structure.

When designing an elastic coating it is necessary to consider that kinematic–dynamic characteristics of a boundary layer on various stages of its formation are essentially different. At the initial transition stages, there are least five initial stages where a 2D wave exists, a quasi-static approach and simplified elastomer can be used [13, 32, 108, 220, 229]. At the following stages and in the turbulent boundary layer interaction between flow and the elastic surface takes place in a wide range of frequencies with dynamic loading. In this case, the surface model must be statistical, for example according to equations (1.10)–(1.13) or [502]. It is also necessary to take into account "bursts" from the viscous sublayer, liquid "patches" falling on the wall, etc. [274, 232, 233, 277].

The foregoing approach to selecting elastic coatings in real conditions can be explained using an example of the flow over a wing (see Figure 1.91). Dotted lines denote borders of sections of a non-controlled uniform elastic surface, points show the modified distributions of loading parameters and new borders of sections after the selection of optimum

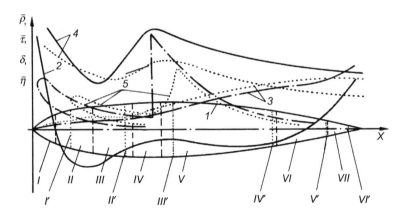

Figure 1.91

Standard distributions of loadings along a wing: 1 — outer contours of a wing, 2 — distribution of pressure $\bar{p}$ on x axis, 3 — boundary layer thickness δ, 4 — shear stress $\bar{\tau}$, 5 — frequency ranges of laminar and turbulent fluctuations $\bar{n}$ in the boundary layer.

characteristics of elastomers for them. The Roman numerals designate numbers of corresponding sections. Borders of sections are determined from following requirements:

- within each section, gradients of parameters of the loading should be constant or small;
- in an area where sections join, parameters should not contain singularities.

The same is valid for bodies of revolution.

The curves presented in Figure 1.91 show that the elastic surface along the wing chord is subjected to different static and dynamic load actions. With constant velocity flow over a wing in section *I*, the pressure gradient $\bar{p}$ is maximal, i.e. pressure sharply decreases, but, remaining positive, it compresses the elastomer. Gradient of shear stress $\bar{\tau}$ is as maximal, which means rapid changing of shear load in the elastomer. In the same section intensive high-frequency velocity fluctuations take place. Because of the maximum gradient of δ and of its minimal value, the heat transfer gradient is maximal. Thus in section *I* the elastic surface must be rather stiff to be stable to the action of high loads and their gradients. With this, the high-frequency velocity and pressure fluctuations are being dumped. The positive effect can be amplified by dividing section *I* into subsections.

In section *II* the above-mentioned loads and their gradients are altered not so intensively and do not reverse sign; the region of unstable oscillation frequencies and the frequencies themselves decrease in this section. Because of this, the mechanical properties of the elastomer must differ from that in the section *I*. For example the compliance of the surface must be increased in comparison with that in section *I*. The $\bar{p}$ load, in contrast with that in section *I*, acts in separating. In section *II* the $\bar{p}$ gradient is positive and gradient reverses sign, i.e. the separating load decreases. In section *III* the gradient $\bar{p}$ is positive and gradient $\bar{\tau}$ is changed sign, which means that loading on separation decreases. The range and frequencies of velocity fluctuations in the boundary layer tend to a minimum, as well.

In section *IV* the $\bar{p}$ gradient practically becomes equal to zero, i.e. the elastomer is subjected to shear loads due to $\bar{\tau}$. With this, the stress $\bar{\tau}$ becomes maximal and its gradient reverses the sign. In addition, due to the transition of the laminar boundary layer into the turbulent boundary layer in the region, the range of load $\bar{n}$ sharply widens and the law of growth of δ and heat transfer coefficient modify.

In section *V* all loads vary weakly and they can be assumed constant. In sections *VI* and *VII*, the load $\bar{p}$ increases, i.e. the separating load increases. Besides these loads, vibration loads must be added, caused by non-steady motion and by vibration of the base and the loads, caused by modification of the environment properties and flow regime.

According to ideas in the paper [29], the complex interactions of disturbances in a boundary layer can be presented as a multivariate system with arbitrary input processes and one output process (see Figure 1.92). All kinds of outer and inner disturbances with input processes X_i, described by frequency spectrum H_i, are summed up in an elastic surface with an outer disturbance N. As a result, at exit the outer frequency response of the elastomer is obtained (see Chapter 3).

The input process x_1 is a frequency load of the entire elastic surface as a single plate. The load x_1 is caused by nonstationary flow in real hydroaeromechanic problems. Fluctuation of an elastomer occurs under the influence of the load x_1 regularly distributed and acting from above.

The input process X_2 is a frequency of load of the entire elastic surface as a single plate due to vibrations and fluctuations of the base—the surface to which the elastic coating is attached. Loading X_2 acts on the elastomer from below and can be either uniform along the plate or with various laws of distribution along the plate.

The input process X_3 is the frequency of quasi-static loads, caused by distribution $\bar{p}$ and $\bar{\tau}$ along the body, under which influence normal and shear loads appear in each section, in addition to vibrating loads X_2. The loads X_3 have a certain distribution law along the plate and act from above.

The input process X_4 is local dynamic frequency loads due to fluctuations of pressure and velocities in the boundary layer. The pressure fluctuations from above determine the sign-alternating normal stresses and the velocity fluctuations induce shear stresses.

The input process X_5 is a frequency of thermal loading acting from above and from below. The properties of elastomers are substantially dependent on temperature. Therefore, energy dissipation of fluctuations and vibrations, depending on the thermodynamic parameters of composite plates, results in a change of temperature of composites. The heat flux direction is downward from the elastomer to the base and upward to the boundary layer. This heat flux is not uniform in different directions because the elastomer consists of sections and because of different intensity of dissipative energy in the transition stages, and so on. In addition, the elastomer is influenced by thermodynamic fields of the boundary layer and of the base. Thus, there are differences in the thermodynamic regime of the flow over different parts; hence, there are additional energies in each section, for example, changed elastomer rigidity and viscosity, and so its frequency characteristics change.

The disturbance N is caused by variable external conditions and flow conditions. If values X_i are expressed in terms of their components, the scheme in Figure 1.92 would be more complicated. The multivariate system, such as is presented in Figure 1.92, is solved by means of the Fourier transformation [111]. In reality, the interaction process is very complex. Therefore, in the presented scheme the elastic surface acts as an integrator of different disturbance spectrum. Because of summation of all the disturbances getting into an elastic surface, one-frequency process of oscillation of an elastic surface occurs at exit (see

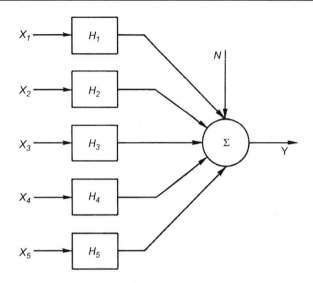

Figure 1.92
Multivariate system with arbitrary input processes and one output process [111].

section 1.12). This process interacts with input processes, i.e. a feedback takes place. For example, a feedback modifies the parameters of base vibrations, the loads $\bar{p}$, $\bar{\tau}$ and N.

Simplified models of real complex interactions are solved by means of calculation of the conditional spectral density and a spectral matrix [111]. Thus, an elastic surface under complex interactions is always subjected to complex load condition. Simultaneously there are various kinds of loadings—compressing, stretching, and shear, unidirectional, sign-variable, static and dynamic loads.

It is practically impossible to construct a model that includes all these loadings. Therefore, it is obviously not possible to develop a mathematical model for the description of a real picture of interaction of a flow with an elastic surface at all kinds of loadings. However, different simplified approaches are possible. In particular, we can use a superposition method. At any interaction, an elastic surface must be hydraulically smooth, monolithic, composite, with appropriate inner structure, and must be controllable to adjust it for optimal interaction.

The hydrobionics approach is promising for understanding the processes of interaction of elastic coatings with a flow, and for designing various characteristics and structures of drag reducing elastic materials.

CHAPTER 2

Types of Elastic Surfaces and Research of their Mechanical Characteristics

2.1 Models of Elastic Surfaces

A major characteristic of the structure of elastic (elasto-damping) materials is the presence in molecules of repeating identical parts. Thus, the length of a macromolecule is 10 times the cross-section size of a molecule, and the molecules are arranged chaotically in the form of a ball [219, 423]. A piece of a macromolecule that moves as a unit is referred to as a segment. In each macromolecule there is a certain order to the arrangement of atoms. The spatial arrangement of atoms, which does not vary as a result of thermal movement, defines the configuration of a molecule [423].

The material which has the greater length of a segment, at its deformation shows resistance to a bend more than a material with a lot of short segments. The segment moves as a single whole. Therefore at the big length of segments it is necessary to make an effort to produce a bend in these segments. If segments are too short for their bend it is necessary to spend a smaller loading as the short segment does not bend and moves as a single whole. The bend occurs due to the intervals between segments.

In this paragraph, and further in the text, the theory of the elasticity of polymeric materials which do not demand the explanatory are stated. It should be known and it is possible to find in any book under the theory of highly elastic materials. For understanding of the received experimental results at a flow of elastic materials it is necessary to know the bases of the theory of elasticity of elastic polymeric materials. Therefore these substantive provisions on examples of their own experimental results in mechanical characteristics of elastic materials are stated. Numerous authors have received results of experimental researches of a flow of elastic plates in air and water streams. However they could not explain the received results as there had been no results in Chapter 2.

In response to a constant deformation and an establishment of balance of external forces and internal stresses, after long action of the imposed forces a spontaneous change of pressure—a relaxation—occurs. Visco-elastic bodies are characterized by one more important property: when the duration of action of external forces increases, the structure alters, including the relative conformation of the molecular chains, their relative positioning, and the supramolecular shapes.

There are different theories on high-elastic strains [12, 219]. The most convenient method of investigation is the modeling of high-elastic strain kinetics, although it describes the behavior of the materials approximately. According to the linear theory of visco-elasticity [371, 423] the material response to loading is a combination of linear viscous and linear elastic motion. A spring is a perfect elastic body. The stress and the strain are connected by the Hooke equation:

$$\sigma = E\varepsilon \tag{2.1}$$

An example of a perfect viscous body is a piston in which the stress and the strain are connected by the Newton equation:

$$\sigma = 3\eta\varepsilon \tag{2.2}$$

Slonimsky proposed a formula in which the Hooke and Newton laws are written in the form of a uniform relation [48]. Combining the elastic and viscous elements, it is possible to construct different models, which can be considered, at least qualitatively, as real bodies. Joining elastic and viscous elements in a series we obtain a combination known as a Maxwell element (see Figure 2.1a, circled by dash-dot line). The stress—strain ratio in such an element can be obtained from the equation:

$$\frac{d\varepsilon}{dt} = \frac{1}{E}\frac{d\sigma}{dt} + \frac{\sigma}{\eta} \tag{2.3}$$

A model with relaxation properties has been proposed by Voigt and Kelvin (see Figure 2.1b, circled by dash-dot line). In this model the elastic and viscous elements are joined in parallel. The differential equation of the model motion is:

$$\sigma = \eta\frac{d\varepsilon}{dt} + E\varepsilon \tag{2.4}$$

The Maxwell model does not take into account the elasticity in polymers due to the unrolling of macromolecules but it satisfactorily describes the relaxation of materials that are strained with a constant rate. The Maxwell model qualitatively characterizes the behavior of a linear polymer, and the Voigt—Kelvin model of a cross-linked polymer. The Voigt Kelvin model does not take into account the fact that the relaxation time depends not only upon temperature and polymer structure but also upon the applied force. In reality, the polymer strain consists of an elastic and viscous part modeled by the Maxwell element and an elasto-plastic part modeled by the Voigt Kelvin element.

To describe the behavior of elasto-plastic materials many rheological models are constructed, which are combinations of elastic and plastic elements. In [219, 336, 448] it has been shown that all mechanical schemes can be reduced to two patterns. The first pattern (see Figure 2.1a) consists of Maxwell elements joined in parallel (model which for the first time has proposed Kun). The second one (see Figure 2.1b) consists of

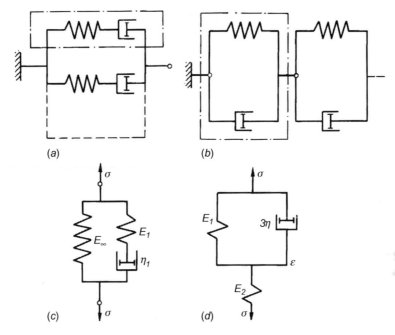

Figure 2.1

Mechanical models of polymer materials: a — Maxwell elements joined in parallel,
b — Voigt—Kelvin elements joined sequentially, c — Dogadkin—Bartenev—Resnikovsky model,
d — standard three-element model.

Voigt—Kelvin elements joined in a series (Alfrey model). The formula of stress relaxation based on the KUN model is as follows:

$$\sigma = \varepsilon \int_0^\infty E(\tau)e^{-i/\tau}d\tau, \tag{2.5}$$

where $E(\tau)$ is the distribution function expressing the part of the Young modules depending on the relaxation time. This part is attributed to those molecular mechanisms that have the relaxation time within τ through $\tau + d\,\tau$ [219]. Based on the model which for the first time has proposed Alfrey, the strain with constant stress and continuous distribution of relaxation time is as follows:

$$\varepsilon = \sigma \int_0^\infty I(\tau)(1 - e^{-i/\tau})d\tau \tag{2.6}$$

where $I(\tau) = 1/E(\tau)$. Different empirical methods of determining distribution functions $E(\tau)$ and $I(\tau)$ [219] are known. Figure 2.1c presents the Dogadkin—Bartenev—Reznikovsky model [219]. It is governed by the differential equation:

$$\frac{d(\sigma - \sigma_\infty)}{dt} = E_1\frac{d\varepsilon}{dt} - \frac{(\sigma - \sigma_\infty)}{\tau} \tag{2.7}$$

Alexandrov and Lasurkin [219] suggested a similar model. Its equation is:

$$\varepsilon = \varepsilon_1 + \varepsilon_2 = \left(\frac{1}{E_1} + \frac{1}{E_2}\right)\sigma \tag{2.8}$$

where $1/E$ is the compliance of the elastomer. Their model corresponds to the standard three-element model of visco-elastic medium (Figure 2.1d):

$$\left(\frac{1}{E_1} + \frac{1}{E_2}\right)\sigma + \left(\frac{\eta}{E_1 E_2}\right)\frac{d\sigma}{dt} = \varepsilon + \frac{\eta}{E_2}\frac{d\varepsilon}{dt} \tag{2.9}$$

A better approximation to real monolith elastomers can be obtained by means of four-element elastic models, which are analyzed in [107, 219]. However, the mathematical description of such models and solution of corresponding differential equations are too complex. In addition, all these models are only an approximation to a real system and do not take into account the molecular structure of elastomers. The more general connection between σ and ε can be stated by means of Volterra equations [107], which present the linear ratio written as follows:

$$\sigma = E\left[\varepsilon - \int_{-\infty}^{t} \Gamma(t-\tau)\,\varepsilon(\tau)\,d\tau\right] \tag{2.10}$$

$$\varepsilon = 1/E\left[\sigma - \int_{-\infty}^{t} K(t-\tau)\,\sigma(\tau)\,d\tau\right] \tag{2.11}$$

where $\Gamma(t-\tau)$ and $K(t-\tau)$ are kernels characterizing the hereditary properties of materials (104).

The kernels in formulae (2.10) and (2.11) are selected for the material under consideration, taking into account its physical properties determined in experiments. There are several kernels suggested by different investigators. Boltzmann suggested the kernel in the form of: $K(t-\tau) = C(t-\tau)$, where C is a constant. Abel suggested as a creep kernel the function $K(t-\tau) = C/(t-\tau)^a$, $0 < a < 1$.

The function was successfully used for calculations of certain materials. Several expressions for the kernels were suggested in [107, 423]. The model description of visco-elastic behavior of polymer materials, taking into account their molecular structures, is suggested in [107], where complex multi-component models are considered. The modules of elasticity and of mechanical compliance are written by expressions of the type:

$$E'(\omega) = \int_{-\infty}^{\infty} H(\ln \tau)\frac{\omega^2\tau^2}{1+\omega^2\tau^2}d\ln \tau \tag{2.12}$$

where $H(\ln \tau)$ is distribution function. The elastic surfaces used in our investigations consist of several layers of an elastomer of different densities separated with more rigid and heavy

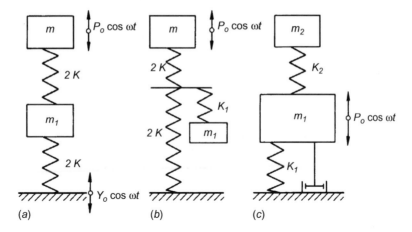

Figure 2.2
Schemes of oscillation of dynamic dampers and of a double cascade of vibro-insulation.

glue membranes, and of layers made of synthetic fiber differing from the elastomers in structure. For constructing such composites the approved approaches of vibration damping can be used [382, 406]. It is the best approximation to real models. For example, the outer film and the middle layer can be considered as a system of two oscillating masses separated by elastic elements. Such schemes are presented in Figure 2.2 [406]. For the scheme *a* (see Figure 2.1a) main dynamic compliance can be expressed by Panovko's equation [423]:

$$\frac{y}{P_0} = \frac{4k - m_1\omega^2}{2k(2k - m_1\omega_2) - m\omega^2(4k - m_1\omega^2)} \tag{2.13}$$

The coefficient of transmission of vibro-insulation (load on the base) is as follows:

$$\frac{P_f}{P_0} = \frac{y}{y_0} = \frac{4k}{2k(2k - m_1\omega^2) - m\omega^2(4k - m_1\omega^2)} \tag{2.14}$$

The anti-resonance frequency is between two resonances and presented by the formula:

$$\omega = \sqrt{4k/m_1} \tag{2.15}$$

The equations of motion for schemes similar to those in Figure 2.2a are also presented in [108]. For the scheme *b* the equations are as follows:

$$m\frac{d^2y}{dt^2} + 2k(y - y_2) = P_0 \cos \omega t$$

$$m_1\frac{d^2y_1}{dt^2} + k_1(y - y_2) = 0, \tag{2.16}$$

$$2k(y_2 - y) + k_1(y_2 - y_1) + 2ky_2 = 0$$

The coefficient of transfer of load on the base is:

$$P_{base}/P_0 = \frac{4k^2(k_1 - m_1\omega^2)}{(k_1 - m_1\omega^2)[(4k + k_1)(2k - m\omega^2) - 4k^2] - k_1^2(2k - m\omega^2)} \qquad (2.17)$$

The main dynamic compliance is:

$$y/P_0 = \frac{4kk_1 - m_1\omega^2(4k + k_1)}{(k_1 - m_1\omega^2)[(4k + k_1)(2k - m\omega^2) - 4k^2] - k_1^2(2k - m\omega^2)} \qquad (2.18)$$

The self-resonant frequencies of oscillations are:

$$\omega^2{}_{1,2} = \omega^2{}_0\{0.25(\omega_{a.res}/\omega_0)^2(2 + \varepsilon) + 0.5(\omega_{a.res}/\varpi_0)$$

$$\pm \sqrt{2[(\omega_{a.res}/\omega_0)^2 - (\omega_{a.res}/\omega_0)^2]^2 + 0.25\varepsilon(\omega_{a.res}/\omega_0)^2} \qquad (2.19)$$

where $\omega_0 = \sqrt{k/m}$; $\varpi_0 = \sqrt{k_1/m_1}$; $\omega_{a.res} = \sqrt{4kk_1/m_1(4k + k_1)} = 2\omega_0\varpi_0/\sqrt{4\omega_0^2 + \varepsilon\varpi_0^2}$

Scheme *a* (see Figure 2.2) effectively and uniformly damps the vibrations. Scheme *b* is used for damping in the range of the above resonance frequencies, and *c* is applied to dissipate resonant frequencies of fluctuations of the main system, its full equations of movement are presented in Karamishkin [382].

Describing the complex multi-element elastic material by the model presented in Figure 2.3a, the values of hardness, inertia and damping are calculated by means of tensor analysis using computers [6, 12, 97, 107, 134]. Oscillation amplitudes of complex elastic coatings (see Figure 2.3b) under the influence of mono-harmonic excitation, when the outer layer is replaced by a system of single oscillators, can be defined by means of the continual scheme of calculation using Wolfson's integral equations [219]. The elastomer behavior under the load acting from above and from below is described by equations (2.13) and (2.14) (see Figure 2.2a).

For scheme *b* (see Figure 2.2b), when a disturbing load also acts from below, the equations of movement is written down in the form:

$$m\frac{d^2y}{dt^2} + 2k(y - y_2) = P_0 \cos \omega_1 t$$

$$m_1\frac{d^2y_1}{dt^2} + k_1(y_1 - y_2) = 0 \qquad (2.20)$$

$$2k(y_2 - y_1) + k_1(y_2 - y_1) + 2ky_2 = 2 \ ky \cos \omega_2 t$$

Klykin [382] presents the equations neglecting outer loads. There are a plenty of experimental works that shown that elastic surfaces essentially reduce vibrations of the base in a wide range of frequencies or temperatures [22, 107, 218, 242, 382, 406, 526].

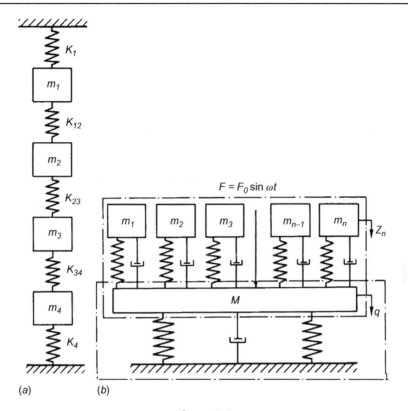

Figure 2.3
Schemes of dynamic dampers of complex, 3D systems.

Theoretical explanations of these effects are given in [107, 218, 406, 526], and have given expressions for determining the damping factor η, logarithmic decrement d, quality coefficient $\omega/2\delta$, resonance peak width Δf, relaxation time $\tau = 1/\delta$, coefficient of space attenuation, amongst others. Klykin [382] has suggested a scheme of low-frequency and high-frequency acoustic vibration dumpers (see Figure 2.4). This uses an elastic gasket *1* as a viscous element (see Figure 2.2), and steel overlays *2* as an inertial element. The high-frequency damper has a steel overlay of small mass and an elastic gasket made of a monolithic material. The low-frequency damper has a heavier steel overlay and the elastic gasket can be porous or weakened with embrasures. Although these weakened areas are localized, by considering them as unit elements and ignoring the loads over the area it is possible to imagine the action of the distributing vibration damper.

Despite the progress achieved in constructing models and theories of elastomer behavior under the applied forces, it should be noted that until now there has been no rigorous mathematical theory of monolithic visco-elastic materials, particularly with the dynamic action of small and two-sided forces. Therefore, experimental investigations on the

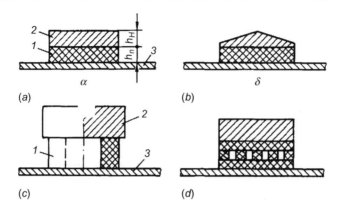

Figure 2.4
Scheme of resonance vibration of damping plates: *a, b* — high frequency, *c, d* — low frequency;
1 — elastic element, *2* — mass of vibration damper, *3* — damping plate.

mechanical properties of such materials will continue to form the basis for both constructing theories and detecting the features of elastomer behavior.

2.2 Mechanical Characteristics of Elastomers

The physic mechanical properties of elastomers can be divided into three main groups: mechanical, physical and technological. The constant equation of motion of elastomers is Young modules, modules of hardness, bulk modules, Poisson coefficient, viscosity, compliance, and so on. The most important independent constants defining the stress state of elastomers are stress $\sigma = P/F$ and elastic strain $\varepsilon = \Delta l/l$. Under small stresses when the relation $\sigma - \varepsilon$ is linear and governed by Hooke's law these constants can be united in the elasticity modules (see formula 2.1). In general, the elasticity module is a function of many variables: the composition of the original mixture, degree of filling, vulcanization regime, and others. The elasticity modules consist of two components:

$$E = E_\infty + E_\tau \tag{2.21}$$

Here E_∞ is an equilibrium modulus, which is determined as $E_\infty = RTn^{2/3}$ and not equal to zero only for cross-linked polymers; R — constant depending on the elastomer chemical structure, T — absolute temperature, n — number of molecular chains in a unit volume. E_∞ is independent of strain form and has the same value at compression or tension as the test piece, but E_∞ depends on absolute temperature and on the degree of molecular linking. The elasticity modules obtained from the linear section of the $\sigma - \varepsilon$ curve can be assumed to be balanced or conventionally balanced. In practice the elasticity module is determined at a very low rate of loading $(2 \cdot 10^{-4}$ m/s) or at long duration of the load, when the relaxation has completed and has no effect upon the elasticity module value. E_τ is the non-balanced relaxation part of the elasticity module, which is also called the instantaneous elasticity module. It is determined as $E_\tau = \lim_{t \to 0} E(t)$. The instantaneous module corresponds to the perfect elastic state of the material. In practice it is determined either at high-rate loads (impact load and so on) or at a low temperature using the temperature–time superposition. The shorter the period of load action, the larger E_τ, the higher the frequency of operating load, and ε decreases at $P = $ const. E_τ substantially depends on the material structure. The relaxation time can be determined from the formula:

$$\tau = \eta/E \tag{2.22}$$

Since E_τ depends on the duration of load action, then according to equation (2.21) real elastomers have a wide set of relaxation times. An important parameter is a bulk module $K = -P/\varepsilon_v$, where $\varepsilon_v = \Delta V/V$ is a 3D deformation. The modules of elasticity, hardness and bulk are connected as follows:

$$\begin{aligned} E &= 2G(1 + \mu) \\ E &= 3K(1 - 2\mu) \end{aligned} \tag{2.23}$$

where μ is Poisson coefficient:

$$\mu = \overline{\varepsilon}/\varepsilon \tag{2.24}$$

$\overline{\varepsilon}$ and ε is relative transversal and relative longitudinal strain respectively. The parameter $\nu = 1/\mu$ is called a transversal number. The change of volume of a cube with the unit side is expressed after deformation by the formula:

$$e = (1 + \varepsilon)(1 - \nu \, \varepsilon)^2 - 1 \approx (1 - 2 \, \nu)\varepsilon \tag{2.25}$$

where e is relative 3D or volumetric expansion. On stretching there is an increase, and on compression a reduction of the cube volume. At uniform loading of an element of the cube a by the stress σ volumetric expansion is equal to triple e:

$$3e = 3(1 - 2v)\varepsilon = 3\sigma(1 - 2v)/E = \sigma/K \tag{2.26}$$

As can be seen, if the above-introduced Poisson coefficient is used a bulky expression is obtained. Therefore, it is preferable to use the lateral number ν as Poisson coefficient μ. From Poisson equation $\frac{\sigma}{\eta}\mu = \frac{1}{2}\left[1 - V^{-1}\left(\frac{dV}{d\varepsilon}\right)\right]$, at $\frac{dV}{d\varepsilon} \to 0$ the coefficient $\mu = 0.5$. Then from formula (2.23) we have: $E = 3\,G$ and $K \to \infty$. For all materials, μ lies in range $0-0.5$. For rubber with compression and dry friction at the ends $\mu = 0.465-0.485$ [336]. The shear stress τ^* is defined by two constants:

$$\tau^* = G\gamma \tag{2.27}$$

where G is shear module, and $\gamma = \frac{\Delta l}{l} = tg\,\beta$ the relative shear of the upper side of the unit cube. For isotropic elastic bodies, there is a relationship that makes it possible to retain as independent only two constants from the three μ, E and G:

$$\varepsilon = \frac{\lambda}{l} = \overline{\tau} \cdot \frac{1 + \mu}{E}$$

$$\tau^* = \gamma \cdot \frac{E}{2(1 + \mu)}$$

$$G = \frac{E}{2(1 + \mu)}$$

(2.28)

Table 2.1 presents the relationships between the elastomer constants [343, 448]; $\lambda = l/l_0$ is the strain in the direction of the stress line; l_0 is the length of the edge of the unstrained cube.

The simplest method of evaluating the elastic properties of an elastomer is to measure its hardness. Hardness is the ability of the elastomer to resist the penetration of an indenter under the action of certain force that causes deformations of stretching, shear and compression. As all these modules are connected we can consider the dependence of hardness upon the Young modules. There are theoretical and empirical equations describing the relatioship between hardness and Young modules for standard elastomers [219, 423, 448].

The reverse parameter of hardness is compliance $I = \frac{1}{E}$, $j = \frac{1}{G}$, i.e. the ability of an elastomer to be deformed under loading action. Elasticity is defined as the part of the expended energy that is being returned after the impact. Typically, it is determined by the ratio between the rebound height and incidence height of the indenter. An important property is creep under a constant force imposed, or a stress relaxation, when the deformation is constant. These parameters can be determined at stretching, compression or

Table 2.1: The main relationships of the elastomer constants.

Constant	Principal Pair				
	(λ, G)	(K, σ)	(σ, μ)	(E, μ)	(E, σ)
Lame constant λ	λ	$K - \dfrac{2}{3}\sigma$	$\dfrac{2\sigma\mu}{1 - 2\mu}$	$\dfrac{\mu E}{(1 + \mu)(1 - 2\mu)}$	$\dfrac{\sigma(E - 2\sigma)}{3\sigma - E}$
Lame constant σ	σ	σ	σ	$\dfrac{E}{2(1 - \mu)}$	σ
Modulus K of volume pressure	$\lambda + \dfrac{2}{3}\sigma$	K	$\dfrac{2\sigma(1 + \mu)}{3(1 - 2\mu)}$	$\dfrac{E}{3(1 - 2\mu)}$	$\dfrac{E\sigma}{3(3\sigma - E)}$
Poisson coefficient μ	$\dfrac{\lambda}{2(\lambda + \sigma)}$	$\dfrac{3K - 2\sigma}{6K + 2\sigma}$	μ	μ	$\dfrac{1E}{2\sigma} - 1$
Young module E	$\dfrac{(2\sigma + 3\lambda)\sigma}{\sigma + \lambda}$	$\dfrac{9K\sigma}{3K + \sigma}$	$2(1 + \mu)\sigma$	E	E

shear of any test pieces used for strain investigations. With these tests, the force (or the deformation) is sustained as a constant for a long period, and the gradual increasing of strain (or decreasing of the force) is measured. The relaxation is the delay of deformation relative the operating load.

For an elastomer being dynamically loaded, an important frequency–temperature characteristic can be obtained by applying the superposition concept, which was first proposed by Alexandrov and Lasurkin, later being developed by Ferry, Tobolsky and Williamson. This concept makes it possible to establish the dependence of the mechanical properties on the frequency of force applied in a wide range of temperatures, which cannot always be tested in a laboratory. The results of tests carried out over a narrow range of frequencies with different temperatures can be adopted for constructing a curve in a wide frequency range, taking into account the equivalence of temperature and frequency action. This approach is presented in detail in [218, 219, 242, 423, 526]. It is to this effect that the curves $G(t)$, constructed at different test temperatures $t°$ and at small values of relaxation time τ, are transposed to the $G(t)$ curve with t, varied in a wide range. First, a curve at $t° = 25°C$, for example, is constructed. This curve is joined with curves constructed at other test temperatures in such a way that the ends of the curves match. In this way a smooth curve is constructed. For matching, the adjustment factor a_T is of prime importance. This factor is a ratio of two relaxation periods at different temperatures. It is determined by the Wilson–Landel–Ferry equation:

$$\lg a_T = -\frac{8.86(t - t_0)}{101.6(t - t_0)} \qquad (2.29)$$

The paper [448] presents formulae for recalculating the frequency necessary for the change in the mentioned modules and time necessary to achieve the same deformations. For the analysis of elastomers under dynamic loading, the nonlinear effects play an essential part [423], which is considered in Section 2.3. It is important to take into account thermodynamic parameters of elastomers interacting with the boundary layer. Density, specific heat, thermal conductivity and thermal diffusivity of some elastomers is presented in [448].

2.3 Methods of Measuring the Mechanical Characteristics of Elastomers

These days elastomers have a wide variety of applications, and standard methods of measuring of their mechanical characteristics rarely meet the requirements of a new task. Therefore, together with standard methods it is necessary to develop new methods of measurement and the corresponding equipment for defining the mechanical characteristics of elastomers. The large number of these methods is explained by the fact that elastomers in real conditions are subjected to loadings of very different times of action—from

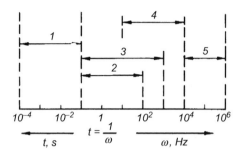

Figure 2.5

Fields of application of different methods for determining mechanical characteristics of rubber over the logarithmic scale of time: *1* − constant rate straining, *2* − free oscillation, *3* − forced non-resonance oscillation, *4* − forced resonance oscillation, *5* − propagation of waves and impulses [423].

microseconds to many years. The fields of application for the different methods of determining mechanical characteristics of elastomers are presented in Figure 2.5.

The nature of the elastic behavior of elastomers, especially complex composites, has not been fully investigated. Equipment and methods are not sufficiently developed for complex dynamic tests to define the laws of fluctuation development inside composites.

Measurements of the mechanical characteristics of dermal coatings of live hydrobionts have shown their essential differences from the characteristics of standard elastomers. The complex composite structure of these live "elastomers coverings" and the analysis of their hydrodynamic value has shown that a new understanding of the role of elastic coatings is necessary. The design of new elastomers should be similar to the skin structure of live hydrobionts. It is necessary to develop corresponding methods of measurement for researching the corresponding static and dynamic characteristics of composite elastomers.

The standard testing methods of elastomers are presented in [336, 423, 434, 448]. The stress−strain relationship is usually investigated with four strain forms: stretching, shrinking, shearing and penetration. A method of testing under stretching and shrinking is given in [336, 434, 448].

With testing in compression we have:

$$\frac{F}{A} = G(\lambda^{-2} - \lambda)S \tag{2.30}$$

where F is compressing force; A is cross-section area of the non-strained test piece; G is shear modulus ({1/3} Young modulus); λ is ratio of heights of compressed and uncompressed pieces; S is the form factor concerning the "barreling" of the test piece, $S = 1 + B\,(d/H)^2$, B is factor depending on modulus of elasticity [336]; d is diameter; H is height of the cylindrical piece.

Hardness is determined by pressing a solid ball into the elastomer to the depth of 0.8 of the ball diameter. The hardness is calculated as follows:

$$F = K_1 G R^{0.65} P^{1.35}$$ (2.31)

For the perfect elastic isotropic elastomer, the depth of penetration of an indenter with a ball tip is expressed by the formula:

$$h = K_2 (F/E)^{0.74} R^{-0.48}$$ (2.32)

where R is the ball radius; K_1 and K_2 are constants.

To define the module of elasticity from measurements of the penetration depth of the indenter, it is convenient to use the formula:

$$D = 615 R^{0.48} \left[\left(\frac{F}{E}\right)^{0.74} - \left(\frac{f}{E}\right)^{0.74} \right]$$ (2.33)

where D is difference in penetration depth under action of total force F and contact force f. The hardness number is calculated as follows:

$$H = F/\pi dh$$ (2.34)

where $R = 9.8 H$; $d = 0.5 \cdot 10^{-2}$ m. The elasticity is determined by the Shore elastomer in relative units. It is made by means of a needle penetrating the sample under consideration, and also by a ball rebound.

Stiffness of an elastomer can be determined by stretching, compressing, torsion (of strips or cylinders), bending, indenter penetration (hardness), or under multiple repeated loading. Stiffness is frequently defined at low temperature. Definition of stiffness of rubber and non-vulcanized rubber mixtures is made according to the resistance of a sample to axial compression by a given deformation and its restoration after unloading. In this case, the stiffness is the elasticity modules divided by the cross-section area of the sample or by its thickness.

It is important to determine the dynamic parameters. For this purpose the methods of free vibration or forced vibrations (resonance and non-resonance) are used, as well as methods based on the wave and pulse propagation properties.

Experiments with periodic loading at the frequency $\omega = 2\pi n$ rad/s are qualitatively similar to the experiments with a constant strain rate at the lead action time $t = 1/\omega$. Two main components of visco-elastic behavior are determined: the real and imaginary components of the complex dynamic module. The first component is the stress component coincident in phase with the strain, and the second is the 90°-shifted stress component (see Figure 2.6). Table 2.2 represents the relationship between the real and imaginary

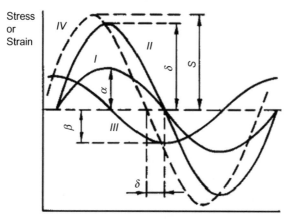

Stress or Strain

Time or Phase Angle

Figure 2.6

Sinusoidal strain and stress: *I* − strain with amplitude *a*; *II* − stress coincided in phase with the strain, with amplitude *b*; *III* − stress, which is shifted by 90° in phase, amplitude *c*; *IV* − total stress (*II* and *III*), with amplitude *d*; δ − loss angle.

Table 2.2: The relationship between real and imaginary components of the modules.

Name	Designation	Young Modules	Modules of Rigidity
Real component of modules	δ/a	E'	σ'
Imaginary component of modules	b/a	E'	σ'
Complex modules	$\dfrac{2}{a} = \dfrac{\sqrt{\delta^2 + b^2}}{a}$	E^*	σ^*
Loss angle tangent	b/δ	$tg\ \delta$	$tg\ \delta$

components of the modules. As it is shown in Figure 2.7, the complex modulus can be expressed by:

$$G^* = G' + iG''$$
$$E^* = E' + iE'' \tag{2.35}$$

and in absolute values by:

$$|G^*| = \sqrt{(G')^2 + (G'')^2}$$
$$|E^*| = \sqrt{(E')^2 + (E'')^2} \tag{2.36}$$

The loss tangent characterizing the dissipative properties of an elastomer and the components of the complex modulus are presented by the formulae:

$$G' = |G^*|\ \cos\ \delta$$
$$G'' = |G^*|\ \sin\ \delta \tag{2.37}$$

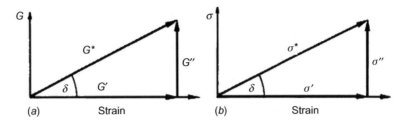

Figure 2.7
Vector picture of the components of modulus (*a*) and stress (*b*) at sinusoid strain.

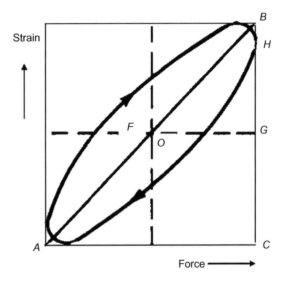

Figure 2.8
Dependence of deformation on force. *F* — static or average deformation (force).

The real and imaginary components increase with decreasing temperature or increasing frequency of cyclic strain. The value *tg δ* varies following a curve that has a maximum corresponding to a certain temperature (when ω = const) or a certain frequency (when $t°$ = const). As a rule, with reduction of cyclic loading, the magnitude of *G'* or *E'* considerably increases while *tg δ* decreases. Therefore, tests should be made with amplitudes of real loads.

The dynamic stiffness *S** is a parameter that is in direct proportion to the force amplitude divided by the strain amplitude, i.e. to *BG/FG* (see Figure 2.8). The dissipation energy in one cycle is presented with the area of the loop $\Delta W = 2\pi BFGsin\delta$. The triangle *ABC* presents the energy consumed for damping the elastic sample. The ratio of the area of hysteresis loop to the triangle *FBG* area, i.e. ratio of irreversible dissipated energy to its amplitude *W*, is called the energy absorption factor: $\Psi = \Delta W/W$.

When a dynamic test is being done, the relationship between the modulus and stiffness areas is follows:

$$\text{shearing} \quad G^* = \frac{S^*H(1 - H^2/36K^2)}{A}$$

$$\text{compressing} \quad E^* = \frac{3\,S^*H}{AS(1 + 2\lambda_0^{-3})} \tag{2.38}$$

$$\text{stretching} \quad E^* = \frac{3\,S^*L}{A(1 + 2\lambda_0^{-3})}$$

Here H is the sample height; $K = \alpha/2\sqrt{3}$ is the inertia radius of square cross-section; A is area of the sample cross-section ($a \times \delta$, a and δ are sides of the section); the value of $H^2/36\,K^2$ can be neglected if $H < 0.5\,a$; S^* is form factor; $\lambda_0 = \dfrac{\lambda_1 + \lambda_2}{2}$; λ_1 and λ_2 are minimal and maximal values of H during compression.

The simplest dynamic test is the determination of elasticity by a ball rebound [336, 448] according to the expression $e^{-\pi tg\delta}$, where:

$$tg\,\delta = \Delta/\pi(1 - \Delta^2/4\pi^2) \tag{2.39}$$

Δ is logarithmic damping decrement:

$$\Delta = \ln(An/A_{n=1}) \tag{2.40}$$

The dynamic elasticity can also be determined by the expressions $(A_3/A_1)^2$ or $e^{-2\pi\ tg\delta}$. Elasticity on the bounce of a ball is defined by means of pendulums of Lupke, Schob, Danlop, Goodyir Hil and by using a trip meter [448]. Elasticity is also defined by throwing a ball from a certain height and measuring the height of the ball rebound [56, 448]. The modern theories [448] give the expression of the time function of elastic displacement α in the contact region at impact:

$$\alpha = \alpha_m\ \sin\ \alpha t$$

$$a_m = v_0\sqrt{\frac{m}{2\pi Rz}} \tag{2.41}$$

$$a = \sqrt{2\pi Rz/m}$$

where v_0 is the ball velocity at impact; m is the ball mass; R is ball radius; a is oscillation amplitude; z is dynamic limit of fluidity calculated according to the formula [448]:

$$z^5 = \frac{h_2^4}{\left(h_1 - \dfrac{3}{4}h_2\right)^3}\frac{mg}{R^3}\frac{10^4}{3^4 4^3 \pi^5}\frac{1}{\left(\dfrac{1-\mu_1^2}{E_1} + \dfrac{1-\mu_2^2}{E_2}\right)^4} \tag{2.42}$$

where h_1 is height from which the ball is thrown; h_2 is rebound height; E_1, E_2, μ_1, μ_2 are Young modules and Poisson coefficients of the material of the ball and sample. Shock stress spectrum in the sample can be obtained by means of the Fourier transform:

$$|S(\omega)| = \frac{2\alpha m}{a[1 - (\omega/a)^2]} \cos \pi\omega/2a \tag{2.43}$$

Let us consider the main methods of determining the dynamic parameters of elastomers.

Free oscillation method: the sample tested, with or without added mass, oscillates at natural or resonance frequency, which is defined by dimensions, and visco-elastic properties of the sample and total mass. From the rate of oscillation damping it is possible to determine the loss tangent. As the amplitude of fluctuation decreases during damping, the dynamic module of elasticity accordingly changes. Therefore, in this way its average value is defined. The Iertsly oscillograph and torsion pendulum are used in testing [336, 423, 434, 448].

The method of forced nonresonance oscillations: the sample is subjected to multiple cyclic strain of sinusoid form, which is forced by eccentric gear, the phase shift angle between the force and strain is measured with a RAPRA device [448], Alexandrov–Gaev device [219], rotation device, Reling device, or the electromagnet excited device [336, 423, 434, 448].

The method of forced resonance oscillations: the elastic sample oscillating with the device (total mass m) is subjected to cyclic force with constant amplitude F_0. The frequency varies until the maximal amplitude A_0 at the resonance frequency n_0 is reached.

The real and imaginary parts of the complex hardness are calculated as follows:

$$2\pi n_0 = \sqrt{S'/m} \tag{2.44}$$

The same devices for making measurements are used as those utilized in the previous method [336, 423, 448].

Methods based on propagation of waves and pulses: elastic wave or pulses are generated in the elastomer by piezoelectric, mechanical, acoustic, ultrasound or other means. Their parameters are registered and analyzed while propagating along the specimen [56, 219, 336, 382, 423]. The velocity of the elastic wave propagation is connected with the elasticity modulus as follows:

$$c = \sqrt{E/\rho} \tag{2.45}$$

$$E = \rho h^2 n^2 \tag{2.46}$$

where c is strain propagation velocity; $\lambda = c\tau$ is wavelength; n is frequency.

The dynamic characteristic of porous elastic materials is determined in the same way as for monolith elastomers [113, 448]. For example, to check the quality of a polyurethane sponge a test is conducted by throwing a steel ball (diameter $1.58 \cdot 10^{-2}$ m, mass $1.63 \cdot 10^{-2}$ kg) down from a 0.46 m height inside a pipe (diameter $3.8 \cdot 10^{-2}$ m) onto the parallel-piped sample (cross-section side 0.1 m, thickness $5 \cdot 10^{-2}$ m). The ratio of rebound height to throwing height in percent is called an elasticity parameter.

The physical and mechanical properties of elastic foam polyurethane (PPU) depend upon the apparent density, dimensions and shapes of pores, composite formula and production process [113]. Most foam polyurethanes with low seeming density are anisotropic and are stronger in the direction of foam lifting. The methods of investigation of composite materials are given the beginning of this chapter [293, etc.]. Mainly rubber–fabric composites have been investigated [113, 448].

2.4 The Apparatus and Devices for Measuring the Mechanical Characteristics of Elastomers

To determine the mechanical parameters of elastomers while they are interacting with a boundary layer the standard devices were used, as well as the devices constructed especially for these investigations.

2.4.1 Apparatus for Measuring the Static Characteristics

The application of elastic coatings for various tasks in hydromechanics demands knowledge and the definition of the mechanical properties of elastomers, i.e. describing behavior under the action of imposed static and dynamic loadings. Using the measurements of a large number of samples of various kinds of elastic materials is justified by that the results obtained, and can promote the establishment of general laws and their specific features. It is necessary to choose an optimum material for each particular hydrodynamic task, and for testing new devices for the measurement of mechanical characteristics of elasomers. The standard devices shown in Figure 2.9 have been used to define the static characteristics. Device VN 5704 is intended for the definition of the hardness of spongy elastomers. The sizes of its plates make it possible to measure products up to 0.7 m wide. The device has three replaceable measuring platforms with areas of 50, 100 and 150 cm^2 and it allows for the defining of the degree of compression in millimeters on loading replaceable cargoes of 2.5, 5.0 and 7.5 kg. The error of the indication is ± 0.01 mm. Device VN 5705ps is used for defining the hardness of rubbers, by measuring the difference of depths of penetration of a ball into the sample under the action of contact and basic loadings with its subsequent transfer into SI units. To increase the sensitivity of the vibrator of this device, only the hardness of solid rubbers is measured with this device. Device VN 5404 is intended for testing only spongy elastomers. The principle by which it

Figure 2.9
Standard devices for measurement of static characteristics of elastomers: *1* — samples of elastomers for research of their mechanical characteristics; devices: *2* — VN-5704, *3* — VN-5705 ps, *4* — VN-5404.

works is same as that of device VN 5704. Thus, the quantity of replaceable cargoes (25−6000 gr) is increased. The sizes of the investigated samples are adjusted for measurement with this device. Samples have the shape of cylinders with a diameter of 37 mm and a height within the limits of 5−45 mm. The standard supposes the various form and sizes of samples for carrying out measurements have the ratio of height to diameter up to 1.5. It is necessary to compare samples of identical shape and size.

2.4.2 Equipment for Measurement Elastomers being Stretched

The equipment developed by Babenko (in 1984) for measuring the properties of elastomers at small stretching under action of small forces is shown in Figure 2.10. It consists of a measuring device, separate block, sound generator GS-56, zero-indicator F-582 and millivoltmeter VS-36. In the separate block there is an electric symmetric bridge circuit (see Figure 2.11). The resistive-strain sensors R_3, R_6 are mounted on the tensometric unit. The zero-indicator output is connected to the millivoltmeter. The marked elastomer strip is fixed in clamps. The shift of the marker defines the deformation of the sample. The lower movable clamp is brought to an extreme upper position so that the sample was in an unloaded state. The upper clamp is connected to the drag balance. By means of potentiometers a minimal (close to zero) value is indicated by the millivoltmeter. Calibration of the drag balance is done by loading the drag balance with standard weights through the hole on top in the frame.

After calibration of the electronic equipment (see Figure 2.10) the strip of the sample of an elastic material 2 has been fixed in a neutral position. On the sample of an elastic material 2 the thin line has been put, located in a transversal direction, concerning moving the sample.

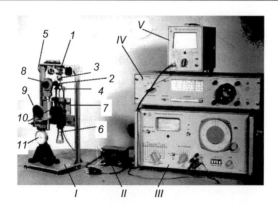

Figure 2.10

Equipment for measuring samples of elastomers at their stretching: *I* — measuring device, *II* — separate block of the amplifier, *III* — sound generator GS-56, *IV* — zero-indicator F-582, *V* — millivoltmeter VS-36. Measuring device I consists of: *1* — strain-gauge beam, *2* — elastomer strip, *3* — motionless clip, *4* — mobile clip, *5* — frame, *6* — micrometric screw, *7* — way, *8* — optical tubes, *9* — micrometric screw, *10* — way, *11* — micrometers.

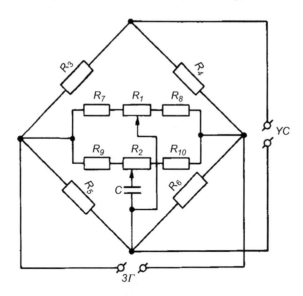

Figure 2.11

Electric circuit of separate block (tensometic bridge).

The sample of the elastomers was stretched by means of the micrometric screw 6. Change of length of the sample has been certain under indications of the micrometric screw and also by means of the optical system which was also fixed to the moving transversal lines on the sample in a longitudinal direction. Simultaneously on an oscillograph pressure in the sample which were defined by means of strain-gage beam 1 were fixed.

The optical tube was moved by means of its microscrew. A transverse marker was established on the bottom part of the investigated elastomer sample. The displacement of the optical tube was defined by means of its micrometric screw, and also by the micrometer *11*. Such threefold control allowed the precise defining of the shift of the elastomer strip. The stress arising in the elastomer strip was defined using the millivoltmeter, and then the magnitude of loading was defined from the calibration graph. Thus, dependences of deformation on loading were identified.

2.4.3 Apparatus for Compression Testing

The device was designed by Babenko et al. (in 1969) and made for use both in modeling experiments and for the measurement of elasticity of the skin of live dolphins. The device is, therefore, designed to be waterproof. The device dimensions are $0.1 \times 0.1 \times 0.04$ m. Figure 2.12 shows a photograph of the elasticity meter with the cover of the device removed. The case and the majority of the detailed parts are made of organic glass and current-insulating materials. Other parts are made of aluminum alloys. A scheme of the elasticity meter is presented in Figure 2.13 [305].

The device functions as follows: the measuring rod *16* bulges projections at certain distance from the plane, which is tangent to the limiters *13*. The rod *16* penetrates until the limit rods *15* touch the surface of the elastomer. The depth of penetration depends upon the elasticity of the surface. As the measuring rod is rigidly connected with the permalloy rod *11*, the depth is registered by measuring the inductance of the coil *8*. Thus, penetration of the rod into the surface and the force on the spring can be identified.

Figure 2.12
Elasticity meter.

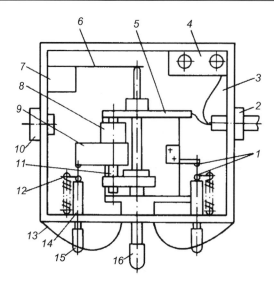

Figure 2.13

Scheme of the elasticity meter: *1* — electro contacts, *2* — hermetic outlet, *3* — frame, *4* — lamp holder, *5* — main gudgeon, *6* — spring, *7* — spring holder, *8* — induction coil, *9* — coil holder, *10* — plug, *11* — permalloy rod, *12* — spring, *13* — limiter, *14* — limit gudgeon, *15* — limit rod, *16* — measuring rod.

Spring *6* is calibrated before and after measurements. The magnitude of the force that presses in the measuring rod to the corresponding depth by means of the spring is defined from the calibrating curve. By processing these data, dependences of the deformation on stress is obtained. The device can be fixed in a frame and moved by means of the micrometric screw, as used in the other devices (see Figures 2.9 and 2.10). The device is universal; it can be used to make measurements of elasticity of elastomers located on finished manufactured products that of the skin or body of live organisms. With this device measurements of the skin elasticity of live dolphins have been made. These measurements had a standard character and were executed according to the requirements of standard devices intended for measuring elasticity of elastomers.

Hands are used to press the measuring rod *16*. Therefore, a special electric system to register the moment of touch was designed to balance the outer loads acting on the measuring rod. This consists of limit rods *15*, springs *12*, contacts *1* and lamps *4*. The limit rods *15*, are touched with the elastic surface, close contacts *1*. The lamps *4* are for registering the nominal position of the measuring rod *16*. The principal electric scheme for this device with the transformer output of elasticity measurement and the scheme of the limiter cut-off is presented in [305]. The oscillograph H-700, power source, sound generator and indicating device were for taking the measurements of elasticity of the surfaces.

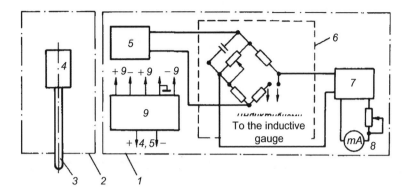

Figure 2.14

Functional scheme of apparatus to measure elasticity of elastomers: *1, 2* — electronic (1) and mechanical (2) parts of the apparatus, *3* — measuring rod, *4* — throttle with movable armature, *5* — RC-generator, *6* — measuring balance, *7* — amplifier of disbalance signal and detector, *8* — measuring device, *9* — power supply.

The design of this device makes it possible to carry out measurements of not only static but also dynamic characteristics. The device can also indicate oscillations of an elastic surface, if the frequency of oscillations is small. However, the measuring rod *16* has weight, which brings errors into such measurements. To reduce the inertial weight of the measuring rod, a new design of the device has been developed for measurement of elasticity of elastic materials. Modernization of this device has resulted from the need to carry out measurements on curvilinear surfaces in the laboratory and field conditions. Changes in the design include the following:

- The measuring rod has been constructed with removable spherical nosepieces of different diameters (2.5; 4; $8 \cdot 10^{-3}$ m). This makes it possible to take measurements at different unit loads on the material without disassembly of the device.
- The device has been made completely hermetical.
- The distance between restrictive cores has been reduced.
- The electronic module has been designed and constructed for off-line working.

The functional scheme of the electronic module is presented in Figure 2.14. The electronic block is made is such a way that it contains three functionally independent parts:

- the generator of sinusoid oscillations;
- the electric bridge scheme for measuring inductance with low quality;
- the electronic scheme to identify absence of balance of the electric bridge.

The parts are assembled as a printed circuit board. The unit can be fed by both an independent source and any suitable one.

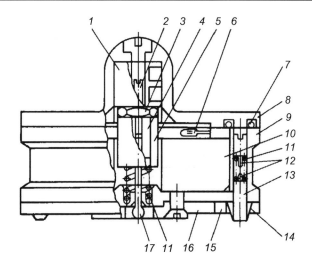

Figure 2.15
Scheme of the gauge — elastometer: *1* — restrictor. *2* — permaloy rod; *3* — check-net;
4 — dielectric brush; *5* — steel cylinder; *6* — light indicator; *7* — seal; *8* — cover; *9* — frame;
10 — inductance coil; *11* — springs; *12* — contacts; *13* — limiting rods; *14* — rubber film;
15 — pressing ring; *16* — support base; *17* — measuring rod.

In addition, a universal compact sealed device (see Figure 2.15) was developed by Babenko [66]. In principal, the device operates in the same way as the device presented in Figures 2.12 and 2.13. If spring *11* is set between the throttle *1* and the check-nut *3*, it operates in exactly the same way as the device presented in Figure 2.13. If spring *11* is set as shown in Figure 2.15, the measuring rod penetrates to the given depth with the force that corresponds to the voltage applied to induction coil *10*. The calibration curve is constructed, which shows the dependence of measuring rod force on voltage applied to the coil *10*. This method shows more accurate results than the method presented in [448]. Changing the inductance of throttle *1* controls the shift of the measuring core.

The compactness of the device (diameter 5.10^{-2} m) means that it is possible to establish it in a cowl and mount it on an elastomer while testing in a flow. The device can be used also for other measurements. For example if the measuring rod slightly supports the base, when carrying out tests it is possible to define parameters of fluctuation of the elastomers surface. For the control of the measurements executed by means of devices resulting from Figures 2.12–2.16. Characteristics of the same elastomers were determined also by means of standard devices. In Figure 2.16 three projections of the device and the sensor for measurement of elastomer elasticity are sketched. In this figure the following designations are used: *1* — case of the device, *2* — cover, *3* — the bottom internal clamping ring, *4* — the bottom external clamping ring, *5* — basic coil of inductance, *6* — screw from Plexiglas®, *7* — the case of the light indicator, *8* — light indicator, *9* — the external metal bush, *10* — the

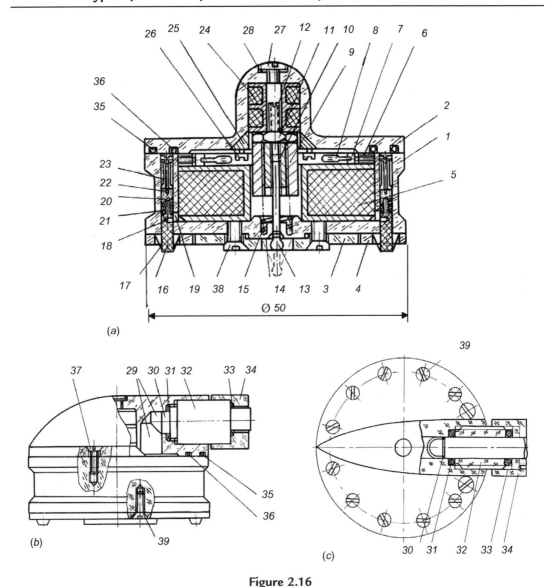

Figure 2.16

Projections of the drawing of the device — sensor for measurement of elasticity of elastomers:
a — main section, *b* — a side view, *c* — view from above.

internal dielectric bush, *11* — a lock-nut, *12* — permalloy rod, *13* — the measuring core, *14*,
17 — a rubber tight film, *15* — basic spring, *16* —restrictive rods (4 pieces), *18*, *19* —
apertures for electrical wire, *20*, *22* — contactors of the restrictive rods, *21* — springs
contactor of restrictive cores, *23* — the plug top of contactor, *24* — a throttle, *25* — collet,
26 — fixing screws, *27* — the screw-bibb for adjusting the permalloy rod, *28*, *31*, *33*, *35*,
36 — rubber sealing rings, *29* — the channel for electro posting, *30* — an electric cable,

32 – the plug for hermetic sealing of the electric cable, *34* – captive nut, *37, 38, 39* – fixing screws.

2.4.4 Apparatus for Measuring the Dynamic Parameters

Apparatus to determine the elasticity by the rebound of a ball

There are a number of methods to determine the dissipative properties of elastomers. The expression to determine the loss coefficient is known [423, 448]:

$$\psi = \frac{\Delta W}{W} = \frac{2\pi}{Q} 2\pi tg\delta \tag{2.47}$$

where ΔW is the energy absorbed by unit volume of the material; W is the energy stored by the unit volume of the material; Q is the parameter characterizing the increased amplitude at resonance.

The impact loading leads to the generation of oscillations in a wide frequency range, as occurs in real complex interaction. The device designed by Babenko (in 1979, see Figure 2.17) by analogy with the one presented in [448] is a hollow cylinder *1* made of transparent polished Plexiglas with a rigidly fixed measuring ruler *2* and viewfinder ring. The cylinder is fixed on the plane curricular base *3* for measuring on plane surfaces, and can be unfixed from the base to measure on curved surfaces. In the upper section of the cylinder, an electromagnet *4* (for holding the ball) and a plumb line (for checking the vertical) are placed. When the electromagnet is de-energized the ball falls onto the surface. The rebound height can be registered by means of the cursor ring, or the photodiode comb set parallel with the cylinder, or by registration of the impact using the piezoelectric transducer mounted under the surface.

The ball can be removed from the cylinder with a rod with a magnet nose. The device is not moved, so it is easy to make several measurements at the same point. Using the balls of different masses *m*, one can vary the energy ΔW. The parameter of damping properties is determined as follows:

$$K = 1 - h/h_0 \tag{2.48}$$

where h_0 is fall height and h is rebound height. For small impact energy, K can be connected to the loss tangent:

$$tg\,\delta = \frac{1}{\pi} \ln \frac{h_0}{h} = -\frac{1}{\pi} \ln(1 - K) \tag{2.49}$$

The apparatus for testing by the method of forced nonresonance oscillations

The device consists of the adjustable clown who rotates by means of the motor. Thus uniaxial compression is formed in time under the sine wave law. Through a rod and spring

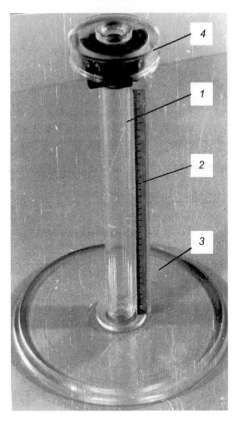

Figure 2.17
Device for measuring dissipative properties of elastomers on recoil balls: *1* — demountable tube from Plexiglas, *2* — measured ruler, *3* — plane curricular base, *4* — demountable case from Plexiglas with electromagnet coil.

(see Figure 2.18) this fluctuation is transmitted to the tested elastomer sample. The maximal compression of the sample does not exceed some percent. Thus, deformation of the sample becomes much less than deformation. Movements are fixed by the movements gauge by a strain sensor and strain measurement equipment.

The elastomer sample is placed between two parallel surfaces. The first surface (the lower one) is a strain gauge and the second is a spring thrust. Two types of calibration were carried out: cargo calibration (with help of draft cargo calibration *10*, see Figure 2.19) and calibration of travel of the sample (with help of gauge of travel *5* and indicator for calibrating the gauge of travel *8*, see Figure 2.19).

When testing was done with the spring *3* (see Figure 2.19), the strain amplitude depended upon the frequency: $\sigma_0 = const$, $\varepsilon_0 = f(\omega)$. When testing was done without the spring *3* (i.e. rod *2* fastened directly to the sample *4*), the stress amplitude depended upon the

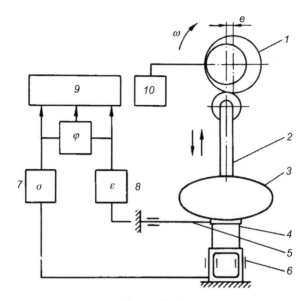

Figure 2.18
Arrangement of the device for measuring the dynamic parameters of elastomers: *1* — eccentric, *2* — rod, *3* — spring, *4* — elastomer specimen, *5* — displacement gauge, *6* — force gauge, *7, 8* — strain amplifier, *9* — two-ray oscillograph, *10* — gauge of turnovers.

frequency: $\varepsilon_0 = const$, $\sigma_0 = f(\omega)$. The gauge signals were registered at the two-ray oscillograph C1-18 and were photographed. While Figure 2.19 shows a photograph of the equipment for carrying out dynamic tests of elastomers developed by Korobov, Figure 2.20 gives an example of photographic data from the screen of an oscillograph [75–77 et al.].

The weighting frequency of the sample was varied by changing the rotation speed of the clown and was fixed on a frequency meter *17* during the moment of photographing. As a drive of the vibrator the direct-current electric motor was used, the speed of rotation of which was regulated by the voltage. A disk with thin slots along the radius was fixed onto the motor. The speed of rotation of the motor was determined using luminodiode, the light of which was registered when a slot in the disk crossed the luminodiode receiver. Impulses of the luminodiode indicated the speed of rotation. The working range of frequencies was $2\ \text{s}^{-1} \leq \omega/2\pi \leq 360\ \text{s}^{-1}$.

In tests, the degree of elastomer deformation varied due to the variation in heights of elastomer samples at constant amplitude of the rod motion, and due to changing the vibrator eccentricity. During dynamic tests the following parameters were defined:

- the complex module of elasticity $E(\omega)$, the real part of which, $E'(\omega)$, is the modulus of elasticity, and the imaginary component, $E''(\omega)\ \pi/2$, shifted on a phase the modulus of losses;
- phase angle $f(\omega)$ between the imposed loading σ and deflection of the sample ε.

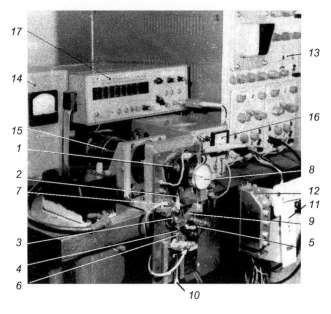

Figure 2.19

Equipment for researching dynamic characteristics of elastomers: *1* — block of clowns with the gauge of speed, *2* — rod, *3* — spring, *4* — sample of elastomer, *5* — gauge of travel, *6* — gauge of force, *7* — additional gauge of travel, *8* — indicator for calibration the gauge of travel, *9* — arm of indicator, *10* — draft cargo calibration, *11, 12* —strain boosters with power units, *13* — two-beam oscillograph, *14* — adjustable power unit of the electric motor, *15* — electric motor of a direct current, *16* — power unit with the amplifier of speed, *17* — frequency meter.

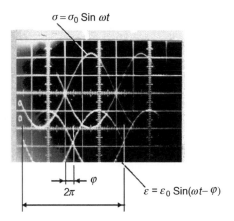

Figure 2.20

Oscillograms of change of applied force and deformation of the sample of elastomer.

If frequency of rotation of the clown is very small, the period of action of force is long and deformations can gain equilibrium values. Change of loading and deformation coincide in the phase. In this case, static tests for compression are carried out on the installation.

For rotation of the clown at high frequency, the period of action of the force is comparable with relaxation time. The stress and deformation do not coincide in the phase. At the beginning of the force action, the highly elastic strain is absent; only elastic strain occurs, which corresponds to Hooke's law. However, its share in the whole strain is very small. Highly elastic deformation due to relaxation develops with a delay so its magnitude mismatches the value of the pressure operating at the instant.

The lag in development of the deformation under action of force leads to that deformation reaching a maximum only when stress has already passed through the maximal value:

$$\sigma = \sigma_0 \sin \omega t$$
$$\varepsilon = \varepsilon_0 \sin (\omega t - \varphi)$$

(2.50)

where δ is the phase difference between stress and strain. (The dependence of strain amplitude on frequency and temperature was discussed in Section 2.3.)

Apparatus for Testing by Forced Resonance Oscillation

The inductance coil *10* (see Figure 2.15) or *5* (see Figure 2.16) being connected to the sound generator and spring *11* (see Figure 2.15) or *15* (see Figure 2.16) being set at the device top, the measuring rod will raise forced oscillation in the elastomer sample. Varying the frequency one can make tests at the resonance oscillations.

Apparatus for Testing by the Method of Wave and Pulse Propagation

The elastic oscillations generated by the indenter falling into the elastomer and on its surface allow for the determining of the intrinsic frequency of oscillation, parameters of surface waves, damping decrement, dynamic modulus of elasticity, its dependence upon the oscillation parameters, amplification factor, etc. Simultaneously, from the time of the ball rebound from the elastomer it is possible to find heights of consecutive rebounds and characteristics of elastomer elasticity depending on the height of the indenter [423, 448, 527]. The apparatus is presented in Figure 2.21 [56, 66]. For holding the steel balls the inductor *4* (see Figure 2.17) was used. The inductor was mounted in the upper part of the glass pipe, which was placed on the test plate.

The ball *13* falls from a certain height and generates the elastic waves. The ball *13* was kept at certain height by means of the electromagnetic coil *12*. By pressing the start button *11*, voltage was fed simultaneously from the block *10* to blocks of the electromagnetic

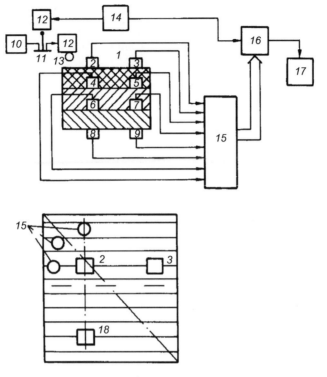

Figure 2.21

The equipment for defining the dynamic characteristics of an elastomer's distribution of waves and impulses: *1* — tested material, *2—9, 18* — piezoelectric detectors, *10* — power unit, *11* — contactor, *12* — electromagnetic actuators, *13* — steel ball, *14* — block of electronic start, *15* — amplifiers, *16* — magnetograph and loop oscillograph, *17* — oscillograph.

actuator *12*. Thus, the current in the block *12* was interrupted, and the steel ball fell down onto the investigated surface, causing elastic oscillations in the material. These oscillations were registered by means of piezoelectric detectors *2—9* and *18*, amplified on the amplifier *15* and transmitted on blocks *16, 17*. The block *16* was connected in parallel to loop oscillograph H-041 and magnetograph H-036. In addition, oscillograph C1-18 (block *17*) was connected in parallel. The signal from the start block turned on oscillograph C1-18, loop oscillograph and magnetograph. All the electronic parts and the technique for carrying out the tests were developed by Kanarsky [56, 66].

The electronic scheme, the basis of which is a biased multivibrator, was developed for controlling the scanning of oscillograph C1-18. The delay in the period of falling of the ball was smoothly regulated and adjusted in order that the scanning started at the moment of contact with the surface investigated. Signals on the oscillograph screen were photographed and then analyzed.

To investigate the anisotropic properties of the materials and the parameters of the surface waves, piezoelectric detectors were fixed outside and inside the composite, at the layer boundaries, in three perpendicular orientations to each other and at certain distances from each other. Such an arrangement of piezoelectric detectors made it possible to extend the number of measuring parameters and to control the parameters while testing and using elastomers. This is important aspect in the development of methods of investigating the properties of a composite and its layers, and for research on ways of increasing the performance of elastic materials capable of friction drag reduction.

A technique for defining the mechanical characteristics of elastic materials has been developed by Kanarsky together with Babenko [66]. The technique consists of generating oscillations inside a material, the parameters of which are measured in three mutually perpendicular directions. In this way the elastic-damping properties of materials are determined; also, the oscillation parameters are registered on the boundary between layers, and the parameters of oscillations are registered simultaneously.

By fading the amplitude of oscillation, damping and nonlinear properties of a material can be defined. The dynamic module of elasticity and anisotropy of a material can be determined by the difference of phases between various signals, and the frequency of natural oscillations of the material in various layers by the oscillation frequency. Dispersion can be defined by comparing frequencies in different places of the material.

When sensors are arranged in longitudinal and transversal directions, measurements of the ball impact are made either at the bisector of the angle, on which sides sensors are located, or sequentially in longitudinal and transversal directions. It is known that in multilayered composite materials oscillations concentrate: near the border of neighboring layers (in wave-guides); on the surface of a composite in the form of surface Rayleigh waves; and in the form of Stoneley and Love waves between layers. Elastic dynamic characteristics of layers of a composite depend on the parameters of oscillations, first of all, on phase and amplitude. Therefore, sensors arranged as described make it possible to identify and constantly control both dynamic characteristics of the composite and its reaction to external disturbances. The sensors allow the constant control of the kind and character of external vibrations and the laws of their dispersion inside the material, including in wave-guides. It enables the researcher to constantly receive information about the complex of mechanical characteristics of the composite depending on temperature and other physical parameters relating to external conditions, and also about variation of the material properties during long-term exploitation. Furthermore, if a composite has a regulation system for its mechanical characteristics, it allows the mechanical properties of the composite to be maintained under various conditions.

2.5 Construction of Elastic Surfaces

With regard to research on the hydrodynamic properties of elastic coverings, the most challenging problem facing engineers is that of drag reduction. Current methods of drag reduction are based in the known representations of a liquid flow about a rigid border. It is obvious that for a successful solution to the problem of flow about an elastic surface it is necessary to know, in an extreme measure, some of the interconnected problems. First it is necessary to understand the structural peculiarities of a liquid flow at a stream on elastic surfaces. It is also necessary to choose and define the peculiarities of the structure of an elastic surface for a specific research task. It is important to investigate the problem of receptivity—interactions of various disturbances near a surface. It is necessary to investigate the influence of the structural and mechanical characteristics of an elastic surface in relation to the problem of receptivity. It is also necessary to explore the structural control and mechanical characteristics of elastic surfaces. There are additionally other peculiarities relating to the problem of an elastic surface flow, and the complexity of solving these problems proves that they are interconnected.

What is known at the present time from analysis of theoretical and experimental research on flows on elastic surfaces is based on the pioneering theoretical work by Gromeco [216] and Kramer's experimental research (in 1957) [311]. The basic conclusions of the analysis consist of the following:

- All the research carried out is not interconnected and does not take into account the above-noted peculiarities of the problem.
- The defining of an optimum design for elastic surfaces has not been studied.
- The flow structure above the elementary designs of elastic has not been investigated (except for work by Babenko, Carpenter and Choi).
- The relationship between elastic surface design and the structure of flow has not been determined (except for work by Kramer and Babenko).
- Through experimental research the basic integral characteristics of the flow on elastic surfaces have been defined.
- In theoretical research basic attention has not been paid to researching hydrodynamic stability (except for work by Amphilohiev, Voropayev and several others).
- The mechanism of interaction of a flow with an elastic surface has not been determined (except for the research by Babenko).
- The problem of receptivity and the problem of control of a flow by means of an elastic surface has not been solved (except for work done by Babenko).

Modern understanding of hydromechanics makes it possible to solve the majority of the specified problems. In relation to experimental or theoretical research, the first question that needs to be asked is: what design of elastic surface should be used for reliable research and, therefore, what mechanical characteristics should the chosen surface have? Obtaining a corresponding result depends on the correct choice of an elastic surface. However, considering the unresolved problems noted above, neither earlier nor modern researchers have been able to answer these questions. The cycle of research for solving these problems is not yet complete.

In relation to this, a combined approach to solving the problem of choosing an elastic surface design for carrying out dependable experimental research has developed. It consists in two components:

- using the experience gained from experimental research performed so far;
- applying the hydrobionic approach.

In relation to the hydrobionic approach, it is thought that during evolution the structure of hydrobionts' dermal coverings have become optimal for each corresponding Reynolds number calculated on the characteristic speed of swimming. Research techniques have been developed and numerous experiments on live hydrobionts have been undertaken. The structure of their skin has also been studied. Based on the results of this hydrobionic research, some initial data for defining the design and mechanical characteristics of elastic materials necessary for solving the above-mentioned problems have been obtained. Furthermore, the hydrobionic approach has provided an understanding of the structure of hydrobiont skin coverings and their hydrodynamic value. There have also been new hydrodynamic problems for studying laws of drag reduction.

The main result of this hydrobionic research is a suggested program of experimental research on the flow on elastic surfaces as follows:

- Research on structure of flow in the boundary layer at all modes of flow, including comparative experiments on rigid and elastic surfaces.
- Research on the problem of interaction of various disturbances on flat and curvilinear rigid and elastic surfaces.
- Research on methods of controlling the characteristics of a boundary layer by means of elastic surfaces.
- Research on the integral characteristics of the flow on elastic surfaces.
- Research of methods of controlling coherent vortical structures in a boundary layer.
- Research on combined methods of controlling the integral characteristics of a boundary layer.
- Development of corresponding designs of elastic surfaces and technologies for their manufacture.

The first of these tasks is to investigate the structure of a boundary layer at all modes of flow on a rigid surface, and then to undertake similar comparative experiments using chosen designs of elastic surfaces. "Comparative experiments" means carrying out experiments using the same setup, under the same experimental conditions, and on the same day. The hydrodynamic stand of small turbulence, (see Figure 1.20, 1.21) has been developed to achieve these tasks. Some results from research on the structure of flow on a rigid surface obtained by means of the tellurium method (see Figure 1.23, 1.24−1.32) and laser anemometer (see Figure 1.72, 1.74) are also described in the previous chapter.

When the initial research on elastic surfaces was undertaken, the results of hydrobionic research on the skin of high-speed hydrobionts were not available. Therefore, the choice of elastic surface design was made on the basis of the available experience of using elastic materials for the absorption of vibrations. In the beginning, experiments focused on the laminar and transition boundary layers. Speeds of flow in the hydrodynamic stand were small, up to 0.5 m/s. Therefore, the following assumptions guided the choice of elastic surface design.

The theoretical analysis of the flow on elastic surfaces provided some recommendations for the choice of material. Preservation of stability of a laminar boundary layer in the flow over an elastic surface is achieved if the hardness of the surface is high. If it is sufficiently high, the match condition is satisfied for the speed of propagation of the surface wave on the elastic surface C_0 and speed of the instability wave from the point of view of Tollmien−Schlichting stability, corresponding to the second branch of the neutral curve. If the rigidity of the elastic surface is not great, speed C_0 should be big enough to prevent occurrence of class *B* waves, and the decrement be small enough to prevent development of Reynolds stress due to class *A* waves, according to the terminology by Benjamin. However, it is practically impossible to find a real elastic material that obeys such theoretical recommendations. The choice of the most dangerous Tollmien−Schlichting wave is also questionable. Detailed results of experimental investigations of hydrodynamic stability on elastic plates are represented in work by Babenko [31, 305]. The basic recommendations are that the phase speed of the surface wave arising in an elastic coating under the action of pulsations of a boundary layer should correspond approximately to phase speed of the basic energy-carrying disturbances in a boundary layer at a corresponding flow regime. The energy of pressure pulsations in a boundary layer should be sufficient to cause fluctuations in an elastic material. If the pulsating energy of a boundary layer is insufficient, the elastic surface behaves like a rigid surface. If this energy is great, depending on mechanical characteristics of the elastic material, there will be oscillation of large amplitude on its surface. Thus, amplitude of oscillation in an elastic material should be small and not exceed the size of the maximum permissible roughness. If the elasticity of an elastic material is small there can be plastic waves of large amplitude on its surface, such as are fixed in the experiments of Gad-el-Hak [195, 197, 198].

There is considerable experience of using elastic materials for the absorption of vibrations and noise. The physical character of the interaction of disturbances and deformations caused by them in an elastic material differs in comparison with the flow in the elastomer's boundary layer. However, results of experiments where elastomers are used for the absorption of vibrations provide some recommendations for the choice of elastic material design: elastomers should have a multilayered design and longitudinal order.

The following design of elastic plates has been chosen, taking into consideration the models of elastic materials described in Section 2.1, the above-stated recommendation and the small flow speeds necessary for researching laminar and transitional boundary layers, and also to provide consistency with results of some previous experimental and theoretical research. Figure 2.22 sets out the design of two kinds of elastic plate. The length of these plates is less than length of the working section of the low turbulence hydrodynamic setup (i.e. 3 m) to ensure their easy installation (see Figure 1.20). The width of the plates is 1–2 mm less than the width of the working section, thereby creating a gap around the edges of plates in order to remove the angular boundary layer along the working section length. The plate (see Figure 2.22a) is designed as a frame consisting of two longitudinal channels connected by staggered rods. The duralumin plate is mounted on that which was used when a thin elastic film was put on the elastic plate.

From four sides the membrane is fixed in the clamping cleat, having clip screws. On one side the membrane is fixed rigidly on the channel to the frame, and on the three other sides the membrane, by means of these clip screws, is fastened to the frame in such as way that it is possible to change the membrane tension. At various distances between the duralumin plate and membrane, various thicknesses of water or air is supplied under the membrane.

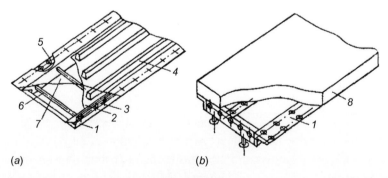

(a)   (b)

Figure 2.22
Design of membrane (*a*) and visco-elastic (*b*) surfaces: *1* — frame, *2* — duralumin plate, *3* — clamping rod, *4* — strips of foamed polyurethane, *5* — clamping plate, *6* — cores of a frame, *7* — membrane, *8* — sheet of foamed polyurethane.

This design makes it possible to investigate so-called "simple" and "complex" membrane surfaces. In the second case, three 12 mm-wide and 11 mm-high propping-up strips are installed between the membrane and plate, which interfere with flutter oscillation in the membrane. The design allows for the changing of the size of these strips, their material and quantity. The top horizontal plane of a shelf of channels slopes, so that the membrane touches the frames at a distance of around 5 mm. Forward and back faces of the frame are made in the form of thin plates established at an angle, so the membrane in this part of frame adjoins to the frame at a width of about 1 mm.

The third kind of design is the frame composed from angle bars and strips with multiple apertures (see Figure 2.22b). An elastic plate made of foamed polyurethane was sewn to the frame. Above the elastic plate a frame made of three duralumin corners was mounted. This frame was used when a thin elastic film was put onto the elastic plate.

Both frames have basic clip screws to establish the elastic plates in the working section horizontally and flush with the surface of the convergent tube in the area of its joint with the plate placed in the working site. Such a design makes it possible to establish a uniform horizontal elastic surface in the working section and to change elastic plates during tests— using a wide range of thicknesses and materials. These designs are well described by the theoretical models presented in Section 2.1. Figure 2.23 shows a photograph of the three kinds of elastic plates.

One of the important mechanical characteristics of an elastic material or membrane is the tension in its surface layer. It is possible to define the tension theoretically by means of a resistive-strain sensor pasted on the surface of an elastic material. However, gluing a resistive-strain sensor introduces essential errors to the results of the measurements due to the glutinous film. A device has been developed (see Figure 2.24) that defines the dependence of the tension on elasticity, measured by the developed elasticity-measuring instrument (see Figure 2.12). The device is made in the form of a rigid frame 1 m long and with a width equal to that of an elastic coating. On the one side the membrane is firmly fixed by a plate put on the frame of the device. On the three other sides the membrane is fixed in a clamping cleat of the same design as for the first kind of membrane coating (see Figure 2.22a). Loads are suspended and are of such a size that equal loading acts on

Figure 2.23
Design of the membrane (*1*) and visco-elastic surfaces: without frame (*2*) and with frame (*3*).

Figure 2.24
Adaptation for definition of dependence of elasticity from a membrane tension.

the unit length of the membrane. Measurement of elasticity of such tense membrane is made and the tension in a unit length of membrane is calculated. The magnitudes of loads can be varied and a calibration curve obtained for tension versus load. With this device it is possible to define similar dependences for monolithic or layered composite elastic materials for thin samples.

Experimental research on the characteristics of the boundary layer on specified designs of elastic coatings has provided new results relating to both rigid and elastic plates. Visualization and measurement of quantitative characteristics has shown essential differences in the structure of flow near an elastic border in comparison with a rigid plate.

Further research was necessary to investigate performance at a wide range of flow speeds and viscosity. A complex of experimental installations and new designs of elastic plates have been developed to accomplish this task [66]. The above-considered designs of elastic surfaces are the simplest and do not have practical value. However, these designs are convenient for comparison with theoretical models of elastomer oscillation. Unfortunately, such designs have a narrow range of frequency characteristics (relaxation times.) They do not perceive range of energy-carrying frequencies of fluctuation of a boundary layer in a spectrum of disturbing movement at the increase of a flow velocity. Dynamic characteristics of a boundary layer lead to an essential increase in the deformation of the elastic material.

The reasoning behind the plastic plate designs are as follows. Elastic surfaces should be of monolithic type and composite. It is known that each kind of a monolithic elastic material has a spectrum of the eigenfrequencies, in which range the elastomer most intensively absorbs the energy of external oscillations. Therefore, composite materials expand the spectrum of fluctuations that are intensively absorbed. The design of elastic coatings should be based on the physics of flow of coherent vortical structures in the near-wall areas of a boundary layer. The character and peculiarities of the development of coherent vortical structures in a transitional boundary layer have been investigated. There is also sufficient understanding of development of coherent vortical structures in a turbulent boundary layer. However, to define the influence of elastic surfaces upon the turbulent flow structure it is necessary to design corresponding elastic surfaces. Therefore, it is necessary to have more monolithic elastic plates with various mechanical characteristics and various kinds of composite elastic plates.

On the basis of the research on the interaction of disturbances in the boundary layer represented in the previous chapter and results of the bionic approach, it became obvious that with an elastic surface of a certain structure it is possible to intensively and purposefully influence the turbulent flow structure and other modes of flow, such as separation flows. By analogy with the structure of high-speed hydrobionts' skin, it was decided to createan orderliness in the longitudinal and cross-sectional directions in the design of elastic materials. Ideas about a wave-guide character of a boundary layer and the mechanism of a complex wave-guide [64] were also used as a basis for the development of elastic coating designs. Again, by analogy to the structure of the skin of dolphins and swordfish, systems by which the temperature and pressure inside elastomers could be actively controlled have also been added to elastic material design. Skin coverings of hydrobionts are supplied by sensor controls, which provide an automatic control system for the mechanical and geometric characteristics of the skin. This is important because the changes resulting from this control can essentially change the structure of flow. When designing long coatings it is necessary to consider the distribution of various power loadings from the environment along the coating length [36]. It is necessary to also consider, during real motion of the body, what external disturbances and disturbances acting from below the body is subjected to. From the case on which the elastic covering is fixed, it will influence the elastic coating. In view of this, it may be necessary to change the design of an elastic coating and its characteristics along the body. The background of flow is also an important consideration for the flow over an elastic coating. It means that the flow over subsequent segments of a coating has a structure that has been changed by previous segments. In each section, the coating will have optimal calculated and designed characteristics and cannot be developed according to the schemes and methods for the flow on a rigid surface.

In considering such a complex multi-purpose problem, it was necessary to develop a design that would allow experimental research of the same elastic plates in various experimental installations and in a wide range of flow speeds. The design of an elastic coating should effectively damp the disturbances acting from the external side of flow and from the rigid case. The elastic coating should have a system to regulate its mechanical and geometrical characteristics and allow mounting sensors of control inside of the coating. Babenko has developed variations of such a design of covering. In the future it will also be essential to develop not only adjustable elastic coatings, but also, by analogy with hydrobionts, active elastic coatings that have a system of automatic control of their dynamic characteristics.

Together with Korobov, Babenko has developed the standard design of the panel to carry out tests of elastic coatings in various experimental installations. Figure 2.25 is a

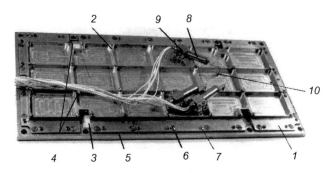

Figure 2.25
Design of the panel for fastening an elastic surface: *1* — panel, *2* — edges of rigidity,
3 —investigated plate, *4* — screws, *5* — framework, *6* — fixing screws, *7* — clip screws, *8* — sensor,
9 — case of the sensor, *10* — mobile axis of the sensor.

photograph of the bottom view of this panel. In this panel a reference rigid plate and elastic plates with various structure and thickness were fixed.

Rigid or elastic plates *3* are fastened to the panel *1* with edges of rigidity *2* by means of the screws *4* located where the edges of rigidity cross. The face sheet (from a flow on a plate in a stream) of the investigated plate presses on the panel *1* of frame *5* via a system of pairs of screws consisting of fixing screws *6* and clip screws *7*. These screws achieve the common plane of the test plate and an external plane of the frame so that there was practically no gap between them (common plane is shown by arrow B in Figure 2.26 (see Figure 2.26)). A base plane is the bottom surface of the frame, which is established on a basic frame of a drag balance (see Figure 2.27 and see Figure 2.26). Various sensors *8* for measuring the characteristics of elastic plates were fastened from below to the panel *1*. Figure 2.25 shows three sensors of mechanotron 6MX-1B, through which fluctuations of the external surface of an elastic plate are measured. The mechanotron (electronic lamp with electrode mechanically controlled from outside) joins in the electric circuit of the direct-current bridge.

In Figure 2.26, the basic frame of drag balance is shown. They consist of two arms connected by a crosspiece. Four screws of the panel with coatings are fastened to this frame. The panel is established flush with the bottom of the working section. For this purpose, for the installation of the measuring panel (the insert) a section is cut out of the end of the working section of the hydrodynamic set-up (between $2.0\,\mu$ and $2.5\,\mu$), in the bottom made of Plexiglas.

Figure 2.26 and (see Figure 2.27), illustrate the fastening of an elastic surface in the panel and the installation of the panel in the hydrodynamic set-up. The frame of panel *17* is fastened to the basic frame *3* with four screws *14* (see Figure 2.27 and also Figure 2.26), which is fixed by means of spring plates *8*. Tangential stress of the flow indicates the friction drag on the plate through the mechanotron or sensitive tensiometric beam *6*. Thus,

Figure 2.26
Photograph strain-gauge suspension brackets in the hydrodynamic stand: *1* — basic frame strain-gauge suspension brackets, *2* — design of the bottom of a working site, *3* — surface of the bottom made from organic glass, *4* — working site of the hydrodynamic stand.

there is simultaneous measurement the kinematic characteristics, by means of a heat loss anemometer, and the integral characteristics, by means of tensiometric beam. The same elastic plates were investigated on the panel *4* in the wind tunnel and in the hydrodynamic channel at greater speeds.

Using the described panel, the kinematic characteristics of the boundary layer over elastic plates have been investigated at various modes of flow. In view of the statements made above, various designs of elastic plates have been developed. All plates were composite and were pasted into a thin duralumin plate *5*. For some plates the layer *1* was made of thin rubber film (see Figure 2.28a, b), for other plates layers or longitudinal strips of electro-isolated conductive graphitic fabric were placed inside (layer *3* in Figure 2.28b, layer *2* or *3* in Figure 2.28c, layer *2* in Figure 2.28e). In the plates presented in Figure 2.28c, layer *2* was made of longitudinal elastic strips, and at variant *d* the external layer *1* was of thin longitudinal strips of the elastic material that was glued. In the plates of variant *e* the external layer *1* is designed in the form of reversed riblets glued to layer *2*.

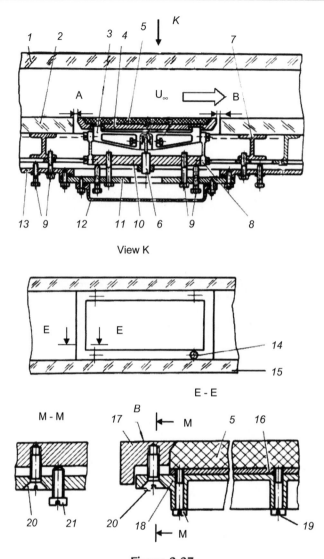

Figure 2.27

The scheme for accommodation strain gauge suspension in the hydrodynamic stand: *1* — cover of the working site, *2* — bottom of the working site, *3* — basic frame strain gauge suspension, *4* — panel for fastening an elastic surface, *5* — elastic plate, *6* — strain gauge beam, *7* — skeleton of the frame of the bottom of the hydrodynamic stand, *8* — spring suspension, *9* — system clip screws, *10* — plate strain gauge suspension, *11* — support plate of strain gauge suspension, *12* — cover strain gauge unit, *13* — case of a working site, *14* — four screws for installation of an elastic insert in a working site, *15* — lateral glasses of a working site, *16* — duralumin plate to which the elastic plate is pasted, *17* — framework of the panel, *18* — panel, *19* — screws for fastening an elastic plate to the panel, *20* — fixing screws, *21* — clip screws.

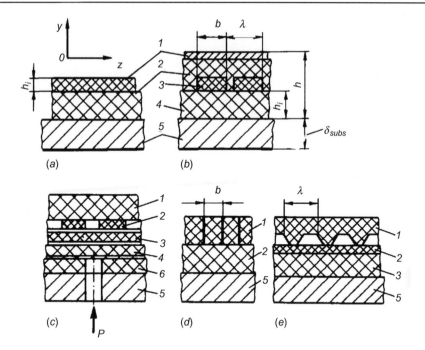

Figure 2.28

Schemes of composite elastic plate cross-sections: *1–4*, and *6* — elastomer layers;
5 — metallic plate.

Thus in one case the layer of turned riblets was mounted streamwise and in others it was mounted across the flow. At low speeds one or two external layers interacted with the flow. At greater speeds, underlying layers also interacted with the flow. The bottom layers played the important role of absorbers, for the reduction of the fluctuations acting on the plate from below and caused by vibration of the installation or the foundation of laboratory. The middle layers also absorbed various external disturbances of large intensity. Tables 2.3 and 2.4 represent geometrical parameters of the elastic composite plates sketched in Figure 2.28.

Plate 2 is shown in Figure 2.29; it is made according to the scheme in Figure 2.28 from foamed polyurethane and also shown in Table 2.3 at numbers 3 and 4. The red conductive wire is connected to the cross-section trunk connecting the conductive strips. In Figure 2.29 fabric is visible as dark color through the rubber film.

Tables 2.3 and 2.4 present the plate structures. All plates, except for 3 and 8*a* in Table 2.4, have inner strips oriented along the flow. Every plate was made in duplicate, the denoted by the number with the suffix "*a*." The first group of elastic plates was investigated first in the hydrodynamic stand of small turbulence [31, 32, 305] and in the flow by means of strain

Table 2.3: Geometric parameters of elastic plates: group 1.

Plate Number	Dimensionless Thickness of Plastic Plates ($10^2 h/l$)	Quantity of Layers	Layer Number	Dimensionless Thickness of i-Layer ($10^3 h_i/l$)	Material of i-Layer	Construction of Plates, Remark Rigidity Above
1, 1a	2	2	1	0.12	RF	CC
			2	19.9	FP-1	CC according to Figure 2.28a
2, 2a	2	2	1	0.12	RF	SAA
			2	19.9	FP-2	SAA, Rigidity above
3, 3a	1.8	4	1	0.12	RF	SAA
			2	6	FP-2	SAA
			3	2	CCF	LS with width $b = 2\,h$ and pitch $\lambda = 2.5\,h$; is pasted with RF
			4	12	FP-2	CC according to Figure 2.28b
4, 4a	1.8	4	1	0.12	RF	CC
			2	12	FP-2	SAA
			3	2	CCF	LS $b = 3.7\,h$; $\lambda = 4.5\,h$, is pasted w/RF
			4	6	FP-2	CC, SAA as plate 3, 3a
5, 5a	1.2	3	2	6	F P-2	CC
			3	2	CCF	CC pasted w/RF
			4	4	FP-2	CC according to Figure 2.28b

6, 6a	1.6	3	2	7	FE-1	CC
			3	2	CCF	LS $b = 2\,h$; $\lambda = 2.5\,h$ pasted w/RF
			4	7	FE-1	CC according to Figure 2.28b
7, 7a	1.6	3	2	7	FL-1	CC
			3	2	CCF	CC pasted w/RF
			4	7	FL-1	CC according to Figure 2.28b
8, 8a	1.8	4	1	6	F-2	CC
			2	2	FP-2	$B = 1.2\,h$; $\lambda = 2\,h$
			4	2	CCF	CC pasted w/RF
			6	8	FL-1	CC according to Figure 2.28c
9, 9a	2.6	5	1	10	FE-2	CC
			2	4	FP-2	$B = 1.2\,h$; $\lambda = 2\,h$
			3	2	FP-2	CC
			4	2	CCF	CC; layers 3, 4 are pasted w/RF
			5	8	FL-1	CC according to Figure 2.28c
10, 10a	1.2	2	1	8	FE-1	For plate 10 $b = 0.85\,h$ 10a: CC
			2	4	FE-1	CC according to Figure 2.28d
11, 11a	1.2	3	1	1	Rough film	CC, roughness in one direction
			2	5	FP-2	CC
			4	6	FP-2	CC according to Figure 2.28b

CC – continuous coating; CCF – conductive fabric; FE – foam elastic; FL – foam latex; FP – foam polyurethane; FP – foam polyurethane; LS – longitudinal strips; PU – polyurethane; RF – rubber film; SAA – same as above.

Table 2.4: Geometric parameters of elastic plates: group 2.

Plate Number	Dimensionless Thickness of Elastic Plates ($10^2 h/L$)	Quantity of Layers	Layer Number	Dimensionless Thickness of i-Layer ($10^3\, h_i/L$)	Material of i-Layer	Construction of Plates, Remark Rigidity Above
1	2	1	2	20	FE-3	CC
1a	1.2	1	2	12	FE-3	CC according to Figure 2.28a
2; 2a				Same as in Table 2.3		
3	2.2	3	1	8	PU-1	CC, bulges and hollows are oriented transverse to flow, $\lambda = 1.8 \cdot 10^{-3}$ m, according to Figure 2.28b. Layers 2 & 3 — the same as for plate 3 & 3a in Table 2.3
			2	2	CCF	
			3	12	FP-2	
5; 5a				Same as in Table 2.3		
6; 6a			Same as in Table 2.3. The outer surface follows to the form of the strips of CCF			
7; 7a				Same as in Table 2.3		
8a		Same as 8; 8a in Table 2.3, but in place of two upper layers the PU-2 layer is glued on the CCF, dimensions of the layer are the same as by PU-1 of plate 3, bulges/hollows are transverse				
9		Same as 9 in Table 2.3, but in place of three upper layers the CC of FE-3 is glued, $h_i/l_E = 8 \cdot 10^3$				
9a				Same as 9a in Table 2.3		
10; 10a				Same as in Table 2.3		
11	1.6	3	1	8	PU-1	CC, bulges-hollows are oriented along the flow, $\lambda = 1.7 \cdot 10^{-3}$ m, $b = \lambda = 0.21\, h$
			2	2	CCF	CC tested w/RF
			3	6	FL-1	CC according to Fig. 2.28d
11a		Same as plate 11, but the third layer — CC of FPPU-3 bulges/hollows are oriented along the flow				

CC — continuous coating; CCF — conductive fabric; FE — foam elastic; FL — foam latex; FP — foam polyurethane; LS — longitudinal strips; PU — polyurethane; RF — rubber film; SAA — same as above.

Figure 2.29
Photograph of various kinds of elastic plates with different methods of regulation of
their mechanical characteristics: *1* — pressure and heating of longitudinal conductive strips;
2 — heating of longitudinal conductive strips; *3, 4* — heating of a conductive continuous layer;
3 — elastic plate established on the panel with a edging.

sensors I and II. Later the plates were modified (see Table 2.4) and investigated mainly in
the airflow. Some plates from this group were investigated also in water flow.

Figures 2.29–2.31 show photographs of some elastic plates made of special devices. Plate 1
is shown in Figure 2.29; it is made according the scheme in Figure 2.28c, basically from
latex—this material having average density, a small amount of closed pores and a smooth
external surface.

Thin layers of foamed polyurethane *6* were pasted to duralumin plate *5* (see Figure 2.28),
which resulted in a thin rubber film with small preliminary stretching from above. From
above the same rubber film and a further a thin layer of latex *4* was pasted. Finally a layer
of rubber film and a continuous layer of a thin conductive graphitic fabric *3* were again
pasted along the whole plate. On the end faces, the fabric was clamped in thin brass grids—
conductive trunks, and onto the end faces of these an electric cable was soldered. On the
above fabrics *3* the thin rubber film was pasted, and on it longitudinal strips of latex *2* were
pasted, onto which the continuous sheet of latex *1* on the size of duralumin plate 5 was
pasted. In Figure 2.29 the latex strip and conductive fabric with the cross-section bus-bar
(dark color) are visible from above on plate 1. Also visible from above is the conductive
wire and an elastic tube in a cavity between longitudinal strips 2, which is the means by
which the liquid moves at constant or variable pressure. Numbers 8 and 8a designate this
plate in Table 2.3.

Figure 2.29 shows plate 3 being assembled on the panel in the way it is established in
experimental installations. The elastic plate 3 corresponds to number 7 and 7*a*, and plate 4
to number 8 and 8*a* in Table 2.3.

Figure 2.30
A photograph of various kinds of elastic plates: *1* — elastic monolithic cylinder with longitudinal conducting delays; *2* — surface with longitudinal cracks for injection of a liquid in a boundary layer; *3* — no controllable plate *10* under the scheme *d*, Figure 2.28; *4* — plate with the conducting delays located on its surface; *5* — plate with narrow conducting strips on its surface.

Figure 2.31
Photograph of the external layer of an elastic plate made according to scheme *e* of Figure 2.28.

In Figure 2.30, plate 3 corresponds to number 10 and 10*a* in Table 2.3. Figure 2.31 is a photograph of the external layer of plate 11 and 11*a* mentioned in Table 2.4 and shown in Figure 2.28e [33].

In Figures 2.32–2.36 schemes of elastic composite plates, which generalize the above-mentioned considerations for the structure of elastic plates, are given. In Figure 2.32

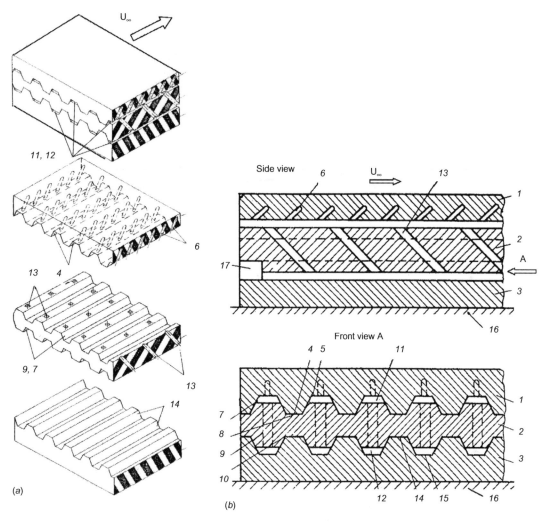

Figure 2.32

Damping covering: *1* — top layer; *2* — middle layer; *3* — bottom layer; *4, 7, 9, 14* — ledges in the top, middle and bottom layers; *5, 8, 10, 15* — corresponding hollows; *6* — blind inclined channels of the top layer; *11, 12* — longitudinal cavities; *13* — through inclined channels of an middle layer, *16* — external surface of a streamline body [33].

the design of the elastic plate consists of three layers, which incorporate the principles used as the basis of other developed elastic plate designs [33].

The basis of this design is the structure of a dolphin skin, which has some layers with different mechanical characteristics, two membranes and two horizontal layers of the circulatory system for the control of skin temperature. The details of the structure and work of this coating are described in the patent [33]. All three layers of this coating have various mechanical characteristics and are connected with each other so there are longitudinal *11*, *12* and cross-sectional *6, 13* channels filled with a viscous liquid. Vertical channels *6* of the external layer also have a variable slope along the body length. Vertical channels of the middle layer *13* have the slope opposite to that of the outer layer. In response to disturbances on the external layer of the coating the system of channels of the external layer bends and works as an absorber. Thus, the liquid from them is pressed out and chokes the top longitudinal channels. On increasing external loading the pressure wave is transferred to the underplaying layer of the coating, leading to an overflowing of the liquid in the second vertical channels and horizontal layers. Thus, throttling increases in channels. It has the effect of damping external disturbances.

The system of connected layers is such that external disturbances cause surface waves in the contact planes of elastic layers. As a result, not only a reduction of energy of external disturbances occurs, but also a redistribution of their energy pulsating components. The design of the coating results in the stabilizing and redistributing of the energy of external spatial disturbances. Thus, the entire coating consists of a closed section, and in each section there are the cross-section horizontal channels promoting the decrement of spatial external disturbances. The weight of the middle layer of the coating is greater than that of other layers. This middle layer plays the role of absorber of intensive disturbances acting not only from the external side of the coating but also from the case on which the coating is fixed.

It is known, that the vibration caused by the work of the power installation is transferred to the case in places of fastening. Regulation of the characteristics of such a coating is carried out by the variable pressure imparted to the liquid filling internal channels. In the case of damage to the external layer, liquid with auto-vulcanizing properties can be used in the channels. To improve a coating's performance, it can be divided into sections. This is done in the way described in Section 2.1, according to laws of distribution of various loadings along a body.

Figure 2.33 outlines the design of an elastic plate that intended for use in the combined reduction [34]. The basic differences from the previous design consist of the following: The external layer *1* is made of a permeable material and, in addition, a permeable thin elastic grid *15* is placed on top. In the external layer, inclined channels *6* have a greater area. The middle layer is thinner, and its cross-section channels *9* are not inclined but vertical. Cross-section channels *12*, which lie in a transverse direction between the second and third layers are placed only in the third layer and closer, and each of these channels is connected to a source of pressure *P*. The internal cavities of this coating are filled with a polymer solution with a concentration 2000 ppm. Along the surface of a body, pressure is distributed according to the body shape: there is overpressure over one part of the body while there is underpressure over other part. It is according to such a distribution of

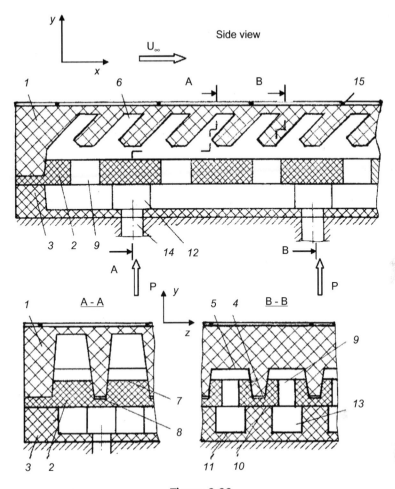

Figure 2.33

Damping covering: *1* — top layer; *2* — middle layer; *3* — bottom layer; *4, 7, 10* — ledges in layers *1, 2* and *3*; *5, 8, 11* — corresponding hollows; *6* — blind inclined channels of the top layer; *9* — through inclined channels of a middle layer; *12* — cross-section channels; *13* — longitudinal channels; *15* — permeable external grid; *16* — external surface of a streamline body; P — the main supply of pressure [34].

pressure that the elastic materials of a covering will be loaded, although there are divisions into separate sections. According to the distribution of pressure along a body, each section of the coating is connected with an internal source of pressure.

Thus, the pressure in each section is such that there is an insignificant difference compared with external pressure and the insignificant part of the polymeric solution is squeezed out through the external permeable elastic layer and the grid on the surface. Alternatively, due to underpressure the polymer solution is sucked from the internal areas of the covering, so that the rough external grid is in the layer of polymer solution. Throttling through internal channels both an external permeable layer and a grid will lead to stretching of the polymer molecules, increasing its efficiency. Such a system of polymer solution injection provides a low consumption. As in the previous design, control of the coating properties is achieved by changing the temperature of the liquid in its channels. In the presence of a corresponding pressure difference, injection or suction of the liquids into the boundary layer occurs in various sections. For example, in initial sections the pressure and release of a polymer solution increases, and in the next sections the polymer solution is partially sucked away. Therefore, this system is economical with polymer solution. The design of the coating means that it is divided into sections both in transverse and longitudinal directions. Thus, in the longitudinal direction it is possible to isolate some longitudinal channels. It is possible to provide variable pressure inside the elastic covering in both longitudinal and in transversal directions according to the magnitude and distribution of external pressure along the body. Thus, there is a control of longitudinal coherent vortical structures of a boundary layer and even greater economy of a polymer solution. Drawings have been developed for manufacturing press-matrixes for this design of elastic coating.

Figure 2.34 illustrates the design of an elastic plate that is an update of the design of plate shown in Figure 2.33 [44]. This new version was developed by Korobov, and in its middle layer it has ledges that enter into corresponding hollows of the bottom layer. In addition, vertical channels on the middle layer are inclined, as well as those in the coating in Figure 2.32.

Figure 2.35 sets out the design of an elastic plate that differs from the previous design (see Figure 2.33) in its two-layer structure; there is no middle layer. Longitudinal strips of a conductive material *3* are located in longitudinal channels *2* of the external layer *1*. On the section edges these strips are connected by cross-section strips *8* to form an

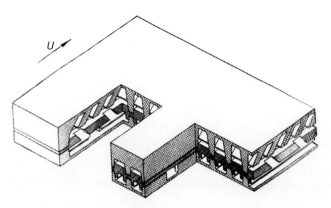

Figure 2.34
Scheme for the updated version of the plate shown in Figure 2.33.

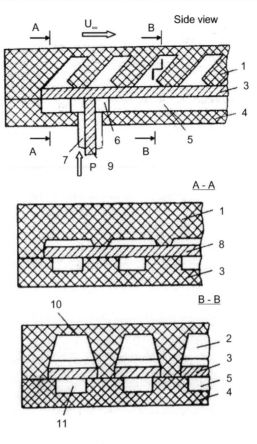

Figure 2.35
Damping covering: *1* — elastic permeable layer; *2, 5* — longitudinal channels; *3* — strips of a current-carrying material; *4* — the bottom elastic layer; *6* — cross-section channels; *7* — tube of pressure supply; *8* — cross-section current-carrying trunks; *9* — electricity cable; *10, 11* — planes to which a permeable material is pasted [44].

electro-insulated conductive section independently connected to the power supply *9*.
The external layer and the strips of conductive material are permeable. In order to increase durability of the coating its cavity can be filled with a wide-meshed porous elastic material. Such a coating can be not divided into longitudinal sections. Its characteristics can be controlled by the corresponding heating of longitudinal conducting strips. As has been shown in the previous chapter, when a body moves in an environment a system of loadings arise on its surface. Therefore, the division of a coating in streamwise sections is provided in this design. In each section it is necessary to regulate the mechanical characteristics in order to achieve distributions along the body of mechanical characteristics of the elastic coating according the existing laws of outer loadings. The way in which mechanical characteristics

are controlled is by changing the pressure in the sections and the temperature (by heating) of strips of a permeable fabric.

All the designs of elastic coating considered above have one common disadvantage— internal cavities that are filled with a liquid and the system of regulating the pressure of the liquid located in these cavities. The problem is that during the maneuvering of a body on which such a coating is located, there is an attached inertial weight, which will influence the covering and change its characteristics.

In the covering shown in Figure 2.34 an attempt to eliminate this disadvantage has been made by filling the cavities with a porous elastic material. Thus, a paste of the porous elastic material is glued only to two planes 10 and 11, which allow the polymer solution to get through the coating to its outer surface. Internal cavities in the coating reduce its durability.

The disadvantage is eliminated in the coating illustrated in Figure 2.36. The design of the coating consists of several layers connected to each other so that in external and adjacent

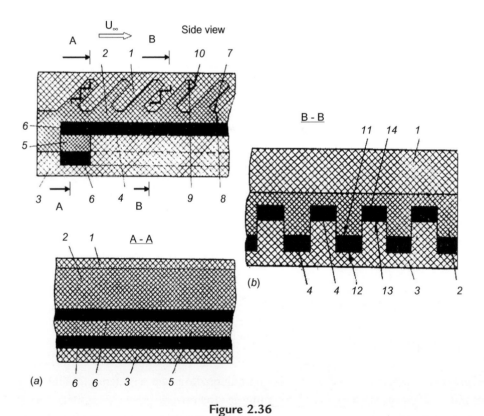

Figure 2.36
An adjustable damping covering: *1–3* — elastic layers; *4* — strips of a conductive material; *5* — bush; *6* — transversal conductive trunks; *7, 9, 11, 13* — ledges; *8, 10, 12, 14* — the corresponding hollows [38].

layers ledges—hollows are located one after another in the direction of flow, and on the adjacent and subsequent layers ledges—hollows are located across the flow. Thus, gaps between ledges and hollows in the bottom layers are filled with a conductive material, for example, fabric or rubber, which forms conductive layers equidistant to the surfaces of the body on which the coating is applied.

On edges the conductive layers have strips of a conductive material running transversal to the flow connected independently to the current source, and each parallel pair of these layers forms isolated sections along the coating. Each layer and section can consist of two conductive longitudinal channels, or any number of channels. On edges of sections, the current is connected to transversal strips either independently to each parallel layer of conductive strips, independently to each section or simultaneously to the entire elastic coating.

One of two parallel conductive layers *4* can be connected to a source of direct current and another to a source of alternating current with a specified or controllable frequency. This results in an electromagnetic field of certain frequency of fluctuation in the material of the coating. The specified design of the coating directs the fluctuations caused by the fluctuations in the liquid flow into the coating in the form of Stoneley and Love surface waves in the area of contact planes of the connection of layers of the coating with each other. These fluctuations propagate inside the coating in the longitudinal direction. However, the design of ledges and hollows located in external layers leads to the attenuation of these surface waves and to the redistribution of the direction of their propagation in external layers in the transversal direction. The specified method of creating electromagnetic fluctuations in the material of the coating will lead to the occurrence of elastic fluctuations in the coating and, as a consequence, to interaction with surface waves. Thereby it is possible to control the parameters of surface waves in the coating and, hence, influence the dynamic characteristics of flow in the near-wall zone. This can be one of mechanisms of influence of elastic coatings upon characteristics of a boundary layer.

The shape of sections of ledges—hollows in the layers of the covering considered can be arbitrary. Depending on its purpose, the elastic coating can be pasted to the surface of a body as shown in Figure 2.36, or onto the external surface of a coating. Densities of layers are arranged such that the density of layer *2* is greatest and the density of the external and internal layers is approximately equal to the density of the environment.

The sizes of controllable damping coatings are defined by the production engineering, the geometrical parameters of the object to which the covering is applied, and the physical features of the fluid flow about the object. Hence, the sizes of the sections are determined for each concrete object according to its external contours, equidistant to its case surface [36].

Controlling the mechanical properties of a coating is achieved by the current through the longitudinal conductive strips *4*, which leads to heating of the layers and changing the mechanical characteristics of the elastic covering according to the control program. Changing mechanical characteristics of elastic coverings along a body length is important for obtaining optimum results in experimental research and field applications. In the initial patents Kramer was guided by the results of research on the external layers of dolphin skin and existing hydrodynamic stability research. In Kramer's fundamental work [317] the hypothesis is put forward that the speeds at which dolphins swim can only be explained by laminarization of the structures of flow in the boundary layer. Based on this hypothesis he performed his theoretical calculations and substantiations for developing the design of his coatings. In Kramer's first three patents [312, 313], his so-called "columellar" coverings are presented, in which the various forms of internal cavity are protected. Later he developed the so-called "ribbed" coatings [314, 316] and monolithic three-layer coating [318]. In each patent the compounding of elastic material layers in the covering is offered. Dependences for the definition of optimum frequency and mechanical characteristics of coatings are given. For constant elasticity of a material along a body length, the thickness of the coating should be varied proportionally to some values of the boundary layer thickness or to the Reynolds number. If the thickness of the coating along a body is constant, in order to obtain optimum characteristics it is necessary to change the elastic and damping characteristics of the covering along the body. Corresponding dependences on the Reynolds number are offered. Proceeding from data existing at that time regarding neutral curve of hydrodynamic stability, Kramer also offered dependences of frequency characteristics of elastic materials of coatings on the Reynolds number.

In Figure 2.37 Kramer's designs for ribbed and monolithic elastic coverings are illustrated (the scale is 1/1000 inch). Laws of distribution of mechanical parameters along the coating considered by Kramer are defined in the authors' designs in each section on the basis of the knowledge about the character of flow structure in a boundary layer and other physical characteristics of loadings of the environment. Modern representations about power-surface waves and other substantive provisions stated above are also considered. Furthermore, the authors offer a technique of mounting sensors inside a coating for controlling the distribution of fluctuations over the material and controlling mechanical characteristics of the coating [56].

Experimental research on the mechanical characteristics of samples of materials used for composite elastic coatings have been undertaken, and various kinds of composite coatings are described in the following sections. Research on the characteristics of the boundary layer over various kinds of composite coatings have also been carried out. Based on the analysis of the obtained results, designs of elastic coverings have been developed for some practical applications.

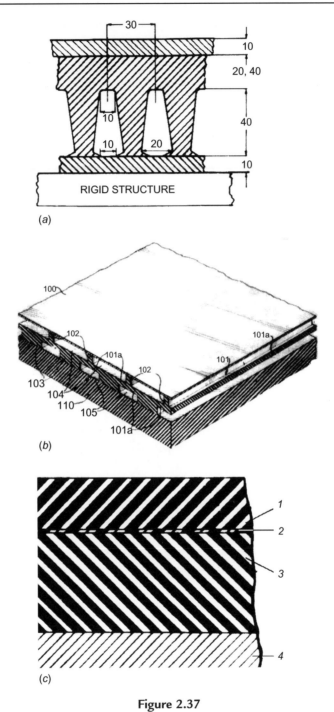

Figure 2.37
Schemes of Kramer coatings: a — ribbed coating; b — coating with longitudinal channels; c — multilayered monolithic coating. Designations in figures are from descriptions of Kramer's patents.

Figure 2.38 shows a new type of coatings for long bodies of revolution (the external layer of the coating ring is absent and the structures of the underlying layer are shown). The elastic coating is placed on the cylindrical part of a body of revolution. It was developed on the basis of research on the receptivity of a boundary layer to various disturbances. Research on the problem of receptivity of the flow over a flat plate, in particular, has been used in the designs of elastic coatings (see Figure 2.28). The results from experimental research on transition and turbulent boundary layer have also made it possible to develop optimum designs of elastic plates, which also model the structure of dermal coverings of high-speed hydrobionts.

Figures 2.30 and 2.39 show photographs of elastic coatings for application of the combined method of drag reduction when conductive longitudinal elements are placed on a surface.

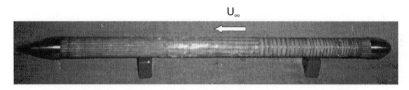

Figure 2.38
New type of elastic covering (external layer of elastic covering is absent).

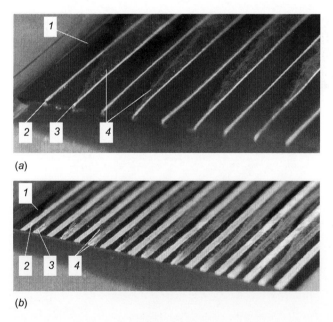

Figure 2.39
Photographs of plates for the combined control of a boundary layer: *a* — plate with wires; *b* — plate with strips; *1* — elastic plates; *2* — wires or strips, the anode; *3* — wires or strips, the cathode; *4* — vertical layer bubbles.

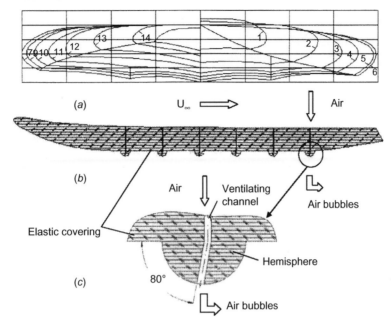

Figure 2.40

Application of the combined method of drag reduction for a board with a sail moving on a water surface: *a* — cross-sections and the form of the body, similar to the form of a sea turtle (front view); *b* — longitudinal cross-section of a body (side view); *c* — cross-section of elements of the case of a body for injection air bubbles in a boundary layer.

In this case, hydrogen bubbles are generated in the boundary layer due to a weak current. When a thin wire was placed on the surface, a thin jet of hydrogen bubbles was generated in the boundary layer, which forms a vertical plane of bubbles in the wake. When narrow strips were placed on the surface of the plate, two parallel vertical planes of bubbles were generated from their opposite sides in the boundary layer.

Figure 2.40 illustrates new kinds of the structure of a body shape that moves on the surface, and the design of a coating for application of the combined method of drag reduction using an elastic coating together with the injection of air bubbles into the boundary layer. The body and elastic coatings were developed by the author together with V. Moros.

In Figure 2.41 the design of an elastic coating for a gliding trimaran is shown. Moros [88, 89] designed the body and elastic covering I. Elastic covering II was developed by Babenko.

Based on the receptivity method a coating has been developed the efficiency of which was checked in towing tests with a model of a three-hull vessel. On the external side of the coating cross-section, numbers mark staggered ledges that are inclined at a positive angle to the flow. When the vessel moves, microcavities are formed behind the protrusions.

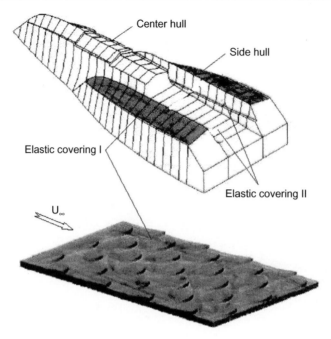

Figure 2.41
Elastic coverings for glider vehicles.

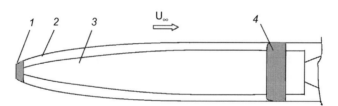

Figure 2.42
Scheme of body movement at cavity flow: *1* — cavitator; *2* — cavity pocket; *3* — body;
4 — stabilizer.

Air bubbles move in the area of local expansion of water flow in the channel between the lateral surfaces of ledges of the coating and are compressed about ledges of the following row under the action of increased pressure. As a result, a gas—water mixture is formed under the bottom of the gliding vessels.

In Figure 2.42 the scheme of body movement at cavity flow is illustrated. The design of an elastic covering has been developed for the stabilization of a body in a cavity.

Based on the analysis of the obtained results, some technical applications for the method of interaction (receptivity) of various disturbances with reference to the considered problem are now discussed. In particular, two ways have been developed to recycle the energy of interaction of the specified disturbances on the border of two environments.

The first way involves the creation of the outer surface of the nasal cone in such a way that specific flat or 3D microvortical formations are generated on its surface; these develop in the shear layer on the liquid—gas boundary and are perceived by the corresponding design of the outer surface of a disk or cone stabilizer located in the tail part of the body. Thus, the specified disk or cone glides on the cavity surface; the system is also supplied with both the known and new combined methods of drag reduction of gliding surfaces.

The second way of energy recycling involves designing the outer surface of the nasal ogival cone and the tail disk or cone in a very different way, while still following the same principles of receptivity. Thus, the gyroscopic moment is given to a body in two ways. In the first case the shape of the nasal cone surface is such that the spiral flow is formed behind its face edge. This flow is perceived by the gliding surface of the tail disk or cone as having a special surface shape similar to the shape of the nasal cone surface. As the tail disk or cone are established motionlessly, under the action of the helicoidal flow they impart rotation to the moving body, therefore the gyroscopic stabilizing moment and flow energy utilization develop. The same effect can be achieved when the gyroscopic moment is realized solely through the special designing of a part of the surface of stabilizers. Through stabilizers, all the variants provide partial accumulation of the energy returned to the flow through the nasal cone. Injection of polymer solutions through the slots in the nasal cone surface or in the area of the face edge is also involved.

A similar scheme of interaction of disturbances in a shear flow in the case of a flat problem leads to the development of new shapes of gliding surface similar to fish scales.

Designs of elastic coverings have also been developed for internal flows. In the previous section a description of patents and experimental results are given together with photographs of the design for a mixer and injector of polymer solutions to which new designs of elastic coatings are applied.

2.6 Main Similarity Parameters

To reveal the basic peculiarities of the interaction of a flow with an elastic surface the modified Voight—Kelvin model can be utilized for simplicity. This model considers the additional tension and weight of an external membrane. The differential equation of motion of a unit element of the model reads as [220, 229, 305, 310, 333]:

$$T_F \frac{\partial^2 \varepsilon_y}{\partial x^2} - M \frac{\partial^2 \varepsilon_y}{\partial t^2} - \eta \frac{\partial \varepsilon_y}{\partial t} - E' \varepsilon_y = P \tag{2.51}$$

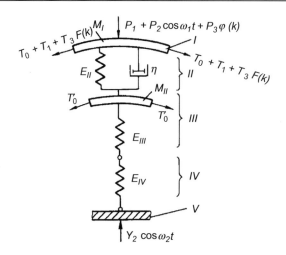

Figure 2.43
Mechanical models of membrane (I, II) and composite (I—IV) of elastic surfaces: 1 — external membrane, II — division of elastomer under the membrane, III — synthetic material, which is glued over a rubber plate, IV — elastic lining; V — base.

The equation concerns only normal surface displacement of the unit element, because it is assumed that the tangential displacement fairly influences the boundary layer parameters.

In the first chapter dolphin skin structure was described, which consists of several layers. Figure 2.43 shows a simplified model of the top four layers of the skin; some models of elastic coatings (see Figure 2.28) have been constructed in a similar way to the structure of these skin layers. When the dynamic loads are rather small the plastic interaction with elastomer takes place essentially in the first and second layers. Therefore, the other layers are considered linearly-elastic. The subscripts in Figure 2.43 denote: 0 — preliminary tension of membranes; 1 — loads caused by distribution of $\bar{p}$ and $\bar{\tau}$ (see Figure 1.89); 2 — vibration loads; 3 — loads caused by fluctuations of pressure and velocities in the boundary layer; $\Phi(\kappa)$ and $F(\kappa)$ — spectra of pressure and velocity fluctuations. The simplified equations are:

$$
\begin{aligned}
\left[T_0 + T_0' + T_1 + T_3 F(k)\right] \frac{\partial^2 \varepsilon_y}{\partial x^2} - (M_I + M_{II}) \frac{\partial^2 \varepsilon_y}{\partial t^2} - \eta \frac{\partial \varepsilon_y}{\partial t} \\
- (E_{II}' + E_{III}' + E_{IV}')\varepsilon_y = P_1 + P_2 \cos \omega_1 t + P_3 l, \ (k) - y_2 \cos \omega_2 t
\end{aligned}
\tag{2.52}
$$

where $\varepsilon = \varepsilon_2 + \varepsilon_3 + \varepsilon_4 = (1/E_{II} + 1/E_{III} + 1/E_{IV})\sigma$

If we know the form and frequency of normal modes, we can write the expression for the response of any point of the composite with coordinate ξ at the moment t:

$$y(\xi, t) = \sum_i \varphi_i(\xi)\varepsilon_{yi}(t), \quad i = 1, 2, 3 \tag{2.53}$$

where $\varphi_i(\xi)$ is the form of the i normal mode, and $\varepsilon_{yi}(t)$ a generalized coordinate, describing the response of the i-mode [29]. So:

$$T_i = \int_0^l \varphi_i^2(\xi)T(\xi)d\xi - \text{generalized tension}$$

$$M_i = \int_0^l \varphi_i^2(\xi)m(\xi)d\xi - \text{generalized mass}$$

$$\eta_i = \int_0^l \varphi_i^2(\xi)\overline{\eta}(\xi)d\xi - \text{damping factor} \tag{2.54}$$

$$E_i' = 4\pi^2 f_i^2 M_i(\xi)\overline{\eta}(\xi)d\xi - \text{generalized rigidity factor, } f - \text{frequency}$$

$$P_i = \int_0^l \varphi_i^2(\xi)P(\xi, t)d\xi - \text{generalized force}$$

and the equation (2.52) takes the form:

$$T_i \frac{\partial^2 \varepsilon_y}{\partial x^2} - M_i \frac{\partial^2 \varepsilon_y}{\partial t^2} - \eta \frac{\partial \varepsilon_y}{\partial t} - E_i'\varepsilon_y = P_i(t) \tag{2.55}$$

According to the scheme in Figure 2.43, we can also write a system of equations of motion of relative displacement components. There is another approach to modeling complex interactions. It is as follows:

- determining the similarity parameters by means of equations (2.55);
- performing model experiments to investigate the physical behavior of the interaction process;
- obtaining the similarity parameters on the basis of experimental data;
- checking these parameters in the integral experiment.

The model approach can be considered now as the most acceptable because as yet no physical investigations of complex interactions have been undertaken; therefore, it is impossible to construct a reliable mathematical model for the problem considered.

When defining the similarity parameters, the laws of complete, partial and conditional similarity must be taken into account. Complete similarity in a model experiment and in real flow over a complex composite takes place if geometric, kinematic and dynamic conditions of the flows coincide. If the similarity of some of the conditions is not satisfied, partial similarity takes place. If the conditions of the experiment do not allow for partial similarity, conditional similarity takes place. If this is the case, the similarity of experimental and real conditions of flow over a composite is considered to be based on general laws of dependence of the investigated parameters upon the selected independent variables. Taking into account the physical picture of the interaction between the flow and the composite (see Chapter 1), it is obvious that complete similarity can only be realized if certain physical experiments are carried out. To determine the dependent variables and the π-parameter 6 we can use the scheme in Figure 2.43 and equation (2.55). Stresses in the composite being summed using parameters of boundary layers—Reynolds number, Cauchy number, thickness of boundary layers, velocity, time δ/U_∞(s), pressure ρU_∞^2 (Pa)—according to the dimensional theory it is possible to write the coefficients in equation (2.52) in dimensionless form [220, 310]. These coefficients determine the π-parameters, which are necessary for the modeling. The dependent variables and π-parameters are presented in Table 2.5 and 2.6. The momentary value of the average velocity $\bar{u}$ is considered as a dependent variable. Some of the similarity criteria were obtained earlier in [220, 305, 310, 333]. In relation to the boundary layer at a flow on elastic coatings the influence of some parameters on $\bar{u}$ consists in the following.

The D parameter is introduced in connection with vibrations and dimensions of the base on which the composite is fixed. The h parameter is introduced in connection with varying thickness along the composite. The real longitudinal curvature leads to the introduction of the R parameter. The local unevenness of the base (bulges and cavities) and dimensions of sections are characterized by parameter l_i. As shown in [305, 310] the kinematics parameters of the boundary layer at a flow of the elastic plate differ from the same parameters at flows over the rigid plate.

It indicates a change in the ratio of characteristic thicknesses of the boundary layer at a flow on elastic plates. To account for these features, parameters δ and δ^* are introduced. Furthermore, these scales are necessary for the definition of their parity with coating thickness h. The parameter G_B characterizes the influence of bend rigidity of the composite or the base on the value of $\bar{u}$. The moment of inertia and the moment of resistance are the additional parameters. Determination of thermal parameters is necessary to control the elastomer properties by heating. The influence of a small degree of heating the elastomer essentially changes its mechanical characteristics. The values of some π-parameters are obvious from the above statements.

The π_6-parameter defines the wave number of unstable oscillations [31, 305] and at other flow regimes it defines the main energy vortices. π_6-parameter can also determine this by wake length of longitudinal irregularities on the surface or inside the composite. The parameters π_7–π_9 are explained in [305]. The parameter π_{10} defines the interaction of the

Table 2.5: Dependent variables.

Variable parameter	Notation	Dimension	Characteristic of Parameter
Instantaneous value of average velocity	$\bar{u}$	LT^{-1}	Dependent variable
Characteristic length	l	L	Geometric parameters of composite and flow
Diameter of base cross-section	D	L	
Thickness of elastic composite	h	L	
Longitudinal radius of base curvature	R	L	
Mutual position of acting parts of the basis	l_i	L	
Boundary layer thickness	δ	L	
Length of boundary layer oscillation wave	λ	L	
Density of elastic composite	ρ_M	FT^2L^{-4}	Kinematics parameters of composite flow
Oscillation mass of composite	M	FT^2L^{-2}	
Viscosity (decrement) of composite	η	FTL^{-2}	
Viscosity of flow	μ	FTL^{-2}	
Density of liquid	ρ	FT^2L^4	
Bond rigidity of base and composite	G_B	FL^{-2}	Dynamic parameters of composite
Elasticity modulus of composite	E	FL^{-2}	
Shear modulus of composite	G	FL^{-2}	
Poisson coefficient	σ	$F^\circ L^\circ T^\circ$	
Frequency of composite oscillations	ω	T^{-1}	
Group velocity of composite oscillations	c_M	LT^{-1}	
Tension of composite	T	FL^{-2}	
Pressure gradient along the base	p	FL^{-3}	
Pressure modulus	$|p|$	FL^{-3}	
Phase angle of pressure fluctuations	φ_n	$F^\circ L^\circ T^\circ$	
Main frequency of pressure fluctuation spectrum	β	T^{-1}	
Base vibration frequency	n	T^{-1}	
Time	t	T	
Composite temperature	$t^\circ m$	Θ	Thermal parameters of composite
Flow temperature	t°	Θ	
Thermal conductivity	K	$L^{-1}T^{-1}\Theta^{-1}$	
Thermal diffusivity	D	L^2T^{-1}	
Specific heat	C	$QFT^2L^{-1}\Theta^{-1}$	

Table 2.6: List of π-parameters.

π-Parameter	Characteristic of Parameter
$\pi_1 = \bar{u}/U_\infty$	
$\pi_2 = l/D$, $\pi_3 = h/\delta$, $\pi'_3 = h/D$, $\pi_4 = l_i/l$, $\pi_5 = U_\infty l/\nu$,	Geometrical parameters, Reynolds
$\quad \pi'_5 = U_\infty/\nu$, $\pi''_5 = U_\infty \delta/\nu$, $\pi_6 = 2\pi\delta/\lambda$, $\pi'_6 = U_\infty R/\nu(\lambda/R)^{3/2}$	and wave numbers
$\pi_7 = \rho_M/\rho$, $\pi'_7 = \rho_M h/\rho\delta$, $\pi_8 = \eta/\mu$, $\pi'_8 = \eta/\rho U_\infty$	Kinematic parameters of
$\pi_9 = \pi'_5 M/\rho$, $\pi'_9 = M/\rho l^3$	composite and flow
$\pi_{10} = \|\rho\|h/G_{bend}$, $\pi_{11} = \rho U^2_\infty/E$	Cauchy number
$\pi_{12} = E/h$, $\pi'_{12} = Eh/\rho D^3\omega^2$,	
$\pi''_{12} = E/h \, (\mu U_\infty \, \pi'^2_5) - \nu/ h(U_\infty \pi_{11})$	
$\pi_{13} = G/E$	
$\pi_{14} = (\pi_{12} / M)^{0.5} \delta/ U_\infty$,	Dynamic parameters of
$\pi'_{14} = (\pi_{12} / M)^{0.5} U^{-1}_\infty \pi'^{-1}_5 = \pi_{14}/\pi''_5 = (\pi''_{12}/\pi_9)^{0.5}$	composite and flow
$\pi_{15} = C_M/U_\infty = (T/M)^{0.5} U^{-1}_\infty$	
$\pi_{16} = (\pi_{15}^2 + \omega^2/\pi_6^2)^{0.5} = (\pi_{15}^2 + \pi'_{14} \pi''^2_5/\pi_6^2)^{0.5}$	
$\pi_{17} = \sigma$, $\pi'_{17} = \Delta/\pi$, $\pi''_{17} = \pi'_8 \pi''_5/\pi_6 \, \pi_9\pi_{16}$	
$\pi_{18} = T/\mu U_\infty = \pi_9 \pi_{15}^2$	
$\pi_{19} = D\sqrt{\omega \rho/\mu}$	
$\pi_{20} = \varphi_n$	
$\pi_{21} = nl/U_\infty$	
$\pi_{22} = U_\infty/\sqrt{gl}$, $\pi'_{22} = U_\infty/\sqrt{g^3}\sqrt{D}$	Strouhal number
$\pi_{23} = \delta/U_\infty$, $\pi'_{23} = U_\infty t/l$	Froude number
$\pi_{24} = t^0_M/t^0$, $\pi_{25} = (t^0 - t^0_M) k/h$	Thermal parameters of composite and flow

pressure field and the elastic surface, and depends on controllable composite parameters and on environment properties; $\pi_{11} - \pi_{14}$ reflect the effect of composite elasticity upon the boundary layer; π_{15}, π_{16} are speeds of propagation of disturbances in a composite; π_{17} relates to absorption properties of composite; π_{18} is the composite tension influence upon the boundary layer parameters; π_{19} reflects the relationship between the inertia properties of the fluctuating pressure field and the viscosity of flow liquid; $\pi_{21} - \pi_{22}$ reflect the vibration of composite and base.

To realize the complete similarity between the model and natural experiments, it is necessary to satisfy the relationship:

$$\pi_{1real} = \pi_{1model} \text{ or}$$
$$\pi_{2r} = \pi_{2m}, \quad \pi_{3r} = \pi_{3m}, \ldots, \pi_{25r} = \pi_{25m} \tag{2.56}$$

Indices denote: m — model; r — natural. If the similarity is partial, some of the relationships in equation (2.56) are correct. If the similarity is conditional, the π-parameter values can be approximately equal. π-parameters characterize geometry, properties and forces. Measurements have showed that π-parameters in natural and model experiments stay constant and are within the limits:

$$\pi_3 \approx 1 - 2$$
$$\pi_6 \approx 16$$
$$\pi_7 \approx 1$$
$$\pi'_7 \approx 1 \div 2 \tag{2.57}$$
$$\pi_9 \approx (1 - 20) \cdot 10^4$$
$$\pi''_{12} \approx (0,2 \div 6,5)10^{-5}$$
$$\pi''_{14} \approx (0,2 \div 5)10^{-5}$$

Some other π-parameters are determined below. They also stay constant. The values of the rest of the parameters must be determined in a number of complex experiments.

2.7 Measurement of Static Mechanical Characteristics of Elastomers

Characteristics of several elastic plates (see Figure 2.28) and the elastomers of which the plates were made were determined using standard devices (VN-5404, VN-5704, see Figure 2.9) and constructed devices (see Figures 2.10, 2.12 and 2.13). The results of the measurements are presented in Figures 2.44–2.48 as a function of force, affecting on the area unit σ, N/m^2, upon the relative strain ε. Overall, 53 specimens of 17 types of elastomer and 32 monolith and composite plates were investigated. The most typical results obtained are presented here.

Measurement of the mechanical properties of the materials began with definition of their density. For the measurement of mechanical characteristics using standard devices, round specimens were made (see Figure 2.9, position *1*). The thickness of the sheets from which specimens were cut out was not identical. Properties of elastomers depend on the thickness of specimen. Therefore, for carrying out measurements of mechanical characteristics, 3–4 specimens of each type elastomer were prepared. The density of all specimens was determined by measuring the weight and volume of the specimen. The weight of specimens was measured using laboratory weights in milligrams, and volume was established using a glass measuring cylinder filled with water. When the specimen was lowered the volume of the displaced liquid was noted. In addition, the volume was calculated from the geometrical size of the specimens. In Table 2.7 results of the measurements are given. The compounding of latex and foam elastics during manufacturing easily changes the density of a material. Therefore, the density of samples of foam latex 3 and 13 (see Table 2.7) and foam elastics differ.

A comparison of densities of two types of rubbers—vacuum (15) and carbonaceous (14)—was made. The results of the measurements are presented as a dependence of relative displacement of the elastomers specimen from a pressure. The displacement of the top

Table 2.7: Density of elastic materials.

Number of Sample	Density (10^2 gr/cm^3)	Material Type
1	4.2	Foam polyurethane
1a	4.6	
1b	4.3	
2	4.2	
2a	4.8	
2b	4.2	
3	41	Foam latex
3a	42	
3b	39	
4	17.0	Foam elastic
4a	21.0	
4b	16.7	
4c	18.5	
5	18.0	
5a	20.3	
5b	19.0	
5c	18.0	
6	28.5	Foam elastic
6a	27.3	
6b	25.4	
7	25.4	
7a	25.2	
7b	21.2	
7c	23.3	
8	25.7	Foam elastic
8a	27.7	
8b	27.0	
8c	26.9	
9	30.6	
9a	32.2	
9b	34.0	
9c	31.6	
10	31.8	
10a	26.4	
10b	26.9	
10c	25.7	
11	29.0	
11a	33.1	
11b	31.1	
11c	31.7	
12	17.5	Foam elastic
12a	17.3	
12b	17.3	
13	45.6	Foam latex
13a	47.0	
13b	46.0	
13c	46.0	
14	130	Rubbers
14a	131.3	
15	143.6	

measuring platform of standard devices under the action of the imposed loading value of relative compression of the specimen or elastic plate was defined from the formula:

$$\varepsilon, \% = \frac{L_0 - L_x}{L_0} \cdot 100 \tag{2.58}$$

where ε is the material deformation, L_0 height of the specimen before weighting, L_x height of the specimen after weighting. Deformation of the specimen was measured in millimeters.

The pressure was determined from the formula:

$$\sigma = P/S, \ gr/cm^2 \tag{2.59}$$

where P is magnitude of compressing force in grams, and S is the area of the specimen in square centimeters, which for all specimens was $\approx 10 \ cm^2$. Results of the measurement are represented in Figures 2.44–2.47 for the specimens according to Table 2.7. For the best representation of the physical parameters of results of measurements, the authors shall present ratios of dimensions of elastomer pressure: 1 gr/cm^2 = 100 N/m^2 = 100 Pa; 1 KPa = 1000 Pa; 1 MPa = 1000 KPa.

In Figure 2.44 results of measurements of the specimens made from foam polyurethane are outlined. Plates were made of the same materials as 1 and 2 according to Table 2.3.

The majority of specimens were made of the same elastomer sheets from which the elastic plates were made. Therefore, for the majority of specimens the thickness did not meet the necessary standards. Because of this, measurements of mechanical characteristics were

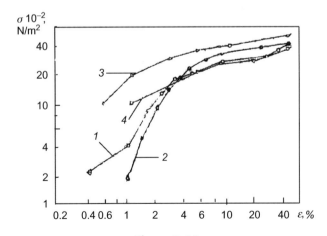

Figure 2.44

Dependence of lengthening on pressure of the elastomer sample: *1* — sample N1; *2*—*1a*; *3*—*2*; *4*—*2a*.

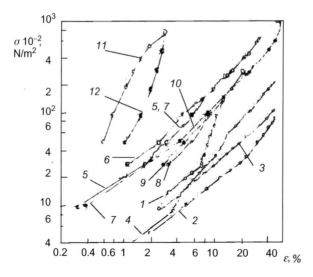

Figure 2.45

Dependence of lengthening on pressure of the elastomer sample: *1*, sample N3 + 3*a* + 3*b*;
2, 4 + 4*a* + 4*b* + 4*c*; *3*, 5 + 5*a* + 5*b* + 5*c*; *4*, 6 + 6*a* + 6*b* + 6*c*; *5*, 13 + 13*a* + 13*b* + 13*c*;
6, 13 + 13*a*; *7*, 13 + 13*b*; *8*, 16 + 16*a* + 16*b* + 16*c*; *9*, 16 + 16*a* + 16*b*; *10*, 16 + 16*b* + 16*c*;
11, 14; *12*, 15.

sometimes performed on stacks of two to three, but sometimes four, specimens. However, for the majority elastomers measurements on a single-layered sample were acceptable. For definition of an error of measurement, the majority of specimens were tested on the devices VN-5404 and VN-5704. The results practically coincided. Some specimens were also tested on the device VN-5705.

Figure 2.45 shows the results of measurements of elastomer specimens the numbers of which correspond to the numbers in Table 2.7. If these are compared with those in Figure 2.44 it becomes evident that the properties of an elastomer can ben divided into four groups. The softest and most pliable are specimens made of foam polyurethane (see Figure 2.44). Thus, the constructed dependences of the pressure on deformation for foam polyurethane can be divided conditionally into three sites in the plane of the plot; in view of that, the results are presented in logarithmic scale. Dependence within the limits of $\varepsilon = 0.4 - 2$ is same as for rigid rubbers (curves 11, 12 in Figure 2.45). The site at $\varepsilon = 2 - 4$ has the same dependences for foam polyurethane specimens as for foam elastic (curves 2−4 in Figure 2.45). Further dependence is characteristic only for this material up to $\varepsilon = 40$. This site corresponds practically to laws of highly elastic deformation, when for small weighting greater deformation follows.

The second group include various types of foam elastic (curves of 2−4 in Figure 2.45), at which there is linear dependence $\sigma = f(\varepsilon)$.

More elastic materials belong to the third group: porous latex (curves of 5–7 in Figure 2.45) and more rigid foam elastic (curves 8–10), which has the same regularity but the material is more rigid. All these curves are located equidistant from curves 2–4 but above in the plane of Figure 2.45.

Even more elastic materials belong to the fourth group of elastomers, and include vacuum rubber (curve 11) and carbonaceous rubber (curve 12). Finely porous latex (curve 1) can be related to the same group, although their elasticity is essentially less than that of the rubbers but law $\sigma = f(\varepsilon)$ is same as for rubbers.

Figures 2.46 and 2.47 represents laws $\sigma = f(\varepsilon)$ for foam elastic 7–12 differing from each other by density (see Table 2.7). The measurements plotted in Figure 2.46 have shown that curves are practically equidistant. In Table 2.7 their density does not differ considerably, however the curves lay on the graph strictly according to density: the greater the density of the specimen, the further up the curve lies, indicating greater elasticity. These measurements testify to the high sensitivity of the measuring devices for static characteristics. The data show that it is not only the structure of a material, but also its compounding structure, that noticeably changes its mechanical characteristics.

In relation to small loadings in the form of pulsations of pressure and speed in a boundary layer, it is already obvious from these data how important the proper choice of elastomer is.

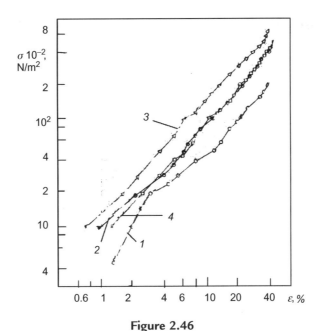

Figure 2.46

Dependence of lengthening on pressure of the elastomer sample: *1*, sample N7 + 7a + 7b + 7c; *2*, 8 + 8a + 8b + 8c; *3*, 9 + 9a + 9b + 9c; *4*, 10 + 10a + 10b + 10c.

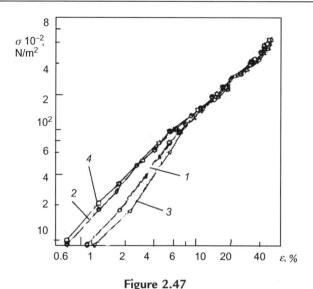

Figure 2.47
Dependence of lengthening on pressure of the elastomer sample: *1*, sample
N11 + 11*a* + 11*b* + 11*c* + 11*d*; *2*, 11 + 11*b* + 11*c* + 11*d*; *3*, 11*a* + 11 + 11*b* + 11*c*; *4*, 12.

Figure 2.47 sets out the results of research on the arrangement order of specimens in a composite pile. At greater loading the results of measurements coincided. However, at small loadings some distinctions are apparent and depend on the configuration of specimens in the pile. Therefore, when designing composite materials it is necessary to consider these data. The results of measuring specimens in combination in a pile according to curves 2 and 4 (see Figure 2.47, foam elastic) have practically coincided with the results of the measurements represented in Figure 2.45 (curves 5–7, foam latex). Curves 1 and 3 in Figure 2.47 coincide well with curves 8 and 9 in Figure 2.45.

The results obtained indicate that measurement of static mechanical characteristics presumes to make a prediction about acceptability of certain materials for tasks under consideration. Figure 2.48 shows results of measurement of elasticity for all elastic plates, the design of which is presented in Table 2.3. Measurements were performed using the device VN-5704 at three arbitrary points on all plates. As the plates were not made using a press—differences in the thickness of plates at the points of measurement may have affected results obtained. Furthermore, practically all the plates were composite and had a design not available to industry. Therefore, it was necessary to establish as much statistical data on the mechanical characteristics of these composites as possible.

In Figure 2.48 the curves are plotted without averaging. It is necessary to bear in mind that it was not possible to maintain the requirements of the standard for the plates. The character of the curves for plates 1 and 2 is the same as for the measurement of mechanical

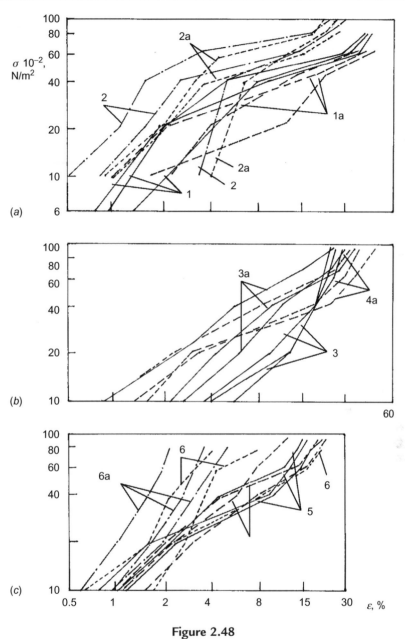

Figure 2.48
Dependence of lengthening on pressure of elastic plates.

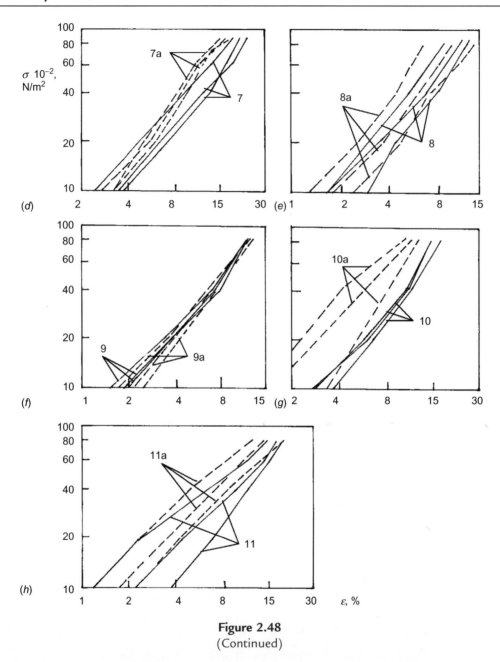

Figure 2.48
(Continued)

characteristics on the corresponding specimen (see Figure 2.44). Plate 2 is made of more rigid foam, and measurements at small values of loading were influenced by insignificant curvature of the aluminum plates on which the elastic plates were pasted. The numbers of curves in Figure 2.48 correspond to numbers of the plate in Table 2.3. The results of the measurements correspond to the design of the plates and the materials of which they are made. In order to gain a greater understanding of the obtained results the authors indicate values for thickness of each plate at three points of measurement, and thickness of the aluminum plate on which the elastomers were pasted. From Table 2.8 it is evident that plates vary in thickness, a consequence of the different designs of plate. Therefore, in order to compare the characteristics of various plates it is necessary to consider their design features and the properties of the materials of which they are made (see Figures 2.44–2.47). Only two plates (1 and 2) were made of monolithic layers of foam polyurethane, the other plates were composite. The analysis of results for these plates (see Figure 2.48a) has shown that the curves differ the most at small pressure. It has been revealed that the character of the curves depends on the initial thickness of a plate at the point of measurement: the thicker the plate, the less elasticity of the plate at the point of measurement. However, with increase of pressure this tendency disappeared. For the other

Table 2.8: Thickness of elastic plates at point of measurement.

Number of Plate	Thickness of Substrate (cm)	Thickness of Plate at Point of Measurement		
1	1.8	10.65	10.8	11.5
1a	1.8	10.28	10.3	10.9
2	1.8	10.05	10.15	10.5
2a	1.8	10.58	10.6	10.7
3	1.8	12.4	12.7	13.2
3a	1.8	10.8	11.1	11.45
4a	1.8	9.4	9.7	10.1
5	2.9	6.7	6.8	7.32
5a	2.9	6.4	6.4	6.8
6	1.8	8.05	8.12	8.44
6a	1.8	8.23	8.67	9.88
7	2.9	7.85	8.2	8.75
7a	2.9	7.0	7.13	7.7
8	2.9	8.6	8.95	9.0
8a	2.9	8.2	8.4	8.45
9	2.9	11.2	11.4	11.45
9a	2.9	11.3	11.7	12.1
10	2.9	6.9	7.1	7.9
10a	2.9	6.8	7.0	7.45
11	2.9	8.0	8.95	9.2
11a	2.9	8.5	8.6	8.6

plates, thickness at the point of measurement did not influence the results. Thickness of the substrate also did not influence the results.

The character of the curves for plates 1 and 2 is the same as for the measurement of mechanical characteristics on the corresponding specimen (see Figure 2.44). Plate 2 is made of more rigid foam polyurethane; therefore, the measurements have indicated that its elasticity is higher than that of plate 1 (see Figure 2.48a).

Figure 2.48b illustrates the results of measurements on plates 3 and 4, which are made of foam polyurethane layers inside of which strips of a synthetic fabric of different widths are located. Plate 3 is thicker and also has narrower strips of synthetic fabric inside. Under smaller loading plate 4 is more rigid, possibly owing to its smaller thickness. At greater loading the mechanical characteristics of these plates became similar. The laws of these plates are similar to those for plates 1 and 2, and the changed slope of dependence $\sigma = f(\varepsilon)$ is caused by the pasted layers of the thin rubber film located inside.

Plate 5 (see Figure 2.48c) is made according to the scheme for plates 3 and 4. The difference is that a continuous sheet of synthetic fabric is located inside instead of separate strips, and the thickness of plate 5 is essentially less. However, it led to a small increase in rigidity of the plate, and law $\sigma = f(\varepsilon)$ has a similar appearance to that in Figure 2.48b.

On the same figure the results of measurements of elasticity for plate 6 are shown, which has same design as plates 3 and 4, but instead of foam polyurethane, foam elastic layers are pasted, the material characteristics of which (curves 2–4, Figure 2.45) belong to the second group. Therefore, the dependences for plate 6 in Figure 2.48C essentially differ from the dependences for plates 3–5 and indicate the increased rigidity of this plate, although the thickness of this plate is greater than that of plates 3 and 4. The characteristics of plate 6 are similar to the characteristics of the corresponding specimens of the foam elastic (curves 2–4, Figure 2.45). However, the elasticity of the plate is higher than in the specimens because inside the plate there are strips of synthetic fabric and rubber film.

Characteristics of all other plates have an identical inclination in the plane of the plots, but depending on the design of plate their rigidity changes. Rigidity is the elasticity of a material divided by its thickness. According to these plots, if at identical values of deformation of a plate the value of σ for each plate is divided by ε and thickness of the plate, the value of rigidity of the plate in the point of measurement is established. If the ratio $\sigma/\varepsilon = E$ is linear it is defined by Hooke's law. At the increase in loadings of dependence become nonlinear. Thus highly elastic and plastic deformations can come when, after the removal of loading, deformation changes with delay. It is important for understanding the analysis of a flow of elastic plates. In this case the behavior elastomer will depend on the size of the hydro dynamic loadings.

Figure 2.48d shows the results of measurements for plate 7, and Figure 2.48e for plate 8. These plates have same design as plate 5. However, they are made of foam latex, the characteristics of which belong to the third elastomer group. The thickness of plates 7, 8 and 6 is approximately identical and is greater than the thickness of plate 5. The elasticity of plates 7 and 7a is much less than that of plates 5, 1 and 2, and is comparable with the elasticity of plate 3. Specimens of latex (curve 1, Figure 2.45) have elasticity similar to that of plate 7, the external and basic layers of which are made of the same material. However, owing to the peculiarities of the design of plate 7, with increased loading the law $\sigma = f(\varepsilon)$ does not vary, while for the specimen of foam elastic the law of elasticity varies and becomes essentially similar to the rubbers.

The design of plate 8 differs moderately from the design of plate 7. On the base layer of plate 8, which is made of the same foam latex as in plate 7a, a continuous sheet of synthetic fabric is glued, to which strips of rigid foam polyurethane are pasted, and between which are cavities for feeding polymer solutions or other liquids. Then there is the external layer on which a rigid foam latex layer is pasted (curves 5−7, Figure 2.45). Despite the air cavities between the strips of foam polyurethane, the rigidity of plate 8 appeared to be higher than that of plate 7 and is similar to the dependences of elasticity of plate 3a at small loadings. It also indicates that the external layer of foam latex was not deformed, and the device measured as for the characteristics of foam polyurethane strips. However, at increased loading the other layers of the composite plate become deformed.

The design of plate 9 differs from that of plate 8 in that a thin layer of a more rigid foam polyurethane is glued on the layer of synthetic fabric, and onto these, strips of the same foam polyurethane are glued. A sheet of foam elastic is pasted to this (instead of latex). The thickness of the plate, therefore, increased to about 12 mm. Foam elastic material belongs to the third group of elastomers. The dependences are the same (see Figure 2.45f) as for plates 7 and 8, but the given design has led to the rigidity of plate 9 being higher than that of plate 8, despite the thickness of plate 9.

Plate 10 is intended to check the "working capacity" of the foam elastic material and is made of continuous sheet. Plate 10a is made of thin strips of foam elastic stuck together, the glutinous seams function as vertical thin elastic membranes—wave-guides, with a step that should be correlated with the longitudinal coherent vortices in transitional or turbulent boundary layers. Both 10 and 10a are made according to the scheme in Table 2.3, and their thickness is in the order of 7 mm. The characteristics of plates 9 and 10 practically coincide, but the elasticity of plate 10a is increased due to the elasticity of the longitudinal glutinous seams.

Plates 11 and 11a are made according to the data in Table. 2.3. They consist of two thin layers of foam polyurethane stuck together. On the outside of these layers, a layer of the skin of the spiny dogfish shark (*Squalus acanthias*) was pasted so that the scales were located along the longitudinal axis of the plate. The thickness of the plates was about 9 mm. The skin of the shark was relatively flexible, but on measuring its elasticity it practically transferred loads to the layers of rigid foam polyurethane. Therefore, the obtained characteristics $\sigma = f(\varepsilon)$ agreed with data for plate 1 up to values $\sigma = 40$. Plates 3−5 also include two layers of rigid foam polyurethane, but the laws of pressure of these plates differ from plates 11 and11a. From the slope in the plane of plots of law $\sigma = f(\varepsilon)$, plates 11 and 11a agreed with data for plates 10 and 10a, although the design of these plates and structure of their materials is different.

Of all the investigated plates, the most rigid are plates 1 and 2 made of sheets of foam polyurethane. The characteristics of sample 7 are optimal for interaction with a boundary layer, they have a linear character of dependences $\sigma = f(\varepsilon)$ and low rigidity. The next best are the parameters of elasticity of plates 10 and 11. From the analysis of the data for plates 8 and 11 it follows that it is possible to design a plate in which, depending on the external loadings, consecutive layers of elastomers will start working.

The non-monolithic character of the samples in a pile and the non-parallel nature of the sides of each specimen explain the spread of measurement points for the samples. The spread of the measurement points for the plates was defined by the non-parallel nature of the plates and their design peculiarities.

For the convenience of carrying out of calculations, some dependence presented above for the specimens and plates are plotted in uniform coordinates. Figure 2.49 presents the curves determined both by standard devices and by the devices constructed by Babenko (see Figures 2.12 and 2.13).

Curves I, II, III and the shaded sector of IV show the results of measurements of the elastic properties of a live dolphin's skin made using the same device (see Figure 2.12) [24, 28, 39]. A satisfactory agreement in results is obtained. Knowing the speeds at which dolphin move and the size of their bodies, it is important to consider the skin characteristics of live dolphins when selecting the design and characteristics of elastic coverings for a given body. It is worth bearing in mind that the parameters of a dolphin's skin are the result of evolution. Figure 2.49 illustrates some averaged data for elastomer specimens in Figures 2.44−2.47 and the results of measurements on polyurethane specimens. Tables 2.8−2.10 represent the corresponding static characteristics of elastomer specimens and numbers of curves in Figure 2.49. Figure 2.49 shows that while the strain is small the tested elastomers obey the linear law. Elasticity modulus is determined from the curves as a tangent of inclination angle of the curves. In the linear region the conditional-balanced modules are in the range $3.5 \cdot 10^4 - 4 \cdot 10^5$ N/m^2, and the instantaneous within the $8.3 \cdot 10^4 - 6.4 \cdot 10^5$ N/m^2 limits.

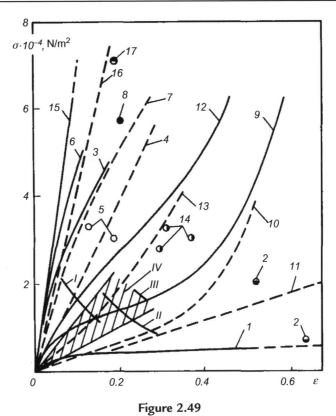

Figure 2.49
Elastic properties of elastomer samples in Table 2.8.

Table 2.9: Elastomer specimen parameters.

Type of Elastomer	Density ρ (kg/m^3)	Number of Pores on 10^{-2} m (ppc)	Elasticity Modulus E_∞ (MPa)	Remarks
FPU-1	40	14−16	0.4	Dry
FPU-1	40	14−16	0.4	Soaked up with water
FE-3	107	30−40	0.49	Closed pores
FE-1	129	30−40	0.56	Opened pores
PU-3A	1050	0.05	8.8	—
PU-3B	1050	0.05	9.0	—
PU-3C	1050	0.05	9.5	—
Rubber film	1370	—	1.8	—

Table 2.10: Static characteristics of elastomer specimens.

Specimen Number	Material	Number of Pores on 10^{-2} m	ρ (kg/m^3)	Curve number in Figure 2.49		
				Elasticity — Instant Value	Conditional — Balanced Value	Measurements by Constructed Device
1	FPU-1	14—16	40	—	1	2
2	FE-1	20—25	129	3	4	5
3	FE-3	20—30	133	—	—	—
4	FE-2	25—45	173	6	7	8
5	FL-1	20—30	250	9	10	11
6	FL-2	15—20	280	12	13	14
7	PU-1	3	1050	15	16	17

It is obvious that the measurements made with the elasticity meter are in satisfactory agreement with the data obtained with the standard devices. Therefore, the above-described device makes it possible to determine the elasticity modulus by two orders faster than with standard devices. For all tested elastomers except for 1, the thickness of the specimen does not influence the strain value if the specimen thickness is $h \geq 3 \cdot 10^{-3}$ m. Taking into account that the thickness of individual layers is higher than h, we can directly compare the results obtained for the specimens and the composite plates.

Figure 2.50 presents the results of measuring the elasticity of the skin of three dolphins and three elastomer samples. The measurements were made using Babenko's elasticity meter (see Figure 2.13). Using the calibration curves constructed at the loading of the measuring core with reference loads, displacements of the core (m) and working loading (N) were defined. By using the calculated area of contact of the measuring core spherical tip with the material under investigation, the specific loading or pressure was defined. The abscissa in Figure 2.50 corresponds to relative deformation $\varepsilon = (l_0 - l)/h$, where l_0 is displacement of the measuring core relative to the coil of inductance on a rigid surface, and l on an elastic surface; h is thickness of the sample or the top layers of a dolphin's skin. The ordinate corresponds to stress $\sigma(N/m)$. The characteristics of the elastomer samples are enumerated in Table 2.11 using the same numbers as in Figure 2.50 (4—6). Elasticity data of other elastomer samples are also provided in Table 2.11 in order to compare them with those of dolphin skin.

Modules of elasticity of the elastomer samples and the skin of dolphins were defined by the ratio $E_{st} = \sigma/\varepsilon'$. Measurements on dolphins were performed along a lateral line. Because the thickness of the skin was taken as the distance between the skin surface and skin muscle it varied from $15 \cdot 10^{-3}$ m in the middle part to $6 \cdot 10^{-3}$ m in the tail part of body [29]. The modulus of elasticity in the head part of a healthy dolphin body changed within the limits of $E_{st} = (2.1-0.96) \cdot 10^5$ N/m^2, of a sick dolphin $E_{st} = (4.7-1.2) \cdot 10^5$ N/m^2;

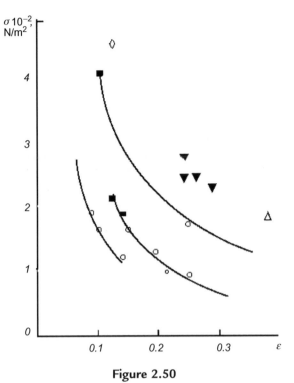

Figure 2.50
Dependence between pressure and deformation of integuments of healthy (*1*), sick (*2*) and
dead (*3*) dolphins, with measurements taken in the areas of the head (I), middle (II) and tail (III)
parts of a body; *4–6* — measurements of elastomer samples.

in the middle part of a body the measurements were $E_{st} = (1.03-0.34) \cdot 10^5$ N/m^2 and
$E_{st} = (0.54-0.37) \cdot 10^5$ N/m^2 respectively, and of a dead dolphin $- E_{st} = (1.7-1.4) \cdot 10^5$ N/m^2.
In the tail part of a dolphin body with the specified state of health, the corresponding data
are: $E_{st} = 0{,}7 \cdot 10^5$ N/m^2, $E_{st} = 0{,}37 \cdot 10^5$ N/m^2, $E_{st} = 4.2 \cdot 10^5$ N/m^2. The skin of a dolphin
that had just died was characterized by a smaller range of skin elasticity along the body
and considerably greater values of the elasticity modulus in comparison with that of a live
dolphin skin. It was discovered that on a healthy dolphin the measurement of elasticity
along its body has a specific character: the elasticity of skin increases in the direction from
the lateral face of the body to its back or stomach. However, on a dead dolphin, it
decreases.

It can be seen from Table 2.11 that elastomer samples with an ample quantity of pores have
elastic characteristics like those of dolphin skin. However, the elastic characteristics of
elastomers that have a density similar to the density of dolphin skin have essentially
different elasticity compared to that of dolphin skin.

Table 2.11: Elasticity of dolphin skin and elastomer samples.

Specimen Number	Density P (kg/m3)	Number of Pores on 10^{-2} m	Thickness (cm)	10^{-4} Est (MPa)	Notes
1	42	20−26	1.0	0.116−0.023	It is measured with the device VN-5704
					It is measured with the elasticity meter (Figure 2.12)
2	300	30−40	0.8	0.035−0.031	It is measured with the device VN-5704
3	1045	6	0.8	0.45	It is measured with the elasticity meter (Figure 2.12)
4	1050	6	0.8	0.45	It is measured with the elasticity meter (Figure 2.12)
5	107	30−40	0.8	0.16−0.06	It is measured with the device VN-5704
				0.13−0.09	It is measured with the elasticity meter (Figure 2.12)
6	107	32	1.0	0.049	It is measured with the elasticity meter (Figure 2.12)
Dolphin	1030	−	1.5−0.6	0.21−0.34	It is measured with the elasticity meter (Figure 2.12)
Rubbers of mark:					
2959	−	−	−	0.98	It is measured wit standard devices (Poturaev)
1847	−	−	−	2.4	It is measured with standard devices (Poturaev, 1975)

Figure 2.51 presents some results of tests of composite elastic plates in the linear coordinates considered earlier in Figure 2.48. The numbers of the curves in Figure 2.51 designate plates the numbers of which correspond to Tables 2.3 and 2.4 (see Table 2.12). The measurements were made using standard devices and the measuring instrument of elasticity developed by Babenko [24]. Each point on the curves is an averaged value from six measurements. Figure 2.51b,c shows the results for elastic plates and the samples of the materials from which they are made. Figure 2.51 shows that the glue layer leads to local strengthening of the elastomer specimens (curves III and VI), which was the material in the outer layers of the composites. The glue layer acts like a thin rigid membrane. Curve 1 corresponds to the plate made of thick homogeneous material FP-1, which is the most compliant. Curve 2 is made of a more solid material FP-2 and is glued over with rubber film. This leads to a reduction in compliance. Curves 4 and 5 (see Tables 2.3 and 2.9) have an equal thickness, but with a different arrangement of the glue layer relative to the outer surface. They have different elastic properties, particularly at high loads. The plate with a thinner outer layer is more elastic. Curve 3 (see Tables 2.3 and 2.9) is

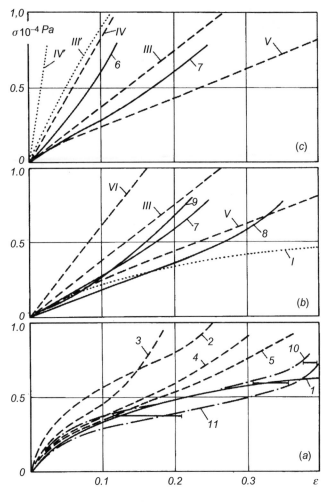

Figure 2.51

Elastic characteristics of elastic plates and elastomer specimens: *1—9* — plate numbers (Tables 2.10 and 2.12); *I, III, V* and *VI* — specimen numbers; *III* and *IV* — numbers of specimens *III'* and *IV'* after aging; *10* and *11* — properties of plate 4 before and after heating. Horizontal lines denote the size of change of the plate parameters with different degrees of heating.

made of the thinnest layers and shows an increase of elasticity. This is a result of both the glue membrane and the rigidity of the layers.

In the region of small strains the plate properties do not differ so much, and this is the very situation of the interaction between the elastic surface and boundary layer. Therefore, the most careful measurements must be carried out for these loads, as must the investigation of the dynamic properties. Comparing the data in Figure 2.51a,b we can see that the plates corresponding to curves 7–9 are made of softer elastomers (see Table 2.13). If the

Table 2.12: Curve designation in Figure 2.54.

Curve Number	a		b		c	
	Material	tg δ	Material	tg δ	Material	tg δ
1	FE-2	0.70	Acrylic plastic	0.82	Acrylic plastic	0.18
2	FE-3	0.59	Standard on a plate	0.18	FPU-2_1	0.62
3	FL-1^*_2	0.46	Silicate glass with a backlash	0.36	FPU-2_3	0.62
4	FL-1^*_2	0.28	Silicate glass	0.13	Plate 2a according to Table 2.3	0.43
5	FL-1_1	0.18	FE-2	0.7	Plate 11a according to Table 2.4	0.31
6	FL-1_2	0.15	PU-1_1	0.39	Plate 11a according to Table 2.4	0.31
7	FL-1_3	0.15	PU-1_2	0.39	Plate 10 according to Table 2.3	0.37
8	FE-2/ FL-1_1	0.64	FL-2	0.14	Plate 10a according to Table 2.4	0.3
9	FL-1_1/ FE-2	0.47	FE-2/ FL-1_1	0.64	Plate 11a with heating	0.58
10	Plate 1 according to Table 2.4	0.56	FE-2/ FL-2	0.64	Plate 10a with heating	0.29
11	Plates 6 and 6a according to Table 2.4	0.51	PU-1/0	0.56	—	—
12	Plate 7a according to Table 2.4	0.36	PU-1/FL-2	0.43	—	—
13	Same	0.36	PU-1/FE-2	0.39	—	—
14		—	FL2/ FE-2	0.14	—	—
15	—	—	Plate 9a according to Table 2.3	0.7	—	—
16	—	—	Plate 1a according to Table 2.3	0.57	—	—
17	—	—	Plate 9a with heating	0.2	—	—
18	—	—	Plate 8 according to Table 2.3	0.14	—	—

Table 2.13: Curve designation in Figures 2.51 and 2.52.

Reference	Number								
Curve number (Figures 2.51 & 2.52)	1	2	3	4	5	6	7	8	9
Plate number (Table 2.3)	1	2	5	3	4	9	.	7	8
Plate number (Table 2.4)	—	—	—	—	—	—	9	—	—

composite is made such that the external layer is more rigid (curves 7 and 9), the general properties of the plate in the field of small deformations are completely defined by the properties of underlying layers (curves 7 and 9, and specimens I and V according to Figure 2.51b).

In this case, the outer layer acts as an elastic membrane, transmitting the stresses to the inner layers. As the load increases the more elastic layers are included in the action. Inclination angles of curves 7 and 9 are increased and approached the angles of the curves of the elastomer specimens (curves III and VI), which were the material of the outer layers of the composites. The inclination angles of curves V and 8, describing the properties of specimen and plate, made of elastomer V, glued over the solid plate, practically coincide. If the outer layer was separated from the inner layers by a rigid membrane (curve 6, Figure 2.51b), the composite response is large and defined by the outer layer (compare with curve 7, Figure 2.51b). Such specific behavior of a composite at its weighting testifies that at dynamic loadings the glutinous internal connections have great value in the complex interactions of different disturbances in a boundary layer.

In Figure 2.51c curves III$'$ and VI$'$ present the specimen parameters measured 7 years after the data was obtained on specimens III and VI and presented in Figure 2.49. The material compliance critically decreases over time.

The heating of plate 4 (see curves 10 and 11, Figure 2.51a and Figure 2.52) has an effect on the elastic parameters investigated. The absolute strain $\Delta \varepsilon = \varepsilon_{hot} - \varepsilon_{cold}$ increases up to 50% within the entire region of stress, and the relative strains $\xi = \Delta\varepsilon/\varepsilon_{cold}$ increase only up to $\sigma = 0.4 \cdot 10^4$ Pa and then sharply decrease with further heating. The heating process was checked with microtermistor MT-54, which showed that the outer surface temperature was slightly increasing. At a flow of elastic plates in real conditions various kinds of static and dynamic loadings operate.

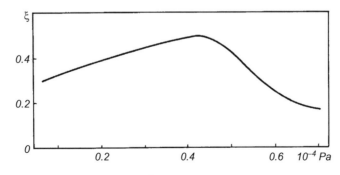

Figure 2.52
Relative change of deformation of plate 4 during heating depending on load value.

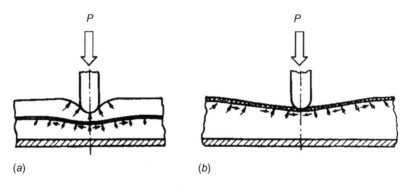

Figure 2.53
The scheme of loading of a composite (*a*) and monolithic (*b*) elastic plate.

Existing methods for calculating hydrodynamic stability or a turbulent boundary layer at a flow on a single-layered monolithic elastic coverings do not take into account the influence of other kinds of loadings in their calculations. Thus, the elementary designs of elastic coverings pay off. Figure 2.53 presents a scheme of loading of two kinds of elastic plates using the method of penetration of the elasticity-meter measuring core into a plate (see Figure 2.12). Figure 2.53*a* illustrates loading of composite plates of the type 7−9 (see Table 2.3). Deformation of the top layer is transferred to the elastic membrane located inside. Thus, the specific loading acting on the bottom layer is much less. If a rigid membrane is located outside of the plate, like for plate 11 (see Table 2.3), or the size of the imposed external loading is insignificant, the scheme of deformation of such an elastic plate is shown in Figure 2.53b. Bending deformation of the external layer causes a system of normal and tangential stresses in the underlying layers, which extend across a greater area in comparison with the originally imposed loading. Thus, the deformation in underlying layers is considerably smaller. If the design of the plate is even more complex, the pattern of deformation of such a composite plate is even more complex.

Existing calculation schemes consider only the deformation of an elastic plate under the action of fluctuation fields of a turbulent boundary layer. If much greater external loadings affect a surface, caused, for example, by a large a vortex in the environment or by the occurrence of vortices due to maneuvering, there will be greater deformations in an elastic coating. The above-presented dependences at static loading of elastic materials indicate that the laws, $\sigma = f(\varepsilon)$, are not linear in the whole range of loading magnitudes. It is related to the properties of the elastic materials, which vary at additional external loadings, and to the presence of additional elastic coatings with other mechanical characteristics. With a change in elastomer rigidity, the characteristics of the elastic waves propagating inside an elastic coating change. These features should be considered in schemes of arrangement. The findings of research on static elastomer characteristics are also contributing to the development of elastic covering that can eliminate adverse influences of various kinds of external loadings.

2.8 Measurement of Dynamic Characteristics of Elastomers

2.8.1 Determination of Elasticity by the Ball Recoil Method

The definition of dynamic characteristics, in particular a tangent of a corner of losses, is achieved by using the technique and equipment discussed in Section 2.4. By means of the device in Figure 2.17 dependences $K(P)$ where $P = mgh_0$, kinetic energy falling from height h, indenter of weight m. The height from which metal balls and cylinders were dropped was a constant. Weights falling on indenters changed and the size P was defined. The size K was defined using formula (2.48) given in Section 2.4. The results of tests on elastomer specimens and plates made of the elastomers are presented in Figure 2.54 as functions $K(P)$. Solid lines denote the results of composite plates made of materials described in Table 2.8. Dash lines denote the tests on the specimens; dash-one-dot lines denote tests on structural materials; dash-two-dot lines denote tests on double-layer elastomers specimens. The designation of the curves at tg δ values, calculated using formula (2.49) for the lowest P, are presented in Table 2.12. Curves $3-7$ in Figure 2.54a present specimens of the same material but with different thicknesses and different pore size. The asterisks in Table 2.10 designate specimens with larger pores, and the indices 1, 2, 3 designate the specimens of $3.5 \cdot 10^{-3}$, $7 \cdot 10^{-3}$ and $10.5 \cdot 10^{-3}$ m thickness. As can be seen from Figure 2.54 the small-pore materials have lower damping properties. Decreasing the thickness causes increasing damping factors, especially with large P values. Sticking the specimen does not give effective results (curves 2 and 10, Figure 2.54). Curves *11* and *12* present plates made of two layers of FE-1 (FE-1 material properties are close to those of

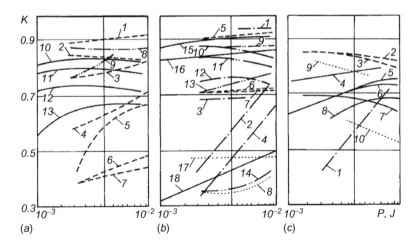

Figure 2.54

Dependence of damping properties of elastomer specimens and composite plates on disturbance energy. Designations are presented in Table 2.12.

FE-3). Curve *11* shows the damping properties of the plate between the current-conducting strips, and curve *12* over them. Structural features of these plates make it possible to obtain different properties in a transversal direction. Curve *13* concerns the plate *7a*, which has outer and inner layers made of FL-1. As was indicated above, thickening of the specimen makes the damping properties worse (compare curves *3* and *4*). Therefore, curve *13* is close to curve *4*. Curves *7* and *8* have the stable high K values.

The influence of a substrate on elastomer properties when falling on its indentor is shown in curve *2* in Figure 2.54b (for tests of a substrate the reference plate made from organic glass was used). The elastic characteristics of silicate glass (curve *4*) and samples of organic (curve *1*) and silicate (curve *3*) glass of a small weight, established with a gap between the substrate have also been investigated. The air gaps in the construction materials, in the same way as elastomers with small pores and high density (type PU-1), increase damping ability. It follows that the curve *11* is located considerably above a curve *6*. The thickening of the sample by 2 times (curves *6* and *7*) did not influence the damping characteristics. On the contrary, for elastomers with many pores, an air gap reduced damping.

In order to model the properties of composite plates the composite specimens were investigated. The specimens were made of the same elastomers and in the same order as the plates were made. The results of research on them are presented in Table 2.10 in the lines containing two materials. For a combined system of elastomer samples such as these, the mechanical parameters are defined by those of the outer layer (compare curves *5, 8* with *10, 14* in Figure 2.54b). The influence of the bottom sample was shown more strongly than that above in terms of energy of disturbance. It is obvious that by carefully selecting layers of material with known mechanical characteristics, it is possible to achieve any kind of dependence K (P)—from linear (curve *9*) up to extreme.

The K values for plate *1a* (curve *16*, Figure 2.54b) and for the specimens (curves *2, 3,* Figure 2.54c) are practically the same. In this case, the specimens and the plate were glued over with rubber film. The structure of plate *8* (curve *18*, Figure 2.54b) is the same as plate *7* (curve *13*, Figure 2.54a) but the external layer differs a little. Thus, the damping parameters of plate *8* are the lowest.

An external layer defines curves *8* and *14*, and when compared with curve *18* shows that FL-2 has a minimal value K and properties of a plate. The outer layer of the plate *9a* (curve *15*) is made of FE-2 (curve *5*). This plate and specimen showed the greatest damping parameters. Heating led to an increase in elasticity and a decrease in damping parameters (curve *17*).

Other parameters of composite plates are presented in Figure 2.54c. Depending on the structure and properties of the elastomer its characteristics depend on thickness to different degrees. So for FL-1 thickness strongly affects the K value (curves *3−7*, Figure 2.54a), and for FPU-2 thickness slightly affects the *K* value (curves *2, 3,* Figure 2.54c). The decreasing *K* and changing the function *K* (*P*) of plate *2a* (curve *4*) made of FPU-2 are explained by the presence of the rubber film glued over it.

K values of plates *11* and *11a* (curves *5* and *6*) are higher than those of the specimens made of the same material as the outer layer of the plates. They are close to the *K* values of the composites constructed from the elastomers in the same order as the plates (curves *12* and *13*). *K* values of plates *10* and *10a* (curves *7* and *8*) are substantially lower than the *K* values of the component elastomer (curve *2*, Figure 2.54a). This is caused by the specific construction of the plates (see scheme in Figures 2.28–2.31).

Heating plate *11a* (curve *9*) and plate *10a* (curve *10*) led to opposite results due to the different properties and construction of the outer layers (see curves *5* and *8*, Figure 2.54c). In the first case the damping properties increase and in the second case they decrease.

When the obtained data was compared with results of [219] it showed that functions $K(P)$ for the elastic and structural materials differed substantially from each other. The tg δ values obtained by the authors for soft elastomers was 0.2–0.8 and for solid elastomers was 0.1–0.2. The obtained *K*-temperature functions, *E* and tg δ values qualitatively agree with the data in [242, 293].

2.8.2 Determination of Elastomer Characteristics by the Method of Forced Non-Resonant Oscillation

Tests were carried out in the test unit shown in Figure 2.19 with a spring ($\sigma_o = $ const; $\varepsilon_0 = \varphi$ (ω)) or without a spring, when a rod was fixed directly to the specimen ($\sigma_o = f(\omega)$; $\varepsilon_0 = $ const). In these experiments the degree of elastomer deformation due to different heights of samples varied at constant amplitude of by moving a rod and changed in eccentricity of the vibrator. The measurement results are presented in Figure 2.55. The parameters of some of the tested specimens are presented in Tables 2.9 and 2.10. The obtained data are confidently agreed with [29]. The compression static modulus of elasticity is determined by step loading, maintained until the relaxation process is over. The dynamic modulus of elasticity, depending on oscillation frequency, can be very different from the static one: for FPU-2 the ratio $E_0/E_\infty = 2$, for FE-3 and PU-3 the ratio is ≈ 3 and ≈ 5 respectively. The tg δ values, determined by this method and by the ball recoil, for FPU-1 are close over a range frequencies of forced oscillations. For FE-3, the values coincide up to a frequency of 100 Hz. Therefore, up to 100 Hz of disturbance loads the method of ball recoil is suitable for determining the averaged parameters of elastomers. Only the method of forced oscillations makes it possible to determine the frequency regions of the most absorbing of the disturbance loads by a material. It is remarkable that samples treated with water leads to a substantial increase of tg δ and to a small decrease in the elasticity modulus.

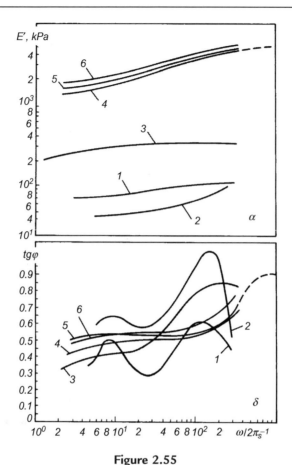

Figure 2.55

Dependence of plasto-elastic properties of elastomers on the loads frequency: *1* — FPU-1; *2* —FPU-1 (soaked with water); *3* — FPU-2; *4* — PU-3 A; *5*— PU-3B; *6* — PU-3C.

2.8.3 Determination of Elastomer Parameters by the Method of Wave and Pulse Propagation [2, 386]

Measurements were made in the test unit shown in Figures 2.17 and 2.21 according to the procedure described in Section 2.4. The technique involved a glass tube inside which an indentor fell to the surface of an elastic plate but did not touch plates, so as not to deform the formation of a superficial wave. The elastic waves occurring on the surface with the ball propagate with a velocity depending on the dynamic elasticity modulus of the material. Figure 2.56 shows oscillograms of signals from two gauges. Value l is a measure of time of elastic wave propagation between the gauges. The testing conditions are presented in Tables 2.11 and 2.4. Knowing the electronic scan velocity and the distance L

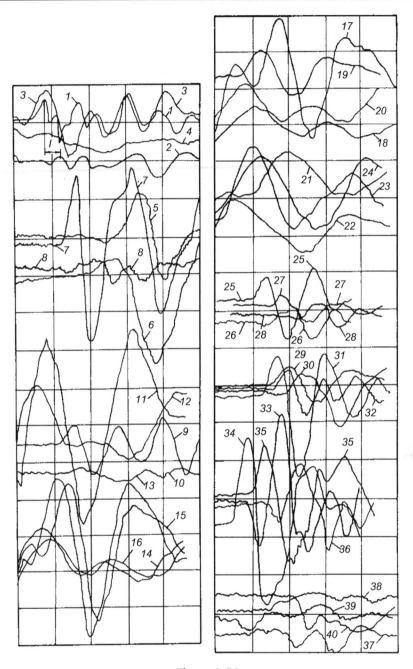

Figure 2.56

Oscillations of the surface of elastomers. Designations of curves are given in Table 2.14.

between the gauges fixed on the tested surface, it is possible to determine the elastic wave propagation velocity. According to formula (2.44) the dynamic elasticity modulus of material is:

$$E = \frac{L^2 v^2}{l^2} \rho \qquad (2.60)$$

Formulae (2.45) and (2.60) are of one type. According to the formula:

$$K_1 = E_{dymam.} / E_{static.} \qquad (2.61)$$

we can determine the anisotropy coefficient κ_1. The anisotropy coefficient κ_2 is calculated in accordance with in longitudinal and transverse directions of the flow:

$$k_2 = E_{long.} / E_{transv} \qquad (2.62)$$

This method was checked by measuring the parameters of the elastic waves for the standard plate, made of acrylic plastic (see curves *1*, *2*, Figure 2.56). These measurements were used as a reference for the analysis of the distribution of fluctuations on the elastic plates. On the surfaces investigated, in response to the falling indenter, for the distribution of fluctuations up to the first gauge the greatest amplitude A depended on frequency of the investigated material's own fluctuation (or the composite). Thus, in Figure 2.56 the first front of a superficial wave and fluctuations following it are fixed. It can be seen that character of fluctuation differs from cleanly harmonious fluctuation. It indicates that real impact causes a spectrum of fluctuations in an elastic plate. Furthermore, on the surface of an elastic plate there are scratches and roughness that also cause distortion of the form of waves. With regard to amplitude and character of the fluctuations fixed by both by gauges, the size of attenuation is defined. The oscillations fixed by the first gauge fade poorly, and those fixed by the second gauge on the form are similar to the fluctuations fixed by the first gauge, but have smaller amplitude. As can be seen from the oscillograms of fluctuations (curves *1, 2,* Figure 2.56), after the third period on a rigid plate the distortion of amplitude caused by nonlinear effects begins.

Figure 2.57 shows a photograph of the development of fluctuations in the standard, and is fixed as a result of both gauges. A large amplification in the signal has been made; therefore, only the first periods of fluctuation are visible in the photograph. On the rigid standard the fluctuation is closer to harmonious in comparison with elastic plate.

The behavior of wave propagation on the elastic surface is clearly different to that on the rigid plate. The fluctuations differ from the harmonious, but their development in time is better than the exponential law describes the standard.

To generate the surface waves in the standard plate it was enough to use the indentor of diameter $d = 2 \cdot 10^{-3}$ m and to register the date with an amplification coefficient $K_a = 10$.

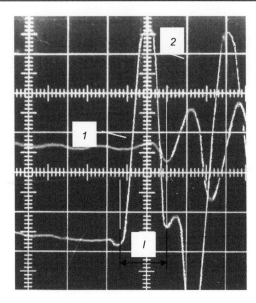

Figure 2.57
An oscillograph of fluctuations on a rigid surface: *1* — data from the first gauge, *2* — data from the second gauge; *I* — scheme of distribution of the front of a wave between the first and the second gauges (curves *1, 2*, Figure 2.56).

For the disturbance being absorbed by the elastomers, these parameters of the indenter had to be increased. The amplification coefficient is graduated in decibels, so the changing of K_a from 10 to 20 means amplifying the signal by 3.16 times, and changing from 10 to 30 means amplifying by 10 times.

To determine the disturbance energy effect upon the oscillation amplitude the experiments were carried out with different balls. The parameters of the balls were as given in Table 2.15.

The elastic plates were divided into two groups. In the first group were the plates with an outer layer made of material with a large number of pores (compliant) and the second group comprised plates with a small number of pores (elastic).

Figure 2.56 (curves *3–6*) presents the oscillation propagation on the surface of plate *5* (Figure 2.28b, Tables 2.3 and 2.4). The outside layer of the plate is a FPU-2 open sheet, which intensively damps the oscillations. The oscillation amplitude A is approximately 15 times less than by the standard plate. The second gauge registered only very small oscillations, which could be detected only by increasing the disturbance energy (mgH) (curves *5* and *6*). The plate, being glued over with rubber film, showed increasing of elastic parameters. So the amplitude A, registered by both gauges on the

Table 2.14: Conditions for carrying out experiments (Figure 2.56).

Curve Number	Plate Number	$H \cdot 10$ (m)	$d \cdot 10^3$, (m)	Gauge Arrangement
1, 2	Standard	1.3	2	Longitudinal
3, 4	5	1.3	10	Longitudinal
5, 6	5	1.45	10	Longitudinal
7, 8	2	1.3	7	Longitudinal
9, 10	2	1.45	10	Transverse
11, 12	1	1.3	10	Longitudinal
13, 14	1	1.3	8	Longitudinal
15, 16	1	1.3	7	Longitudinal
17, 18	1	1.3	6	Longitudinal
19, 20	1	1.3	7	Longitudinal
21, 22	1	1.45	10	Transverse
23, 24	6a	1.3	7	Longitudinal
25, 26	10	1.3	7	Longitudinal
27, 28	10	1.45	7	Transverse
29, 30	10a	1.45	10	Transverse
31, 32	11a	1.3	10	Longitudinal
33, 34	11a	1.3	10	Longitudinal with a
35, 36	11a	1.3	10	rod
37, 38	11a	1.3	8	Longitudinal with a rod
39, 40	11	1.3	10	Same, $h = 70$ mm water
				Same, h = 70 mm oil

Table 2.15:

Parameter	Range of Balls					
$10^3 \, d$ (m)	2.0	5.3	6.0	7.0	7.9	9.9
$10^3 \, \rho$, (kg/m^3)	7.8	10.1	7.6	7.8	7.74	7.77
$10^3 \, m$ (kg)	0.33	0.6	0.88	1.4	2.0	3.96

plate *2*, was increased compared to the amplitude of plate *5* by 1.2 times (compare curves *5* and *6* with *7* and *8*).

On plate *2* made from a monolithic homogeneous material, pasted with an outer layer of thin rubber film, the parameters of fluctuation have been measured in longitudinal and transverse directions. Amplitude *A* in the transverse direction is 4 times less than in the longitudinal direction. This means the longitudinal directed disturbance energy propagates more intensively. Therefore, the plate construction makes it possible to orient the disturbance growth in a certain direction.

The more rigid plates (*1, 1a* and *9*) are made of the same material—FE-3—but of different thicknesses (see Table 2.4). The oscillations of the surface of plate *1*, which has the greatest

Table 2.16: Conditions for carrying out experiments (Figure 2.62).

Plate Number	Section of Figure 2.62	Curve Number	l (m)	H (m)	$d \cdot 10^3$ (m)	Remarks
Standard (acrylic plastic)	a	1	0.2	0.13	2	The acrylic plastic is fixed in a frame
		2	0.2	0.13	2	Acrylic plastic not fixed
2	b	1	0.2	0.13	7	The gauges are arranged in transverse direction
		2	0.1	0.13	10	The gauges are arranged in longitudinal direction
		3	0.1	0.145	7	Same
2a	c	1	0.2	0.13	6, 7, 8, 10	Same
		2	0.2	0.13	6	Water above plate, h = 22 mm. Rod
		3	0.2	0.13	6, 7, 8, 10	Same, h = 70
5	d	1	0.1	0.13	6, 7, 8, 10	The gauges are arranged in longitudinal direction
		2	0.2	0.13	7	Water above plate, h = 22 mm. Rod
		3	0.2	0.13	8, 10	Same

thickness, depend on load intensity (curves *11−18*, Figure 2.56). The behavior of the oscillation remains equal with all masses of the indentor. With reduction in energy of the disturbances the amplitude of fluctuation *A* decreases. In all cases, in view of the value of coefficient of amplification K_a size *A* of plate *1* is more than 2.6 times that of plate *2*. On plate *1* the development of fluctuations on a surface of plate in longitudinal and cross-section directions has also been checked (curves *19−22*). The design of a plate 1 monolithic is a fixed anisotropy of properties in different directions. It tells us that at carrying out of measurements in transversal direction parameters d and H were more than at carrying out of measurements in a longitudinal direction (see Table 2.14). On this plate, anisotropy of properties in various directions was also found: in a longitudinal direction the size *A* is 2−3 times more than in the cross-section direction. In the latter case, fluctuations are fixed in an antiphrasis, which is caused by more essential attenuation of a wave on distance *L*.

Figure 2.58 shows a photograph of the development of fluctuation on the surface of an elastic plate *1* in a longitudinal direction. The elastic layer of plate *1a* is thinner than that of plate *1*. Therefore, plate *1a* has a higher rigidity and amplitude of oscillation. The outer layer of plate *9* is of the same material as plates *1* and *1a*, but has the smallest thickness. For plate 9 the rigidity of the layer is the greatest but amplitude *A* is the lowest because the plate is composite (whereas *1* and *1a* are one-layer plates). The anisotropy of properties is absent.

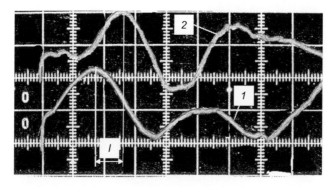

Figure 2.58
An oscillograph of fluctuations on elastic plate *1*: — data from the first gauge, *2* — data from the second gauge; *I* — scheme of distribution of the front of a wave between the first and the second gauges (curves *19, 20*, Figure 2.56).

The elasticity of plates *6, 6a, 10* and *10a* is still higher. Curves *23, 24* in Figure 2.56 present the oscillation behavior of plate *6a*. The value *A* for plate *6a* is 1.7 times lower than for plate *1* because of its composite structure. The anisotropic properties of plates *6* and *6a* are lost because of the construction of inner layers. The anisotropy of plates *10* and *10a* is lost due to the construction of outer layers. They are made of agglutinate strips of the same material as plates *6* and *6a*. The strips were turned by 90° on plates *10* and *10a* (properties of porous materials depend on the direction of frothing). With oscillation propagating in a transverse direction, its amplitude decreases and the phase shift increases up to 180°. Plate *10* has the most marked anisotropy of its properties (curves *25−28*). The oscillation amplitude in the longitudinal direction is 2 times less than for plate *1*, and in the transverse direction it is 2.7 times less than in the longitudinal direction. The strip spacing of the outer layer of plate *10a* is 2 times less than for plate *10*. The value *A* of plate *10a* in the longitudinal direction is a little less, and in the transverse direction is 1.5 times less than for plate *10*.

In Figure 2.59 the photograph of the development of fluctuation on a surface of elastic plate *10* in a transverse direction is shown. Plates *7, 7a* and *8* (outer layer made of FL) showed a rapid damping of oscillations.

Plates with an outer layer of low-porosity material are of the most interest. These are plates *11, 11a, 3* and *8a* (see Table 2.4). Plates *11* and *11a* have a longitudinally ordered inner surface and plates *3* and *8a* have a transversally ordered inner surface. Figure 2.56 shows the surface oscillations of the plate *11a*. Plate *1* has the maximal value of amplitude among the compliant plates of the first group. For plate *11a* the amplitude of fluctuation 1.6 times less than that of plate *1*. The first group of compliant plates includes plates made of, or

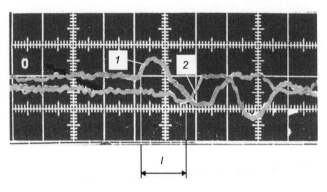

Figure 2.59
An oscillograph of fluctuations on elastic plate *10*: — data from the first gauge, *2* — data from the second gauge; *I* — scheme of distribution of the front of a wave between the first and the second gauges (curves *27, 28*, Figure 2.56).

having in their design, the material FPU. Other plates conditionally relate to the second group of elastic plates.

In order to define the influence of the density of a liquid in which a body with elastic coverings moves the device used for carrying out of these experiments was updated. The indentor fell not on an elastic plate, but onto a special rod established vertically by means of rubber extensions. To avoid additional distortions, a light from a thin tube with a tip made of organic glass was used. In spite of this, the oscillation amplitude of plate *11a* in this experiment was increased by 1.25 times. The heated plate *11a* (curves *35* and *36*) had increased compliance and the value *A* decreased by 1.3 times (compare curves *33, 34* and *35, 36*). It is not clear why the influence of the indentor on a plate through a core increases the amplitude of fluctuation. This requires additional research to reflection.

To compare the obtained data with results of elastic plates tested in a hydro-chamber the plates were fixed in a case that was filled with water or another viscous liquid, and tested. The thickness of the layer of liquid above the plates was 7 cm. The oscillation parameters changed and the value *A* decreased by 8.5 times. The oscillation parameters of plate *11* were the same as for plate *11a*. The reduction of fluctuation in this case relates to the damping properties of the viscous liquid located above the elastic plate. From the character of the oscillograms it follows that the fluctuation in elastic plates is nonlinear and fading. Such fluctuations lead to the occurrence of the attached liquid, which starts to change with a delay in phase in comparison with an elastic plate. It also leads to the reduction in amplitude of fluctuation of elastic plates.

In Figure 2.60 a photograph of the development of fluctuation on a surface of elastic plate *11a* in a longitudinal direction is shown. With regard to such photographs, owing to

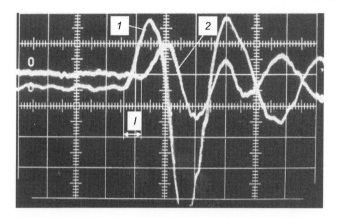

Figure 2.60
An oscillograph of fluctuations on elastic plate *11a*: — data from the first gauge, *2* — data from the second gauge; *I* — scheme of distribution of the front of a wave between the first and the second gauges (curves *31, 32*, Figure 2.56).

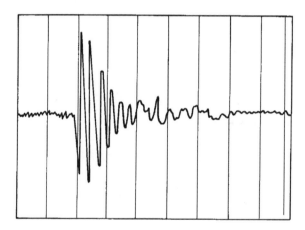

Figure 2.61
Surface oscillation oscillogram obtained by means of a mirror-galvanometer oscillograph. The vertical lines are the time marks.

the limited area of the monitor screen it is impossible to track the dynamics of development of fluctuation in time. It is convenient to run down the time development of oscillations by means of mirror-galvanometer oscillograph (see Figure 2.61). The natural oscillation frequency and logarithmic decrement were determined with this type of oscillogram. The free oscillation of an elastic surface can be expressed by equations such as:

$$y = A \ e^{at} \ \sin(\beta t + \varphi_0) \tag{2.63}$$

2.8.4 Definition of the Complex of Mechanical Characteristics of Elastomers

When performing ball recoil methods, is important to know the height of the first rebound of the ball. However it is also important to determine the following recoil heights, as this makes it possible to determine the dynamic elastic-damping properties of elastomers. From oscillograph data the sizes of impulses of consecutive impacts of an indentor with a probationer plate are established, as is the time between impacts, and this defines the consecutive heights for the rebound of an indentor. The ball recoil heights are presented in Figure 2.62a. The dashed lines note the time trajectory of the indentor. The unfixed plate absorbs the indentor's kinetic energy more intensively than the fixed one. This is in agreement with the data presented in Figure 2.54. Such determining of the absorbing properties of elastic plates can be considered as an approximate modeling of the interaction of a boundary layer with an elastic surface. Figure 2.56 presents the disturbance developing in the plate with unit indentor shock. The boundary layer pressure pulsations cause similar surface waves in the plate. The waves interact with the following oscillations caused by the flow pressure pulsations. The experiments with the falling indentor are approximately reflecting the interaction mechanism: the first shock generates the elastic waves, interacting with the waves generated by the following shocks. In addition, the data in Figure 2.62 are analogous to the damping factor. The data make it possible to determine the dependence of absorbing value of disturbance energy upon the indentor fall height, and the group of the plate.

Measurements were made on all the elastic plates. In the tests the following parameters were changed: mass of the falling ball (diameter d), the falling height H, thickness of the water layer above the plate h', and the distance between the shock point and the location of pressure gauge. The gauge was arranged either along or transverse to the longitudinal construction layers of the elastomers. The ball dropped either onto the surface or onto the rod fixed vertically on the plate's surface.

The rubber film practically did not change the absorbing properties of the plate. The water layer compresses the plate, so the elasticity increases and the absorbing properties are decreased (curves *2* and *3*, Figure 2.62b, d). In this case, the rubber film decreases the absorbing properties. Anisotropy is absent. Plates *1, 1a, 9* and *9a* are made of more elastic materials compared to the above-considered plates. However, in contrast with the data presented in Figure 2.56, the absorbing properties were increased compared to plates *2, 2a, 5* and *5a*. In addition, the thinner plates (*1a* and *9a*) showed greater absorbing properties. The function $H(K)$ is similar but in contrast with plates *2* and *5*, with the increasing of H, the absorbing properties can be either increased or decreased.

The layer of water h' also reduces the absorbing properties of plates, and increasing the thickness of the water layer further decreases the absorbing properties. According to Figure 2.56, anisotropy of properties depending on a direction of accommodation of gauges

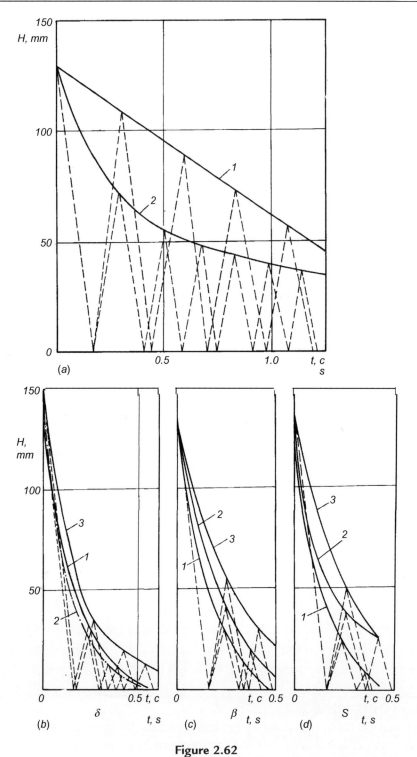

Figure 2.62
Height of indentor recoil at its consecutive impacts with a standard plate and elastic plates (see Table 2.16).

has been revealed. The elasticity of plates *10, 10a, 6* and *6a* has this to a greater degree. The obtained data correspond well with the results given in Figure 2.56. The absorbing properties of these plates are less than for plates *1, 1a, 9* and *9a* and are comparable to those for plates *2* and *5*. The absorbing properties of these plates are independent of indentor mass and depend upon H and h'. The absorbing properties in this case increase as the h' increases.

The absorbing properties of plates *7, 7a* and *8* are the lowest, due to the high elasticity of the material and to the rigidity of the lower layers.

Plates of the second group, *11, 11a, 3* and *8a*, differ in their mechanical parameters and construction (see Table 2.4). In the external layer of plates *11* and *11a* there is orderliness similar to that of plate *10* (conditionally named "turned riblets.") Therefore, it has appreciable anisotropic properties, and in a transverse direction the absorption of impact is poorer. The absorbing properties of plate *11* are practically independent from H and depend on h', as do plates *2a* and *5* (see Figure 2.62c,d). The greater the indentor mass, the less disturbance energy is absorbed. When the water layer is replaced with an oil layer the absorbing properties increase sharply. This agrees well with data in Figure 2.56 (curves *39* and *40*). Heating the plate leads to a decrease in the absorbing properties. Increasing H and m leads to a decrease in energy absorption. The outer layer of plate *11a* is less elastic than that of plate *11*. This may be the reason why the anisotropy is less and the absorbing parameters greater than for plate *11*. When the indentor falls onto the rod fixed on the plate, the absorbing parameters are decreased. The value of H of the first recoil is the same as for the standard plate (see Figure 2.62a) because the kinetic energy of the indentor is completely transferred to the rod without loss. Heating plate *11a* without the water layer causes an increase of compliance and absorbing parameters depending on indentor mass. The water layer practically changed nothing, i.e. the absorbing properties of heated plate *11a* are the same in air and water. When the indentor mass is increased, the absorbing properties are decreased.

The outer layer of plate *3* is made of solid PU and the outer layer of plate *8a* of soft PU. Therefore, the absorbing properties of plate *3* are decreased and those of plate *8a* are increased. Only these plates were equally absorbing of any disturbances irrespective of disturbance parameters. Absorbing characteristics of plates of the second group were the same as plates *1* and *9* and the maximal values of absorption of disturbances are for plates *8a* and *9a*.

The comparison of test results obtained by wave and pulse propagation and by ball recoil methods is presented in Table 2.17. The surface oscillation amplitude A and the ball recoil height H are parameters characterizing the same physical properties of materials. The table shows that the results obtained by the both methods are agreed. There are differences for plates *1, 9* and *7, 8*. The surface waves propagating in the plates mainly depend on the

Table 2.17: Comparison of test results according to wave propagation and ball recoil materials.

Data in Figure 2.56	Data in Figure 2.62
Plate 5: The value A is decreased approximately by 15 times compared to the standard. *Plate 2*: Value A is more than 1.2 times that for plate 5. Value A in longitudinal direction is 4 times greater than that in transverse direction (anisotropy). *Plates 1, 9*: A is by 2.6 times higher than that for plate 2. Longitudinal A is by 2–3 times higher than transverse A. *Plate 10*: A is 2 times less than for plate 1 and reached about same as at a plate 2. Anisotropy: longitudinal A is 2.7 times less than the transverse. *Plate 11a*: A is 1.6 times less than for plate 1 and 1.6 times higher than for plate 2. For the plate being heated, the A decreases by 1.3 times and by 8.5 times with water layer $h' = 70$ mm. Anisotropy is registered.	*Plates 5, 2*: The value H is decreased by 3–5 times compared to the standard. Plate 2 has H that is slightly higher than that for plate 5. Anisotropy of plate 2 is absent. H depends on kinetic energy of indentor $W = mgH_0$. Water layer h' increase H and decreases the absorbing properties. *Plates 1, 9*: H is 1.5 times less than for plate 2. For thinner plates (1a, 9), the value for H is less, and depends upon W and h' as for plate 2. Anisotropy is registered. *Plates 10, 10a*: The value H is increased in comparison with H for plates 1, 9 and reached the same as that for plate 2. H depends on W (mainly on H_0) and on h'. In contrast with plate 2, H decreases with h' increasing. Anisotropy is the largest. *Plates 11, 11a*: H is 1.8 times higher than for plate 2. The transverse absorption is less than the longitudinal one (anisotropy). H depends on W (on which h' depends, as plates 2a, 5): H increases with m increasing. Oil layer leads to H decreasing by 3–5 times.

properties and construction of outer layer and to a lesser extent on the deeper layers. The absorbing properties depend on all layers and their combination. Therefore, both methods supplement each other and make it possible to investigate different mechanical properties of elastomers.

According to measurements the logarithmic decrements and absorbing factors were calculated by:

$$\delta_i = \ln\left(\frac{A_i}{A_{i+1}}\right), \tag{2.64}$$

$$K_a = \frac{h_i - h_{i+1}}{h_i} 100, \tag{2.65}$$

where A_i and A_{i+1} are amplitudes of the preceding and the following wave in accordance with data in Figures 2.56–2.62, h_i and h_{i+1} are the fall height and the recoil height of the indentor in accordance with Figure 2.62. As is seen from Table 2.18 the surface waves in the tested plates are irregular.

Therefore, the calculated decrement of attenuation is different. Figure 2.56 shows that the oscillation amplitude of the standard plate rather increases along the plate and then becomes

Table 2.18: Logarithmic damping decrements.

Plate Number	$\delta_1 = \ln(A_1/A_2)$	$\delta_2 = \ln(A_2/A_3)$	$\delta_3 = \ln(A_3/A_4)$
Standard plate	0.42	0.14	0.18
5	0.82	−0.53	−
5	0.34	−0.14	0.18
2	1.8	−	−
2	0.34	−0.5	0.3
1	0.54	0.6	0.14
1	0.5	0.54	0.5
1	0.43	0.5	−
1	0.37	0.6	0.18
1	0.54	0.7	0.18
1	1.1	0.7	−
6a	0.83	0.34	−
10	0.47	0.4	1.1
10	0.45	0.16	0.7
10a	0.75	−0.26	0.85
10a	0.58	−0.23	1.6
11a	0.65	0.29	0.59
11a	0.56	0.13	0.56
11a	0.6	0.15	−
11a	1.2	0.3	0.6
11a	0.98	0.3	0.5
11a	1.1	0.79	−
11a	0.45	−0.47	0.1
11	0.74	3.1	−

of saw form, characterizing the nonlinear effects. Because of this, the value δ_2 is negative and δ_3 is small. The decrement of attenuation of the first fluctuation of plate 5 has grown, but the character of development of disturbances is the same as for the standard. The difference is that attenuation occurs to increase the energy of the impact more intensively.

The surface film of plate 2 leads to an increase in absorption if the disturbance is small. At increasing impacts the absorption becomes the same as that of the standard plate and plate 5. Therefore, the material (FPU) cannot be effective in damping the disturbance energy.

The laws of absorption of impacts by elastic plates are that δ_2 is less than δ_1, but at the third wave δ_3 increases, i.e. after a certain decreasing of damping a sharp increasing of damping occurs. It is also necessary to take into account the surface wave amplitude. Therefore, the value K_a is a supplementary parameter. K_a calculated with following recoils is presented in Table 2.19.

At has been shown that all the elastic plates (but particularly the plates of the first group) absorbing the disturbance energy. Using formula (2.61) the logarithmic decrement of

Table 2.19: Damping coefficient of elastic plates.

Plate Number	k_1	k_2	k_3	k_4	$\delta_1 = \ln (A_1/A_2)$	$\delta_2 = \ln (A_2/A_3)$	$\delta_3 = \ln (A_3/A_4)$
Standard Plate	17.7	15.7	20.4	34.5	0.18	—	—
5	81.4	85.0	—	—	1.6	—	—
2	76.5	64.5	54.5	—	1.46	1.0	0.79
2	77.2	37.2	48.0	—	1.41	0.42	0.64
1	84.1	78.7	—	—	1.97	1.6	—
1	84.8	77.8	—	—	—	—	—
1	83.2	78.9	—	—	—	—	—
6a	54.76	44.6	44.5	38.1	0.83	0.69	0.47
10	61.2	58.0	60.8	59.0	1.0	0.83	0.87
10a	67.51	51.8	54.9	57.3	1.0	0.8	0.7
11a	77.4	62.2	60.4	—	1.3	0.83	0.69
11a	69.8	26.0	67.4	—	1.1	0.69	—
11a	60.01	64.9	65.1	—	1.1	0.69	—
11	49.6	—	—	—	—	—	—

attenuation δ_i is calculated. The same formula has been used to calculate quasi-logarithmic decrements of attenuation, when instead of values of fluctuation amplitude according to the gauge indications, in this formula corresponding heights h_i of consecutive indentor bounces were substituted. The values of logarithms obtained (see Table 2.20) come nearer to values of δ_i (see Table 2.21) only at small energy of disturbances, i.e. at subsequent indentor bounces. Thus, it is possible to define approximately the decrement of elastomer attenuation by measuring the subsequent indentor rebounds.

Comparison of the obtained laws relating to the distribution of fluctuations in elastomers and laws of indentor bounce from elastomers leads to the conclusion that the reaction of elastomers with various characters of disturbances of movement is different. Depending on the intensity of the disturbances acting on elastomer, there are fluctuations of various characters and a degree of nonlinearity in the material. In response to the action of intensive disturbances (i.e. a large indentor) the elastomer will bend, which will influence the character of the development of fluctuations in the elastomer, caused, for example, by earlier pulsations of a boundary layer. It reasonable to assume that except for pulsations of a boundary layer, the liquid's lumps fall on the elastomer. These liquid lumps act in the form of large vortices from the external border of a boundary layer or in the form of large external vortices, for example, from the roughness of a surface or arising from maneuvering a body.

Such liquid lumps act in a similar way to a falling indentor, and will bend the elastomer and be rejected by an elastic surface, sometimes up to their full dissipate. However, the action of such diverse disturbances will cause in an elastomer the development of a spectrum of disturbances. It is necessary to investigate the further development in elastomers of a spectrum of fluctuations under diverse and non-uniformly scaled loadings.

Table 2.20: Damping characteristics of materials.

Plate Number	$<\delta>$	$<K_a>$	f (Hz)
Standard Plate	0.42–0.49	47.7	639
5	0.68	80.0	78
5a	0.58	79.4	99
2	0.66	78.2	81
2a	0.6	76.0	83
1	0.75	81.0	214
1a	0.69	84.6	246
9	0.7	79.7	137–153
9a	0.9	87.2	75
6	0.6	54.0	260
6a	0.64	59.1	247
7	0.52	52.7	98
7a	0.6	63.0	74
8	0.61	40.1	65
10	0.59	68.2	200
10a	0.75	57.3	231
11	0.7	65.6	185
11a	0.69	75.7	239
3	0.64	62.5	169
8a	0.68	75.7	233

It is necessary not only to consider it, but also to analyze the behavior of elastomers in real conditions.

Values δ and K_a depend on many parameters: H_0, m, h° amongst others. Therefore, it is very difficult to determine them unambiguously for a composite or an elastomer. We can only determine their limits when the parameters are constant. The averaged values δ and K_a were calculated from different conditions of the tests (see Table 2.20). Such averaging is relative but it makes it possible to determine the statistical parameters and potentialities of materials. It was found that the averaged values δ and K_a corresponded to the data in Table 2.12.

The method of wave and pulse propagation was also used to determine some other parameters of elastomers (see Table 2.21). For comparison, in Table 2.22 the parameters obtained in Section 2.7 are also presented.

In Table 2.21 the mechanical characteristics of plates are presented, necessary for analysis of their hydrodynamic properties. In addition to the modulus of elasticity and anisotropy coefficients, the values f and C are of significance. The frequency parameters are important to determine the resonance interactions of elastomers with frequency parameters of the boundary layer. Tables 2.21 and 2.22 show that the values f, δ and K_a depend on many parameters of outer layer, its construction and that of the whole composite. The value f

Table 2.21: Mechanical parameters of elastic plates presented in Table 2.4.

Plate Number	Density ρ (kg/m³)	Quantity (pores/cm²)	Frequency f (Hz)	Phase Velocity C (m/s) Longitudinal	Transverse	Static Modulus of Elasticity 10^{-4} E_{st} (MPa)	Dynamic Modulus of Elasticity 10^{-4} E_{long} (MPa)	10^{-4} E_{trans} (MPa)	Anisotropy Factor $k_1 = E_{long}/E_{st}$	$k_1 = E_{long}/E_{trans}$
Standard plate	1190	–	639	1230	1230	1.77	1.77	1.77	1	1
5	4	20–25	78	29	29	0.35	0.353	0.353	1	1
5a	42	20–25	99	43–48	40–47	0.53	0.7–1.0	0.7–0.9	1.3–1.9	1
2	42	20–25	81	47	53	0.3–1.4	0.9	1.15	1	0.8
2a	42	20–25	83	–	48	0.3–1.2	1.23	0.98	1	1.25
1	107	32	214	64.3	58.6	0.49	4.3	3.6	8.85	1.2
1a	115	36	246	66.6	48.6	0.49	5	2.67	10	1.87
6	107	30–40	260	70.7	67	0.59	0.536	0.48	0.9	1.1
6a	107	30–40	247	60.4	61	0.59	3.76	3.9	7	0.97
9	107	32	137–153	51.7	50.5	0.49	2.8	2.7	5.7	1.04
9a	173	45–55	75	87.4	40	0.39–5.7	13	2.7	2.3–3.3	4.8
7a	200	40–60	74	43.6	–	0.35	3.82	–	11	–
8	300	30–40	65	123	–	4.5	44	–	9.8	–
10	410	50–55	200	52.7	84.1	0.35	0.98–3.9	28.4	2.8–11	0.14
10a	410	50–55	231	52.7	90	0.35	0.98–3.9	32.4	2.8–11	0.13
11	1050	5–10	185	61	66	8.8	39	45.6	4.3	0.85
11a	1045	4–9	239	49.7	68.2	5.9	25.3	47.6	4.3	0.53
3	1050	4–7	169	40	23	3.7	15.7	5.5	4.1	2.9
8a	930	4–7	233	63.2	52.7	7.85	36.5	25.3	4.65	1.45

Table 2.22: Mechanical parameters of elastic plates presented in Table 2.3.

Plate Number	Density ρ (kg/m^3)	Quantity (pores/ cm^2)	Static Modulus of Elasticity 10^{-4} E_{st} (MPa)	Dynamic Modulus of Elasticity 10^{-4} E_{dyn} (MPa)	Anisotropy Coefficient $k_1 = E_{dyn}/E_{st}$
1	40	14−16	0.47−0.8	−	−
1a	40	14−16	0.43−0.9	−	−
2	42	20−25	0.3−1.4	1	1
2a	42	20−25	0.3−1.16	1.23	1
3	42	20−25	0.17−0.2	−	−
3a	42	20−25	0.37−1.1	−	−
4	42	20−25	0.65−1.0	−	−
4a wet	42	20−25	0.5−1.0	−	−
5	42	20−25	0.3−0.5	0.37	1
5a	42	20−25	0.4−0.64	0.7	1
6	107	30−40	0.57−1.6	0.536	9
6a	107	30−40	0.55−0.87	0.384	7
7	200	40−60	0.26−0.4	−	−
7a	200	40−60	0.3−0.35	3.8	12
8	300	30−40	0.3−4.6	−	−
8a	300	30−40	0.4−0.8	−	−
9	173	45−55	0.28−0.33	13.2	27
9a	173	45−55	0.4−5.8	13.2	23−33
10	410	50−65	0.98−0.4	11.2−29	7.2−11
10a	410	50−65	0.98−0.4	11−33	7.2−11
11	42	40	0.27−0.89	−	−
11a	42	40	0.4−0.86	−	−

must be taken into account to predict the application of plates, and reducing hydrodynamic drag. The phase velocity C of disturbance propagation in elastic plates must be compared to the velocity of disturbance propagation in the boundary layer. The C value also depends on many parameters of the composite and until now it has not been possible to determine its connection with mechanical properties of plates.

2.9 Oscillations and Waves in Composite Elastomers

For tasks vibro-absorbing homogeneous single-layered or composite designs of elastomers are considered. Their frequency characteristics are investigated within the practical limits of vibrations that will be encountered, basically, in a range of units and tens of kilohertz. For the majority of hydro−aerodynamic tasks, the range and intensity of loadings is very different. It has been shown that some of the designs of elastic coverings discussed have specific ordered structures that will essentially change the behavior of the elastomers under action of enclosed loadings. According to the representations put forward about interaction in a boundary layer of various impacts, the character of interaction of various external disturbances in real conditions of a flow on elastomers will be complex and ambiguous. Much research on the interaction in a boundary layer of various disturbances has been undertaken [310]. Based on the obtained results it is necessary to study the physical representations of the interactions of elastic coverings with disturbances acting both from a boundary layer and from the case on which the elastic covering is fixed.

As a first approximation it is necessary to find out the order of size of the loadings acting on the elastomer from an average boundary layer. It is known that energy of pulsations of pressure p' on a wall is connected with speed of stream U_∞ dependence:

$$\frac{2\sqrt{(p')^2}}{\rho U_\infty^2} = \alpha_\tau \lambda, \tag{2.66}$$

where α_τ is the coefficient of proportionality (for a smooth wall $\alpha_\tau \approx 2.1$), and the coefficient is defined from a parity:

$$\lambda = \frac{2\tau_0}{\rho U_\infty^2} = \frac{2\overline{u'v'}}{U_\infty^2}, \tag{2.67}$$

where τ_0 is tangential pressure on a wall, and $\frac{2\overline{u'v'}}{U_\infty^2}$ is Reynolds stress expressed through longitudinal u' and cross-section v' components of pulsating speeds. Measurements have shown [256, 303] that for a rigid plate $\lambda = 0.0046$, for elastic plate $\lambda = 0.0034$. For two values of speed 0.6 m/s and 10 m/s are received accordingly for rigid $\sqrt{(p')^2} = 0.83 \cdot 10^2$ and $4.83 \cdot 10^2$ Pa (Pascal) and for an elastic surface $\sqrt{(p')^2} = 0.61 \cdot 10^2$ and $3.7 \cdot 10^2$ Pa. Similar calculations for a rigid plate have been defined $\sqrt{(p')^2} = 3.3 \cdot 10^2$ Pa [41]. If it is necessary to re-count the coefficient of transformation $a_\tau \lambda$, obtained from measurements on dolphins [438], for speed of 10 m/s, at passive swimming $\sqrt{(p')^2} = 4.45 \cdot 10^2$ Pa, and at active swimming $1.95 \cdot 10^2$ Pa. Romanenko [438] and Shlanchauskas [464] estimate the scale of dissipative vortices as approximately 1 cm^2 and the height from which clods of liquids "fall" on a wall as size $y/\delta \approx 0.15 - 0.25$. Then using calculations of thickness of a boundary layer for a rigid surface it is possible to estimate the energy of pulsations of pressure at the chosen speeds accordingly for rigid surface $P_1 \approx 3.3 \cdot 10^{-7}$ J and $P_2 \approx 9.66 \cdot 10^{-5}$ J and for an elastic surface $-P_1 \approx 2.4 \cdot 10^{-7}$ J and $P_2 \approx 7.1 \cdot 10^{-5}$ J (with J meaning joule). If one accepts the distribution of thickness of a boundary layer on the body of a dolphin such as it is calculated in work Babenko [72, 73]

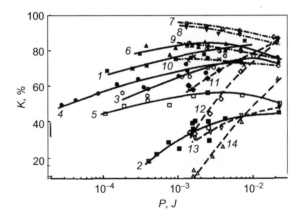

Figure 2.63

Dependence of the damping coefficient on the energy of perturbation (potential energy of the falling balls): *1– 6* — correspond to elastic materials (see Table 2.19); *7, 8* — healthy dolphin; *9* — ill dolphin; *10* — dolphin just after death; *11* — corresponds to an unfixed steel strip; *12* — silicate glass; *13* — wooden plate; *14* — organic glass.

estimations for a floating dolphin have yielded following results: at $\sqrt{(p')^2} = 4.83 \cdot 10^2$ Pa energy of disturbances movement is equal $P = 0.996 \cdot 10^{-4}$ J; at $\sqrt{(p')^2} = 1.95 \cdot 10^2$ Pa; $P = 0.39 \cdot 10^{-4}$ J.

Measurements of damping coefficient on live dolphins [72, 73, 77] have been made. The results obtained are compared to similar results from measuring the damping properties of some samples of elastomers and construction materials (see Figure 2.63). Parameters and designations of curves are given in the Table 2.11 (see Section 2.7). From Figure 2.63 it can be seen that dissipation energy of dolphin skin essentially differs from other materials. Damping of disturbances by healthy dolphin skin (curves *7* and *8*) is maximal and it is essentially above other materials. The skin of a sick dolphin (curve *9*) is less effective at damping disturbances than that of a healthy dolphin. Measurements on plates made of FPU and FE (curves *1* and *4*) at disturbance energy $P = 5 \cdot 10^{-3}$ J show that the damping coefficient K becomes comparable with K of a sick dolphin's skin.

Approximately half an hour after the death of the sick dolphin measurements were taken (a curve *10*). It appears that from the character of this curve and its value, at death dolphin skin becomes more similar to artificial elastomers and commensurable with data for PU (curves *3* and *4*). The basic differences are that the skin of a healthy dolphin possesses maximal damping properties that are practically constant in a wide range of sizes of disturbing loadings. Thus the maximum of absorption of disturbance energy is located in area of values $P < 10^{-3}$ J while for elastomers at this value P damping properties decrease, and the maximum K is located at $P > 10^{-2}$ J.

The measurements have shown that the maximum absorbing ability of skins of live dolphins is located in a range of the maximal values of energy of pulsations of a transitive boundary layer, which, as measurements have shown is $1.5-2$ times above pulsations of a turbulent boundary layer. Samples of elastomers have a maximum of absorption at essentially great values of energy of P. From here, two indirect conclusions follow. For the flow on the skin of live dolphins a boundary layer that has distinct characteristics from that of a boundary layer on a rigid surface is formed. The second conclusion is that homogeneous elastomers cannot compete with the damping characteristics of live dolphin skin. It is necessary to study the peculiarities of the skin structure of live dolphins and to model its hydrodynamic functions.

The above-presented explanations of the obtained results are based on the theory of effective modules. In accordance with the theory the mechanical properties of a composite are considered as the properties of a homogenous substance with an effective modulus that can be determined in terms of elasticity modules of the components and by means of parameters describing the structure of the composite [6]. This applies to static and quasi-static loads that cause the flexural waves in elastomers. However, under the action of dynamic loadings, the surface of monolithic single-layered materials develop stationary (harmonious) and non-stationary (under action of impulses) waves of other kinds [2, 6, 132, 257, 386, 459], including normal, superficial and shift waves.

Dynamic loads in complex elastic composite plates (see Figure 2.28) bring about such waves. The dynamic response of a reinforced composite depends on the correspondence between the wave propagation direction and the reinforcing direction. The reinforcing components work as wave-guides. If the waves are propagated perpendicular to the reinforcing, the reinforcing elements are obstacles, and reflective of the disturbance propagation [2, 6, 132, 484]. It has been revealed that dispersion depends on the form of irregularities and what is more, irregularities of the cross-section section close to the rectangular [21, 495]. The surface and 3D waves go over into each other, and the Raleigh waves are intensively scattered by a lattice with a scale equal to the disturbance wavelength.

The mechanical oscillations in composite materials are characterized by attenuation, dispersion and nonlinearity. It is important to note the other properties of waves in composites [6, 386]. The phase velocities and frequencies of harmonic waves depend on the parameters of the elastic material structure as well as on plate thickness. The non-stable disturbances can be presented as a superposition of sinusoid waves, using the Fourier integrals.

When distribution of waves occurs perpendicularly to reinforcing elements, the composite elastomer works as the wave filter that selectively passes or reflects periodic waves. Some undesirable dynamic effects (for example, increase in pressure at interaction of the reflected and refracted waves) can be reduced through appropriate design of the structure of a composite. It has been established that the level of dynamic pressure very much depends on the quantity of layers, their thickness and the properties of materials. It is possible to pick up these parameters to reduce the dynamic loadings in a composite. Therefore, the composite, depending on its construction, formula and orientation, can be used as a wave-guide, repeller, resonator, concentrator or filter [6, 495].

An other important property of composite elastic materials is that under action of dynamic loadings, composite superficial waves of different types extend. As has been determined in [495] the surface waves are the elastic waves, propagating along the free surface or along the boundary between the solid body and other medium, and fading at removal from the boundary. There are two types of surface waves—vertically or horizontally polarized. The most usual waves encountered are Raleigh waves (vertical polarization). Their energy is concentrated mainly in the layer with thickness $\lambda - 2\lambda$, and the phase velocity $C_R \approx 0.9C$, where C is phase velocity of the plain cross-wave.

At the boundary of two solid media, when their densities and elasticity modules do not strongly differ, a Stoneley wave can originate. The Stoneley wave consists of two Raleigh waves (one in a layer), located at the both sides of the boundary at a distance λ. In addition, the shear (Love) waves (horizontal polarization) can originate. Their phase velocity is between the transverse wave velocities in the neighboring layers. The Stoneley waves usually originate when the thicknesses of layers differ substantially.

Much interest surrounds the superficial waves along the border of a composite with a liquid, as this place most essentially represents the interaction of pulsating fields speeds of a boundary layer with superficial waves of an elastic composite. It is likely that along this border two types of wave are propagated: damping Raleigh waves and non-damping waves with vertical polarization. The damping wave has a damping factor equal to 0.1, i.e. when $x = 10\ \lambda$, the wave damps by l times. Phase velocity of the non-damping wave is less than one of phase velocity in the liquid and of longitudinal and transverse wave velocities in the solid body. The non-damping wave consists of a faint non-homogeneous wave in liquid and two high non-homogeneous waves in the solid body (longitudinal and transverse). The wave energy and particle motion is mainly located in the liquid. The dynamics of a liquid (boundary layer) complicates the picture of waves in an elastomer.

The composite elastic plate (see Figure 2.28) consists of longitudinally oriented reinforcing components. If these components are located transverse to the approach flow, i.e. transverse to the disturbance propagation of a boundary layer (plate *3* and *8*, Table 2.4) the reinforcing components work either as repellers or as filters. In this case, the boundary layer disturbances are either increased and repelled in the flow or, in contrast, permeate into the composite and quickly become damp.

If the reinforcing components are oriented along the flow, the boundary layer disturbances cause oscillations propagating in the composite near the wave-guides formed by the reinforcing components. Energy of impacts will be spent on dissipation and excitation of mechanical oscillations in the elastomer: partly in the damping oscillations in wave-guides (Stoneley and Love waves), partly in surface waves at the outer boundary of the composite. In the latter case, superficial waves will directly cooperate with disturbances of a boundary layer, then will strengthen or damp. If the outer surface is made of transverse microcomponents, the surface oscillations, as shown in [21], mean that superficial fluctuations will dissipate and be reflected either by them or to be transformed to volumetric oscillations of the composite. Thus, through the design of a composite it is possible to reduce and localize disturbances of a boundary layer, especially by taking into account the structure of flow at different stages of transition.

Plates *8* and *9* have oscillation concentrators for a direction of flat disturbances of a boundary layer or any other disturbances in longitudinal wave-guides. All the plates having such a system of regulation can change the mechanical characteristics and properties of wave-guides.

Thus, due to the structure and orientation of reinforcing elements in an elastic composite plate concerning the direction of the basic flow (or a vector of disturbances of a boundary layer) it is possible to achieve a certain effect on the interaction of a stream with an elastic plate as follows:

- The boundary layer disturbances can be intensively damped through the absorbing of disturbance energy by the elastic composite plate, for example the oscillation caused by boundary layer pressure and shear fluctuations. Oscillation damping can be realized inside of composite plates through purposeful movement of fluctuations on wave-guides with attenuation.
- The oscillation energy can be concentrated near the wave-guides to decrease the interaction of harmonics into the composite, to decrease the negative effect of inner waves repelling in wave-guides and oscillation resonance and to influence the character of distribution of oscillations near the wave-guide.
- It is possible to achieve local intensive interaction of an elastic plate with a boundary layer. For this purpose an elastic plate is used to form longitudinal coherent vortices systems in a boundary layer. The elastic plate near to an external surface has a design in the form of longitudinal wave-guides.

- It is possible to intensively strengthen the disturbances extending in a boundary layer. For this purpose the disturbances are reflected from an elastic plate in a boundary layer through the plate's design in which the external layer reinforcing elements are located perpendicularly to the direction of distribution of disturbances in a boundary layer. In this case a disturbance of a boundary layer does not cooperate with a superficial wave in an elastic plate.

CHAPTER 3

The Turbulent Boundary Layer over Elastic Plates

3.1 Experimental Equipment and Methods of Measurement

The measurements in this chapter were executed together with Dr M. V. Kanarsky. The measurements of the turbulent boundary layer were carried out on the lower wall of the working chamber of a low-velocity open wind tunnel [66, 256]. The dimensions of the closed working chamber were $3 \times 0.5 \times 0.2$ m, the range of working velocities was $0-18$ m/s, the degree of turbulence at:

$$U_\infty < 10 \text{ m/s is } T_u = 1.7\%, \text{ and at } U_\infty > 10 \text{ m/s } T_u = 2\% \qquad (3.1)$$

A scheme showing the position of the insert, on which the elastic plate was fixed, is shown in Figure 3.1. There was a rectangular aperture in the bottom wall of the pipe, in which the tensometric insert was mounted flush with the internal surface of the working section. The detailed structure of this insert is described in Section 2.5 and Figures 2.25–2.27 [256, 285, 286]. Four vertical thin elastic rectangular plates (5) were attached to the insert. The other ends of these plates were fastened to a thick duralumin plate (6). Two wooden bars (7) were attached to this plate. On the bars, two duralumin corners were fixed. On their edges, the corners had four pairs of clip screws (9), which allowed the establishment of an elastic plate flush with the wall of the working section of wind tunnel with accuracy within 0.01 mm. Simultaneously, by means of normed metal linings, the insert was fixed in the aperture of the pipe wall with 0.2 mm gaps on the forward and lateral sides, and up to 1 mm gaps on the rear side.

The dimensions of the insert plate were 0.528×0.246 m and of the elastic plate were 0.48×0.2 m. The standard plate was made of 10 mm-thick acrylic plastic. To prevent the deformation of the elastic surface due to underpressure in the flow, the region of the working chamber was made airproof. It also prevented the velocity profile distortion due to air suction through clearances. To measure the drag of the insert, the mechanotron 6MX-2B (8) was used. Its mobile anode was fastened to the insert, and the case to plate 6.

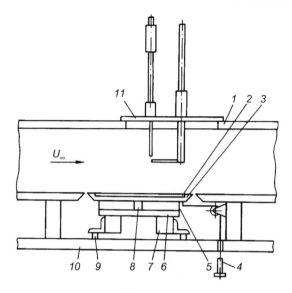

Figure 3.1

Arrangement of sensors in the working chamber of wind tunnel: 1, working chamber; 2, removable plate; 3, tensometric insert; 4, calibration unit; 5, elastic components; 6, plate; 7, supports; 8, mechanotron; 9, clip screws; 10, frame of the working chamber; 11, plate with thermo anemometer sensors.

Figure 3.2 shows the locations of the turbuliser and coordinates x sections (l, ll, lll, llla) in which measurements of a turbulent boundary layer characteristic are executed. It was found that the turbulent boundary layer originates practically just across the turbuliser.

The turbulent boundary layer characteristics were measured with a DISA thermo anemometer. The sensor 55A25 measured the average longitudinal fluctuation velocity. The X-shaped sensor 55A38 measured the Reynolds stresses and transverse fluctuations. To determine the 3D correlation coefficient, the correlator 55D70 was used. The root-mean-square stress fluctuations were measured with a 55D35 voltmeter, and the averaged value was measured with a 55D30 voltmeter. A low-voltage spectrum analyzer C4-44 performed the spectrum analysis of velocity fluctuations. The thermo anemometer wire sensors were fixed in holders, which could move the "vertical" sensor with an accuracy of 0.01 mm and the "horizontal" sensor with that of 0.005 mm (Figure 3.1). Both holders were mounted on the plate, which could be moved along the investigated surface.

The technique used for carrying out the measurements consisted of the following: the investigated elastic plate was mounted. Simultaneously the mechanotron and thermo anemometer sensors were calibrated. The sensors were calibrated in the DISA special

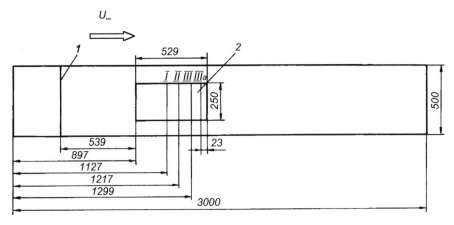

Figure 3.2

Working chamber of a wind tunnel (upper view): 1, turbulizer; 2, insert; I, II, III, IIIa, sections for measurement of boundary layer characteristics.

calibrating tube 55D41 by means of an inclined manometer F-7462. The Nusselt number was determined as a calibration function of the Reynolds number, calculated with the wire diameter as a characteristic linear dimension:

$$Nu = A + B\,Re^n \tag{3.2}$$

The function can be approximately presented as an output voltage against the velocity of ongoing flow:

$$E^2 = A + B\,U^n \tag{3.3}$$

where A and B are constants. The value n is taken to be 0.45 [181, 445, 501]. This gives the most agreement with experimental data within $0.02 < \text{Re} < 44$.

The mechanotron sensor 6MX-2B was calibrated by means of calibration device 4. After the calibration mechanotron an aperture in frame 10 of the working site was pressurized. Through this aperture passes a cable of the calibration unit 4. The thermo-anemometer sensor moved along coordinates x, y in a longitudinal plane which included a longitudinal axis of a working site. The average speed was measured. At the given speed of a flow measurement a plate resistance was spent. Then a wedge was mounted flush in the gap between the back edge of the insert and the aperture in the bottom wall of the working site, such that the forward edge of the insert was close up against the wall. It prevented rocking the insert, which was fixed on elastic suspension brackets.

The kinematic characteristics of the turbulent boundary layer were being measured in sections (see Figure 3.2), with two free stream velocities. The velocity profile was measured with a vertical sensor (see Figure 3.1) on the left. While moving the sensor down in the viscous sublayer, the wall influence distance was measured by a pulse counter. The velocity profile was determined twice—while the gauge was moving down and then up. Simultaneously the longitudinal fluctuations of velocity were being measured.

At the distance where fluctuations were maximal, the spectrum analysis was undertaken. The Reynolds stresses and velocity transversal fluctuations were measured by an X-shaped sensor, fixed in a holder (see Figure 3.1) on the right. The vertical sensor was mounted along the tunnel axis for registering the average speed of flow.

3.2 Velocity Profiles of Average Speed over Elastic Plates

The calculations showed that the most of the π-parameters of the tested plates in the work velocity range were not optimimal. Therefore, the positive effects revealed in the experiments are explained by the structural and kinematic-dynamic principles of interaction between a flow and an elastic surface (see Chapter 1), rather than by the elastic-damping properties of the elastic plates. Correspondence between disturbance structure in viscous sublayer and elastic plate structure is characterized by the parameters:

$$\lambda_z^+ = \overline{\lambda}_z u_\tau / \nu \tag{3.4}$$

$$\overline{y}^+ = \overline{y}\, u_\tau / \nu \tag{3.5}$$

As has been shown [467], the distance between Kline stripes equals twice the wavelength λ_z of longitudinal vortices in the viscous sublayer.

The dynamic correspondence between a disturbance motion and elastic surface is characterized by the parameters:

$$T^+ = Tu_\tau^2 / \nu \tag{3.6}$$

$$f^+ = (T^+)^{-1} \tag{3.7}$$

Here and below, the effectiveness of elastic plates means a set of positive attributes, indicating the reduction of hydrodynamic friction drag in comparison with that of a rigid standard plate.

The measurements were being made at 12 elastic plates in the following order: standard plate 1, 1a, 5, 9, 2a, 11a, 10a, 7, 3, 8a, and 11. The mechanical parameters of the plates are presented in Table 2.19, their structures in Table 2.4, and free flow velocities at the measurement sections in Table 3.1. Reynolds numbers on the standard plates in section III varied within limits $7.7 \cdot 10^5 \div 1.5 \cdot 10^6$. As it can be seen from Table 3.1, the measurements were being made with two average values of free flow velocity, 9.87 and 16.42 m/s.

Table 3.1: Range of work velocities for tests of the elastic plates.

Plate Number	Section I			Section II			Section III		
	U_∞	u_τ	u_τ/U_∞	U_∞	u_τ	u_τ/U_∞	U_∞	u_τ	u_τ/U_∞
	m/s	m/s	10^2	m/s	m/s	10^2	m/s	m/s	10^2
Standard plate	10.15	0.46	4.5	9.62	0.488	5.1	10.0	0.475	4.75
	15.98	0.744	4.66	16.99	0.778	4.6	17.06	0.774	4.53
5	9.88	0.488	4.94	–	–	–	10.07	0.44	4.4
	17.04	0.758	4.4	–	–	–	16.94	0.746	4.4
11*a*	10.13	0.425	4.2	–	–	–	9.9	0,439	4.4
	17.67	0.758	4.3	–	–	–	17.43	0.701	4.02
11	–	–	–	–	–	–	9.9	0.414	4.8
	–	–	–	–	–	–	16.37	0.558	3.4
8*a*	–	–	–	–	–	–	9.94	0.418	4.2
	–	–	–	–	–	–	15.54	0.652	4.2
3	–	–	–	–	–	–	9.93	0.398	4.0
	–	–	–	–	–	–	15.64	0.734	4.7
1	9.92	0.418	4.2	10.27	0.377	3.67	10.09	0.447	4.4
	15.53	0.595	3.83	15.69	0.562	3.58	15.86	0.549	3.5
1*a*	9.36	0.414	4.4	9.904	0.394	3.94	9.36	0.373	3.99
	15.69	0.578	3.7	15.69	0.545	3.5	17.01	0.68	3.99
9	9.49	0.371	3.9	–	–	–	9.77	0.412	4.2
	16.12	0.691	4.3	–	–	–	15.92	0.68	3.9
2*a*	–	–	–	–	–	–	9.58	0.451	4.7
	–	–	–	–	–	–	16.17	0.562	3.48
10*a*	–	–	–	–	–	–	10,07	0,42	4.17
	–	–	–	–	–	–	15.81	0.75	4.74

Besides π-parameters characterizing mechanical properties of elastic plates (see Section 2.5), and parameters $(3.3) \div (3.6)$, there is one more parameter determining the effectiveness of complex interactions, namely, roughness factor K_s [424, 445]:

$$K_S^+ = K_S u_\tau / \nu \leq 5 \tag{3.8}$$

$$K_{\lim} = K_S U_\infty / \nu \leq 100 \tag{3.9}$$

where K_s is roughness scale.

The roughness of the investigated plates ($K_s < 0.2$ mm) was less than allowable, except for plate 5.

Besides "static" roughness, there is "dynamic" roughness, caused by flexural vibrations of the plate [186, 187, 229] or by local vibrations under action of the boundary layer pressure fluctuations [287, 288]. These parameters are presented in detail elsewhere [16, 20, 290] and in Chapter 1. In the considered measurements, the "dynamic" roughness was also under the allowable limit.

The obtained velocity profiles on the standard plate agree well with other sources [123, 127, 130, 169, 283, 328, 362, 381, 385, 408, 424, 433, 524] and follow the 1/7 power law:

$$\frac{u}{U_\infty} = \left(y/\delta\right)^{1/7} \tag{3.10}$$

The experimental points of elastic plates lie under the standard plate curve as if increasing the value $1/n$ in velocity distribution law [256]. Velocity profiles on elastic plates, plotted in linear scale in the form $u/U_\infty = f\left(y/\delta\right)$, have smooth bends typical for complex interactions between different disturbances in the boundary layer.

In Figure 3.3, the averaged velocity profiles have been plotted in logarithmic coordinates to make it possible for detailed investigation of the fluid behavior in immediate proximity to the wall. Experimental points on the standard plate (series *a*) lie on curves *2* and *3*, depending on the mean flow velocity. The vertical lines denote the spread of experimental points. The obtained data agree well with data from elsewhere [435]. Curves *1* and *2* describe the flow in the near-wall region and have equal slopes. Different Reynolds numbers, as well as the turbulence level in the wind tunnels, cause their difference. Curve *3* corresponds to a larger Reynolds number and to lesser thickness of the viscous sublayer. Its slope in the near-wall region differs from that of curve *2*. Therefore, the velocity profiles allow the determining of the viscous sublayer thickness and the profile slope in the near-wall region. The value of this slope and the metrological kind of the profile shape are estimated in the series of measurements, *b*. Curve *9* denotes the Blasius profile obtained in the hydro-chamber with low turbulence scale (the first stage of transition) on rigid and elastic plates (see Chapter 1 and [305]). The profile of the heated elastic plate remains linear (curve *10*). The Blasius profile changes as turbulence increases (curve *5*); linear function comes to a parabolic function (II stage of transition). As the velocity increases and stages of transition pass, the profile remains parabolic but become closer to the turbulent profile; curve *6* corresponds with IV stage of transition, curve *7* to VI stage of transition. It should be noted that the profile (curve *4*) [435] is similar to the transitional profiles. In the end, the profile in turbulent boundary layer (curve *8*) and viscous sublayer becomes linear again, but with another slope comparing to curves *9, 10*.

These data make it possible to characterize the velocity profiles on the elastic plates. In Figure 3.3 the plates are combined as follows:

- groups *c, d* present results of measurements on the foam polyurethane plate (c) and on the foam polyurethane pasted over with elastic film (d);
- groups *e, f, g, h* present results of measurements on compliant plates;
- groups i, *j, k, l* are for elastic plates. The plate 10*a* can be considered as an elastic and rough plate due to its structure and properties.

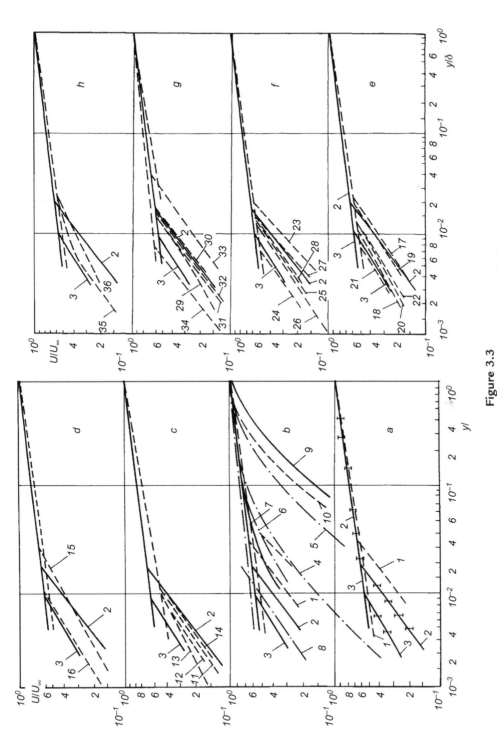

Figure 3.3

Distribution of averaged longitudinal velocity across the boundary layer thickness on: the standard plate (a), the rigid plate (b), and on elastic plates 5 (c), 2a (d), 9 (e), 1a (f), 1 (g), 10a (h), 11a (i), 11 (j), 8a (k), 3 (l): 1, 4 – data from [435] for turbulent and transitional boundary layer; 2, 3 – measurements with slow and high flow velocities; 5–8 – measurements in a hydro test chamber at different stages of transition with U_∞ equal to 0.1; 0.18, 0.27 and 0.6 m/s [310] and $\varepsilon > 0.1\%$; 9 – the same with $U_\infty = 0.08–0.15$ m/s and $\varepsilon < 0.05\%$ on rigid and elastic plates; 10 – the same as 9, but the membrane surface is heated [305]. Other notation is given in Table 3.2.

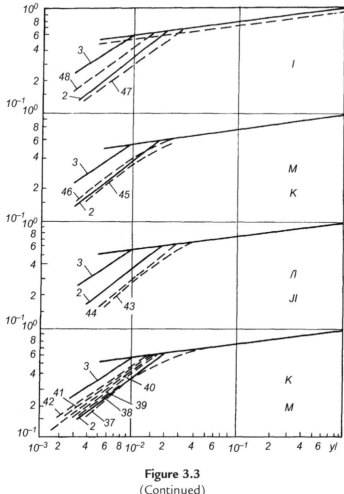

Figure 3.3
(Continued)

The important characteristic is the thickness of the viscous sublayer, which allows for understanding a physical pattern of flow in a turbulent boundary layer, and is necessary for closure of various semi-empirical theories. Nising [493] offered a formula to calculate the viscous sublayer thickness, based on the analogy of the viscous sublayer with a laminar boundary layer:

$$\overline{\delta}_{lam} = 4.64 \left(\frac{\nu x_0}{u_0} \right)^{1/2} \tag{3.11}$$

where x_0 is the length of the viscous sublayer before burst or destruction, and u_0 the velocity on the outer border of the viscous sublayer at the destruction. Some methods of definition of viscous sublayer thickness are described in the literature [142, 325, 435]. The simplest of them are:

- according to the vertical coordinate of maximum longitudinal velocity fluctuation;
- according to the point of intersection of the viscous sublayer profile and the logarithmic profile.

Based on the dimensional theory, von Karman offered a simple formula to determine viscous sublayer thickness:

$$\overline{\delta}_{lam} = \alpha \frac{\nu}{\sqrt{\tau_w/\rho}} \tag{3.12}$$

According to Nikuradse, the factor α is 11.6. In accordance with theoretical conclusions, the turbulent motion is characterized by two empirical constants: $\aleph = 0.41$ and $\alpha = 11.5$, which are determined by the formula:

$$\alpha = \frac{\overline{\delta}_{lam} u_\tau}{\nu} \tag{3.13}$$

The value α can be also determined from the relationship:

$$u^+ = \alpha - 2.5 \ln \alpha + 2.5 \ln y^+ \tag{3.14}$$

where u^+ and y^+ are taken in the point of intersection of profiles.

Recently obtained experimental data show that $\alpha = 10 \div 12.5$ [177] and $\aleph = 0.3 \div 0.41$.

Let us calculate, according to equation 3.13, the constant α for the investigated plates. The value of δ_{lam} was calculated based on experimental data in Figure 3.3, according to the point of intersection of the viscous sublayer profile and logarithmic one. The value of u_τ was determined according to Clauser method [142, 441]. Calculation results are presented in Table 3.2. The obtained law $\delta_{lam}/\delta = f$, the Reynolds number calculated on thickness of loss of an impulse is completely in agreement with [181, 501].

Measurements on the solid plate showed that $\alpha_{aver} = 11.55$. But the velocity being lower, $\alpha > \alpha_{aver}$, while the velocity being higher, $\alpha < \alpha_{aver}$. It occurs because with increasing U_∞ the value of u_τ increases more slowly than the value of δ_{lam} decreases.

In most experiments on the flow over elastic plates, the value of α increases due to the growth of thickness of the viscous sublayer, because speed u_τ decreased. Value δ_{lam}/δ has accordingly increased.

In the flow over the foam polyurethane plate (group c), the value of coefficient α was substantially increased due to roughness. Velocity profiles in the viscous sublayer become parabolic, as at VI stage of transition (curve 7, Figure 3.3). The roughness effect being

Table 3.2: Thickness of viscous sublayer, according to Figure 3.3.

Plate Number	Number and Designation	Section Number	u_τ (m/s)	$10^4\,\delta_{lam}$ (m)	$10^2\,\delta$ (m)	$10^2\,\delta_{lam}/\delta$ (m)	$\alpha = u_\tau\,\delta_{lam}/\nu$
Standard plate,	2	I	0.46	3.8	2.51	1.51	11.73
group *a*	3	I	0.744	2	2.21	0.9	9.99
	2'	II	0.488	4.1	2.78	1.47	13.4
	3'	II	0.778	2.2	2.63	0.84	11.5
	2"	III	0.475	3.7	2.93	1.33	11.8
	3"	III	0.774	2.1	2.75	0.76	10.9
5 , group *c*	11 12	I	0.488	4.7	3.32	1.42	15.39
	13	I	0.758	2.6	3.4	0.76	13.23
	14	III	0.44	4	4.1	0.98	11.8
		III	0.746	3.7	3.6	1.03	18.5
2*a* , group *d*	15	III	0.451	6.7	3.71	1.8	20.3
	16	III	0.562	2.8	3.76	0.75	11.24
9 , group *e*	17	I	0.371	4.2	2.31	1.82	10.46
	18	I	0.691	2.1	2.18	1.01	9.74
	19	III	0.412	4.4	2.73	1.61	12.17
	20	III	0.628	3.4	2.67	1.27	14.33
	21	III *a*	0.594	2.3	2.94	0.78	9.2
	22	III	0.562	3.5	3	1.17	13.2
1*a*, group *f*	23	I	0.414	3.9	2.57	1.51	10.84
	24	I	0.578	2.3	2.75	0.83	8.9
	25	II	0.39	3.1	2.74	1.12	8.1
	26	II	0.545	3.4	2.64	1.3	12.4
	27	III	0.373	3.7	2.93	1.2	9.26
	28	III	0.68	3.7	2.96	1.24	16.9
1 , group *g*	29	I	0.418	4.5	2.7	1.67	12.6
	30	I	0.595	4	2.73	1.5	15.97
	31	II	0.377	4.3	2.96	1.45	10.88
	32	II	0.562	4	2.93	1.37	15.1
	33	III	0.447	4.7	3.42	1.37	14.1
	34	III	0.549	1.64	3.28	0.5	6.04
10*a*, group *h*	35	III	0.42	2	3.83	0.52	5.64
	36	III	0.75	4.2	2.66	1.6	21.14
1.1 *a*, group *i*	37	I	0.425	4.4	2.49	1.76	12.6
	38	I	0.758	3.6	2.51	1.4	18.3
	39	III	0.439	3.7	2.89	1.28	10.9
	40	III	0.701	3.9	2.82	1.38	18.35
	41	III *a*	0.672	3.2	2.92	1.1	14.4
	42	III	0.635	3	2.825	1.06	12.8
11, group *j*	43	III	0.414	6.3	2.9	2.17	17.5
	44	III	0.558	4.6	2.82	1.63	17.23
8*a* , group *k*	45	III	0.418	4.6	2.93	1.6	12.9
	46	III	0.652	3.8	2.82	1.35	16.6
3, group *l*	47	III	0.398	5.1	2.95	1.73	13.62
	48	III	0.734	3.1	2.87	1.08	15.3

eliminated (group *d*), the velocity profile becomes similar to that of a standard plate and coefficient α is substantially increased due to good mechanical parameters (the properties of this plate were the closest to optimal).

In groups *e, f, g,* as well as in groups *c, d* (see Figure 3.3), elastic plates with outer layers of solid plate were tested. The outer layer thickness of elastic plates of group *f* was 1.5 times as much as that of group *e*. The outer layer thickness of group *g* was 2.5 times as much as that of group *e*. These plates were tested to investigate the mechanical parameters effect upon the complex interactions. In accordance with criteria of similarity, the effectiveness of the interaction depended upon the Reynolds numbers: coefficient α depended on the location of the measurement point and mean flow velocity. In group *e*, the value of α increased only in sections III and III*a*. In group *f*, the compliance of the plate increased due to increasing thickness. However, essential changes were not revealed: the coefficient α increased only in sections II and III at greater speed. Only in group *g* did the value α increase in all sections, particularly in sections II and I.

The effectiveness of the compliant plates substantially depends on their mechanical parameters. Slight changes in mechanical parameters led to appreciable changes in effectiveness. To eliminate the disadvantage, the compliant plate was structured to generate below a set of longitudinal vortices (see Chapter 1 and Figures 2.28–2.31). The testing of this plate (10*a*) showed a substantial increase in δ_{lam} and α. The velocity profile became a transition one (parabolic).

Due to the increasing of elasticity, plate 10*a* can be classified among the elastic plates, which have to be analyzed in accordance with the principles and criteria in equations (3.4) to (3.7). The determination of these criteria is presented in Table 3.3. The value of λ_z^+ and the frequency of bursts from the viscous sublayer f_b^+ for the standard plate is taken as in these sources [118, 136, 153, 177, 225, 274, 435]. From equations (3.6) and (3.11) we can obtain the period of regeneration of viscous sublayer:

$$T^+ = \left(\frac{u_\tau \overline{\delta_l}}{4.64\nu}\right)^2 \tag{3.15}$$

In equation (3.11) δ_l is the thickness of viscous sublayer, including the laminar and buffer layers, and was determined by intersection of the curves of velocity profile in the turbulent core and buffer zone. The values of δ_{lam} are, therefore, larger than in Table 3.2, and it is more convenient to determine them by the velocity profiles expressed in universal coordinates. The value of T^+, calculated for section I and for $U_\infty = 10.15$ m/s on the rigid plate, is $T^+ = 215$; and $f^+ = 4.65 \cdot 10^{-3}$. The values agree well with the known data.

In Table 3.3, for the compliant plates, values for λ_z^+ are taken as for the standard plate. The values of λ_z^+ are underestimated as can be seen from the measurements of correlation coefficients. The value of f_b^+ was determined from the results of measurements of natural

Table 3.3: Parameters of flow in the viscous sublayer.

	U_∞ (m/s)	λ_z^+	$10^2\,\lambda_z$ (m)	$10^3 f_b^+$	f (1/s)	Re^{**}
Plate Number				**Section I**		
Standard plate	10.15	80−100	0.26−0.32	3.5−4.1	49.7−058.2	1662
	15.98	80−100	0.16−0.2	3.5−4.1	130−152	2304
5	9.88	80−100	0.24−0.3	4.88	78	2140
	17.04	80−100	0.16−0.2	2.02	78	3794
11a	10.13	48.5	0.17	19.72	23	1645
	17.67	86.5	0.17	6.2	239	2894
11	−	−	−	−	−	−
	−	−	−	−	−	−
8a	−	−	−	−	−	−
	−	−	−	−	−	−
3	−	−	−	−	−	−
	−	−	−	−	−	−
1	9.92	80−100	0.29−0.36	18.2	214	1748
	15.53	80−100	0.2−0.25	9.0	214	2766
1a	9.36	80−100	0.29−0.36	20.1	246	1569
	15.69	80−100	0.2−0.26	11.0	246	2815
9	9.49	80−100	0.32−0.4	14.8	137	1430
	16.12	80−100	0.17−0.2	4.3	137	2292
2a	−	−	−	−	−	−
	−	−	−	−	−	−
10a	−	−	−	−	−	−
	−	−	−	−	−	−
Plate Number				**Section III**		
Standard plate	10.0	80−100	0.25−0.31	3.5−4.1	53−62	1912
	17.06	80−100	0.15−0.19	3.5−4.1	141−165	3061
5	10.07	80−100	0.27−0.34	6.0	78	2676
	16.94	80−100	0.16−0.2	2.1	78	3968
11a	9.9	50.1	0.17	18.5	239	1867
	17.43	80.0	0.17	7.25	239	3207
11	9.9	47.2	0.17	16.1	185	1872
	16.37	63.7	0.17	8.85	185	3012
8a	9.94	80−100	0.29−0.36	19.9	233	1901
	15.54	80−100	0.18−0.23	8.2	233	2859
3	9.93	80−100	0.3−0.37	15.9	169	1910
	15.64	80−100	0.16−0.2	4.67	169	2929
1	10.09	80−100	0.27−0.33	16.0	214	2251
	15.86	80−100	0.2−0.27	10.6	214	3394
1a	9.36	80−100	0.32−0.4	26.3	246	1789
	17.01	80−100	0.17−0.22	7.9	246	3283
9	9.77	80−100	0.29−0.36	12.0	137	1739
	15.92	80−100	0.19−0.24	5.2	137	3773
2a	9.58	80−100	0.26−0.33	6.1	83	2319
	16.17	80−100	0.21−0.27	3.9	83	3965
10a	10.0	112.8−141	0.4−0.5	19.5	23	2516
	15.8	201−252	0.4−0.5	6.1	23	2743

vibration frequency of elastic plates, because the bursts of viscous sublayer are very affected by the oscillations of the elastic surface of plates. For elastic plates, the values of λ_z^+ were being determined in accordance with the structure of the outer layer (see Table 2.4) and the values of f_b^+ by measuring the dynamic properties of the plates (see Table 2.19).

The results presented in Table 3.3 show that interaction effect on plates 5 and 2a takes place only because of their mechanical properties. Thus, such compliant plates can be successfully used with low Reynolds numbers (see Figure 1.42). The compliant plates 1, 1a, 9, and 10a have such values of f_b^+ that they could, in accordance with the pre-burst modulation hypothesis [137, 138], effectively interact with boundary layers by regulation of the burst frequency from the viscous sublayer. But this mechanism, due to the large compliance of the plates, did not work. It can be seen from the fact that large values of f_b^+ (see Table 3.3) correspond to low values of α (see Table 3.2).

The effectiveness of the compliant plates with raised elasticity can be increased by structural interaction. So, the raised value of α corresponds to $\lambda_z^+ = 201-252$. The value of λ_z^+ of polymer solutions has also increased by up to 200 times [385]. Such a trend is also registered with high negative gradients of pressure [183].

All elastic plates (see Figure 3.3, groups *i, j, k* and *l*) have an increased values of α by up to two times in comparison with the standard plate. The value of λ_z^+ was either the same as for the standard plate or 1.5−2 times less. The value of f_b^+ was 1.5−5 times higher. It is obviously that the effectiveness of these plates is caused by the "pre-burst" modulation (the value of f_b^+ increased). The principle of structural interaction is also of substantial importance. Therefore, the maximal increasing of the viscous sublayer thickness was registered for plate 11, which had the value of λ_z^+ 2 times less than the standard plate. But, taking into account that λ_z of coherent vortical structures in the viscous sublayer of the standard plate is 2 times less than the Kline stripes, we can see a good correspondence of λ_z to the distance between the longitudinal structures in the elastic material.

From this follows that λ_z of an elastic plate structure must be either equal to λ_z of longitudinal vortices in a viscous sublayer or 2 times higher, to correspond to "peaks" and "valleys" of vortices.

The obtained results are presented in Figure 3.4 in universal semi-logarithmic coordinates. The experimental points in the viscous sublayer are clustered along the curve $u^+ = y^+$, and in turbulent core between the solid curve:

$$u^+ = 5.61 \lg y^+ + 4.9 \tag{3.16}$$

and the dashed curve [272]:

$$u^+ = 2.5 \ln y^+ + 5.5 \tag{3.17}$$

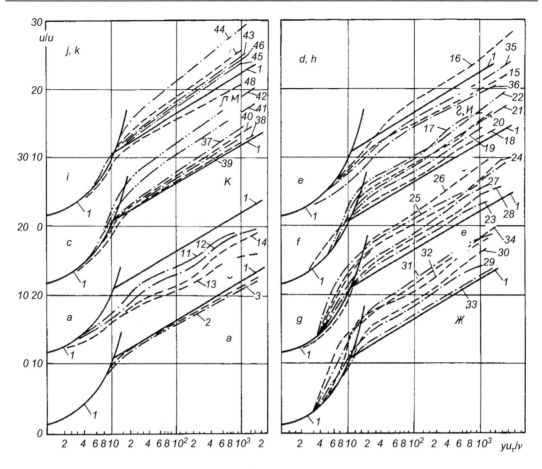

Figure 3.4
Distribution of mean velocity across the boundary layer. Designations are as for Figure 3.3.

In [49, 325, 425, 435, 441, 498] on the basis of dimensional theory and taking into account the condition of matching profiles in the viscous sublayer and turbulent core of the boundary layer, the formulae of the universal profile of velocity were derived:

$$u^+ = A \ln y^+ + B$$
$$u^+ = \frac{1}{\aleph} \ln y^+ + \alpha - \frac{1}{\aleph} \ln \alpha$$
$$u^+ = \frac{2.303}{\aleph} \lg y^+ + \alpha - \frac{2.303}{\aleph} \lg \alpha$$

(3.18)

where α is calculated according to (3.13).

After calculating the values of coefficients $A = 2.303/\aleph$ and $B = \alpha\,(2.303/\aleph)\,\lg\alpha$ (α is taken from Table 3.2) we can get, for example, from formula (3.16) for section I of standard plate at $U_\infty = 10.15$ m/s:

$$u^+ = 5.611\ \mathrm{g}y^+ + 5.7 \tag{3.19}$$

It is seen from Figure 3.4 and Table 3.2 that the constants $\aleph$ and α are varied even for the rigid plate. So the coefficients A, B in the formulae (3.18) vary. It is a result of the fact that the formula (3.18) is derived based on the two-layer model, without taking into account the buffer layer. In [177], the limits of A and B along rigid plates, in pipes and channels are presented.

Recently some attempts were made to obtain a formula for the mean velocity profile taking into account the buffer layer. The equations of velocity profiles in universal coordinates are presented in Table 3.4.

The velocity can also be written as follows [142]:

$$\begin{aligned}
u^+ &= \varphi_1(y^+) + [\textstyle\prod(x)/\aleph]\,\omega\,(y/\delta) \\
\varphi_1 &= (1/\aleph)\ln y^+ + c
\end{aligned} \tag{3.20}$$

where with $y^+ \geq 50$, $C = 5.0$. The parameter Π is a function of x and the function $\omega(y/\delta)$ at zero pressure gradients can be written as follows:

$$\omega(y/\delta) = 2\,\sin^2\left[(\pi/2)y/\delta\right] \tag{3.21}$$

If $Re^{**} > 5000$, than $\Pi = \mathrm{const} = 0.55$. If $Re^{**} < 5000$, the parameter Π varies according to the law presented in [142]:

$$\textstyle\prod = 0.55\,[1 - \exp\,(-\,0.243)\,\zeta_1^{1/2} - 0.298\,\zeta_1] \tag{3.22}$$

where $\xi_1 = (Re^{**}/425 - 1)$.

In the paper [142], it is shown that, if $Re^{**} < 6000$, the values of $\aleph$ and C (see Table 3.4) are not constant and can be described as follows:

$$\aleph = 0.4(Re^{**}/6000)^{-1/8} \tag{3.23}$$

$$c = Re^{**1/8}\,(7.9 - 0.737\ln|Re^{**}|) \tag{3.24}$$

The value $\aleph$ can be also calculated as follows:

$$\aleph = 0.4 + \frac{0.19}{1 + 0.49(Re^{**}10^{-3})^2} \tag{3.25}$$

According to [514], at $425 < Re^*600$ $\aleph = 0.0013\ Re^{**} < {}^* - 0.362$. For the standard plate, the values of Re^{**} in the measurement sections (see Table 3.2) were: 1662; 2304; 1745;

Table 3.4: Equation of velocity profile of the turbulent boundary layer.

Region of Boundary Layer (BL)	Location in Section of BL	Equations	Author (Researcher)
Laminar sublayer (LS)	$0 \le y^+ \le 5$	$u^+ = y^+$	Schlichting [445], Prandtl [445]
	$0 \le y^+ \le 4$	$u^+ = y^+$	Daily [156]
	$0 \le y^+ \le 5$	$u^+ = Re_\tau(\eta + 1/2\Phi\,\eta^2)$, $Re_\tau = u_\tau\,\delta/v$, $\Phi = (\delta/\tau)$ $\partial p/\partial x + u_\tau\,\delta/v$, $\eta = y/\delta$	Fedjaevsky [181]
	$0 \le y^+ \le 5$	$u^+ = Re^*_{\delta lam}\, y/\delta_{lam}$	Nikitin [381], Sherstjuk [462]
Buffer layer (BL)	$5 \le y^+ \le 30$	$u^+ = 5\,1ny^+ - 3.05$	Daily [156]
	$5 \le y^+ \le 30$	$u^+ = 11.5\,1ny^+ + 5$	Karman [445]
	$10 \le y^+ \le 70$	$u^+ = 11.7\,1ny^+ - 17$	Schlichting [445]
LS and BL	$0 \le y^+ \le 13.2$	$u^+ = y^+ - 22.86 \cdot 10^{-3} y^{+2}$	Povh [425]
	$0 \le y^+ \le 26$	$u^+ = \int_0^y dy^+/[1 + B(1 - e^{-B})]$, $B = C^2\, u^+\, y^+$	Spolding [493]
	$0 \le y^+ \le 20$	$u^+ = 2\int_0^y \left\{1 + \sqrt{4k^2 y^+[1 - exp(-y^+/a)]^2 + 1}\right\}^{-1} dy^+$	Fedjaevsky [181]
	$0 \le y^+ \le 27.5$	$u^+ = 14.54\,\tanh(0.068y^+)$	Davies [157]
Logarithmic law (LL)	$y^+ \ge 13.2$	$u^+ = 2.5\,1ny^+ + 5.5 - 36.08/y^+$	Povh [424], Repik [435]
	$y^+ \ge 30$	$u^+ = 5.75\,\lg y^+ + 5.5$	Karman [445]
	$y^+ \ge 27.5$	$u^+ = 2.5\,\lg y^+ + 5.5$	Rao, Povh [441, 424]
		$u^+ = 5.5\,\lg y^+ + 5.45$	White [514]
	$y^+ \ge 60$	$u^+ = 5.75\,\lg y^+ + 5.24$	Laufer [493], Loytsjansky [362], Schlichting [445]
	$y^+ \ge 70$	$u^+ = 5.75\,\lg y^+ + 5.7$	Fedjaevsky [181], Goldstick [206], Klebanoff [272]
	$y^+ \ge 30$	$u^+ = 5.75\,\lg y^+ + 4.2$	Daily [156]
		$u^+ = a - 2.5\,1n\,a + 2.5\,1n\,y^+$	Fedjaevsky [181]
	$y^+ \ge 20$	$u^+ = 5.61\,\lg y^+ + 4.9$	Zhukauskas [522], Reynolds [433],
		$u^+ = 5.75\,\lg y^+ + 5.1$	Klein, Smith [274]
	$y^+ \ge 26$	$u^+ = 2.781\,ny^+ + 3.8$	Spolding [177]
	$y^+ \ge 25$	$u^+ = \varphi(y^+) + [\Pi(x)/\aleph]\,\omega(y/\delta)$	Couls [153]
	$y^+ \ge 30$	$u^+ = Re^*_{\delta l}\,(1.15\,\lg y/\delta_l + 1.5 - 0.5\,\delta_l/y)$, $u^+ = Re^*_{\delta l}$ $[1.15\,\lg y^+ + 1.5 - 1.15\,\lg Re^*_{\delta l}) - 0.5\,Re^*_{\delta l}v/u_\tau y]$	Nikitin [381]
	$y^+ \ge 20$	$u^+ = 6.2\,\lg y^+ + 4.5$	Kont-Bello [283]
		$u^+ = 2.6\,1n[1 + 0.39(y^+ - 7.8)] + 7.8$	Millionschikov [424]
		$u^+ = 5.31\,\lg y^+ + 5.8$	Repic [435]
LS + BL + LL	$y^+ \ge 0$	$u^+ = 2.5\,\ln(1 + 0.4\,y^+) + 7.8[1 - e^{-y^+/11} - (y^+/11)\,e^{-0.33\,y+}]$	Spalding [177]
	$y^+ \ge 0$	$u^+ = 2\int_0^y dy^+/[1 + (1 + 0.64y^{+2}C^2)^{1/2}]$ $C = 1 - exp\,(-y^+/26)$	Van-Drist [445]
	$y^+ \ge 0$	$f(y^+) = y^+ exp[Kg\,(\pi, y/\delta)]$	Goldstein [207]
	$y^+ \ge 0$	$u^+ = 6.5\,\tan^{-1}(y^+/6.5) + 2.5\,\ln[1 + (y^+/6.5)^2]^{1/2}$	Sill [466]

2916; 1912 and 3061. So the formula (3.19) of the measured velocity profile in Section I can be expressed by means of Couls' parameter Π according to formulae (3.20)–(3.25) as follows:

$$u^+ = 4.91 \lg y^+ + 6.15 + 1.62 \sin^2\left[(\pi/2)y/\delta\right], \; \aleph = 0.48 \tag{3.26}$$

Extending the concepts of [181], the formulae to determine the averaged velocity profiles in the viscous sublayer and in the region of logarithmic law are presented in [381]:

$$\frac{u}{u_{*\overline{\delta}}} = \mathrm{Re}_{\overline{\delta}}^* - y\overline{\delta} \tag{3.27}$$

$$\frac{u}{u_{*\overline{\delta}}} = \mathrm{Re}_{\overline{\delta}}^* - \left(1.15 \lg \frac{y}{\delta} + 1.5 - 0.5\frac{\overline{\delta}}{y}\right) \tag{3.28}$$

where $u_{*\overline{\delta}}$ is the friction velocity at the upper boundary of the sublayer and $\overline{\delta}$ is thickness of the sublayer with linear law of velocity distribution. This value is less than δ_{lam} in Table 3.2 and must be approximately equal to $\delta_{lam}/2$. The parameter $\mathrm{Re}_{\overline{\delta}}^*$ is a proportionality coefficient determining the thickness of the near-wall layer δ_{lam}. To determine the unknown values in (3.27) and (3.28), we can use the nomographic chart presented in [381]. In order to do it, one has to plot the experimental points in coordinates $\lg u = f(\lg y)$ on the chart. The paper [381] presents also other functions, where the parameter δ_{lam} is a thickness of water layer in a river or channel. Thus, they can be used to determine the kinematic characteristics of such flows.

The formulae (3.27) and (3.28) can be used in other ways. One can plot the velocity profile in dimensional coordinates and determine the value of $\overline{\delta}$ as the value of y in the point, where the profile begins to diverge from linear law. The value of u_τ, obtained by other methods, almost does not differ from $u_{*\overline{\delta}}$. Therefore, the parameter $\mathrm{Re}_{\overline{\delta}}^*$ is equal to u/u_τ at $y/\overline{\delta} = 1$. The error of this method is higher than for the method in [381]. However, the approximate method is very simple. Its accuracy can be increased by correcting the value of $\mathrm{Re}_{\overline{\delta}}^*$ in (3.28) to achieve correspondence with experimental points. The formula (3.28) obtained in this way for the same conditions as (3.18) are as follows (the values of y and δ are in mm):

$$u^+ = 5.754\left(1.15 \lg \frac{y}{0.15} + 1.5 - 0.5\frac{0.15}{y}\right) \tag{3.29}$$

The equation (3.29) slightly overestimates the experimental curve because of difference between the u_τ and $u_{*\overline{\delta}}$.

Comparison of formulae (3.16), (3.19), (3.26) and (3.29) shows that experimental points better correspond to formulae (3.16), (3.26) and (3.29) are lower then the lines plotted according to formulae (3.19), (3.29) and (3.26). The high values of y^+ — (3.29) also gives overestimated values in comparison with experimental values. The formulae given in [425] (see Table 3.4) are also convenient to use.

In [274, 277] the experimental profiles of the turbulent boundary layer in a water flow were brought to dimensionless form in two ways: by using the value u^* derived from the modified Clauser method, and by using the value u_τ, determined from the distribution of velocity near the wall. Results of measurements presented in papers [274, 277] and brought to dimensionless form by using the dynamic velocity u^* correspond to the formula (3.16) and agree with our data for the rigid standard plate. At $y^+ > 100$, our data differ from the results in [274, 277]. As it is shown in [49, 274, 277, 408, 441, 524], the difference can be explained by the Reynolds number effect upon the velocity profile, by negative pressure gradient caused by the thickening of the boundary layer over the walls of the wind tunnel, and by high turbulence level of the mean flow.

At positive gradients of pressure, the structure of speed near the outer border of the boundary layer rises upwards, and at negative gradients it goes downwards. The results received on the elastic plates (see Figure 3.4) have shown that the gradient of pressure was the same as at the flow on the standard. Thus, the distinction of the forms of velocity profiles on rigid and elastic plates was caused by specific features of the flow on elastic plates and the mechanism of complex interactions of various disturbances in the boundary layer.

The analysis of velocity profiles in Figure 3.4 is carried out in the same order. The rigid plate (see Figure 3.4a) and the rough elastic plate 5 (group b) were the standard plates. The obtained velocity profiles are typical for rough plates [142, 181, 381, 462, 473] and, as it follows from (3.8), the roughness of the plate increased with the increasing of U_∞.

When the flow velocity was low (see Figure 3.4d), the rubber film pasted over the elastic plate 5 caused the decreasing of roughness, and when the velocity was high, the roughness increased. However, at $y^+ > 30$, due to elasto-damping properties effect upon the parameters of the boundary layer, the velocity profile lies under the profile of the standard plate.

Comparing the groups *e, f, g* presented in Figures 3.3 and 3.4 we can conclude that the effectiveness of the compile plates is proportional to the value ΔB of deviation of velocity profile from that of the standard plate. The flow over plate *10a*, which had a rough surface, gave the curves laid under the standard as in the group *c*.

In groups *i, j, k, l,* the measurement results of the elastic plates are presented. Growth of size ΔB characterizes a positive effect. Based on this it is possible to conclude that plates *11a* and *11* obtained better results than plates *8a* and *3* (the best results are obtained on plate *11*). Comparing their mechanical (see Table 2.19), structural and dynamic (see Table 3.3) properties, one can conclude that the better results are caused by structural and dynamic properties more than by mechanical ones. Therefore, the natural oscillation frequency of plates *11a, 11* and *8a* is higher than of plate *3*. The plates *11a* and *11,* the outer layers of which are longitudinally structured, have a lower anisotropy coefficient K_2 than the plates *8a* and *3*, i.e. the disturbance damping is more intensive in the longitudinal direction than in transverse direction. All this promoted intensive generation of longitudinal vortices in the viscous sublayer over the plates *11a* and *11,* and their development was more stable. In addition, the generation of turbulence decreases due to the mechanism of pre-burst modulation.

As can be seen from Figure 3.4 the effectiveness of elastic plates depends on different factors; the mechanical properties depend on the local Reynolds number. That is why the curves in the groups differ so significantly from each other.

Behavior of curves in the region of logarithmic law can be explained either by the changing of the pressure gradient along the plate or by the changing of the constant $\aleph$. In all experiments, the slope of the curves changed only because of changing $\aleph$. It follows from the fact that the pressure gradient did not change along the standard and elastic plates, and the curve slope on the standard plate decreased due to outer turbulence. It should be noted that in most cases the turbulence level at the outer boundary of the boundary layer over the elastic plates was higher owing to oscillations of the outer layers of the plates.

There is one more essential difference in the form of the velocity profiles of the flow on elastic plates: the experimental points on the elastic plates, unlike the standard plates, lie to the left of the laminar profile in the viscous sublayer. This fact, as well as the law of location of experimental points in the region of logarithmic law, can be explained taking into account the complex interactions in the boundary layer (see Section 1.14) in the following way. As it follows from the analysis of velocity profiles presented in Section 1.11 and in [310], the vortex generators amplify the longitudinal vortices of the boundary layer and transform the Blasius profile into a transitional profile. The same can be shown with heated membrane surfaces (see [310] and Figure 3.3) and with negative pressure gradients [435].

The velocity profiles on the elastic plates (see Figure 3.4) in the laminar sublayer and in buffer layer are exactly the same as the profiles in [310], when the longitudinal vortical disturbances were being inserted in the boundary layer. This means that in the specified areas of the boundary layer on elastic plates longitudinal vortical systems are formed, the sizes and intensity of which lead to a thickening of the viscous sublayer and changing of kinematic parameters of the boundary layer in the near-wall area. This, in turn, influences the energy balance of the boundary layer, which causes the change in the value of $\aleph$ in the region of logarithmic law of velocity profile.

Thus, it has revealed that there is a phenomenon of self-structuring of stable longitudinal vortices in the near-wall region of the turbulent boundary layer and in the transitional boundary layer over the elastic plates. The mechanism of the phenomenon depends, in particular, on the behavior of surface wave propagation in elastic plates (see Chapters 1 and 2).

To define the velocity profile analytically by means of (3.17), it is necessary to know two values, $\aleph$ and α. The value of α is determined from Figure 3.3. Let us write (3.17) in the form:

$$u^+ = \frac{1}{\aleph} \ln \frac{y^+}{\alpha} + \alpha \tag{3.30}$$

Now we have to select the value of $\aleph$ to correspond to the experimental points. For plate *11* at high velocity (see Figure 3.4j, Tables 3.1 & 3.2) the formula (3.16) in the region of logarithmic law will be as follows ($\aleph = 0.33$; $\alpha = 16$):

$$u^+ = 6.98 \lg y^+ + 7.6 \tag{3.31}$$

It can be noted that with low Re** the value of $\aleph$ on the rigid plate can rise up to 0.48, and on the elastic plate $\aleph$ can decrease to 0.33. Similar decreasing of $\aleph$ is registered in [225, 385] during the investigations of polymer solutions.

In [425], the expressions are obtained for the calculation of the boundary layer profiles in weak solutions of polymers:

$$
\begin{aligned}
0 \leq y \leq \delta_{lam} \quad & u^+ = y^+ - G_1(y^+)^2 \\
\delta_{lam} \leq y \leq \delta \quad & u^+ = \frac{1}{\aleph} \ln y^+ + B + \frac{G_2}{y^+} + \frac{\prod(x)}{\aleph}\omega \\
\text{where} \quad & G_1 = \frac{1}{3\alpha\overline{A}}\left(1 - \frac{1}{2\aleph\alpha\overline{A}}\right) = \frac{1}{2\overline{\delta}_{lam}} \\
& B = \alpha\overline{A} - \frac{1}{\aleph}\ln\alpha\overline{A} - \frac{1}{2\aleph} \\
& G_2 = \frac{2\alpha\overline{A}}{3}\left(\frac{1}{\aleph} - \alpha\frac{\overline{A}}{2}\right)
\end{aligned}
\tag{3.32}
$$

$\overline{A}$ is anisotropy coefficient of dynamic viscosity.

In this paragraph the empirical method of calculation of an averaged longitudinal structure of speed of a boundary layer results in the injection of small concentrations of polymer solutions. Values of the empirical constants necessary for carrying out calculations are given. The method is described in more detail in [425]. In [142] the explanatory parameter is given. In Russia and Ukraine some monographies and many articles in which various semi-empirical methods of calculations of kinematic parameters of a boundary layer are developed at the injection of polymer solution are published. In Chapter 7 one such method is given.

The profiles on the elastic plates with a positive effect (see Figure 3.4) are described by the equation (3.32) with $\overline{A} = 1.1 - 1.3$. The measurements of mechanical parameters of these plates (Table 2.19) showed that anisotropy coefficient K_2 is within the same limits but reverses its sign. Such coincidence of $\overline{A}$ and K_2 verifies the hypothesis that anisotropy of properties of elastic plates in transverse direction leads to anisotropy of properties of fluid, as well as to the generating and sustaining of longitudinal vortices. The fluid anisotropy takes effect not in the changing of viscosity, but in anisotropy of properties of fluctuating velocity field and the near-wall shear. Thus, the following procedure can be suggested to calculate the boundary layer. The value of K_2 is determined by mechanical measurements, and α determined by measuring the velocity profiles, using the data in Figure 3.3. It is known that B = 4.9. Having defined values α and B, we can find $\overline{A}$ and compare to K_2. By means of K_2, we make a correction of $\overline{A}$. After that, we can use the calculation method presented in [142, 169, 181, 232, 424, 441, 501, 503]. Experimental research of the turbulent boundary layer on elastic plates, as well as the transitional boundary layer, has revealed a change in the ratio of characteristic linear scales in comparison with rigid plates. For example, all elastic plates increase the values of δ_{lam} and δ. Only the elastic plates generating longitudinal vortical systems from below somewhat decreased δ.

3.3 Profiles of Fluctuation Velocities

The fluctuation velocities give more information about the complex interactions of disturbance motions in the boundary layer. Figure 3.5 shows the values of longitudinal fluctuation velocities on the rigid and elastic plates. The vertical lines denote the measurement error. Some data showed in Figure 3.5 are presented in [256]. The data are agree well with those presented in other research [169, 177, 272, 283, 325, 362, 381, 425, 435, 441, 473, 493, 524].

The logarithmic *x*-coordinate axis allows for defining the laws of fluctuations in the viscous sublayer more clearly. It can be seen from Figure 3.5 that at the flow on a rigid plate the intensity of velocity longitudinal fluctuation is maximal near the boundary of the laminar sublayer. The increasing of mean flow velocity leads to the decreasing of fluctuations, in agreement with [381]. The increasing of the Reynolds number, calculated with $U_\infty = const$, was not followed by substantial decreasing of maximal longitudinal pulsations because the measurement sections were located close to each other.

Analyzing the results presented in Figure 3.5, we must take into account that the intensity of vortices in the near-wall region of the boundary layer increases due to the high turbulence level in the wind tunnel [48, 54, 105]. It causes a response in the elastic materials according to their structure and mechanical properties. The compliant plates absorb, more or less, the energy of the pulsating field and the elastic plates generate 3D disturbances depending on the energy of the pulsating field and the structure of the outer

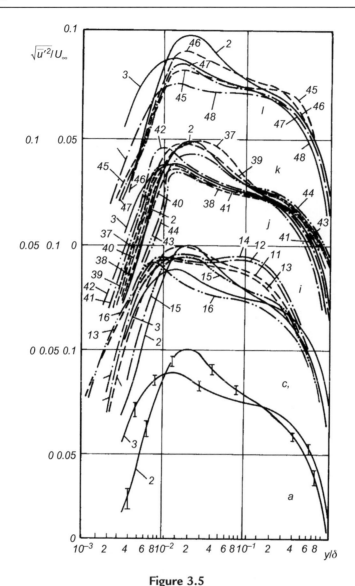

Figure 3.5

Distribution of longitudinal pulsating velocity across the boundary layer over the rigid and elastic plates. Designations are as for Table 3.2.

layer of the plates. The disturbances arising in elastic plates interact with pulsating field of the boundary layer and modify it.

The pulsating characteristics of rough elastic plate *5*, as well as the averaged velocity profiles, can be measured. The pulsating curves of the rough plates have no clear maximum (curves *11−14*). It is strongly stretched by y/δ in the region of the maximal values.

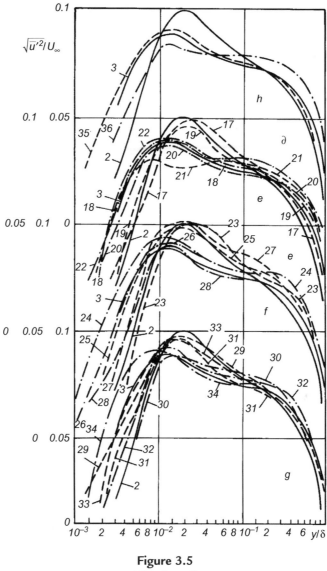

Figure 3.5
(Continued)

The velocity and Reynolds number being increased, the second maximum is formed (curves *12* and *14*). Thus, the first maximum is located closer to the surface than at the rigid standard. The pulsating curves of plate *5* pasted over with elastic film are the same as for a rigid plate.

The pulsating field is influenced by the oscillations of the plate surface. Because of this, the location of the maximum of the longitudinal velocity fluctuations does not characterize the thickness of the viscous sublayer, as it does on the rigid plate.

The groups *e, f, g, h* present the results of measurements of longitudinal fluctuations on the compliant plates. Let us consider the dependence of effectiveness of compliant plates upon their mechanical parameters, in particular upon parameters determined by the thickness of plates. The thickness of plate *9* (group *e*) is the least (4 mm). Its pulsating profiles practically coincide with those of the standard plate. The thickness of plate *1a* (group *f*) is more (6 mm) than of plate *9*. The maximal values of its pulsating velocity are the same as for the standard plate and the form of profiles are changed: the maximum is extended by *y*. The profile at $U_\infty, = 9$ m/s in the region of maximal pulsations has a wide direct strip— similar to that in complex interactions in a transitional boundary layer [134]. The conditions of the experiments were similar: the plate surface oscillated in the longitudinal direction with an amplitude of 0.5 mm and a frequency of 10 Hz. It is remarkable that the friction drag was very low in this case.

Plate *1* was the thickest (10 mm, group *g*). The value of maximal fluctuations did not decrease. At a low velocity in section III the thickness of the viscous sublayer was slightly increased.

In group *h*, characterizing plate *10a*, in spite of its increased roughness compared with plates *1, 1a* and *9*, the fluctuation maximum decreases and the profile is transformed closer to the transitional profile. It is caused by generating disturbances in a boundary layer of plate *10a* from below in the form of longitudinal coherent vortical structures. The value of λ_z increased (see Table 3.3). The decreasing fluctuations can be explained by the complex interactions between the 3D longitudinal vortices in the viscous sublayer and the vortices generated by plate *10a*.

The largest change in longitudinal pulsations was registered in the flow on elastic plates in groups *i* and *j*. There, all three positive factors are indicated:

- pulsation maxima decreased;
- values of maxima have moved away from the wall to the thickness of a boundary layer (i.e. the value of α increased) and stretched on *y*;
- profiles of fluctuation velocities became similar to those in a transitional boundary layer at the last stages of transition, when systems of longitudinal coherent vortical structures were generated in the boundary layer (see Chapter 1).

The pulsations, unlike the averaged velocities, according to complex interactions principles (see Chapter 1) were decreased more by plates *8a* and *3* (groups *k* and *l*), which have transverse structured outer layers. The effectiveness depended on the Reynolds number and increased at higher flow velocity, when the value of λ_z^+ increased and the value of f^+ decreased, and parameters of similarity approached their optimum values. The effectiveness of pulsating characteristics of plates *11a* and *11* depended on the values of λ_z^+, and of plates *8a* and *3* on the value of f^+. The values of λ_z^+ and f^+ were the same, or twice as high than at the rigid standard. The results indicate the structural interaction and pre-burst modulation.

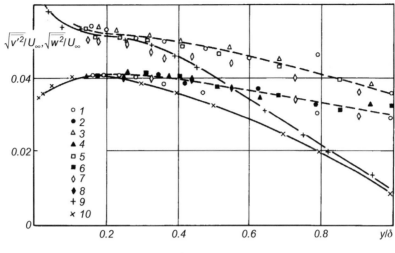

$\sqrt{\overline{v'^2}}/U_\infty, \sqrt{\overline{w^2}}/U_\infty$

○ 1
● 2
△ 3
▲ 4
□ 5
■ 6
◇ 7
◆ 8
+ 9
× 10

Figure 3.6

Transversal velocity fluctuations in the turbulent boundary layer over a rigid plate: *1, 3, 5, 7, 9* — velocity fluctuations *v; 2, 4, 6, 8, 10* — velocity fluctuations *w; 1–4* — section I; *5–8* section III; *1, 2 — U_∞* = 10.15 m/s; *3, 4* — 15.98 m/s; *5, 6* — 10 m/s; *7, 8* — 17.06 m/s; *9, 10* — measurements [109, 190].

The distribution of root-mean-square values of transversal velocity fluctuations in the boundary layer over the rigid plate can be seen in Figure 3.6. It is well agreed with the measurements in a channel [177, 272, 283]. The transversal fluctuation velocities measured in [272, 435] differ from the results of our measurements because of the different experimental conditions, especially the high turbulence level of the flow in the wind tunnel. In our experiments, the measurements near the wall were limited by the design of the X-shaped gauge.

Figure 3.7 presents the results of measurements on elastic plates of three kinds. The compliant plate slightly influences the transverse fluctuations in accordance with the structure of the outer layer, and only near the surface do the transverse fluctuations slightly decrease.

The transverse pulsations near the surface of the compliant plate *10a* generating the 3D longitudinal vortices systems increase, but at *y/δ* > 0.4 they decrease. The same picture is seen on the elastic plate *11a*. Elasticity of plate *11a* is higher than elasticity of plate *10a*. Therefore, the amplitude of fluctuations of the surface of plate *11a* is larger, and it influences the most part of the boundary layer thickness. Therefore, longitudinal vortical structures over plate *11a* have greater height than those over plate *10a*.

The transversal fluctuations *v'* decrease on the entire thickness of the boundary layer. Qualitatively the dependences of fluctuation velocities on the boundary layer thickness have similar characteristics as in tests with solutions of polymers [222, 223, 385, 425]. The form of fluctuations velocities on the rigid standard is defined by the degree of turbulence of the

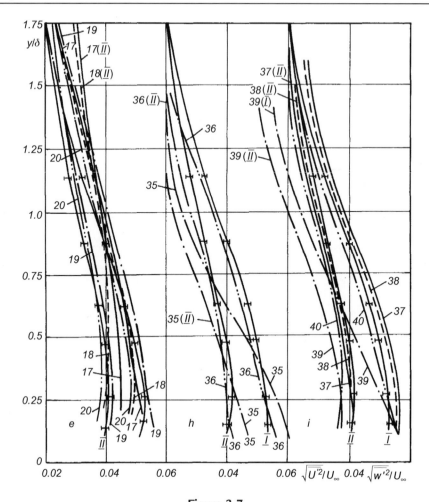

Figure 3.7

Distribution of transversal velocity fluctuations in the boundary layer over the rigid and elastic plates. The solid lines I and II denote fluctuation velocities v and w on the standard plate. Designations are as for Table 3.2.

flow in the channel. Measurements have shown that on the external border of boundary layer $y/\delta = 1$, fluctuation velocities had the following values: $\sqrt{\overline{u'^2}}/U_\infty \approx 0.014 - 0.016$, $\sqrt{\overline{v'^2}}/U_\infty \approx 0.03$, $\sqrt{\overline{w'^2}}/U_\infty \approx 0.035$. Moving the sensors of the hot-wire anemometer to the channel axis, these sizes decreased, so at $y/\delta = 1.5$ the corresponding values were established: $0.007 - 0.008$; 0.02; 0.02. The magnitudes of fluctuations velocities continued to decrease up to the channel axis. These data correspond to [1, 169, 177, 462]. The elastic plates, depending on their structure and properties, influence the flow up to $y/\delta > 1.5$. Therefore, the plates 2 and 5 do not influence the outer region of the boundary layer.

In Table 2.5 the structure of the investigated elastic plates is given. Compliant plates 1, 1a are executed from a monolithic sheet of foamed polyurethane, pasted over from above with a thin rubber film. Measurement of the mechanical characteristics of these plates and samples of material from which these plates are executed have shown that such plates are more likely to be relevant to elastic plates. The rubber film and a layer of glue make an external part of these plates rigid and elastic. The plate 9 is a composit and is executed from several continuous layers of a material and the internal layer has a conductive layer which is also isolated by a thin rubber film. Therefore such a design of an elastic plate is also rigid and elastic. These plates seem as though they reject the pulsations of a boundary layer for the external area of thickness of a boundary layer. The physics of this process are given in 1.10. In general they were all received by experimental kinematic characteristics of a boundary layer and it is necessary to consider them in conformity with the data results from parts 1 and 2.

At the same time compliant plate 10a and elastic plates 8, 8a, 11 and 11a have a specific structure of an external layer which generates longitudinal CVSs in a boundary layer. Their pulsating characteristics differ essentially from the characteristics of Compliant plates 1, 1a and 9 according to a problem with the susceptibility of the character of flow compliant in elastic plates.

Fluctuation characteristics of compliant plate *10a* and of elastic plates *8, 8a, 11*, and *11a* decrease in the region of $0 < y < 1.5$. To approximately calculate the fluctuating characteristics on the rigid smooth plate, the empirical relations can be used [381].

3.4 Velocity Field in the Near-Wall Region

The processes of dissipation and generation of energy in a turbulent boundary layer are most intensive in the near-wall layer. Let us consider in detail the above-described experimental results near the streamline surface. In Figure 3.8, the velocity profiles on the standard rigid and elastic rough plates and on typical compliant and elastic plates can be seen. The designations used in Figure 3.8 are same as in Figures 3.3 and 3.4 and in Tables 2.4, 3.1 and 3.2. The straight line A corresponds to identical numbers on the coordinate axes and defines the slope of velocity profiles near the wall in Laufer's experimental data [441] for the rigid wall. The results obtained on the rigid plate practically coincide with data in [283, 441,177]. The Reynolds numbers were $0.77 \cdot 10^6$ in our data and $1.2 \cdot 10^6$ in data from [283]. The results of measurements with Re $= (0.77 \div 0.87) \cdot 10^6$ are well agreed with Laufer's data [177, 441] (Re $= 0.5 \cdot 10^6$ and $5 \cdot 10^6$). However, while in [177, 441] the velocity profile lies lower with increasing Re, in our investigations the velocity profile in the near-wall region lies higher than in [177, 441]. This is related to the higher turbulence level of the mean flow.

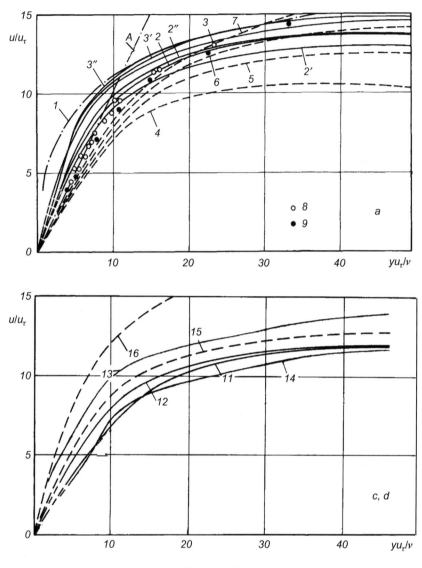

Figure 3.8
Distribution of mean velocity in the near-wall layer depending on different Reynolds numbers on the rigid and elastic plates. Designations are as for Table 3.2. Curves *4–7* – calculations of Eeinstein and Li [111] for $T^{+} \cdot 10^{5}$ equal to 1.35; 3; 5.6; 7.5; curves *8, 9* – Laufer's data [111] for Re $= 5 \cdot 10^{5}$ and $5 \cdot 10^{6}$.

The velocity profiles on the rough elastic plate *5* (curves *11–14*, Figure 3.8c,d) agree well with data from Einstein and Li [493]. The velocity profiles of plate *5* due to roughness are below experimental data for the rigid standard (see Figure 3.8a) and Laufer's data. As soon as the rough elastic material has been pasted over by a thin rubber film (plate *2a*, curves *15*

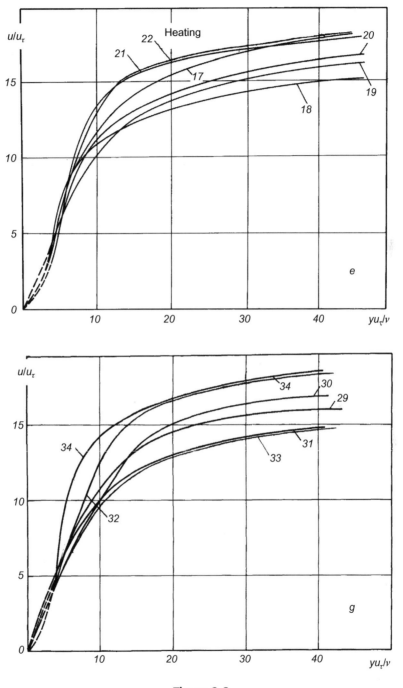

Figure 3.8
(Continued)

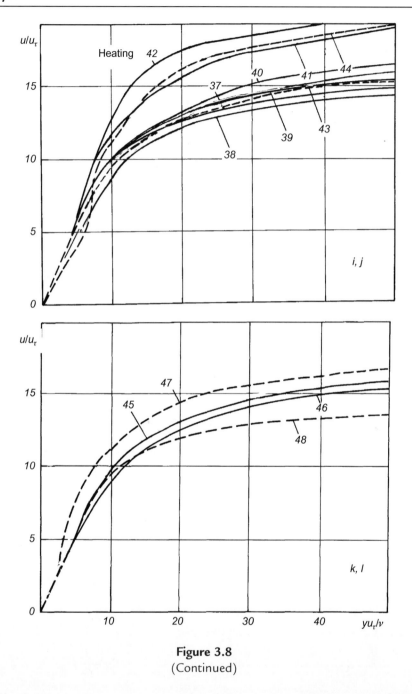

Figure 3.8
(Continued)

and *16*), the velocity profiles became similar to those of the rigid standard. At greater speed (curve *16*) parameters of the elastic plate *2a* changed the structure of velocity, which began to differ from the standard and took the form typical for elastic plates.

The profiles on the compliant plate *9* (curves *17–22*, Figure 3.8e) are located much higher than on the standard plate. At the boundary of the laminar sublayer near $y^+ = 4$–5, lock in velocity profiles have inflections. δ_{lam} and appear the same as at the reference rigid plate. Plates *1a* and *1* have similar profiles with points of inflexion (curves *29–34*, Figure 3.8g). The velocity profiles of both plates are the same, and differ from plate *9* in that they lie higher in comparison with the standard. The inflexions of the profiles in the region of $y^+ = 5$ can be explained by the generation of longitudinal vortices or by oscillations of the plate surface induced of the oscillations of the boundary layer. In the region of laminar sublayer the velocity profile slope is same as on the standard plate or decreases showing that δ_{lam} increases. The inclination increases at $y^+ > 5$, i.e. far from the wall, so friction does not increase and the "flow rate" of liquid in the boundary layer increases.

The velocity profiles in the flow of plate *10a* are same as on plate *2a* till $y^+ = 5$. Therefore, in contrast with Figures 3.3 and 3.4, the longitudinal velocity profile over the plate *10a* in these coordinates does not have the characteristic features. It occurs because in the near-wall region the positive factor of generated vortices is compensated by the negative factor of roughness of plate *10a*.

As it can be seen from the text above, the elastic plates are of the most interest. The velocity profiles of these plates have a number of similar properties (see Figure 3.8i,j,k,l). When $y^+ < 5$, the velocity profiles lie under the standard, i.e. the thickness of the viscous sublayer δ_{lam} is increased. At $5 < y^+ < 10$, the elastic plates' velocity profiles intersect the rigid plate velocity profiles, i.e. they have the less pronounced inflexions than on compliant plates (see Figure 3.8c,d,e,g). At $y^+ > 10$, the velocity profiles lie higher than the standard profiles that reflect the increased "flow rate" of liquid in the boundary layer. The velocity profiles in section III practically coincide for all plates. However, as the velocity increases the profiles in this section on plates *8a* and *3* (transverse lamination of the outer layer of plates) lie lower than at low velocity, and for plates *11a* and *11* (longitudinal stratified structure) they are higher, compared with the standard plate.

This indicates that the effectiveness of the elastic plates depends on the Reynolds number and on the plate structure. Inflections of velocity profiles at $y^+ = 5$ take place, but their intensity is less than at compliant plates, and these inflections are smooth and stretched up to $y^+ = 10$. In this sense, the best is the velocity profile over the elastic plate *11* (curve *44*, Figure 3.8j), where at greater speed the profile is characterized by an appreciable inflection and greater "flow rate" of the liquid in the boundary layer. It can be noted that the inflexions of the velocity profiles reflect the stages of transition of the boundary layer, in which the longitudinal vortices were being generated.

Velocity profiles have been measured while regulating the mechanical characteristics of elastic plates. Such regulation was carried out by heating the current-carrying fabric located inside of plates. The size of the current was increased gradually so that on the surface of the elastic plates there was no thermal boundary layer. Interesting results were obtained for the compliant plate *9* and an elastic plate *11a* in section III*a* at greater stream speeds. For plate *9*, $U_\infty = 16.3$ m/s without heating and $U_\infty = 16.1$ m/s when heated. For plate *11a*, these measurements were 19.4 m/s and 19.45 m/s respectively. These conditions correspond to curves *21* and *22* (Figure 3.8e) and curves *41* and *42* (Figure 3.8i). Heating the compliant plate did not lead to appreciable changes. At the same time, heating the elastic plate essentially influenced the characteristics of the boundary layer.

The "discharge" in a boundary layer is understood as a concept—it is the specific interaction of a stream and an elastic plate so that the structure of speed changes and at the same speed of the basic stream the discharge in the field of a boundary layer increases. It indirectly causes a reduction of friction by a wall. However the physical picture of flow in a boundary layer can also differ in a number of features, analysis of which will be dealt with later.

Heating of the elastic plates has led to an appreciable excess of a profile of longitudinal velocity in the near-wall area. It is known that profiles of velocity with such excesses testify to a reduction of friction near to a wall. And the increase in the form of a curve of a profile far from the wall testifies t an increase of the velocity due to the reduction of friction by the wall. Such physical process can be named conditionally as the increase of the charge of a liquid through a boundary layer. The mechanism of the phenomenon is clear from a position of the problem of susceptibility at consideration of the interaction of the CVS of a boundary layer with a CVS generated in a boundary layer of a streamline wall which has fluctuations caused by the pulsations of pressure of the boundary layer. Other features of interactions of elastic plates with a boundary layer will be considered below.

According to the data in Figure 3.5, there are no substantial differences between the values of maximal pulsations of the standard plate and those of compliant plates. Whereas, according to the data in Figure 3.9c,d the longitudinal fluctuation velocity on the compliant plates, as well as the averaged velocities, substantially increases due to substantial increase of the value of u_τ. The maximum points of profiles of velocity fluctuation are closer to the surface and closer over plates *1* and *1a* than over the plate *9*. A similar pattern was observed in the transitional boundary layer. All complaint plates have inflexions of the fluctuation velocity profiles at $y^+ = 5$, like the averaged profiles.

The thermo anemometer sensor could not be placed closer to the surface. Therefore, all curves of averaged and fluctuation velocities are extrapolated at $yu_\tau/\nu < 5$ in the origin of the coordinates. The shape of the fluctuation profile of plate *10a* changed and became similar to the profile on the rough plate.

The shapes of fluctuation velocity profiles of elastic plates are the same as over compliant plates, but fluctuations are decreased and the points of maximum are shifted away from the

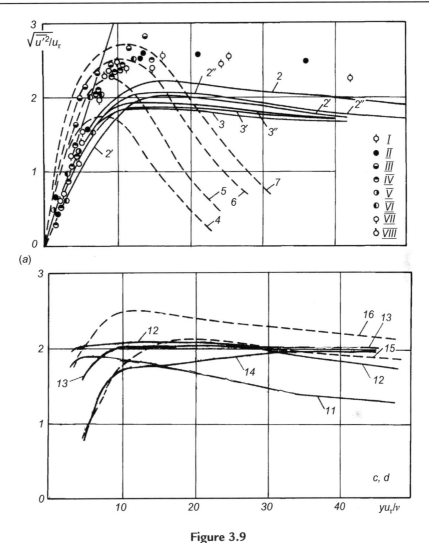

Figure 3.9

Distribution of velocity longitudinal fluctuations in the near-wall layer over the rigid and elastic plates. Curves I-VII are Laufer's data [71] at Re = $6.2 \cdot 10^5$. Other designations are as for Figure 3.8.

surface. Fluctuations on elastic plates are higher than on the rigid plate; only on plate *3* at high velocity U_∞ was the longitudinal fluctuation velocity drastically decreased.

Thus, there is an inflection point in the longitudinal averaged and fluctuation velocities nondimensional by friction velocity in the near-wall areas. Maxima of fluctuation velocities are increased in comparison with the standard. Experiments with polymer solutions have showed a similar phenomenon. The behavior of pulsations in polymer solutions was similar to that in flows over elastic plates.

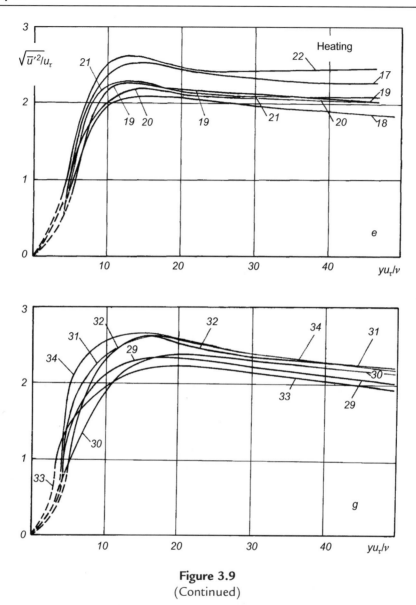

Figure 3.9
(Continued)

In addition to measurements of average velocity, measurements of fluctuation velocity have been executed at by heating two plates, *9* (compliant) and *11a* (elastic). Profiles of average velocity over the compliant plate did not essentially change at heating; at the same time there was an essential variation over the elastic plate. Profiles of fluctuation velocity at heating both plates (curves *21*, *22* and *41*, *42*) changed essentially—fluctuation increased across the entire thickness of the boundary layer. Based on these results it is possible to state that heating and other methods provide efficient control of the characteristics of the boundary layer on elastic plates.

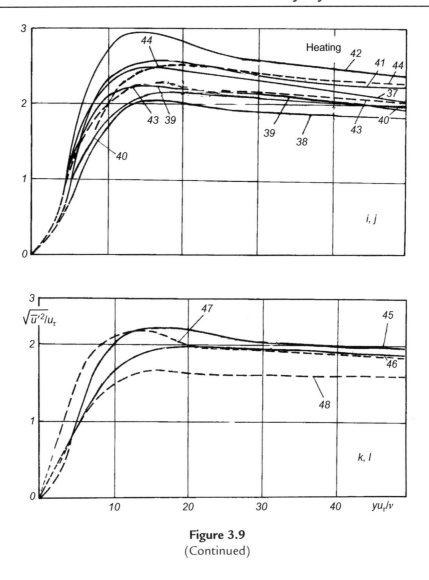

Figure 3.9
(Continued)

Recently the investigators paid greater attention to the distribution of the wall shear (du/dy) in the near-wall layer, i.e. fluctuations of averaged velocity gradient near the surface. Measurements [176, 203] showed that the shape and values of fluctuations of longitudinal velocity and longitudinal near-wall shear are the same near the wall.

In the paper [170], it is shown that $\sqrt{\overline{u'^2}}/u_\tau \to dy^+$ at $y \to 0$, i.e. near the surface $du'/dy^+ = \text{const} = a = 0.24$. According to Srinivasan [177] $a = 0.25$, according to [225] $a = 0.32$ and according to [177, 225] $a = 0.3-0.35$. The profile of velocity longitudinal pulsations over the rigid plate in our measurements agrees well with the profile presented in

[170]: at $y^+ = 2$, it is seen as a light inflexion in the fluctuation velocity profile and, thus, the tangent of the slope angle is $a = 0.2-0.24$. At the flow on elastic plates, thickness of the viscous sublayer increases. Therefore, the inflexion point in the fluctuation velocity moves away from the wall and is situated at $y^+ = 5$. Thus, the value of parameter a at $y^+ > 5$ increases for compliant plates to 0.37, and for elastic plates to 0.31.

These changes in values of a and coordinate y^+ of the inflexion point in the velocity profile are provided at elastic plates by the essential change of compliance of the external layer. Thus $a = 1-1.5$ means that the profile of longitudinal velocity is intensively laminarized and, behind the inflexion point at $y^+ > 5$, a increases. However, since the increase of fluctuations of the near-wall shear, leading to increasing τ, takes place in the layer located far from the surface, it does not influence the friction on the wall.

According to [225] the value of a decreased from 0.34 to 0.2 in the flow of a polymer solution. The value of a of the transversal shear on a rigid wall is $0.1-0.13$ and decreases to 0.07 in the flow of a polymer solution [225]. So, the equating of a in longitudinal and transverse directions on elastic plates corroborates Rotta's hypothesis [441]: liquid in the shear flows tends to decrease anisotropy.

In Figure 3.10 we represent the distribution of transversal fluctuation velocities across the boundary layer on a rigid plate. The distribution is of the same magnitude and behavior as in [177, 225, 473], but differs from the data in [170, 223]. The solid curves—vertical and dash—are the transversal fluctuation velocities. The profiles of the transverse and longitudinal fluctuation velocities completely coincide with data in [283] at x/D = 40.

All the above-mentioned results in this section were received by means of the wire sensor. It allowed making measurements near the surface, beginning on average 0.1 mm above the rigid plates. On elastic plates, the sensor could be placed even closer to the surface because it was possible to deepen the wire sensor holder in the elastic surface. To measure the transversal velocities, an X-shaped sensor was applied. Its design did not allow placement closer to the surface. Therefore, measurements began at 4 mm. The experimental data represented in Figures 3.7 and 3.10 began at small velocities approximately at $y/\delta \approx 0.15$ and $y^+ \approx 120$, and at greater velocities at $y^+ \approx 200$. Thus, it was not possible to investigate the influence of elastic plates on transversal velocities near the surface. Designs of elastic plates and their mechanical characteristics determine the influence of plates on the entire boundary layer thickness. It is shown that such influence varies depending on the type of elastic plate (compliant or elastic). Therefore, the results in Figures 3.7 and 3.10 are of interest from the point of view of the influence of various types of elastic plates not only on the near-wall area but also on the entire thickness of the boundary layer.

While on the rigid plate the transverse fluctuation velocities practically do not depend on the Reynolds number, plate 9 (Figure 3.10e, the worst from the point of view of drag reduction) showed a dependence on the Reynolds number. The values of v' and w' increased

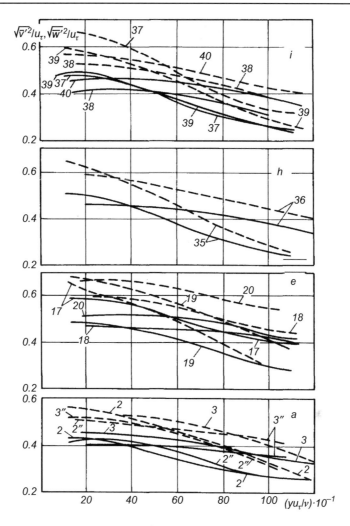

Figure 3.10

Distribution of velocity transverse pulsations in a near-wall layer over rigid and elastic plates. Designations are as for Table 3.2.

by 1.5 times compared with that on the rigid plate. The velocity profiles changed: v' and w' increased with increasing y^+.

The distribution of transversal velocities across the boundary layer in section III of plate *10a* (see Figure 3.10h) is practically the same as that of plate *11a* (see Figure 3.10i). The values of v' and w' decrease in comparison with those of plate *9* but they remain slightly higher than that of the standard plate.

Both cases (Figures 3.7 and 3.10) allow us to conclude that complex interactions cause the redistribution of the velocity fluctuation field, increasing the transversal component.

It might be the result of oscillations of elastic plate surface, generation of longitudinal vortices or changing in vortex sizes and in Reynolds stresses in the near-wall region.

To calculate the fluctuation velocities on the rigid plate, we can use the formulae from [381]:

$$\sqrt{\overline{u_m'}^2}/u_\tau \approx \{-0.5\,[\lg(10^3\lambda)]+1.36\}^3 \tag{3.33}$$

$$\sqrt{\overline{u_m'}^2}/U_\infty \approx \left(\sqrt{\overline{u_m'}^2}/u_\tau\right)\sqrt{\lambda/2} \tag{3.34}$$

$$\sqrt{\overline{v_m'}^2}/U_\infty \approx \left(\sqrt{\overline{v_m'}^2}/u_\tau\right)\sqrt{\lambda/2} \tag{3.35}$$

where $\lambda = 2\tau_0/\rho U_\infty^2$

At $0 < y < 2\delta_{lam}$

$$\sqrt{\overline{u_m'}^2} = \sqrt{\overline{u_m'}^2}(y/2\overline{\delta}_{lam}) \tag{3.36}$$

where $2\overline{\delta}_{lam} \approx \delta_{lam}$

According to [225]:

$$\sqrt{\overline{w'^2}}/u_\tau = 0.077y^+$$
$$\sqrt{\overline{u_m'^2}}/u_\tau \approx 0.28y^+ + 0.014y^{+2} \tag{3.37}$$

According to [435, 473], at $0 < y < \delta_{lam}$:

$$\sqrt{\overline{v_m'^2}}/u_\tau \approx 0.01y^+$$
$$\sqrt{\overline{w_m'^2}}/u_\tau \approx 0.1y^+ \tag{3.38}$$

The papers [177, 283, 381] present the discrepant data concerning the maximal values of results obtained: $\sqrt{\overline{u_{max}'^2}}$ and c_f decrease with increasing Re. The measurements gave the following results for the standard plate:

$$\sqrt{\overline{u_{max}'^2}}/u_\tau \approx 2.3 \tag{3.39}$$

for low velocities:

$$\sqrt{\overline{v_{max}'^2}}/u_\tau \approx 0.85 \div 0.9$$
$$\sqrt{\overline{w_{max}'^2}}/u_\tau \approx 1.06 \div 1.17$$

for high velocities:

$$\sqrt{\overline{u'^2_{\max}}}/u_\tau \approx 1.9 \tag{3.40}$$

and values of v' and w' can be calculated by the formulae similar to (3.39).

For all elastic plates we have (in accordance with the above indicated gradation):

$$\sqrt{\overline{u'^2_{\max}}}/u_\tau \approx 2.1 \div 2.6$$
$$\sqrt{\overline{v'^2_{\max}}}/u_\tau \approx 0.84 \div 1.04 \tag{3.41}$$
$$\sqrt{\overline{w'^2_{\max}}}/u_\tau \approx 1.07 \div 1.38$$

At high velocities we have for the plates *3*, *8*, and *10a*: $\sqrt{\overline{u'^2_{\max}}}/u_\tau \approx 1.7 \div 2.0$, i.e. less than by the standard plate. It can be seen from (3.41) that mainly the longitudinal and transversal components of fluctuating velocity increase.

The presented results show that the mechanism of interaction between the boundary layer and the elastic plates is very complex and depends upon many factors. According to Section 1.6 and Figure 1.6, it can be expected that the frequency spectrum of oscillations of an elastic plate surface will be the key information parameter. The spectrum characteristics are the frequency distribution of energy, the form and amplitude of the surface oscillation.

As it is shown in Sections 2.6 and 2.8, the form of the spectrum is determined through the self-resonant frequency of the elastic plate oscillation. Therefore, it is analyzed as the dependence of maximum $\sqrt{\overline{u'^2_{\max}}}$ on the coefficient K_w of the elastic surface oscillation frequency (Strouhal number). The results presented in Figure 3.11 have shown that the magnitude of $\sqrt{\overline{u'^2_{\max}}}$ at elastic plates depends on character of the interaction of a boundary layer with the plate and the speed of the mean flow. It is revealed that at elastic (*11* and *11a*) and compliant (*2a*, *5*, *9*) plates the dependencies of fluctuation velocities on the fluctuation frequency coefficient K_w of the surface are identical. It is possible to consider the received law to be universal, as far as K for the standard calculated from the frequency of the viscous sublayer regeneration agrees with the general law.

The distinctive behavior of function $\sqrt{\overline{u'^2_{\max}}} = f(K_\omega)$ for plates *3* and *8a* can be explained by the transverse lamination of their outer layer, which generates the 2D disturbances of the type of Tollmien–Schlichting wave. It is shown in Section 1.4 and [310] that the complex interaction in the boundary layer of 2D and 3D (natural) disturbances causes an intensive decrease in longitudinal fluctuation velocity. The curves III and IV corroborate the conclusion.

So at low velocities the minimal longitudinal fluctuations take place with:

$$K_w = 2.8-3 \tag{3.43}$$

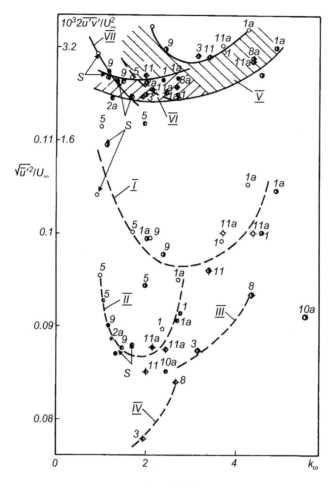

Figure 3.11
Dependences of the maximal fluctuation velocity and Reynolds stress upon the frequency coefficient of oscillation of the elastic plate surface: I, III, V — low velocities; II, IV, VI — high velocities; I, II — compliant plates; III, IV — elastic plates; I–IV — distribution of $\sqrt{\overline{u'^2_{\max}}}$; V–VII — distribution of C_f; VII — standard plate. Designations are as for Table 3.2. S — standard plate.

and at high velocities with:

$$K_w = 1.8 - 2 \qquad (3.44)$$

Smaller values of these are characteristic for plates *8a* and *i*, as well. The experiments showed the connection between fluctuations of near-wall shear stresses and pressure pulsations and coherent vortical structures in the near-wall region of turbulent boundary layer [177]. It should be remembered that elastic plates possess the property of spontaneous formation of longitudinal vortices structures. Furthermore, the design of plates with reversed riblets in the outer layer generates such structures in a boundary layer. In connection with these properties, a correlation of $\sqrt{\overline{u'^2_{\max}}}$ and $|\overline{u'v'}|$ (see Figure 3.11) is found. In spite of the dispersion in experimental points, we succeeded in resolving two regions of values of $|\overline{u'v'}|/U_\infty$: for low (*V*) and high (*VI*) velocities. The values of K_w are determined according to the minimum values of $|\overline{u'v'}|/U_\infty$, coinciding with (3.43) and (3.44). Besides, the area *VI* smoothly matches the area *V*, and the law $|\overline{u'v'}|/U_\infty = \varphi(K_\infty)$ has other characteristics for the rigid standard.

The tested elastic plates were not optimal for the given conditions of the experiment. Therefore, the minimal values of $|\overline{u'v'}|/U_\infty$ are identical for the standard and elastic plates. From Figure 3.11 it follows that in order to reduce the longitudinal fluctuation velocities it is necessary to carefully choose the mechanical characteristics and structure of elastic plates according to the specified criteria of similarity. It is possible to use the generalized data in this Figure for predictions of efficiency of elastic plates. Knowing operational speeds of a body, it is possible to calculate the energy-carrying range of the spectrum of the boundary layer fluctuations, and to define the optimal elastic material of the plate from the data given in Figure 3.11.

To calculate the turbulent boundary layer, it is important to determine the length of mixture *l*. Some different formulae to calculate the value of *l* on a rigid plate are presented in [169, 241, 325, 362, 424, 435, 441] at $\aleph = 0.4-0.41$. In our tests we did different calculations to determine the value of $\aleph$ using formulae in [181, 232, 441, 445]:

$$0 \leq y^+ \leq 40 \; l = (-0.1 \, y)^2 \, u_\tau \tag{3.45}$$

$$y^+ \geq 40 \quad l = \aleph y \tag{3.46}$$

$$l = v' \Big/ \left| \frac{\partial U}{\partial y} \right| \tag{3.47}$$

$$l^2 = u'v' \Big/ \left| \frac{\partial U}{\partial y} \right| \frac{\partial U}{\partial y} \tag{3.48}$$

$$l/\delta = 0.14 - 0.08(1 - y/\delta)^2 - 0.06(1 - y/\delta)^4 \tag{3.49}$$

The results of the calculations are presented in Figure 3.12. The calculations according to (3.45) for the rigid plate in the section I (see Figure 3.2) at low velocity brought to the curve *2*, which is well agreed with [138] (curve *10*) by $y^+ = 30$. The point "*3*" marks the results according to (3.47), and "*4*" according to (3.48). The results are well agreed with each other and with results according to (3.45). The curve "*1*" is the result obtained from (3.46). The natural spread of points "*3*" and "*4*" caused by the errors of measurement of the components in formulae (3.47) and (3.48) does not allow the determining of $\aleph$ exactly. But

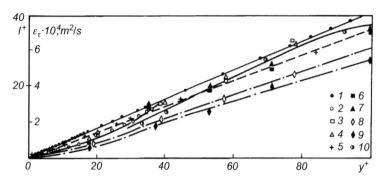

Figure 3.12

The distribution of value l^+ (*1–7, 10*) and ε_τ (*8, 9*) in the near-wall regions of the turbulent boundary layer over rigid (*1–4, 8*) and elastic *11a* (*5–7, 9*) plates; *10* — measurement l^+ [138].

for the rigid plate up to $y^+ = 100$ (3.46) can be used with $\aleph = 0.4 -0.41$ as a first approximation. At $y^+ < 40$ the distribution l across the near-wall layer is governed by (3.45) or even more complex relations, in particular the Van Driest formula [181]. Distribution of size by way of mixture l in the near-wall areas calculated on experimental results with the application of various formulas is well agreed with [138] for a rigid standard. Experimental results for an elastic plate 11a have sown that distribution l in the near-wall area is similar to the reverence but the size at l of an elastic plate is less than at the standard.

Calculations for plate *11a* according to (3.47) (point "*6*") and according to (3.48) (point "*7*") are well agreed with each other and averaged by the curves *5* according to (3.46) with $\aleph = 0.36$. This value of $\aleph$ corresponds to the value obtained for plate *11* according to the method presented in Section 3.2 ($\aleph = 0.33$). In this case, one can also use the formulae (3.45) up to $y^+ = 40$ because (3.46) cannot be used in the viscous sublayer due to complex interactions in the boundary layer. The value of l in this region can be calculated by formulae given by Vasetskaya and Iosilevith [177]:

$$l = ky\{1 - \exp[c\varphi/(\partial u^+/\partial y^+)]\} \tag{3.50}$$

Figure 3.12 presents the distribution of the turbulent viscosity coefficient in the near-wall layer according to [441]:

$$\varepsilon_\tau = |\overline{u'v'}|/(|\partial U/\partial y|) \tag{3.51}$$

$$\varepsilon_\tau = -\overline{v'}l = l^2(|\partial U/\partial y|) \tag{3.52}$$

It can be seen that the turbulent viscosity in the boundary layer on the elastic plate *11a* (curve *9*) is less than on the rigid plate (curve *8*). Similar results were obtained when polymer solutions were being investigated [223, 425]. The values of ε_τ and l lie under the straight lines up to $y^+ = 30$ and can be better described by the formulae presented in [181, 441].

3.5 Energy Balance of a Turbulent Boundary Layer over an Elastic Plate

The set of equations determining the balance of kinetic energy for the three components of fluctuation velocity is presented in [181, 232, 362, 424, 441, 473]. The main flow energy is being transferred to the fluctuation motion through the turbulent shear stresses. It manifests itself mainly in changing the longitudinal fluctuation velocity, which transfers energy to the transversal components through turbulent fluctuations of pressure. From Section 3.4, based on the character of change of fluctuation velocities it is possible to estimate, in the first approximation, a degree of change of energy balance of the turbulent boundary layer at the flow on elastic plates.

The existing instrumentation does not allow measuring all components of energy balance of the turbulent boundary layer. We analyze the components of energy balance which can be determined by thermo anemometer. The equation of energy balance of a 2D turbulent boundary layer is as follows [232, 381, 441]:

$$
\begin{aligned}
&-\overline{u'v'}\frac{\partial U}{\partial y} - \frac{1}{2}\frac{\partial}{\partial y}(\overline{u'^2} + \overline{v'^2} + \overline{w'^2}) - \frac{1}{\rho}\frac{\partial}{\partial y}\overline{v'p'} + \frac{1}{3}U\frac{\partial}{\partial x}(\overline{u'^2} + \overline{v'^2} + \overline{w'^2}) \\
&+ \frac{1}{2}V\frac{\partial}{\partial y}(\overline{u'^2} + \overline{v'^2} + \overline{w'^2}) - \upsilon\left(\overline{u'\nabla^2 u'} + \overline{v'\nabla^2 v'} + \overline{w'\nabla^2 w'}\right) = 0
\end{aligned}
\tag{3.53}
$$

The terms of this equation have the following physical meaning:

- 1st − generation of fluctuation turbulent energy by the averaged motion;
- 2nd − turbulent diffusion of kinetic energy of fluctuation motion;
- 3rd − turbulent diffusion of pressure fluctuation energy;
- 4th − convection of the turbulent energy from averaged motion in the longitudinal direction;
- 5th − convection of the turbulent energy from averaged motion in the transverse direction;
- 6th − the dissipation of turbulent fluctuation energy.

The simplified equation does not involve the convective terms. The equation of energy balance in the near-wall region becomes [441]:

$$
\frac{1}{\rho}\frac{d(\tau_0 U)}{dy} = -\overline{u'v'}\frac{dU}{dy} + \upsilon\left(\frac{dU}{dy}\right)^2
\tag{3.54}
$$

It can be seen that the energy is transferred from the outer part of the boundary layer into the near-wall region with the velocity $\tau_0 U/\rho$.

The maximum production of turbulent energy can be determined as follows [441]:

$$\left(\overline{u'v'}\frac{dU}{dy}\right)_{max} = \frac{u_\tau^4}{4v} \tag{3.55}$$

The equation of turbulent energy balance is brought to the dimensionless form using the relations v/u_τ^4 and δ_{lam}/u_τ^3 [381, 441].

Based on the obtained experimental results for distributions of averaged and fluctuation components of velocities across the boundary layer thickness, utilizing formulae (3.53) and (3.54) it is possible to calculate the majority of the components of turbulent energy balance of the boundary layer. To define the contribution of turbulent energy production into the equation of turbulent energy balance it is necessary to measure the distribution of turbulent friction stress across the boundary layer thickness. Figure 3.13 presents the results of measurements on the rigid plate of distribution of turbulent friction stresses across the boundary layer. The obtained results differ from the data in [122, 187, 272] due to higher turbulence levels, the difference in Reynolds numbers and experimental conditions.

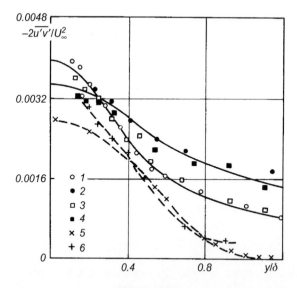

Figure 3.13

Distribution of shear stresses in the turbulent boundary layer over the rigid plate with $U_\infty = 9.9$ (*1*), 16.2 (*2*), 10.0 (*3*), 18.4 m/s (*4*); *1, 2* — I section, *3, 4* — III section, *5* — Klebanoff's measurements [190], *6* — Blick's measurements [159].

In [186, 272], the boundary layer was investigated on a plate located in a wind tunnel, and in our experiments it was located on the bottom of a wind tunnel. That is why our results agree better with [283, 223], where the investigations were being made on the walls of a wind tunnel. It should be noted that Reynolds stresses in wind tunnels reduce to zero not at the boundary of the boundary layer, but at the tunnel axis.

As measurements were carried out at low Reynolds numbers, the Reynolds stress for the rigid plate close to the wall is higher and corresponds to the data in [181, 207, 514].

Depending on the flow velocity, there are two laws of distribution of Reynolds stresses across the boundary layer thickness that can be explained by design features of a wind tunnel: distribution of velocity on the cross-section of a pipe and the degree of turbulence depending on the flow velocity. At greater velocity, Reynolds stresses near the wall are less than at small flow velocities. Moving away from the wall, Reynolds stresses decrease more slowly. This corresponds to the data of Figure 3.5.

Figure 3.14 presents the results on the elastic plates. The laws of Reynolds stresses distribution across the boundary layer differ from that on the rigid plates. At a flow of elastic plates there is a reorganization of structure of the entire boundary layer, so the energy balance, ratio and interrelation of its components change.

The distribution of $\overline{u'v'}$ on y/δ over the compliant plate *9* (curves *17–20*, Figure 3.14e) changed only near the surface. The values of $\overline{u'v'}$ decreased. However, the clear correlation between the value of $\overline{u'v'}$ and free flow velocity is not revealed, unlike on the standard plate.

When 3D disturbances were generated from the elastic surfaces into the boundary layers over plates *10a* (curves *35–36*, Figure 3.14h) and *11a* (curves *37–40*, Figure 3.14i), Reynolds stresses decreased on the entire thickness of the boundary layer, although not so intensively as in measurements [187, 122, 186]. At high velocities near the elastic surface, the value of $\overline{u'v'}$ decreased with the same intensity as for the measurements in the papers mentioned. Table 3.5 lists the values of shear stresses on the smallest distance from surface on which it was possible to install the X-shaped sensors. The values in the table can differ from the corresponding values on plots because averaged values of shear stresses are indicated in the plots.

It can be seen from Figures 3.13 and 3.14 that the production of fluctuation energy is concentrated in the near-wall region. In addition, at elastic plates, Reynolds stresses near the wall appeared less than at the standard. The maxima of shear stresses moved substantially away from the wall: that can be explained by the thickening of the viscous sublayer. The maximum on the rigid plate also moved away from the wall in the case of positive pressure gradient [441]. In this case, in contrast with elastic plate, the shear stresses increase.

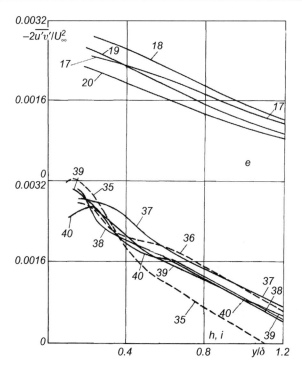

Figure 3.14

Distribution of shear stresses across the boundary layer over elastic plates. Designations are as for Table 3.2.

Table 3.5: Values of shear stresses near the surfaces.

Plate Number	Place of Measurement (Number of Section)	Velocities (m/s)	$10^3\left(-2\overline{u'v'}/U_\infty^2\right)$
Standard	I	9.9	3.8
Plate	I	15.8	3.47
	III	10.0	3.6
	III	18.4	3.3
9	I	9.85	2.44
	I	15.4	2.98
	III	9.66	2.86
	III	17.1	2.49
10a	III	9.75	3.25
	III	17.4	2.8
11a	I	10.3	2.8
	I	16.6	3.04
	III	10.4	3.14
	III	17.6	2.44

Figure 3.15 presents the distribution of Reynolds stresses, which are brought to dimensionless form using the friction velocity. The results for the standard plate are well agreed with [177, 223, 283, 327, 441, 473]. In some papers, the dependence of turbulent shearing stresses on the Reynolds number is not revealed.

The behavior of Reynolds stresses distribution across the boundary layer, in dependence on Reynolds, numbers obtained in our experiments agree well with the data presented by Laufer [441] and Chang [177]. When comparing the results obtained at the standard with

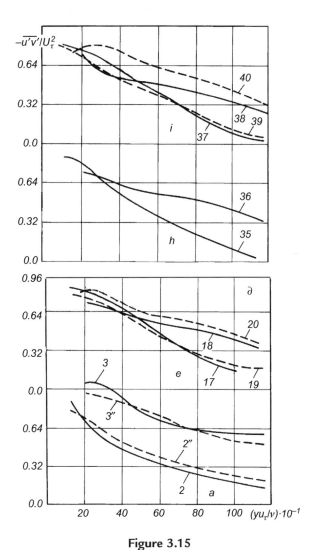

Figure 3.15

Profiles of turbulent shear stresses in the boundary layer over the rigid and elastic plates. Designations are as for Table 3.2.

those in Figure 3.15 for elastic plates, two types of dependencies are found out at different flow speeds. Thus, the revealed distinctions are more essential than those in Figure 3.14. At small flow velocities Reynolds stresses on elastic surfaces match the data for the standard. Only over plate *9* (curves *17, 19*) do Reynolds stresses increase in comparison with the standard (curves *2, 2″*). At the standard, Reynolds stresses in all measurements on the boundary layer thickness increase with increasing velocities. For all the kinds of elastic plates investigated, at greater flow velocities, up to $y^+ = 600$, Reynolds stresses practically do not increase in comparison with small flow velocities, while on the standard, they increase essentially. Therefore, at greater flow velocities up to the specified values of vertical coordinate all elastic plates have less Reynolds stresses than at the standard.

Due to the sensor design, it was not possible to measure closer to the surface. However, on the standard there is the tendency to increase Reynolds stresses at approaching the surface up to the value equal 1. According to Laufer [441], reductions of Reynolds stresses will begin after $y^+ = 50$. In our tests, we could not approach the standard below $y^+ = 123$. At a flow on elastic plates maxima of Reynolds stresses are revealed at $y^+ = 250$ and below. Thus, at a flow on elastic plates, not only do Reynolds stresses decrease, but also their maximal values locate essentially higher, in comparison with the standard. Thus, the elastic surface decreases the turbulence production and, in addition, shifts Reynolds stresses maxima away from the surface. These conclusions are consistent with the inflections of velocity profiles near the wall and with thickening of the viscous sublayer.

The Reynolds stresses decrease was weaker than in experiments with polymer solutions [223, 325, 327, 425]. It can be explained by non-optimal properties of the plates for the given conditions.

Reynolds stresses in the near-wall region of the boundary layer over a rigid plate can be calculated according to [473]:

$$0 < y < \delta_{lam} \quad |\overline{u'v'}|/u_\tau = 6 \cdot 10^{-4}(y^+)^3 \tag{3.56}$$

As is shown in [441], the term $-\overline{u'v'}(\partial U/\partial y)$ in the energy balance equation characterizes velocity, with which longitudinal component increases energy under the influence of the shift of average velocity $|\partial U/\partial y|$. Similarly, the term $\overline{v'^2}\,(\partial U/\partial y)$ characterizes velocity of production of $-\overline{u'v'}$. The redistribution of the turbulent boundary layer energy is caused by the velocity gradient effect and, thus, reacts against the turbulent anisotropic distortion. Therefore, it is possible to imagine that physically the energy balance in the boundary layer is determined mainly by the wall effect, which distorts the isotropic turbulence of the free flow. The stresses in the boundary layer are caused by these distortions of isotropy, and the turbulent boundary layer tends to reduce anisotropy. Therefore, we can expect that any action decreasing the wall effect will lead to a decrease in anisotropy. All computation procedures for a turbulent boundary layer utilize the relations of Reynolds stresses and kinetic energy of the fluctuation motion. The [k − ε] model of turbulence based on the equations of turbulence energy and its dissipation rate is a very commonly used model especially for flows of polymeric solutions [177, 493]. Figure 3.16 presents the experimental profiles of kinetic energy of turbulent fluctuation motion and the ratio between the turbulent friction stresses and the energy. The distribution of these parameters on the rigid plate is well agreed with [272, 441, 492]. We also presented the curves for plate *11a* (other elastic plates have the same curves).

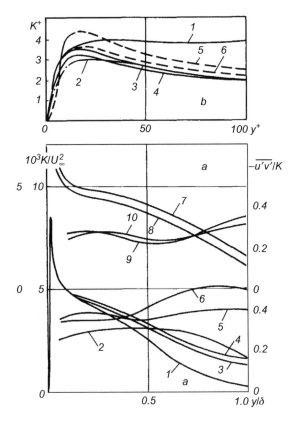

Figure 3.16

The profiles of kinetic energy of turbulent pulsating motion and the relationship to turbulent stress of friction in the boundary layer over the rigid and elastic plates (*a*), and in the near-wall region (*b*): *a* − measurement over the rigid plate according to data in [272] of values of k/U_∞^2 (*1*) and $-\overline{u'v'}/k$ (*2*); our data: for low (*3, 5*) and high (*4, 6*) velocities on the rigid plate and on the plate *11a* for low (*7, 9*) and high (*8, 10*) velocities; *b* − measurements on the rigid plate (*1* − [492], *2* − [177]; our data: *3* − for low velocities, *4* − for high velocities) and on plate *11a* (*5*− for low velocities, *6* − for high velocities).

In comparison with the rigid plate, the maximal value of k on the elastic plate decreases and its coordinate on the boundary layer thickness increases. The value of k increases all across the boundary layer. As at a flow on elastic plates, Reynolds stresses decrease, the relation $-\overline{u'v'}/k$ essentially decreases due to the fact that turbulence production decreases and kinetic energy of fluctuation motion increases. It indicates a change of ratio between terms of the turbulent energy balance equation. We should note the direct correspondence between magnitudes of $-\overline{u'v'}$ and k which, in the opinion of Harsh [228], is universal for turbulent flow. It is known, that on a rigid plate at a positive gradient of pressure k/U^2_{∞} increases, and its maximum is shifted away from the wall [441].

The data presented in Figure 3.16 qualitatively agree with the measurements data presented in [111, 228] and with results of experiments with polymer solutions. The increasing of k^+ at coordinate y^+ corresponding to the maximum of the turbulence energy is not so essential at the flow on elastic plates, as in tests with polymer solutions. Hassid [327] noted the decreasing of k^+ near the wall. As can be seen from Figure 3.16b, such a decrease is caused by the inflexion in the fluctuation velocities. The values of $\sqrt{\overline{u'^2_{max}}}/u_{\tau}$ increase on the elastic plates, as in [327].

Hassid calculated [147] the turbulence scale as follows:

$$\Lambda = ck^{3/2}/\varepsilon \tag{3.57}$$

where ε is dissipation rate of turbulence energy for unit mass, $C_k = 0.46$ for Newtonian liquid. The value of C_k is selected to satisfy the condition $\varepsilon^+_{\tau} \approx \Lambda^+$ in Newtonian liquid in the region of velocity logarithmic law. Since Λ varies inversely with ε, we have $\Lambda \approx \delta_{lam}$. Theoretical and experimental investigations of the boundary layer of polymer solutions show that the turbulent scale increases all across the viscous sublayer, and that Λ is proportional to δ_{lam}. Investigations on the elastic plates (see Table 3.2), as well as the investigations in non-Newtonian liquids showed the increasing of Λ and δ_{lam}. Indeed, at a flow on elastic plates the turbulence scale has increased on all the near-wall areas of the boundary layer, following from (3.57). Based on the mentioned data, according to equation (3.54), it is possible to define the distribution of production and dissipation of the turbulence energy in the near-wall region of the boundary layer over a rigid plate. Figure 3.17 illustrates the production of the turbulence energy in the near-wall areas on a rigid plate determined from the formula offered in [111].

$$P^+ = \frac{-\upsilon\overline{u'v'}}{u^4_{\tau}} \cdot \frac{dU}{dy} \tag{3.58}$$

Figure 3.18 presents the distribution of dissipation rate of the energy of the averaged motion:

$$D^+ = \frac{\upsilon^2}{u^4_{\tau}} \cdot \left(\frac{dU}{dy}\right)^2 \tag{3.59}$$

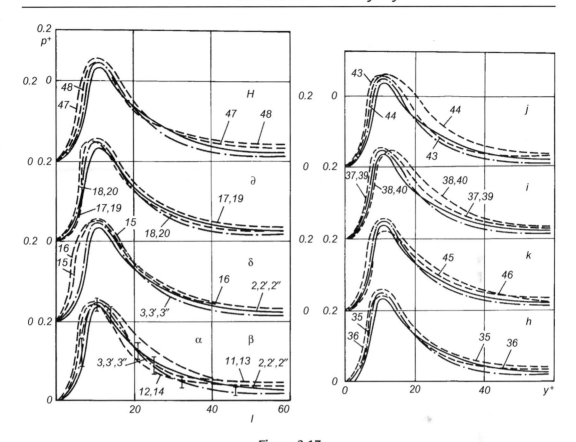

Figure 3.17

Distribution of velocity generation of turbulent energy in the boundary layer over the rigid and elastic plates. Designations are as for Table 3.2.

The obtained data are well agreed with the data by Laufer [441] and Klebanoff [272] who brought the P and D parameters to dimensionless form, using the value of U_∞.

They determined the following values of P and D:

$$P = \frac{\delta \overline{u'v'}}{U_\infty^3} \cdot \frac{dU}{dy} = 0.037$$

$$D = \frac{\delta v}{U_\infty^3} \cdot \left(\frac{dU}{dy}\right)^2 = 0.15$$

(3.60)

The curves P and D intersect according to Laufer at $y^+ \approx 11.5$, and at $y/\delta \approx 0.04$ according to Klebanoff.

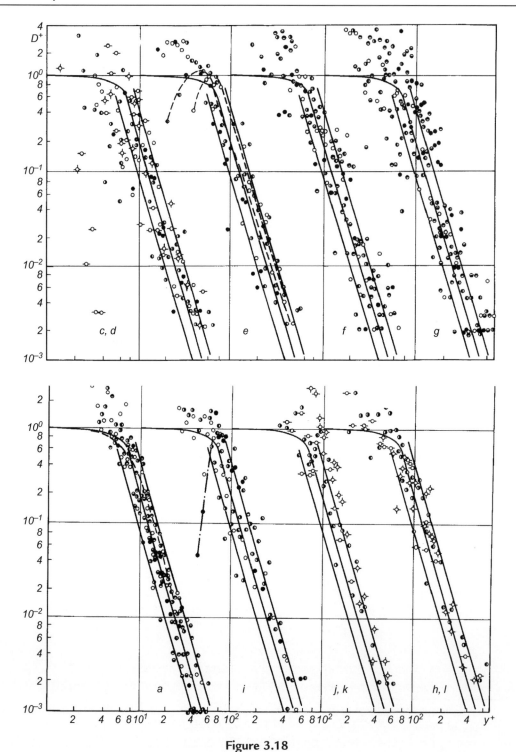

Figure 3.18
Distribution of dissipation rate of turbulent energy in the near-wall region of rigid and elastic plates. Designations are as for Table 3.2.

The curves of energy dissipation are constructed in logarithmic scale to make them more convenient to analyze. The points above $D^+ = 1$ are obviously reflective of the errors in calculation of velocity gradients and the value of u_τ near the wall. Taking into account that near the wall u_τ decreases to zero and the velocity gradient remains to be a finite quantity, the curves presented in Figure 3.18 better correspond with the real conditions than those in [272, 441].

The two outermost solid lines in Figure 3.18a mark the spread of the experimental points for the rigid plate. The middle solid line agrees well with data by Laufer [441] (broken line). On all other figures for elastic plates these data are drawn as three solid lines for the rigid standard plate.

Therefore, the data presented in Figures 3.17 and 3.18 verify that the maximum of turbulent energy production is located at the outer boundary of the viscous sublayer and the energy dissipation is concentrated in narrow layer $0 < y^+ < 15$. Figures 3.5 and 3.9 showed that the maximum of the longitudinal fluctuation velocity is located at $y/\delta = 0.15-0.2$ or $y^+ = 12-15$, which is very close to the location of the maximum of turbulent energy production. As follows from the equations of turbulent energy balance, the value of longitudinal fluctuations depends on the Reynolds stress and the mean velocity gradient [441]. Therefore, we can, as a first approximation, estimate the changing of turbulent energy production in accordance with the value and shape of the profiles of longitudinal fluctuation velocity.

Proceeding from the profiles of longitudinal fluctuation velocity (see Figures 3.5 and 3.9) one can expect that the maximal value of turbulence energy production on the elastic plates does not change substantially (on the compliant plates they slightly decrease, while they slightly increase on the elastic ones). The same conclusions can be drawn from the analysis of the profiles of shear stresses (see Figure 3.15), which decrease near the surface of elastic plates, and increase with moving away from the surface in comparison with the rigid plate. The turbulent energy production increases because of the increasing velocity gradient. This conclusion is corroborated by Figure 3.17: the maximal values of P^+ on all plastic plates do not change, but move in the direction of y^+. The turbulent energy production on the rough plate *5a* decreases after the maximum and increases near the wall. Such behavior remains in the case of the plate pasted over with a rubber film (plate *2a*). At the compliant plate *9* (see Figure 3.17C) differences from the standard are appreciable after P^+_{max}: depending on flow velocity, P^+ increases or decreases. The plates generating the longitudinal vortex or 2D disturbances are of the most interest. The plates generating the longitudinal vortices (*10a, 11, 11a*) show a slight shift of the curves to the right (particularly at high velocities) compared to the standard plate. The turbulence generation decreases at $P^+ < P^+_{max}$, at $P^+ > P^+_{max}$ increases, and P^+_{max} is located at greater y^+. At the plates generating 2D disturbances (*3, 8a*) various results are obtained. The behavior of plate *8a* is the same. The turbulence energy on plate *3* increases at $P^+ < P^+_{max}$ (similar to the compliant plates). Thus, production of turbulent energy at a flow on elastic plates does not change essentially in comparison with the standard, and is redistributed on y^+, as if it moves above on the boundary layer thickness, that well correlates with fluctuations of the outer surface of elastic plates.

The experiments with polymer solutions showed other results. With a solution of potassium oleate [426, 474], the value of P^+ substantially decreases all across the boundary layer and the value of D^+ increases. Solutions of polyakrylamid showed the decreasing of turbulent energy dissipation rate at $y \approx 0.3\delta$ [248].

In the authors' opinion, the energy balance of the turbulent boundary layer over elastic plates should not change essentially in comparison with energy balance of over the rigid standard. The processes of momentum and energy transfer across the boundary layer depend on flow conditions near the wall. Thus, there are two ways to influence the turbulence anisotropy in the boundary layer. The first requires decreasing the extraction of energy from the averaged motion. This can be done either by "smoothening" of the outer boundary of the boundary layer or by decreasing bursts from of the viscous sublayer.

However, such a statement does not mean a reduction of the turbulent energy production. For example, for organization of longitudinal vortical systems in the viscous sublayer (plates *10a, 11, 11a*) value P^+ increases. Thus $-\overline{u'v'}$ can either increase or not change. In reality "smoothening" of the outer boundary can be done by decreasing the burst from viscous sublayer.

The second way of reducing anisotropy is thought to be the most promising. It is based on the fact that the turbulent energy balance and amount of energy extracted from the averaged motion remain constant. Only magnitudes of components of the energy balance equation are redistributed so that an increase of turbulence energy production should be accompanied by its increased dissipation, and the specific contribution of components of energy should be redistributed across the boundary layer thickness. In this case, a thickening of the viscous sublayer and moving of P^+_{max} above along y^+ must be followed by enlargement of the vortices near the wall and decreasing anisotropy. It means that the dissipation must increase. Note that both specified ways of anisotropy reduction are realized in the near-wall areas where anisotropy is most essentially pronounced.

Let us consider the value $\int_{y^+=1}^{y^+=100} D^+ dy^+$ that characterizes the turbulence dissipation (see Figure 3.18). While flowing over plates *5* and *2a*, the dissipation substantially decreases only near the wall. The dissipation on the plate *2a* at $y^+ \to 0$ tends not to 1, but to essentially smaller values, so that at $y^+ = 9$ we have D^+_{max} (i.e. in the region where the velocity profiles have inflexions).

The dissipation integral on compliant plates *9, 1a* and *1* (see Figure 3.18c,d) increases according to the plate effectiveness: at plate *9* D^+ is the same as at the standard plate, and at plates *1* and *1a,* depending on the velocity, D^+ becomes higher than at the standard plate.

At the elastic plates (*11, 11a, 10a*) generating 3D disturbances in a boundary layer, D^+ has increased (at a plate *10a* to a lesser degree due to the roughness of its surface). The spread of experimental points has essentially decreased. An inconsistent result is obtained at plates *8a* and *3*, as well as for the research of P^+: dissipation has increased for plate *8a*, and remained same at as at the standard for plate *3*. The invariance of components of the turbulent energy balance at plate *3* is caused by its mechanical characteristics, in particular its considerable rigidity. Despite of this, according to principles of complex interactions of different disturbances (See Chapter 1), longitudinal fluctuation velocities noticeably decreased at plates *8a* and *3*.

Thus, two features are discovered. At decreasing y^+ the value of D^+ tends not to 1, as in Laufer's data on a rigid plate [441], but to essentially smaller values indicating the location of maximum D^+_{max} at $y^+ \approx 5-6$ at compliant plates, (i.e. on the border of the laminar sublayer), and at $y^+ \approx 8-9$ at elastic plates. Values of P^+ and D^+ move away from elastic plates. The area in which viscous and inertial forces are counterbalanced shifts away from the wall. Immediately at the wall dissipation decreases, and increases at $y^+ > 5-9$, that can be caused by fluctuations of the surface of elastic plates.

The second feature relates to the fact that at elastic plates, at which turbulence production $\int_{y^+=0}^{y^+=60} P^+ dy^+$ has increased, dissipation $\int_{y^+=0}^{y^+=60} D^+ dy^+$ is simultaneously increased. It confirms the hypothesis about redistribution of components of energy balance across a boundary layer due to the tendency of a flow to reduce the anisotropy of turbulence. As soon as the decelerating influence of a wall upon a flow is weakened, there occurs reorganization of vortical systems in a boundary layer and the anisotropy of turbulence decreases.

3.6 Correlation and Spectral Parameters of a Turbulent Boundary Layer over an Elastic Plate

To develop the models of turbulence based on the empirical constants, it is necessary to know the time or space scales of turbulence determined in statistical investigations. In addition, features of distributions of fluctuation and averaged velocities across the boundary layer over the elastic plates, as well as features of distribution of the components of turbulent energy balance cannot be explained only based on "one-point" measurements.

Below the authors present the results of investigations of statistical characteristics of the boundary layer. Two thermo anemometers [42, 257] obtained the double spatial correlation coefficients. One sensor was mounted so as to be stationary and the other could be moved along the axis y. The correlation functions were being measured at different velocities in two sections relative to the longitudinal coordinate x (I and III sections, Figure 3.2), at different distances from the surface. The stationary sensor was located at y equal to 1 and 10 mm. Parameters of the flow are listed in Table 3.2.

The curves in Figure 3.19 are presented in the normalized form:

$$R_{u'u'(0,r,0)} = \frac{\overline{u_1' u_2'}}{\sqrt{\overline{u_1'^2}} \ \sqrt{\overline{u_2'^2}}} \tag{3.61}$$

The data obtained on the rigid plate agree well with [180, 283, 433]. The shape of the correlation coefficient curve indicates the wide range of the vortices dimensions. It is known that the curve shape, if the distance between the sensors is very small ($r \to 0$), characterizes the micro scales of the dissipation vortices if they are not less than the sensor size [130, 433, 473]. The integral scale or macro scale of turbulence, which characterizes the largest size of vortex in the point of measuring, can be calculated by the formula:

$$L = \int_0^\infty R_{(0,r,0)} dr,$$

$$L = \int_0^1 R_{(0,r,\delta,0)} d\left(r/\delta\right) \tag{3.62}$$

The last formula does not take into account the negative regions of the correlation functions, which decrease the value of L and characterize the reversed compensating flow due to the liquid continuity.

On a rigid plate, as well as in data from [180, 283], the vortex sizes increase with the increase in distance from the wall. It verifies the cascade process of vortex transport to the surface [130, 441]. At $U_\infty = $ const the correlation coefficients slightly decrease as x increases. At a change of Reynolds number the correlation coefficients increase due to increasing speed, because the relative coordinate of the motionless sensor changes $\tilde{y} = y/\delta$.

It should be noted that, according to [223, 225, 283, 441], $R_{u'u'(r,0,0)} >> R_{u'u'(0,r,0)} > R_{u'u'(0,0,r)}$, i.e. the vortex size in the longitudinal direction is much larger (almost 40 times) than in the transverse direction. The hypothetical vortices can be imagined as strongly extending along an axis $0x$, with greater size along $0y$ as along $0z$.

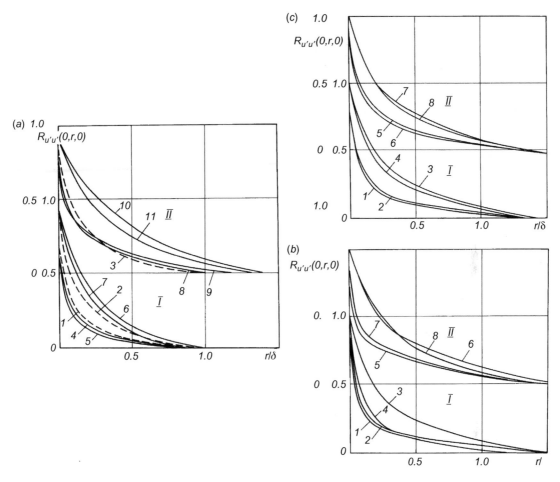

Figure 3.19

Double-space correlation of longitudinal fluctuation velocities in the flow over the rigid (*a*) and elastic (*b, c*) plates at different distances from the wall depending on U_∞ and *x*. Dash curves — data of [129] at $\tilde{y} = 0.033(1)$, $0.0405(2)$, $0.4(3)$. *a*: *I* — low velocities (*4* — first section, Figure 3.2, $U_\infty = 10.15$ m/s; $\tilde{y} = 0.0398$; *5* — third section, 10 m/s; 0.0273; *6* — first section, 10.15 m/s, 0.398; *7* — third section, 10 m/s, 0.341), *II* — high velocities (*8* — first section, 15.98 m/s; 0.043; *9* — third section, 17.06 m/s; 0.0364; *10* — first section, 15.98 m/s, 0.43; *11* — third section, 17 m/s, 0.364), *b*: plate *9*: *I* — low velocities (*1* — first section; 9.49 m/s; 0.0433; *2* — third section, 9.77 m/s; 0.0366; *3* — first section, 9.49 m/s; 0.433; *4* — third section, 9.77 m/s; 0.366), *II* — high velocities (*5* — first section. 16.12 m/s; 0.046; *6* — third section, 15.92 m/s; 0.0375; *7* — first section, 16.12 m/s; 0.046; *8* — third section, 15.93 m/s; 0.375). *c*: plate *11a*: *I* — low velocities (*1* — first section, 10.13 m/s; 0.04; *2* — third section, 9.9 m/s; 0.0346; *3* — first section, 10.13 m/s; 0.402; *4* — third section, 9.9 m/s; 0.346), *II* — high velocities (*5* — first section, 17.67 m/s; 0.04; *6* — third section, 17.43 m/s; 0.0355; *7* — first section, 17.67 m/s; 0.4; *8* — third section, 17.43 m/s; 0.355).

Figure 3.19b,c presents the correlation coefficients on the compliant plate 9 and on the elastic plate 11a near the surface (for the standard plate at $y+ \approx 30$ (low velocities) and at $y+ \approx 50$ (high velocities), for plate 9 at $y^+ = 25$ and 45, for plate *11a* at $y^+ \approx 30$ and 50 accordingly). The measurements of correlation coefficients on the rigid plate, as can be seen from Figure 3.4, were made in the region of logarithmic law of the mean velocity, and on the elastic plates at $y^+ = 25-30$, i.e. in the region of the outer boundary of the buffer region.

The correlation coefficients increased on both types of elastic plates. Comparing corresponded values of $\tilde{y}$ on rigid and elastic plates, it is possible to conclude that such an increase of correlation coefficients indicates an increase of $\tilde{y}$ on elastic plates. However, this is not the only reason. The increase of correlation coefficients near the elastic plates occurs owing to specific features of their interaction with the flow, in particular owing to a changed balance of turbulent energy on the entire thickness of the boundary layer.

The value of $R_{u'u'}$ on both types of elastic plates increased at $y^+ = 25-30$, i.e. the integral scales of turbulence and the sizes of largest vortices near the surface increased. However, the laws of these changes at elastic plates *9* and *11a* are different. At the compliant plate *9* $R_{u'u'}$ increased basically due to the growth of large vortices (at $r/\delta = 0.1-1$) and much less due to increase in the sizes of small-scale vortices (at $r/\delta = 0.1-1$). The size of large vortices increased by 1.5 times and positive correlation coefficients were registered up to $r/\delta = 1.5$. In other words, energy of large vortices and their statistical time of existence near the surface increased.

In contrast with that, on plate *11a*, which generates the system of small-scale longitudinal vortices, the increase of $R_{u'u'}$ at $y^+ \approx 30$ is caused basically by the increasing energy of small-scale and medium vortices (at $r/\delta = 0-0.5$), and the influence of large vortices is less. Hence, near plate *11a* the sizes and energy of small and medium vortices increased.

In both cases (plate *9* and plate *11a*) the exchange of energy between the vortices of different sizes changed. The increase of the transverse size of vortices indicates the decreasing turbulence anisotropy near the surface. These laws remain at velocity increase and in the region of logarithmic law of the velocity distribution (at $y^+ = 50$).

The obtained data showed that the main change in energy balance and turbulence scales over elastic plates takes place in the near-wall region, up to the outer boundary of the buffer layer. These changes propagate onto the boundary layer and even to the free flow due to the surface oscillations.

In experiments with polymer solutions [223, 327], the decreasing of the correlation coefficients $R_{u'u'(0,r,0)}$ is registered at $y^+ = 80$. In the paper [225], it is shown that in the near-wall layer the transversal scale increases almost by 7 times compared with a rigid plate.

From the obtained results, it is possible to draw conclusions that the conditions leading to decelerating effect of a wall promote the reduction of anisotropy. It is reflected in the alignment of longitudinal and cross-sectional (due to an increase in the last) vertical and transversal scales. The vortex becomes symmetrical in a cross-section, instead of an extended shape in vertical direction, as on the standard. Hence, values of the correlation coefficient describing the cross-sectional sizes of vortices should increase in such cases that were registered in the tests near elastic plates.

The statistical analysis of turbulent flows can also be carried out by means of investigations of the spectrum characteristics. Usually a 1D spectrum is analyzed, which leads to underestimation of the energy of 3D spectra.

We can write the relationship between the correlation and spectrum function by means of Fourier mutual cosine-transformations:

$$R(r) = \int_0^\infty E(k) \cos krdr$$

$$E(r) = \frac{2}{\pi} \int_0^\infty R(r) \cos krdr \tag{3.63}$$

The 1D spectra with respect to wake number $E(k)$ were being measured by a thermo anemometer. A sensor was located in the point of maximal fluctuation velocity along $\tilde{y} = y/\delta$. The measurements were being made in three sections along the x coordinate (see Figure 3.2). The measurement results were normalized according to the formula:

$$\int_0^\infty \frac{E(k)}{\overline{u'^2}} dk = 1 \tag{3.64}$$

where $E(k) = \frac{F(\omega)u}{2\pi\Delta f}$ are spectral functions; Δf is the infiltration band-pass filter width; $k = 2\pi f/u = 2\pi/\lambda$; f and λ are frequency and length of the wave harmonic; u is mean velocity in the measurement point.

Figure 3.20 presents the spectra of longitudinal fluctuation velocities on rigid and elastic plates. The character of change of spectral density with regard to wave numbers for a rigid plate corresponds to Klebanoff's measurements [272] for the internal area of boundary layer. The curve *2* at $\tilde{y} = 0.005$ is located closer to Klebanoff's curve *3* at $\tilde{y} = 0.0011$, and the curve *1* at $\tilde{y} = 0.013$ is between the curve *4* at $\tilde{y} = 0.05$ and the curve *3*. At the same time, curves *1* and *2* limit the spread of experimental points; experimental points at small speeds of the flow are situated basically along curve *1*, and along curve *2* at greater speeds. Measurements of spectral density are executed near the surfaces at $\tilde{y} = 0.005-0.16$, that corresponds in universal coordinates to $y^+ = 7-13.5$. Hence, at smaller values the

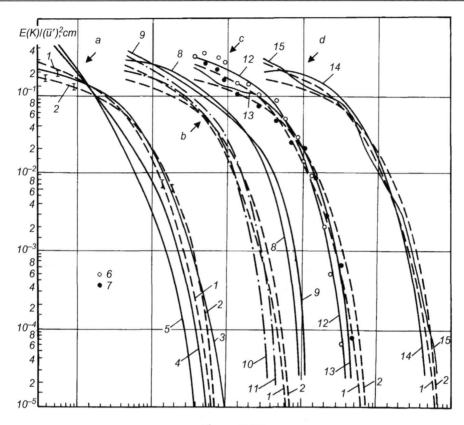

Figure 3.20

Spectral function of the longitudinal pulsating velocity on the rigid (*a*) and elastic (*b-h*) plates: *a*: 1, 2 — integrated curves for low and high velocities in I–III sections according to Figure 3.2 at *u* and $\tilde{y}$ equal to 5.36 m/s and 0.016; 4.2 m/s and 0.0115; 4.98 m/s and 0.013; 7 m/s and 0.007; 7.58 m/s and 0.005, 7.8 m/s and 0.005; *b*: plate 5 — curves 8, 9 for low and high velocities in I–III sections at *u* and $\tilde{y}$ equal to 4.2 m/s and 0.012; 3.49 m/s and 0.0055; 6.08 m/s and 0.006; 5.825 m/s and 0.0067; plate 2*a* — curves 10, 11 at *u* and $\tilde{y}$ equal to 4.34 m/s and 0.0135; 6.55 m/s and 0.0053; *c*: plate 11*a* — curves 12, 13 at *u* and $\tilde{y}$ equal to 4.54 m/s and 0.014; 4.57/ m/s and 0.01; 7.57 m/s and 0.009; 7.5 m/s and 0.0096; section III*a* — curve 6 — 8.68 m/s and 0.008; curve 7 — 7.98 m/s and 0.007 (heated plate); *d* (III section): plate 11 — curves 14, 15 at *u* and $\tilde{y}$ equal to 4.25 m/s and 0.0135; 7.16 m/s and 0.012. Curves 3–5 — measurements [190] at $\tilde{y}$ equal to 0.011; 0.05; 0.02; *e*: plate 1 — curves 16, 17 at *u* and $\tilde{y}$ equal to 4.68 m/s and 0.01; 4.99 m/s and 0.0126; 4.7 m/s and 0.01; 7.4 m/s and 0.0117; 7.64 m/s and 0.01; 8.074 m/s and 0.008; *f*: plate 1*a* — curves 18, 19 at *u* and $\tilde{y}$ equal to 4.06 m/s and 0.0128; 4.6 m/s and 0.0073; 4.04 m/s and 0.009; 7.04 m/s and 0.007; 6.86 m/s and 0.007; 7.17 m/s and 0.0078; *g*: plate 9 — curves 20, 21 at *u* and $\tilde{y}$ equal to 4.42 m/s and 0.016; 3.89 m/s and 0.01; 6.86 m/s and 0.006; 6.73 m/s and 0.007; section III*a*: curve 6 — 6.32 m/s and 0.004; curve 7 — 6.43 m/s and 0.007 (heated plate); *h*: plate 10*a* — curves 22, 23 at *u* and $\tilde{y}$ equal to 4.33 m/s and 0.0135; 6.35 m/s and 0.005.

measurements were done in the viscous sublayer, and at great values on the external border of the buffer layer.

One-dimensional spectra of fluctuation velocity characterize contributions of different scale vortices into the energy spectral density. Therefore, we can rationalize the obtained spectra as follows. At $y^+ = 7-13.5$, the large-scale vortices with the largest λ are registered that corresponds to low values of $k = 0.3-3$. These vortices characterize the extraction of energy from the averaged motion. The inertia vortices are registered at $k = 4-20$, and the dissipative vortices at $k = 20-70$.

According to the cascade process of energy transport, there is an energy flux from large-scale vortices to the low-scale vortices, which dissipate, and their energy near the surface goes over into the heat. The contributions of different scales vortices existing in a turbulent boundary layer into the spectral function depend on the vertical coordinate. As it can be seen in Figure 3.20a, the distance from the wall being increasing (curves *4* and *5*) the energy of large-sized vortices increases and of dissipative ones decreases compared to those in the viscous sublayer (curves *1*, *2* and *3*). According to normalization of scales of vortices, in the specified areas there are certain dependences of spectral density on wave numbers [168, 232, 233, 272, 330, 433, 441, 473]. As it is shown in [433], in the near-wall region we have the law $E \sim k^{-1}$. Klebanoff and Laufer registered it, in particular on the rigid standard points of the curves *2* and *4* (high velocities) corresponded to the law at $k = 1.5-10$ and the curve *4* at $k = 0.3-2$. According to [283], the law $E \sim k^{-5/3}$ describing the inertial sub-domain of intermediate scales was registered only at $y^+ = 15-80$ over narrow range of wave numbers. In our measurements, the law was registered mainly at $y^+ = 13.5$, for example on the rigid plate of the group *a*, curve *1*, low velocities at $k = 4-20$.

At high values of k, which correspond to the region of viscous dissipation, the law $E \sim k^{-7}$ takes place, as it is shown in [441]. Our measurements at $k > 20$ agree with the Klebanoff's data. Thus, reference curves *1* and *2* are characterized by that in the area of energy-carrying vortices ($k = 0.3-4$) experimental points are located along curves $E \sim k^{-1/6}$ and $E \sim k^{-1/3}$ accordingly at low and high speeds of flow. The same differentiation is observed in the field of inertial scales ($k = 4-20$), when at low velocities the law $E \sim k^{-5/3}$ is valid, and $E \sim k^{-1}$ at high velocities. In the region of dissipative vortices, the law $E \sim k^{-7}$ takes place. Such differentiation of the curves is determined by the coordinate y of the measurement point.

In [473] it is shown that transport of energy from components u' to components v' and w' is especially intensive in the high-frequency part of the spectrum, at greater k.

The group *b* characterizes the law $E(k)$ on the elastic plates *5* and *2a*. The increasing of energy of vortices of all sizes on the foam polyurethane plate can take place only if the generation of turbulent stresses is increased and the dissipation of turbulent energy decreased. These measurements correspond to the measurements of the fluctuation velocities, as well as of the energy balance (see Figure 3.17 and 3.18a). The spectrum on the foam polyurethane plate pasted over with elastic film became similar to the standard plate spectrum due to the decreased roughness. But, due to mechanical properties of the elastomer, spectral density of the energetic vortices increases in comparison with the rigid standard, while it decreases in the dissipative region.

According to the system analysis, let us consider, at first, the compliant plates. The least effective of them, plate *9*, is investigated in the group *g* (see Figure 3.20). The spectrum density of the longitudinal fluctuation velocity slightly increased at all wave numbers compared with the standard plate. In the inertia vortices region, only the law $k^{-5/3}$ takes place. In the energy-carrying vortices region, the function *E(k)* changed at low velocity. It corresponds to measurements of the correlation coefficient and energy balance. Similar results were obtained during the investigations of the most effective compliant plate *1*, group *d*.

An unexpected result was obtained on plate *1a* (group *f*): the energy of all vortices substantially decreased, particularly in energy-carrying vortices at high velocities.

The same picture was observed at the measurement of profiles of longitudinal fluctuation velocities (see Figure 3.5f). It is possible to explain it only by the following: the plate surface performed small tangential oscillations influencing the near-wall shear fluctuations. In the specified range of y^+, spectra of the near-wall shear fluctuations practically coincide with spectra of fluctuations of longitudinal velocity [171], consequently, the spectrum *E(k)* was changed. Tangential oscillations of the elastic plate were caused by tangential stress on the wall. It made the insert fixed on elastic membranes moving along the flow (see Figure 2.27). After a burst from the viscous sublayer, tangential stress on the wall decreased, elastic deformation of membranes returned the suspension bracket with the elastic plate into the initial position. The self-oscillatory mode of the suspension bracket occurred when the frequency of bursts from the viscous sublayer coincided with the natural frequency of the entire insert. Fluctuations of the suspension bracket occurred due to extraction of energy from the main flow directed on the formation of the near-wall shear fluctuations in the boundary layer. Such resonant oscillations of the elastic plate led to resonant energy absorption of this layer over the entire wave range of the longitudinal fluctuation velocity. The plate tangential oscillations were then eliminated due to its rigid jamming, so the revealed anomaly has not been investigated in the experiments. Therefore, further research of other kinematic characteristics of the boundary layer on this fixed plate has not found any abnormal phenomena.

The revealed phenomenon can be explained by the fact that mechanical characteristics of the given elastic plate have allowed the causation of fluctuations of its outer surface, which have been concurrent in the resonance with the frequency of bursts from the viscous sublayer. These resonant fluctuations have coincided with natural fluctuations of the elastic suspension bracket of the insert. As a result, the insert got into a self-oscillatory mode at a frequency concurrent with the frequency of bursts from the viscous sublayer. Because of the onset of self-oscillation, there was a so-called pre-burst modulation effect, the physical aspect of which is stated by Bushnell [20, 137, 138]. Thus, friction drag of the plate is essentially decreased.

The revealed feature also makes sense in relation to the complex interaction of disturbances (see Chapter 1). At resonant oscillations of the plate, a flat wave is generated, which interacts with disturbances of the boundary layer in the form of longitudinal vortices. At such an interaction, there is an intensive reduction of longitudinal fluctuation velocities [310].

The group h (see Figure 3.20) presents the results of measurements of the spectral density of homogeneous compliant in plate $2a$ and plate $10a$ (plate $10a$ generates the longitudinal vortices). The results showed the increasing of large-scale vortices and decreasing of energy of the dissipative vortices at high velocities of flow. At low velocities, the reverse takes place. At low values of wave numbers, it can be explained by the relative location $\tilde{y}$ of the thermo anemometer sensor. The decreasing of the dissipative vortical energy can be explained by the absorption of vortices by the elastic surface. Similar effects are registered in experiments with polymer solutions [222, 223]. The experiments on plate $10a$ showed the attributes indicated longitudinal vortices in the boundary layer [310]: there were maxima in the region of energy-carrying vortices and minima in the inertia wave number region.

These attributes are clearly noticeable in the spectra on the elastic plate $11a$ (group c) in all sections at high and low velocities of the flow. According to principles of complex interactions (See Chapter 1), depending on the intensity of longitudinal vortices generated from below, bends and maxima of the curve are different: bends are most appreciable in section III at greater speed and in section IIIa at the regulation of mechanical characteristics of the plate by heating. They become apparent depending on sensor location relative the longitudinal vortices.

Spectra of two types of elastic plates have also been investigated: 8 and $3a$, which generate 2D disturbances, and 11, which, in addition to plate $11a$, generates 3D disturbances. Both types of plates showed the increasing of the energy of large-scale vortices and the decreasing of the dissipative vortical energy. But the attributes indicating coherent structures in the boundary layer in the form of longitudinal vortical systems were detected only over plate 11 (see Figure 3.20d).

Fortuna [225] presented the spectral density of longitudinal near-wall shear depending on frequencies while investigating the polymer Separan AP-30. The spectrum form changed compared to the solvent. The energy increased, as in [310], at small values of wave numbers and valleys appeared in the inertia region of the spectrum. The correlation factors increased in transversal direction. The coefficient of skin friction in Couette's flow decreased by 34.6% and 64.5% due to the Separan AP-30 effect. These data agree well with the results obtained on the plates generating the longitudinal vortices.

The decrease in spectral function of longitudinal fluctuation velocity in the dissipative region can be explained by more intensive transport of energy to the components v' and w', which leads to enlargement of the dissipative vortices and to a decrease of anisotropy in the near-wall region.

CHAPTER 4

Fluctuations of an Elastic Surface in a Turbulent Boundary Layer

4.1 Apparatus for Research on Fluctuations of an Elastic Surface

According to [16, 165, 186, 257, 518] the amplitude of vibration of an elastic surface is less than 20 μm. This creates certain difficulties for the measurement of fluctuations of the surface of an elastic plate. For measurements with high accuracy, the optical method was chosen, the basic advantage of which is a non-contact definition of the amplitude of a surface motion [177, 165]. This method involves the following. The light spot is incident on the surface through the focusing device. The photo detector receives the reflected light. Depending on the distance of the source and the detector from the surface, the light spot is defocused and the detector receives the different amount of the light energy. The apparatus is presented in Figure 4.1. The light from the power supply (1) and light source (2) is incident on the surface (14) through the light guide (3) and focusing device (4). The reflected light gets into the receiver (7), and then through the light guide (8) it gets into the photo multiplier tube (9). The signal (in voltage form) through the amplifier (10) is registered by the direct current voltmeter (11), the mean-square voltmeter (12) and tape recorder (15). In the given apparatus, the authors used the photo multiplier tube PhEU-2, the direct current and mean-square voltmeters DISA 55D30 and 55D35, and the tape recorder MP-1.

The focusing device (4) and receiver (7) are made in the form of a tube with a 10 mm diameter. Lenses for focusing a light beam on the investigated surface and the light reflected from it on the face surface of the light pipe have been built in the focusing and receiving tubes. To increase the accuracy of the measurements, a thin mirror elastic foil with a diameter of the order 1 mm was pasted on the investigated surface. The focusing device and receiver were fastened on the same holder (6), which was connected with the coordinate device necessary for calibrating the measuring device. The holder (6) was not established on a hydrodynamics stand and had independent fastening.

This arrangement made it possible to avoid the influence of vibrations of the hydrodynamics stand on the indications of measuring equipment. The coordinate device was connected with a measure, which measured displacement of the holder together with

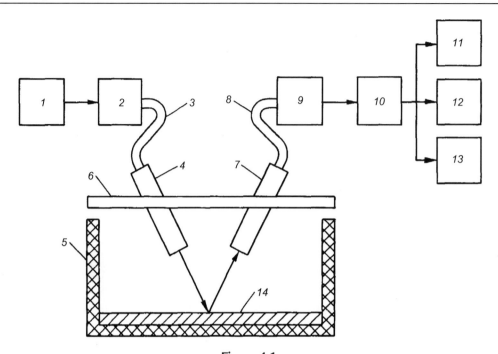

Figure 4.1
Scheme of the equipment for measuring the oscillation of an elastic surface.

the focusing and receiving lenses fixed on the holder, with an accuracy of 1 μm. The calibration function of the output voltage against the distance from the surface was linear.

While these measurements were being carried out, the pressure fluctuations were investigated by the methods presented in [258, 488]. There were vertical orifices of 1 mm in diameter in the elastic plates. Sensors of pressure pulsations were fastened from below on the elastic plates on the case of the tensometric inserts instead of mechanotrons (see Figure 2.25). Measurements were being made on the same plates as in the airflow experiments (see Section 2.5).

4.2 Investigation of Fluctuations of an Elastic Surface

The investigations of oscillations caused by the flow were made on two types of plates: compliant *2a* and elastic *11*. The measurements were done at the hydrodynamics stand at a flow velocity of 0.6 m/s in section III (see Chapter 1). The test section design and the experimental procedure are described in detail in [30, 305].

To indicate the interference level, the target signal from the displacement-measuring instrument was also registered in the flow over the reference rigid plate and at zero flow

Table 4.1: Amplitudes of oscillations of the elastic plate surface.

Type of Elastic Plate	$\sqrt{a'^2}$, μm			$\left\langle \sqrt{a'^2} \right\rangle$, μm
	Measurement Number			
	1	2	3	
Plate *2a*	12	17	15	14.7
Plate *11*	18	21	25	21.3

velocity (i.e. with the gate regulating the flow velocity closed) with the pump working. Each plate was measured three times. Fluctuations of the surface displacement in the form of the voltage were registered on the tape multichannel tape-recorder MT-1. The results of measurements of the root-mean-square values of the surface fluctuations are presented in Table 4.1.

The displacement amplitude for plate *2a* (see Table 2.4, PPU with film) was found to be less than that of plate *11* (PU). It can be explained by the resonance properties of plate *11*, which has higher values for the natural frequency of oscillations and modulus of elasticity (see Table 2.19).

In theoretical investigations of the interaction between a flow and elastic plates, Betchov and Kriminale calculated the complex coefficient of rigidity connecting the intensity of the pressure fluctuations $\sqrt{p'^2}$ with the amplitude $\sqrt{a'^2}$ of the surface oscillations [186, 490]:

$$Z = \frac{\sqrt{p'^2}}{\sqrt{a'^2}} \tag{4.1}$$

where

$$Z = S - iarc - m\alpha^2 c^2 + T\alpha^2 \tag{4.2}$$

Based on formula (4.1) it is possible to estimate the magnitude of surface fluctuation [257]. To do it, we can use the formula from [441, 488, 445, 516, 535]:

$$\frac{2\sqrt{p'^2}}{\rho U_\infty^2} = d_\tau \lambda \tag{4.3}$$

where d_τ is proportionality factor (for a smooth surface $d_\tau = 2.1$ and λ is determined by the formula:

$$\lambda = \frac{2\tau_0 / \rho}{U_\infty^2} \frac{2\overline{u'v'}}{U_\infty^2} \tag{4.4}$$

According to Figures 3.13 and 3.14, we specify the value of λ for rigid plate equal to 0.004 and for the elastic plate as 0.0032. Hence, for $U_\infty = 0.6$ m/s, the value of $\sqrt{p'^2}$ calculated by formula (4.3) is 1.5 N/m^2 for the rigid plate, and 1.2 N/m^2 for the elastic plate.

The approximate values of root-mean-square amplitudes of the surface displacement calculated by formula (4.1) are 8.1 μm for plate *2a*, and 0.5 μm for plate *11*.

The calculated and measured values for plate *2a* are well agreed. For plate *11*, the values are substantially different. In the paper [112] the other expression for the compliance is presented:

$$Z^{-1} = [\sigma\omega^2 - T\alpha^2 - S\alpha^2 - \rho_w g - \sigma_s(\rho_w, \alpha, c, H, v)]^{-1} \tag{4.5}$$

A detailed analysis of boundary conditions is carried out in [32, 504, 310]. It is obvious that the amplitude of an elastic plate fluctuation influences not only pressure fluctuations of the boundary layer and frequency properties of the surface. Of essential importance is the susceptibility of the boundary layer; in particular, under the action of rather weak fluctuations of pressure in a flow there can be a resonant increase in amplitude of the surface displacement, or conversely, its essential reduction (see Chapter 1, with regard to susceptibility).

The assumption about the resonant character of the interaction of surface oscillations and the boundary layer fluctuations is confirmed by values of f and f^+ (see Table 3.3) and research on the spectral characteristics of these fluctuations (see Figure 4.2). In Figure 4.2a, spectra of the surface fluctuations of the rigid plate made of Plexiglas (standard) are shown. With the pump working and the gate closed ($U = 0$, curve *2*), there is an increase in relative amplitudes at frequencies of 12 and 20 Hz, which could be caused by oscillations of the entire strain-sensor suspension. It is important to remember that the measuring insert is fixed with four elastic elements (see Figures 2.26, 2.27 and 3.1).

The power-line noise causes the increase in amplitude of curves *2* at a frequency of 50 Hz. A certain increase is observed at the second and third harmonics of the power-line noise, at frequencies of 100 and 150 Hz. When the pump is on and the gate is closed, curves *2* characterize the vibrations of plates. At a flow over a plate, under the action of pressure fluctuations, oscillations caused by the basis of elastic plates and pressure fluctuations in the boundary layer begin to interact, resulting in spectra of the surface fluctuation change (curve *1*). The mechanism of such interaction of fluctuations speaks on the basis of a problem of susceptibility (see Section 1.11 of Chapter 1 and [47, 55, 59, 61 and 310]). On a rigid plate, the above-noted vibrating and power-line noise induced fluctuations still exist, but their contribution to the spectrum energy decreases. Due to these imposed fluctuations, spikes appear in the curve at 30 and 75 Hz. Energy of fluctuation of the plate surface caused by the flow increases in comparison with absence of a flow (4.2 a curve *2*).

According to the similarity criteria (see Section 2.6), plate *11* (see Figure 4.2b) is not optimal and remains rigid at a flow speed of $U_\infty = 0.6$ m/s, as well as in the airflow. Therefore, the character of curves *1* and *2* are similar to the curves on the rigid standard. However, the amplitude of surface fluctuations of elastic plates is always larger than that of a rigid surface (see Table 4.1). Therefore, amplitudes and energy of surface fluctuations in curves *1* and *2* of plate *11* are increased in comparison with the standard over the whole

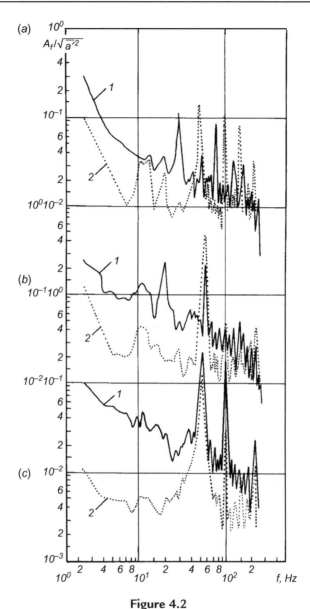

Figure 4.2

Oscillation spectra of the surface of rigid (*a*) and elastic (*b* and *c*) plates: *11* (*b*) and *2a* (*c*); $1 - U_\infty = 0.6$ m/s, $2 - U_\infty = 0$ (the pump works).

range of fluctuation frequencies. Thus, pressure fluctuations of the turbulent boundary layer (curve *1*) essentially increase the amplitude of surface fluctuations.

According to the similarity criteria, plate *2a* is more effective at low velocities. Therefore, its vibro-damping properties are the best (curve *2*, see Figure 4.2c). Since the compliance of

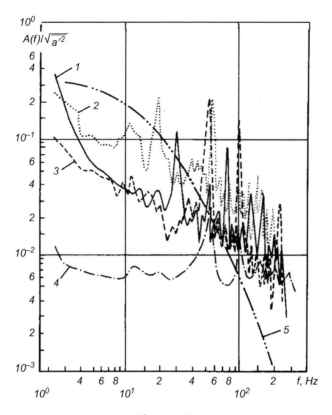

Figure 4.3

Comparison of surface oscillation spectra of rigid (*1*) and elastic (*2* and *3*) plates: *11* (*2*) and *2a*
(*3*) at $U_\infty = 0.6$ m/s, *4* — plate *2a* at $U_\infty = 0$ (the pump works), *5* — spectrum of longitudinal
velocity fluctuations of plate *11*.

plate *2a* and its oscillation amplitude is higher than that of the rigid plate, curve *1* and *2*
differ from each other more than over the rigid plate. Due to good vibro-damping
properties, the peaks in curve *1* are smoothened over in plate *2a*.

Comparison of the data obtained is shown in Figure 4.3. Plate *11* has the greatest energy of
surface fluctuations. Its spectrum of fluctuations (curve *2*) has the classical form typical of
pulsations of longitudinal velocity in a boundary layer (curve *5*).

In the field of low frequencies, the fluctuation spectrum of plate *2a* (curve *3*) is less than at
the rigid standard (curve *1*). In other parts of the frequency range, these curves practically
coincide. The vibro-damping properties of plate *2a* in the absence of a flow (curve *4*)
remain constant practically over the whole frequency range. This is probably why the
minimal coefficients of friction drag (see Chapter 5) were revealed in the water flow over
plate *2a* in the low-turbulence hydrodynamic channel.

Thus, taking into account the kinematic parameters of the boundary layer over the elastic plates and the data presented in Figures 4.2 and 4.3, it suggests that the process of interaction between vibro-damping surfaces and a flow have a complex character. It can be reasoned that the main mechanism of interaction of a boundary layer with compliant plates is the absorbing of the fluctuation energy [285, 287, 289, 502−504]. For elastic plates, the main mechanism is the pre-burst modulation effect [137].

We can see from Figure 4.2 (curves *1* and *2*) that the oscillation amplitudes of a surface are essentially influenced by the pressure fluctuations of a turbulent boundary layer, which correlate well with the longitudinal velocity fluctuations [176, 172, 535] and with the near-wall shear fluctuations [171, 177]. Therefore, it is interesting to compare the spectrum of surface displacements and the spectrum of pressure fluctuations (see Figure 4.4), especially as a method of determining the pressure fluctuations by the displacement fluctuations, as has been proposed in [165].

The sensors of pressure fluctuations in our tests were less sensitive to vibrations. Therefore, spikes at frequencies of 12 and 20 Hz are insignificant, and the basic spikes on the rigid standard are caused by the power-line noise. Comparing Figures 4.2 and 4.4 (series *a*) we can conclude that the pressure fluctuations are well correlated with the displacement fluctuations. Curve *3* in Figure 4.4 (the pump not working) is a metrological reference. It lays above all other curves because the denominator of the ratio $p(f)/\sqrt{\overline{p'^2}}$ decreases in the absence of disturbances.

Figure 4.5 presents the comparison of pressure fluctuations spectra on the rigid and elastic plates. At a flow of elastic plates, these quantities correlate more poorly (curves *1, 4* series *b*). Instead of two spikes at frequencies of 12 and 20 Hz caused by the vibration of the hydrodynamics stand at the pump working, spikes at frequencies of 15−17 Hz were registered by sensors of pressure pulsations on elastic plates.

While, in Figure 4.3 the energy of the plate displacement spectrum of plate *11* is greatest, in Figure 4.5 the energy of pressure fluctuations on this plate is least. Such a contradiction is because spectra of surface displacement characterize vibro-damping properties of an elastic plate, and fairly reflect its hydrodynamic features. It is obvious that the compliant plate *2a* only absorbs the cross-section pulsations of velocity and pressure. Its amplitude of surface fluctuations is less than at plate *11*, and its vibro-damping characteristics greater. Pulsations of pressure reflect the character of the boundary layer structure, which changes in the flow on elastic surfaces. Spectra of pressure fluctuations better characterize the variation of kinematic characteristics of the boundary layer on elastic surfaces. From here it follows that there were more essential changes of the boundary layer structure over the elastic plate *11*.

According to the results described in Chapters 1−3, the authors believe that plate *11* generates 3D disturbances, which enter the boundary layer in the form of longitudinal

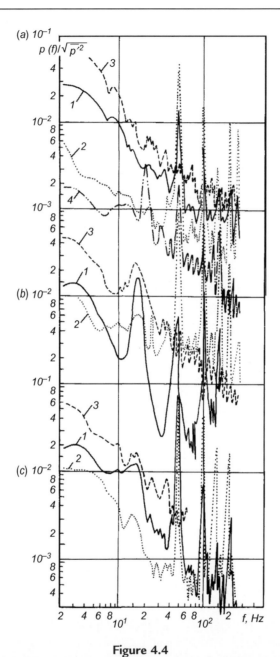

Figure 4.4
Spectra of pressure fluctuations on the rigid (*a*) and elastic (*b* and *c*) plates: *11* (*b*) and *2a* (*c*):
1 − U_∞ = 0.6 m/s, *2* and *3* − U_∞ = 0; *2* − the pump works, *3* − the pump does not work,
4 − spectrum of the plate motion (plate *11*).

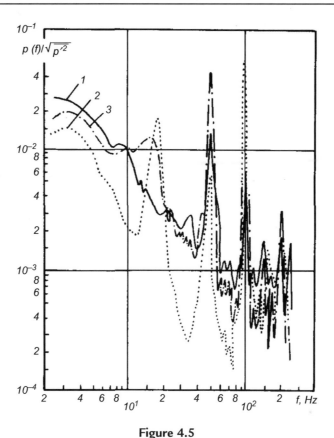

Figure 4.5

Comparison of spectra of pressure fluctuations on rigid (*1*) and elastic (*2* and *3*) plates *11* (*2*) and *2a* (*3*) at $U_\infty = 0.6$ m/s.

vortical systems. It was revealed that this leads to increasing oscillations of the surface of plate *11* in comparison with plate *2a*. Nevertheless, the result is increased transversal components of fluctuation velocity of oscillation, reduced anisotropy of turbulence and, hence, reduced pressure fluctuations. This agrees well with the measurements of the fluctuation velocities (see Figures 3.5j and 3.7), the profiles of kinetic energy (see Figure 3.16) and Willmarth's formula [535]:

$$\sqrt{\overline{p'^2}}/q = 0.0035 \tag{4.6}$$

and Kraichman's formula [117, 216]:

$$\sqrt{\overline{p'^2}}/\tau \approx 6 \tag{4.7}$$

The best normalization of spectra of the near-wall pressure fluctuations is presented in [172]. The decrease in pressure fluctuations in the flow over elastic plates in the region of low frequencies observed in Figure 4.5 agree with the measurements in elastic tubes [165].

4.3 Structure of a Turbulent Boundary Layer over Elastic Plates in Water

In Chapter 3 we presented the results of investigations on the turbulent boundary layer over elastic plates made in the wind tunnel. Calculations of the similarity parameters showed that the mechanical parameters of the plates were not optimal to stabilize the boundary layer. One of the similarity parameters requires the elastomer density to be approximately equal to the flow density. The density of the investigated plates was substantially lower than air density and comparable with water density (see Tables 2.7, 2.9 and 2.10). In order to estimate the reliability of the similarity criteria, and also the usefulness of elastic surfaces in various environments, the characteristics of the turbulent boundary layer on the same plates (see Figures 2.29−2.31) were investigated in water.

In Figure 4.6, you can see the mean velocity profiles on the rigid and elastic plates. Reynolds numbers in air and water flows were approximately equal. However, due to lower turbulence of the main flow in the hydrodynamics stand, the profiles in water flow are not fully developed. In the inner region their shape is typical of the turbulent boundary layer at low Reynolds numbers. Measurements were made in sections I, II, III, indicated in Figure 3.2. The panel-insert is shown in Figure 2.25 (bottom view) and in Figure 2.29, position 3 (top view). The origin of the distances of the specified sections where the measurements were done relate to the edge of the panel, which was installed in the working section of the hydrodynamics stand at a distance of 2 m from its start (see Figures 2.26 and 2.27). In Figure 3.2 the system of coordinates of sections I, II and III result from carrying out measurements in a wind tunnel. Measurements of the fluctuation of elastic surfaces are lead in the hydrodynamic stand, the system of coordinates which results in Figure 1.73. Therefore section I in Figure 3.2 corresponds to section III in Figure 1.73. It is necessary for considering the calculation of Reynolds numbers in air and water flows and by comparison of kinematics characteristics of a boundary layer on elastic plates (Chapter 3), certain in a wind tunnel with fluctuations of a surface of the elastic plates measured in a water flow (Chapter 4). The boundary layer thickness in water increases by approximately 1.5 times, so plate 5 became hydraulically smooth at $U_\infty = 0.6$ m/s. The velocity profile of plate 5 is fuller compared with that in the wind tunnel, but it differs little from that on the standard plate. The profiles on the other elastic plates are practically of the same form as that on plate 5. The difference lies in the pronounced inflections of the profiles at $y/\delta < 0.2$, which indicate the generating of longitudinal vortices in the boundary layer. The longitudinal fluctuation velocity profiles are also presented in Figure 4.6. On the rigid plate the spread in experimental points in the outer region of the boundary layer is the same as in airflow. Because of the sizes of the wedge-shaped sensors, it was not possible to come closer to the surface. The maximal values of the fluctuation velocity are registered near the wall. The profiles of fluctuation velocity on plate 5 practically coincide with that on the

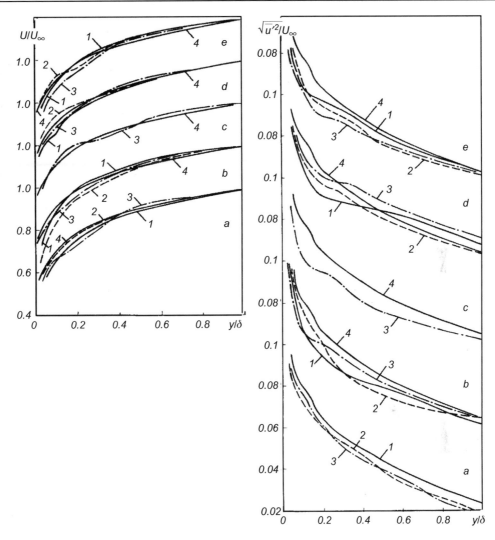

Figure 4.6

Distribution of longitudinal averaged and fluctuation velocities in the turbulent boundary layer over the rigid (*a*) and elastic (*b–e*) plates 5 (*b*), *2a* (*c*), *11a* (*d*), and *11a* (*e*) in water flow: *a*: curves *1–3* — section I–III at U_∞ = 0.61, 0.63, 0.61 m/s; *4* — measurements in air flow (Figure 3.4); *b*: *1* — 0.62 m/s; *2, 3* — 0.6 m/s; *c*: *3* — 0.62 m/s; *d*: *1* — 0.617 m/s, *2* — 0.6 m/s; *e*: *1, 2* — 0.6 m/s, *3* — 0.61 m/s.

standard plate. This confirms the conclusion about the reduction of roughness of this plate in water due to the maximum allowable roughness being increased. On the other elastic plates, the fluctuations of longitudinal velocity substantially decreased at $y/\delta < 0.3$, as it is seen in Figure 4.6. In the outer region of the boundary layer over plate *11a* the longitudinal fluctuations increase; this is related to increasing surface oscillations. The decrease in fluctuations on the elastic plates is in agreement with [122]. The shapes of the fluctuation

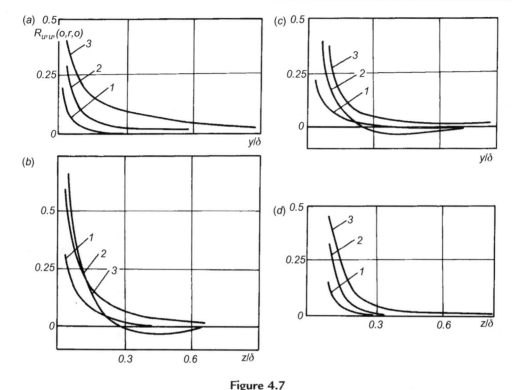

Figure 4.7

Cross-correlation coefficient of longitudinal fluctuation velocity on the rigid (*a, b*) and elastic plate *2a* (*c, d*): *1–3* for *a-y/δ* = 0.0405, 0.122, 0.222; for *b* – 0.0135, 0.135, 0.27; for *c* – 0.0135, 0.297, 0.297; for *d* – 0.0135, 0.0135, 0.23.

profiles on plates *11* and *11a* (see Figure 4.6) indicate the generation of the longitudinal vortices in the boundary layer (compare with data in Chapter 1 and Figure 3.3).

In Figure 4.7 the transverse correlation coefficients on the rigid plate and on plate *2a* can be seen. The coefficients of the transverse correlation between the longitudinal components of the velocity fluctuations are presented as follows:

$$R_i = \frac{u_1'(\xi)u_1'(\xi + x_i)}{\sqrt{\overline{u_1'^2(\xi)u_1'^2(\xi + x_i)}}} \tag{4.8}$$

where $i = 2$ when the moveable sensor was moved past the immovable one along the Oy axis, and $i = 3$ when it was being moved along the Oz axis. The curves in Figure 4.7 characterize the correlation coefficients at different distances from the motionless sensor until it reaches the rigid plate y, when the second sensor moves along the Oy axis. The distance between sensors before the test begins was 1 mm. The obtained results are qualitatively agreed with Figure 3.19a. The integral scales in the water turbulent boundary layer are less than that in air.

In series *b*, the correlation coefficients are shown for the case when the moveable sensor was moved along the Oz axis. The pre-experiment distance between sensors was 1.3 mm. The obtained data agree with data in [111, 168, 180, 192, 232, 233, 272, 376, 433, 473]. From a comparison of measurements in series *a*, *b* it follows that up to $y/\delta \leq 0.2$–the vortex size in a transverse direction in the near-wall region on the bottom of the hydrodynamics stand is larger than in a vertical direction, and at $y/\delta > 0.2$–when conversely the vortex becomes extended more strongly in the vertical direction. In other words, in the cross-section the vortex has the triangular-like shape.

The second conclusion is that the vortex has periodicity in the transverse direction. So, at $y/\delta = 0.27$ negative values of the correlation coefficient were obtained, which indicates the compensating reversed flow of the liquid due to its continuity. From here it is possible to estimate the spacing and wavelength of the longitudinal vortical system in the near-wall region of the turbulent boundary layer on the standard plate: $\lambda_z \approx 0.02$ m, i.e. 2.5 times less than at the linear stage of the boundary layer transition [310].

On plate *2a*, the vertical size of the vortex increases up to $y/\delta \leq 0.2$ due to increasing thickness of the viscous sublayer, and decreases at $y/\delta > 0.2$. The negative values of the correlation coefficients registered at $y/\delta = 0.3$ indicate multi-layered longitudinal vortices. These data are well agreed with the results of investigations into velocity profiles in complex interactions [47, 58, 61, 310]. We can change the vortex sizes by regulating the mechanical parameters of the elastomer (compare curves *2, 3* in Figure 4.7c). In transverse a direction, the vortices have appeared wider in the near-wall areas (there is no reversed flow and periodicity on *z* over this plate) and it has kept this size at $y/\delta > 0.2$.

Thus, the longitudinal vortices of triangular shape in cross-section appear in the near-wall region on the rigid plate (standard). The elastic surface, due to its oscillations, transforms the vortices into a multi-layered vortical system.

Near the surface, the shape of the vortex in cross-section became rectangular and wider in the direction Oz. On the next layer ("second floor,") the system of vortices also became rectangular, but with smaller sides. So, we can corroborate the conclusions of Chapter 3 about the decreasing of turbulence anisotropy near an elastic surface, the changing in the energy balance across the boundary layer and the modifying of the disturbance structure near an elastic surface.

The lifetime of vortices in the turbulent boundary layer can be evaluated according to autocorrelation coefficients, which are calculated by the formula:

$$R_i(\tau) = \frac{\overline{u_i'(t)u_i'(t+\tau)}}{\sqrt{\overline{u_i'^2}}} \tag{4.9}$$

where $i = 1, 2, 3$ denotes the longitudinal, normal and transversal axis; τ is time lag.

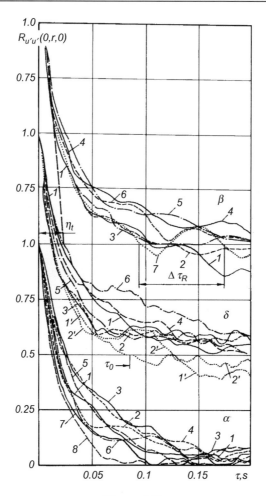

Figure 4.8

Autocorrelation coefficients of longitudinal fluctuation velocity on the rigid (a) and elastic plates
$2a$, 5 (b) and $11a$ (c) in section III (according to Figure 3.2): a: for $\tilde{y}_1 = 0.23$ curve $1 -$
$\tilde{y}_2 = 0.276$; $2 - 0.32$; $3 - 0.3$; $4 - 0.46$; $5 - 0.55$; $6 - 0.645$; $7 - 0.78$; $8 - 0.92$;
b: plate $2a$: for $\tilde{y}_1 = 0.227$ curve $1 - \tilde{y}_2 = 0.273$; $2\ 0.32$; $3 - 0.37$; $4 - 0.55$; $5 - 0.64$; $6 - 0.77$;
plate 5: for $\tilde{y}_1 = 0.222$, curve $1' - \tilde{y}_2 = 0.27$; $2' - 0.31$; c: plate $11a$ (I section) for $\tilde{y}_1 = 0.278$
curve $1 - \tilde{y}_2 = 0.33$; $2 - 0.39$; $3 - 0.44$; $4 - 0.56$; $5 - 0.67$; $6 - 0.78$; $7 -$ for $\tilde{y}_1 = 0$,
$\tilde{y}_2 = 0.278$.

Figure 4.8 illustrates the results of measurement of autocorrelation coefficients of
longitudinal fluctuation velocity on rigid and elastic plates $2a$, 5 and $11a$. The quantity
$\tilde{y}_1 = y_1/\delta$ designates the coordinate of the motionless sensor, and $\tilde{y}_2$ of the movable sensor.
Figure 4.8 shows the data when the movable sensor is moved from the wall to the external
border of a boundary layer. The initial value of the factor of autocorrelation on the rigid

plate is measured at $\tilde{y}_2 = 0.276$ (curve *1*). In the process of motion of the gauge from the wall, the autocorrelation coefficients increase, beginning from $\tilde{y}_2 = 0.55$ (curve *5*) they are compared to values of R_τ of curve *1*, and further (with growing *y*) decrease so that have the least values at $\tilde{y}_2 = 0.92$. The presented results are well agreed with the data in [180, 283, 473]. They can be explained by the fact that the autocorrelation coefficients characterize the timescale of turbulence, i.e. the lifetime of the most large-scale vortices:

$$T = \int_0^\infty R_\tau d\tau \qquad (4.10)$$

In accordance with the cascade process of energy transfer, the lifetime of large-scale vortices increases in the beginning with increasing distance from the wall. Near the outer boundary of the boundary layer the autocorrelation coefficients decrease. In this region, the large-scale vortices appear and move very quickly to the wall. On the other hand, the outer boundary of the boundary layer is intensively mixed—this is caused by low-scale dissipative vortices.

The dot lines in Figure 4.8b show the autocorrelation coefficients on the plate *5* (the open sheet foamed polyurethane). The roughness of the plate in the water flow was not practically noticeable. So, the lifetime of the large-sized vortices at $\tilde{y}_2 = 0.27$ (curve *1'*) and 0.31 (curve *2'*) was substantially decreased compared with the standard plate. The regions with negative values of R_τ indicate the reversed motion (reversed compensating flow).

According to [433, 473], the period of oscillation of the autocorrelation coefficients $\Delta \tau_R$ and the value of τ_0 (interval before the first transition of R_r through zero) are connected with the fluctuation frequency ω_m, at which spectral density reaches a maximum, by the expression:

$$\Delta \tau_R = 2\pi / (\omega_{mi} \tau_0) = \pi / \omega_m \qquad (4.11)$$

The size τ_0 also characterizes the timescale of energy-containing vortices. Hence the energy balance of turbulent boundary layer has changed on plate *5* at the same distance, as at the standard. This occurs owing to a thickening of the viscous sublayer.

From formula (4.11) we obtained the value of the dominant frequency for the standard plate $\omega_m = 17.94 \ s^{-1}$ and for the plate *5* $\omega_m = 39.25 \ s^{-1}$. These values are well agreed with the data in Table 3.3: if we wrote formula (4.11) in the form $\tau_0 = 2\pi / \omega_m$, the obtained frequency of the energy-carrying vortices $(78.5 \ s^{-1})$ practically coincides with the natural frequency of oscillations of plate *5*. So, it can be stated that the natural frequency of oscillation of the elastic plate influences the structure of the boundary layer up to $y/\delta \approx 0.3$.

The vortical structure of the boundary layer over the plate *2a* in water also differs from that on the standard plate. The dependence of the autocorrelation coefficient on time for curve *1* at plate *2a* is the same as at the standard. Moving away from the wall (curves *2* and *3*), the autocorrelation coefficient decreases for plate *2a* (but increases at the standard). Then at $\tilde{y}_2 = 0.55$ the value of R_τ becomes same as at $\tilde{y}_2 = 0.273$ (curve *1*), which coincides with parameters on the standard. But further the value of R_τ at plate *2a* decreases again, and only after $\tilde{y}_2 = 0.77$ (curve *6*) R_τ increases. Such behavior of R_τ across the boundary layer over the elastic plate can be explained by two factors: the thickness of the viscous sublayer and the surface oscillations leading to generation of a multi-floor system of longitudinal vortices.

At a flow of Figure 4.8, plate *11a,* in a boundary layer, 3D indignations are generated from below in the boundary layer. The plate has the greatest frequency of natural oscillation characteristics. Figure 4.8 shows the greatest changes of timescales of turbulence for the plate *11a.* So at $\tilde{y}_2 = 0.28-0.44$ (curves *1–3, 7* in Figure 4.8c) the integral timescale remains constant, i.e. substantially less than on the standard plate. Chapter 1 describes the results of experimental research into the characteristics of the boundary layer on a rigid plate while introducing disturbances in the form of longitudinal vortices by means of the developed mechanical vortex generator. The data obtained in section I are metrological for the analysis of the autocorrelation coefficient for plate *11a* [310] represented in Figure 4.8. The shape of curves *1–3* (see Figure 4.8C) corresponds to metrological data and indicates that longitudinal vortical structures were generated in the near-wall areas. In comparison with the standard, the viscous sublayer became thicker; therefore, large vortices in the boundary layer are located higher, in the area of the movable sensor (curves *4* and *5*).

According to formula (4.11), the frequency of dominant oscillations increased compared to the standard late, similar to plate *5*. The tangent line plotted in Figure 4.8 as a dash line on the abscissa axis indicates on the segment η_t, which characterizes the time micro scale [180, 433, 473]. It is seen that on plate *11a* the micro scale increased compare to the standard plate. It means that the lifetime and sizes of dissipative vortices increased. The obtained data agree with the conclusions of Chapter 3.

Figure 4.9 represents the autocorrelation coefficients in the near-wall region. On the rigid plate, the integral scale L is the same as in Figure 4.8 up to $\tilde{y}_1 = 0.16$, then it begins to decrease in the buffer zone and at $\tilde{y}_1 = 0.092$, i.e. at the outer boundary of the viscous sublayer, it becomes equal to its value on the outer boundary of the boundary layer. That indicates the decreasing in the lifetime of the energy-carrying vortices and increasing L of the dissipative vortices. It agrees with the measurements of the spectra.

On plate *2a* at $\tilde{y}_1 = 0.091$ the values R_τ are same, and at $\tilde{y}_1 = 0.182$ they are less in comparison with the standard. The difference between plate *2a* and the standard is noticeable not near the wall, but starting with the buffer area and above.

In contrast, the values of R_τ near the wall on plate *11a* are greater than on the standard plate, which corresponds to the values of η_t (see Figure 4.8).

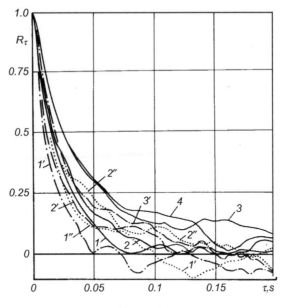

Figure 4.9

Autocorrelation coefficients in the near-wall region on rigid (solid curves) and elastic plates:
2a (dash-dot curves) and *11a* (dash curves). Rigid plate: *1* $-\tilde{y}_1 = 0.092$, *2* $- 0.14$, *3* $- 0.16$,
4-0.184; plate *2a*: *1'*$-\tilde{y}_1 = 0.045$, *2'* $- 0.091$, *3'* $- 0.132$; plate *11a*: *1"* $- \tilde{y}_1 = 0.0466$,
2" $- 0.134$.

Both types of elastic plates generated the longitudinal vortices that can be seen from the presence of the regions of negative values of R_τ. The frequency of bursts on plate *11a* is higher than on the plate *2a* according to Table 3.3.

For more complete study of the structure of the turbulent boundary layer, spatial−time correlations have been measured in the transverse direction (see Figure 4.10). The double correlation is defined as the correlation tensor (covariation) formed by two components of fluctuation velocity in two points at different t [433, 441, 473]:

$$R_{ij}(x, r, t) = \overline{u_i(x)u_j(x + r, t)} \tag{4.12}$$

where i and j denote different directions; x is the vector determining the location of the point in space; t is time of the delay of observation (time lag).

In the experiments discussed, the second sensor in the beginning moved upwards relative to the motionless first sensor, and then was pulled down to the first sensor and became motionless; then the first sensor was pulled down to the wall. As is seen from Figure 4.10, the time lag at which the correlation coefficient becomes maximal, depends on the distance r_y between the measurement points. The maximal correlation coefficients were registered with a positive time lag, when the sensor was moving to the wall (r_y is negative), and with

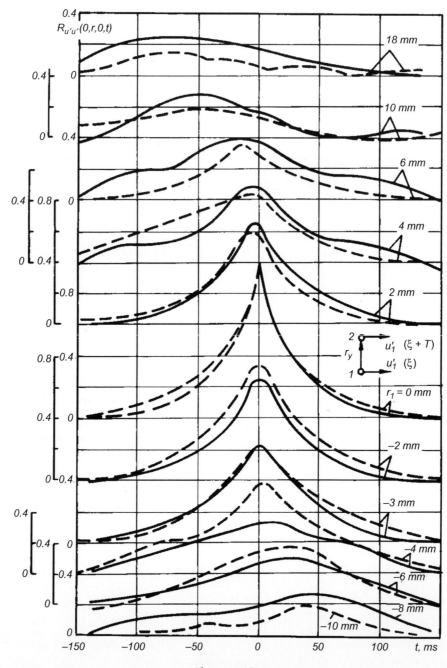

Figure 4.10

Distribution of spatial−time correlation coefficients in the turbulent boundary layer over the rigid (solid curves) and elastic *11a* (dash curves) plates. On the rigid plate the immovable gauge is located at $y = 10$ mm, $\delta = 43.3$ mm, $y/\delta = 0.23$, $U_\infty = 0.61$ m/s; on the elastic plate: $\delta = 36$ mm, $y/\delta = 0.28$, $U_\infty = 0.617$ m/s.

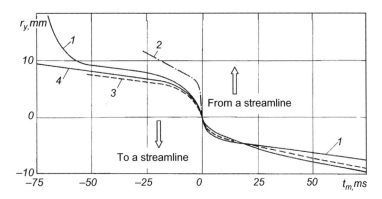

Figure 4.11
Change in time delay, corresponding to maximal correlation factor, depending on the distance between the gauge on the rigid (*1*) and elastic (*2—4*) plates 5(*2*), 11 (*3*) and *11a*.

a negative time lag when the sensor was moving away from the wall (r_y is positive). The value of $R_{r\tau}$ is maximal at $r_y = 0$.

Hence the structure of the vortices is inclined in such a manner that the external part advances towards the internal located close to the wall. Such orientation of vortices indicates continuous distortion of the turbulence field due to the distribution of average velocity across the boundary layer thickness. These results correspond to the diagrams of the boundary layer shown in Chapter 1. The inclination of the vortex is caused by energy exchange, particularly by the production and dissipation of the turbulent boundary layer near the wall, and by the physical properties of the flow at the outer boundary of the boundary layer. These data on the rigid plate in water are in agreement with the data obtained in the wind tunnel [180, 441].

The measurements were also made on the elastic plates (*5, 11* and *11a*). The measurement conditions are the same as for Figures 4.8–4.10. The dash curves in Figure 4.10 present the space–time correlations on plate *11a*. The inclination of vortices is similar to that on the standard plate.

The difference in slopes of the vortices can be estimated from Figure 4.11: the dependence of r_y (the distance between the measurement points) on t_{max} (time lag corresponding to maximum of the correlation function) were plotted. The slope of a vortex indirectly characterizes the balance of turbulent boundary layer energy. There is no substantial changing in the vortex slope on the elastic plates because the elastic plates were not optimal in the similarity parameters. The time lag indicates the time it takes for the vortex to move from the first sensor to the second one, so the data in Figure 4.11 characterize the vortex velocity at different distances from the wall. The line $r_y = 0$ corresponds to coordinate $y/\delta = 0.5$.

It can be seen that at plate *5* the vortex velocity or its slope increased in the outer region of the boundary layer. At the same time, the vortex slope at plates *11* and *11a* in the external boundary layer decreased. In the near-wall areas, the vortex velocity and inclination at plate *11* fairly differed from the standard, and at plate *11a* these quantities decreased. Therefore, the vortex slope on plates *11* and *11a* decreased across the boundary layer, which indicates changing energy balance and a decreasing frequency of bursts. The vortex slope decrease and vortex speed reduction corresponds to the flow rate increase in the boundary layer or to the reduction of friction on the wall and can be defined by increased amplitudes of the surface fluctuations, essentially, the increase in thickness of the laminar sublayer.

Figure 4.12 presents the spectral characteristics of the boundary layer over the elastic plates in water flow. The spectra on the rigid plate are well agreed with Klebanoff's data: curve *1* at $\tilde{y} = 0.074$ better coincides with curve *4* at $\tilde{y} = 0.05$, and curves *2* and *3* with curve *5*, accordingly. Some decreasing of the spectral function in the region of small wave numbers compared with Klebanoff's measurements in air flow can be explained by the parameters of the experimental unit and by the properties of the turbulent boundary layer at low Reynolds numbers.

The spectral function of the longitudinal fluctuation velocity on plate *5* barely differed from that on the standard plate due to the roughness decrease in the water flow. The difference, as in air (see Figure 3.20), consists of the increased fluctuation energy at all wave numbers compared with the standard plate. On plate *2a*, the obtained data coincided with the data obtained in air flow (see Figure 3.20): in the field of energy-carrying vortices the spectral density increased in comparison with the standard, and in the field of dissipative vortices it decreased.

In comparison with experiences in air, the increase of energy is more significant, thus the efficiency of elastic plates in water is increased. The results obtained on plate *11* are the same as that on plate *2a*. Thus the spectral density in the field of dissipative scales differs slightly from the reference measurements, and in the field of energy-carry scales does essentially more. They are also agreed with the results of measurements in the air flow (see Figure 3.20d, curve *14*), but are more noticeable. At $k = 1-3 \text{ cm}^{-1}$, we registered the strong splashes of energy on plate *11*, which is connected with the natural frequency of the plate oscillation.

It has been shown that, as in a transitional boundary layer [310], the amplitudes of fluctuations of the spectral curve decreased, but their frequency increased. This can be explained by the stabilization of the boundary layer by the elastic plates and the increasing of their natural frequencies compared with the standard plate.

As well as in Figure 3.20c, spectra on plate *11a* are the most different from that on the standard plate (see Figure 4.12d). This is due to both its structure and the mechanical

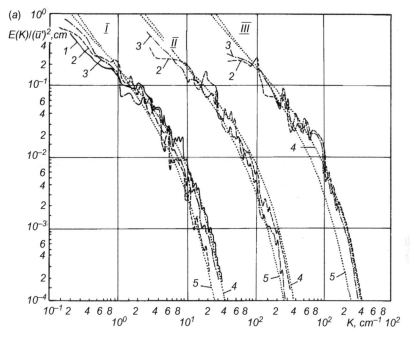

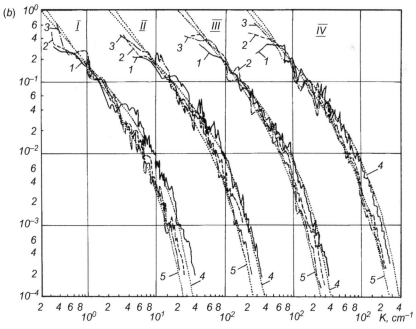

Figure 4.12
(Continued)

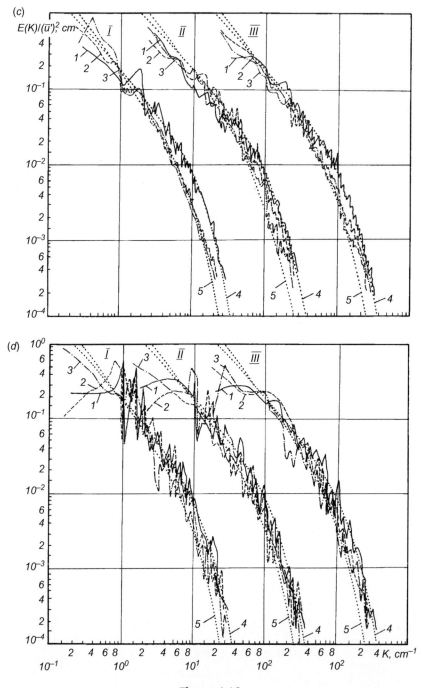

Figure 4.12
(Continued)

properties of the plate, but primarily the frequency and amplitude of surface oscillations (see Section 4.1). Similar to plate *11*, strong spikes are registered in the region of inertial scales. In contrast with plate *11*, there was an essential reduction of energy of longitudinal fluctuations in comparison with the standard in the field of energy-carrying scales at $\kappa < 1 \text{ cm}^{-1}$. These data are well agreed with metrological measurements of spectral density at susceptibility of the boundary layer to longitudinal vortices [58, 61, 310]. The data indicated the generation of complex multi-layered vortices in the boundary layer of plate *11a*.

The above-presented findings allow for the following conclusions to be made:

1. The measurements of the surface displacement of elastic plate correlate well with the measurements of pressure fluctuations. Despite of the increased amplitudes of fluctuation of elastic plates in comparison with the standard, the changed structure of the boundary layer on elastic plates has led to a reduction of energy in pressure fluctuation spectra.

2. The effectiveness of elastic plates depends on many parameters. One of the most important parameters is the parameter characterizing the relationship between the liquid density and elastomer density.

3. The turbulent boundary layer structure changes across the whole boundary layer including its outer region.

4. The amplitude—frequency parameters of oscillations of elastic plates surfaces strongly influence the structure of the disturbance motion.

5. All parameters of the disturbance motion strongly change in the boundary layer over elastic plates. It is important that the longitudinal vortices are generated in the boundary layer.

6. The results of measurements in the water flow are similar to that in the air flow, but the difference from the standard plate is more pronounced in the water flow.

Figure 4.12

Spectral function of the longitudinal fluctuation velocity in the water flow over the rigid (*a*) and elastic plates *5* and *2a* (*b*), *11* (*c*) and *11a* (*d*). I, II, III — section; 4, 5 — experiments [190] at $\tilde{y}_0$ is equal to 0.05 and 0.2. *a*: I — $U_\infty = 0.63$ m/s, $1 - \tilde{y} = 0.074$, $2 - 0.246$, $3 - 0.49$; II — $U_\infty = 0.608$ m/s, $2 - 0.229$, $3 - 0.457$; III — $U_\infty = 0.61$ m/s, $2 - 0.23$, $3 - 0.46$; *b*: I–III plate *5*: I — $U_\infty = 0.62$ m/s, $1 - \tilde{y} = 0.065$, $2 - 0.217$, $3 - 0.435$; II — $U_\infty = 0.6$ m/s, $1 - 0.065$: $2 - 0.217$, $3 - 0.435$; III — $U_\infty = 0.6$ m/s, $1 - 0.067$, $2 - 0.222$, $3 - 0.444$; IV — plate *2a*, third section: $U_\infty = 0.62$ m/s, $1 - \tilde{y} = 0.068$, $2 - 0.227$, $3 - 0.455$; *c*: plate *11*; I — $U_\infty = 0.6$ m/s, $1 - \tilde{y} = 0.067$, $2 - 0.232$, $3 - 0.446$; II — $U_\infty = 0.6$ m/s, $1 - \tilde{y} = 0.067$, $2 - 0.224$, $3 - 0.448$, III — $U_\infty = 0.61$ m/s, $1 - 0.064$, $2 - 0.215$, $3 - 0.43$; *d*: plate *11a*: I — $U_\infty = 0.617$ m/s, $1 - \tilde{y} = 0.083$, $2 - 0.278$, $3 - 0.555$; II — $U_\infty = 0.6$ m/s, $1 - 0.073$, $2 - 0.242$, $3 - 0.484$; III — $U_\infty = 0.61$, $1 - 0.07$, $2 - 0.247$, $3 - 0.5$.

4.4 Interaction of Different Disturbances in a Boundary Layer over Elastic Plates

The results of the experimental research represented in the first part of this book have shown that the ordered coherent vortical structures (CVSs) exist in the near-wall areas of the turbulent boundary layer over a smooth monolithic plate. CVSs in the turbulent boundary layer on rigid plates have been investigated for decades. Schemes of such structures are given in Figures 1.17 and 1.47. The difference between CVSs on rigid and elastic plates relates to the shape and size of the structures. In addition, longitudinal multi-layered vortical structures are formed in the flow on elastic plates (see Sections 4.2 and 4.3).

From the point of view of susceptibility problems (See Chapter 1), the above-presented results can be explained as follows. The pressure fluctuations of the boundary layer over elastic plates cause Rayleigh waves (RW) on the plate surface. In Section 2.8, photos of RW on rigid and various kinds of elastic plates are presented. RW on a rigid plate has almost a sine-wave character. Scratches and sinuosity of a surface cause small differences from the purely harmonious kind. On all kinds of elastic plates, RW are nonlinear, which is defined by the specific molecular structure of the elastic plates (see Chapter 2). Pressure fluctuations of the boundary layer on a wall are caused by liquid chunks falling onto the surface from various distances, including due to the cascade process described by Bradshaw [130]. In real conditions of the flow on an elastic surface, there will be statistical set of RW, which will interact with each other and with waves reflected from borders of the plate. Under the action of the shear field of the boundary layer, the RW system will have a prevailing direction of propagation along the direction of mean flow. In this chapter, spectral curves of elastic surface fluctuations in a boundary layer and fields of pressure fluctuations on the surfaces of plates are presented. Such spectra are caused by complex interaction of the flow and the elastic plate. It is possible to assume that multi-layered longitudinal vortical structures above elastic surfaces will constantly interact with the statistical field of RW displaced downstream.

The analysis of the structure of dermal coverings of high-speed hydrobionts (see Chapter 1) has shown that there are both longitudinal and transversal structures in the skin coverings inducing various kinds of disturbances in a boundary layer. Furthermore, active fluctuations of the surface of a dolphin's skin have been revealed. In relation to this, various kinds of elastic analogues of hydrobionts skin coverings have been made that have both longitudinal and transversal structures in the outer layers. The elastic plates having a longitudinal structure in the outer layer led to occurrence of systems of longitudinal Stoneley and Love waves the in plates under action of fluctuation fields in the boundary layer (see Section 1.12) [515]. It has been supposed that such designs of plates would promote the formation of a complex wave-guide, which would be complex both at the near-wall part of the boundary layer and the outer part of the elastic coating (see Chapter 1).

In connection with this, it is possible to imagine that interaction of uniform elastic plates with a flow occurs in the form of quasi-plane RW appearing in the elastomer and moving downstream with the system of longitudinal multi-floor vortices in the boundary layer of elastic plates. Plates that have the outer layer structure in the form of transversal irregularities generate flat wave structures in the boundary layer under the action of flow, which interact with the system of longitudinal multi-layered vortices in the boundary layer over such kinds of elastic plates. Plates that have the outer layer structure in the form of longitudinal irregularities generate longitudinal vortical structures in the boundary layer under the action of flow, which interact with the system of longitudinal multi-layered vortices in the boundary layer over such kinds of elastic plates.

Such interactions can be simplistically represented as an interaction between the plane wave and 3D disturbances, and as an interaction between two 3D disturbances with different geometrical parameters. The results of such interactions on the rigid plate in model experiments are described in detail in [47, 55, 57−61, 310].

As well as on a rigid plate, complex interactions of various disturbances in a boundary layer have been investigated. At a flow on elastic plates, the field of velocity has been measured at various stages of transition, including the turbulent boundary layer when flat and 3D disturbances were introduced into the boundary layer. Some of the results obtained are noted here [310].

The results of interaction between the disturbances depend on the coordinate y at which the plane disturbance is introduced (see Figures 1.73 and 1.74). The maximal fluctuations were registered not in the region of critical layer (as on the rigid plate), but under the critical layer. This can be explained by the decreasing of the vertical coordinate of the critical layer. Furthermore, the closer the vibrating strips are to the surface, the larger the induced vibrations caused by the wall. For the plane disturbances being introduced into the boundary layer on a rigid plate, the transition came earlier. The complex interactions on elastic plates led to more intensive stabilizing of the boundary layer than at interaction only with 3D disturbances.

When introducing 3D disturbances in the boundary layer from above and from below the wall, the fullness of averaged velocity profiles increased near the wall, and the shape of profiles smoothed out above the top edge of the vortex generator. Introduction of 3D disturbances from above the wall was carried out by means of vortex generators shown in Figure 1.75. Introduction of 3D disturbances from below the wall was carried out by means of designs of the plates represented in Figures 2.28d, 2.30 and 2.31. Interaction of 3D disturbances led to suppression of the system non-stationary small-scale multi-layered vortices arising at and generating only from below, and led to the formation of more ordered and large-scale vortices.

At the flow on elastic plates, the generation of 3D disturbances above the surface of elastic plates leads to their resonant interaction with disturbances existing in the boundary layer, as well as at the flow on a rigid plate. Such interaction occurs in agreement with structural and kinematic—dynamic principles. This means that the greatest stabilization of a boundary layer will occur at similar energy and sizes of 3D disturbances arising at natural transition and generated from above and from below the surface.

Below we present the results of investigations on spectral and correlation parameters at complex interactions, mainly at such velocities of transitional boundary layer when longitudinal vortical structures were formed in natural conditions.

The method of researching a boundary layer reaction to introduced disturbances initially involved the visualization of the flow in the boundary layer at different stages of natural transition: this was performed in the hydrodynamic stand of small turbulence. Special attention was given to photographing the visualized velocities profiles $U(z)$ (see Chapter 1) along the working section in the direction x, at various values of $0 < y < \delta$ and U_∞. Then, velocities profiles were measured in characteristic sections on z (z_1 in 'peaks', z_2 in "valleys," and z_3 in intervals between) by means of laser anemometer at consecutive stages of natural transition [310], and spectral and correlation characteristics [258] were obtained by means of a thermo anemometer. Results on the rigid plate were standard for research of susceptibility of the boundary layer to different disturbances.

The 3D disturbances were introduced into the boundary layer in two ways. In the first, the lattice of vortex generators (see Figure 1.75) was established upstream of the measurement point inside the boundary layer, but above the plate surface [310]. The scale of disturbances in this case was determined by the distance between the vortex generators, λ, and by their sizes (1 — small vortex generators, 2 — medium, 3 — large). The sizes of the vortex generators were: for B1, height $3 \cdot 10^{-3}$ m and length $12 \cdot 10^{-3}$ m; for B2, $5 \cdot 10^{-3}$ m and $15 \cdot 10^{-3}$ m; for B3, $7 \cdot 10^{-3}$ m and $18 \cdot 10^{-3}$ m. The second method consisted of generating the 3D disturbances by the elastic surface. In this case, the parameters of disturbances depended on the structure of the plate, its mechanical properties and the spacing z_i, at which the mechanical properties change along z.

The spectral characteristics were being measured in a hydrodynamic stand at Reynolds numbers of $2.2 \cdot 10^5$, $4.4 \cdot 10^5$ and $1.3 \cdot 10^6$. These values of *Re* correspond to the transition stages, which are caused by Tollmien—Schlichting waves, longitudinal vortices and by generation of developed turbulence accordingly. The thermo anemometer gauges were located at distance $4 \cdot 10^{-3}$ m from the surface, i.e. in the region of the most intensive interaction between the introduced and natural disturbances (in the area of fluctuation

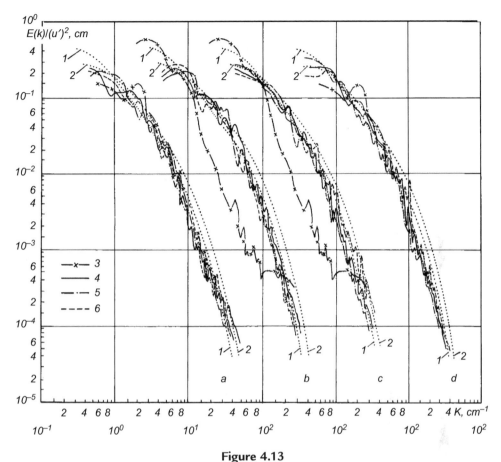

Figure 4.13

One-dimensional spectra of longitudinal fluctuation velocity on the elastic plate *2a* (*a, b*) and *11a* (*c, d*) carrying in the boundary layer the longitudinal vortices: *1, 2* — turbulent spectra on the rigid plate at y/δ equal to 0.21 and 0.063 [190]; *3* — spectra on the rigid plate when 3D disturbances were generated from above [134] at $z = z_*$; the given measurements are made at $z = z_1$ (*4*), $z = z_2$ (*5*), $z = z_3$ (*6*).

velocities maxima along *y*), and at a distance of $5 \cdot 10^{-2}$ m from the vortex generators. The spectra were being measured at three values z_1, z_2 and z_3, as well as velocity profiles.

The spectral curves characterizing the susceptibility of the boundary layer to 3D disturbances on the rigid plate are presented in Chapter 1 and papers [58, 310]. In Figure 4.13, curves *1* and *2* represent classical turbulent spectra on a rigid flat plate [272], and curve *3* is the spectrum of the boundary layer of a rigid plate subjected to 3D disturbances introduced by means of lattice vortex generators. The 3D disturbances have changed the kind of spectrum on the rigid plate. Comparison of curves *1* and *2* with curve *3* reveal, for following differences, and at a distance of $5 \cdot 10^{-2}$ m behind the vortex generators

Table 4.2: Conditions of experiments in Figure 4.13.

Series of Experiments	a	b	c	d
Plate	2a	2a	11	11
Vortex generator	2	3	3	2
$\lambda \cdot 10^3$ (m)	8	12	12	8
$z_* \cdot 10^3$ (m)	–	–	1.8	1.8
$U_\infty \cdot 10$ (m/s)	2.2	2.2	2.2	2.0

below on a flow. For curve *3* there are: 1) valleys in certain parts of spectrum, particularly at large wave numbers; 2) clear maxima at low frequencies; 3) abrupt spikes at certain values of k; 4) different "smoothness" of the curves depending on the transition stage [310].

The results of measurements of spectral characteristics when the 3D disturbances generated from above the boundary layer interact with the disturbances generated from the surface are presented in Figure 4.13. Groups of the curves making consecutive series of measurements, are shifted relative each other on the abscissa axis by one cell of the scale grid. In groups *a* and *b* we investigated the complex interactions of 3D disturbances in the boundary layer over the smooth isotropic elastic plate *2a*, and in groups *c* and *d* over the anisotropic plate *11*, with 3D disturbances generated from below with the wavelength $z_* = 1.8 \cdot 10^{-3}$ m. Conditions of the measurements are listed in Table 4.2.

All characteristic properties that were present when testing simple interactions on the rigid plate remained the same for the testing of complex interactions of 3D disturbances, but the effects were "smoothed." A decrease of valleys in the inertial sub-domain of the spectrum ($0.8 \text{ cm}^{-1} < k < 3 \text{ cm}^{-1}$) corresponds to the shape of the averaged and fluctuation profiles. The profiles change by z [310], much weaker than on the rigid plate. This indicates less developed longitudinal vortices or a change in sizes and shape of the vortices. From this it follows that, at complex interactions, there is not only a redistribution of energy of disturbances between spectral components, but also a change of the shape and structure of vortices due to the properties of the surface and features of interaction of the 3D disturbances generated from above and from below.

The complex interactions differ from simple ones in the following characteristic properties of spectral parameters of boundary layer:

* the intensity of high frequency disturbances decreases;
* the spectral curve in the inertia region is "fuller;"
* the number of maxima increases and they are less pronounced.

Other attributes are the same. Changing the parameters of the disturbing movement or application of other plates does not change the specified general character and distinctive attributes of the spectral distribution of disturbances energy.

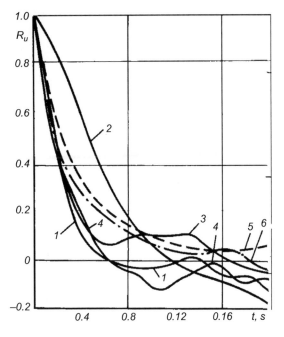

Figure 4.14

Autocorrelation coefficients of longitudinal fluctuation velocity on different plates: *1, 2* — rigid plate, vortex generator B2, $\lambda_z = 8 \cdot 10^{-3}$ m, $U_\infty = 0.2$ m/s, z_3 and $U_\infty = 0.1$ m/s, z_1; *3* — plate *2a*, B2, $\lambda_z = 8 \cdot 10^{-3}$ m, $U_\infty = 0.22$ m/s, z_1; *4* — plate *11*, B3, $\lambda_z = 12 \cdot 10^{-3}$ m, $U_\infty = 0.2$ m/s, z_1; *5, 6* — turbulent boundary layer over the rigid plate and the plate *11a* without vortex generators, $U_\infty = 0.6$ m/s.

The autocorrelation coefficients R_u are presented in Figure 4.14. It is known that the integral timescale (see Figure 4.10) defines the average time of action of turbulence in the given point. It is determined by the length scale L_0 and velocity scale u: $T \sim L_0/u$.

As is shown in [269], the second maximum of the autocorrelation curve corresponds to the period of bursts from the viscous sublayer. Based on this, it is possible to conclude that the turbulent flow on the elastic plate *11a* (curve *6*) is accompanied by reduction of size T in comparison with the flow on a rigid surface (curve *5*). It indicates the increase in the frequency of bursts from the viscous sublayer or, in other words, a change in the energy balance of the turbulent boundary layer [40]. It corresponds to the results presented in [47].

In the viscous sublayer of the turbulent boundary layer, the stage of development of longitudinal vortical systems precedes the stage of bursts from the viscous sublayer (see Figures 1.17 and 1.47) [510]. Proceeding from the analogy between the disturbance development in the viscous sublayer and in transitional boundary layer [40], we can assume that second maxima of correlation curves *1, 3, 4* indicate a collapse of longitudinal vortices in the transitional boundary layer. This assumption is confirmed by the shape of curve *2* without a second maximum, which corresponds to the initial stages of transition of a boundary layer.

A reduction of U_∞ up to 0.2 m/s (curves *1, 3, 4*), i.e. in the transitional boundary layer, should naturally increase the timescale. However, introducing 3D disturbances irrespective of the type of surface at t < 0.03 s leads to equality of autocorrelation coefficients in these cases (curves *1, 3, 4*) and in the case of natural turbulent flow on an elastic plate (curve *6*). The greatest difference in autocorrelation coefficients takes place, as well as in Figure 4.13, in the field of intermediate scales: 0.04 < t < 0.1 s. For further reduction of flow velocity over a rigid surface and the same parameters of disturbances, there is an increase in coefficient R_u until $t \approx 0.08$ s (curve *2*). Here, the influence of parameters of undisturbed flow is more important than the introduced disturbances. It is necessary to note that at artificial generation of vortical disturbances, negative values of the autocorrelation coefficients R_u were observed since the time lag of $t \approx 0.06$ s. This fact, as well as valleys of spectral curves in the inertial sub-domain, serve as a confirmation of the presence of 3D disturbances in the form of a system of longitudinal vortices in the boundary layer.

Definition of the structure of the disturbing movement in the case of natural transition to turbulence and at the introduction of 3D disturbances into a boundary layer enables the calculation of the parameter describing dimensionless scale of regular 3D disturbances, $\lambda^+ = \lambda_* u^*/v$, where u^* is the friction velocity, and v the kinematic viscosity coefficient.

For the viscous sublayer of the turbulent boundary layer over the rigid plate $\lambda^+ = 80-100$ and the value of u^*/U_∞, depending on the Reynolds number, changes within the limits $(3.6-4.7) \cdot 10^{-2}$ (in the present investigations, the value was $(4.5-5) \cdot 10^{-2}$) [117]. Three-dimensional deformation flat waves (see Chapter 1) have been revealed as early as in the laminar boundary layer at natural transition. Introduced in the boundary layer, flat disturbances rapidly turn into 3D ones (see Section 1, 1.12). According to measured parameters λ_*, values of λ^+ have been calculated.

At the initial stages of natural transition $u^*/U_\infty = (2-3.5) \cdot 10^{-2}$ and $\lambda^+ = 90-130$. At different stages of transition on the rigid plate (simple interactions) [47], the values of u^*/U_∞ changed within the same limits and λ^+ was 50–60 at $U_\infty = 0.1$ m/c (Tollmien–Schlichting flat waves), 70–100 at $U_\infty = 0.2$ m/s (Benny–Lin vortices) and 100–140 at $U_\infty = 0.6$ m/s (turbulent boundary layer). The value of λ^+ depends on both the transition stage and the scale of the introduced 3D disturbances.

In the turbulent boundary layer over elastic plates, the value of u^*/U_∞ was less than on the rigid plate and the value of λ^+ greater ($\lambda^+ = 120-200$). In the transitional boundary layer with complex interactions $\lambda^+ = 70$. It is close to the lower value of this parameter for a rigid plate. This can be explained by the substantial decrease in u^* due to a reduction of tangential stress, which is caused by a modification of the shape of the velocity profile in the boundary layer. In some cases of interaction of the disturbances in a transitive boundary layer of the elastic plates layer we have value $\lambda^+ = 10-30$. It is essentially less than at the introduction of the 3D disturbances on a rigid plate.

Thus, the used plates decrease the effect of introduced 3D disturbances on the boundary layer. The structure of the disturbing motion changes: the scale of longitudinal vortices decreases in the transitional boundary layer and increases in the viscous sublayer of the turbulent boundary layer. The region of intensive interactions of the disturbances is concentrated inside the layer $0 < y/\delta^* < 1.5$. On the basis of the measured spectral and correlation characteristics, we can determine the dynamic properties of the disturbing motion, i.e. the regeneration period of the longitudinal vortices in the viscous sublayer of the turbulent boundary layer, or the reversed quantity/frequency of bursts f.

According to Figure 4.14, $T = T_1 = 0.22$ s in the turbulent boundary layer over the rigid plate (curve 5), and $T_2 = 0.16$ s for plate *11a* (curve 6). In the transitional boundary layer, $T_3 = 0.14$ s for the rigid plate (curve *1*), $T_4 = 0.14$ s for plate *2a* (curve *3*), and $T_5 = 0.15$ s for plate *11* (curve *4*). Dimensionless parameters T^+ calculated by the formula $T^+ = Tu^{*2}/v$ have the following values: $T_1^+ = 198$; $T_2^+ = 92$; $T_3^+ = 4.7$; $T_4^+ = 14.8$; $T_5^+ = 9.5$. The value of T_1 is well agreed with [269], where the values vary within the limits $220-400$, depending on y^+. The decrease of the value of T^+ for elastic plates can be explained by oscillations of the elastic surface under the influence of the disturbance motion [256]. It increases the frequency of bursts from the viscous sublayer. The decrease in T^+ in the transitional boundary layer is connected with the decrease in the value of u^*. The difference between T_4^+ and T_5^+ is caused by the mechanical properties of the elastic plates.

The parameter T^+ can also be calculated by the formula $T^+ = U_\infty T/\delta^*$. From this formula we obtained: $T_1^+ = 3$; $T_2^+ = 2.2$; $T_5^+ = 0.8$. For the rigid plate, T^+ varies within the range $4-10$ [269]. Therefore, the obtained results lead to the following conclusions:

- The mechanism of development and structure of the disturbing motion in the transitional boundary layer and in the viscous sublayer of the turbulent boundary layer are characterized by identical features and are of the same nature, as has been assumed in [40].
- The receptivity of a boundary layer substantially depends on the mechanical properties and structure of the surface, and on the parameters of the introduced disturbances. At that, there are substantial changes of all kinematic characteristics, including the profiles

of averaged and fluctuation velocities by y and x, the spectral and correlation characteristics.

4.5 Boundary Layer over a Controlled Elastic Plate

It is known that the mechanical properties of elastomers strongly depend on temperature. For this reason it was proposed in [32, 38, 56, 186, 305] to vary the elastomer temperature to control the interaction between the flow and surface.

In order to consider the interaction of elastomers with a flow in terms of susceptibility of a boundary layer to various disturbances, or in terms of a complex wave-guide (see Chapter 1), the regulation of rigidity or viscosity of elastomers by means of heating or cooling, for example, allows for changing the process of interaction of disturbances in the boundary layer. It is known that hydrobionts use this method of reflectory (see Chapter 1). Depending on the design of composite elastic plate (see Chapter 2), by means of heating, it is possible to change its rigidity or its damping properties. This can change the susceptibility to disturbances, depending on its form, and the wave-guide properties of the material.

To check these points, the measurements of kinematic and dynamic parameters of the boundary layer over elastic plates with different structures were carried out with use of a heat-loss anemometer. The measurements were made in air and water while heating the elastomers.

Experiments in water were made in the hydrodynamic stand with controllable turbulence. The construction and experimental procedure are described in detail in Chapter 1 and [30, 305]. The elastic plate 6 (according to Table 2.3) was arranged in the insert (see Figure 2.25) at the end of the work section (see Figures 1.73, 2.26 and 2.27). The composite plate 6 (see Figure 2.29) consisted of two layers of foam elastic material FE-1 (thickness of each layer was $4 \cdot 10^{-3}$ m), one of which was glued onto the aluminum plate $(2 \cdot 10^{-3}$ m thick). There was a conductive fabric, CCF, between the layers. On the end faces, strips of this CCF fabric were clamped in the transversal trunks made of thin brass grids. Conductive wires were soldered to these trunks. Thin rubber film, to which the fabric was pasted, was glued onto the first layer of FE-1. Over the CCF fabric, the same rubber film was pasted for hermetic sealing, to which the second layer of FE-1 with a smooth outer surface was pasted.

When the current was flowing through the CCF fabric, the elastic plate was heated. Layer of FE-1 had good heat-insulating properties; therefore heat was kept inside the plate. It is known that mechanical characteristics of elastomers essentially depend on their temperature. Therefore, the properties of the elastic plate 6 changed at heating.

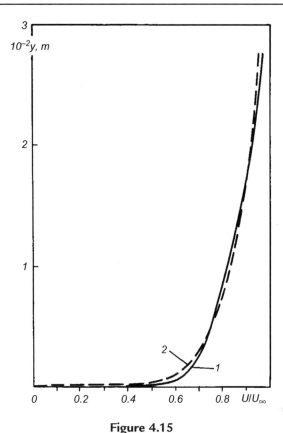

Figure 4.15

Longitudinal averaged velocity profile on the elastic plate *6* (see Table 2.3) without heating (*1*) and with heating (*2*).

Two thermocouples MT-54 (temperature measurement accuracy 0.01°C) were used to measure the temperature of the water and the elastic surface. For the measurements in air and in water, the current voltage was 19.2 V and the current strength 5.9 A. The temperature of the water did not change, while the temperature of the outer surface of the elastomer increased by 0.1°C. Hence, we can say that the kinematic parameters of the boundary layer were modified only due to the change in mechanical properties of the elastic plate.

The measurements were being made in section III*a*, at distance of 0.485 m from the leading edge of the insert. Two heat-loss anemometer sensors were used to measure the transversal correlation coefficients of longitudinal fluctuation velocity. The velocity field and the spectral characteristics were measured using one sensor.

Figure 4.15 presents the typical profiles of longitudinal averaged velocity in semidimensionless coordinates. The main change occurred in the near-wall region and at the boundary between the buffer region and turbulent boundary layer core.

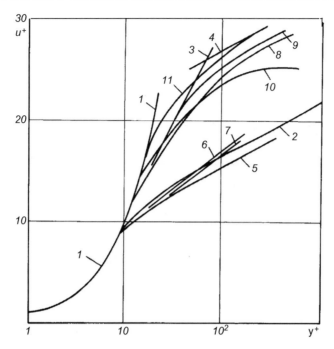

Figure 4.16
The influence of heating duration on the stabilizing properties of plate *6*: velocity profile on the rigid plate [164] in the viscous sublayer (*1*) and in the turbulent boundary layer core (*2*), the law of the flow in buffer layer (*3*) and in turbulent boundary layer core (*4*) in a polymer solution flow. The rigid plate (*5*) and elastic plates (*6, 8, 10*) without heating and elastic plates with heating (*7, 9, 11*).

The velocity profile in the near-wall areas on the heated plate became less "full," indicating the decrease in friction. During measurements, it was necessary to consider a change of heat conductivity of the elastic wall and on the heat-loss anemometer sensor near the surface. Inessential variations of velocity profiles near the wall can influence conclusions about the efficiency of the elastic surface.

Figure 4.16 represents the profiles of averaged velocity measured over plate *6*, at different parameters of heating. The curve *1* presents the velocity profile in the viscous sub-layer $u^* = y^*$, curve *2* the law $u^* = 2.5 \ln y^* + 5.5$. Accordingly, one of the laws of flow in the boundary layer with a polymer solution injection in the buffer layer (*3*) and turbulent core (*4*) can be written down as follows:

$$u^+ = 11.7 \ln y^+ - 17$$
$$u^+ = 2.5 \ln y^+ + 5.5 + \Delta B \tag{4.13}$$

The results of the measurements on the rigid plate (curve *5*, $u_\tau = 3 \cdot 10^{-2}$ m/s) are well agreed with known data and lie a little below curve *2*, as it is usual for rough plates.

The obtained law (curve *5*) can be explained by the increased level of turbulence at the end of the working section at a velocity of 0.6 m/s. Curve *5* is the standard for comparison.

The hydrodynamic stand is the closed pipe in which energy of the pump is partially transferred to flow due to friction on the impeller. As a result, the temperature of the water increased by approximately 1°C during one experiment series. This was because the heat exchanger in the hydro-chamber was not in operation. At the beginning of the experiment on the elastic plate (curve *6*), the temperature of the water was 20.4°C, and it was 21.5°C at the end. In the experiments with heating the elastic plate (curve *7*), the temperature was 21.5°C and 22.5°C. At such a change of temperature the mechanical characteristics of the elastic plate changed, but it did not influence the characteristics of the boundary layer. The difference of the velocity profiles from the reference is insignificant. However, the slope of curve changed (i.e. the value of Karman constant $\aleph$ changed) indicating a thickening of the viscous sublayer. For curve *6*, $u_\tau = 2.7 \cdot 10^{-2}$ m/s, and for curve *7*, $u_\tau = 2.65 \cdot 10^{-2}$ m/s. To evaluate the sensor holder configuration effect upon the measurement results, the followed measurements (curves *8* and *9*) were made with the complex holder. The starting temperature was 23.7°C and the ending temperature 24.7°C. For curve *8*, $u_\tau = 1.9 \cdot 10^{-2}$ m/s, and for curve *9*, $u_\tau = 1.8 \cdot 10^{-2}$ m/s. These measurements were taken just after the pervious series. During this time, the elastic plate warmed up and its compliancy increased, so the obtained results were similar to that in polymer solutions.

To verify the results of the first series, the experiments were repeated, but the starting temperature was 18°C, and the ending one was 19.3°C for the plate without heating and 20.7°C for the experiments with heating. These experiments gave the most substantial results (curves *10* and *11*) comparing both the rigid plate and the heated plate. The shape of curves indicates the thickening of the viscous sublayer. For curve *10*, $u_\tau = 2.08 \cdot 10^{-2}$ m/s, and for curve *11*, $u_\tau = 1.8 \cdot 10^{-2}$ m/s.

Despite the identical conditions of the experiments, essential differences were found out at re-testing. This means that in the initial experiment the elastic plate was well warmed, and, therefore, in the area of the heating strips the properties of the elastic material were essentially changed. As a result, the properties of the material in the transverse direction changed, with the step equal to the spacing of the heating strips. Thus, steady systems of longitudinal vortices with λ_z corresponding to λ_z of the natural boundary layer in the place of measurement were induced in the boundary layer. According to principles of susceptibility, there were effective interactions of these vortical systems. At heating, the efficiency of the elastic plate increased in all cases. The obtained data correlated with the measurements of fluctuation velocities (see Figure 4.17). The conditions of the experiments were the same as in Figure 4.16. As it is shown in Figure 4.17, the fluctuation velocity in the first series only slightly changed at heating the plate, and it decreased in the second and third series. In any case, the maximum of the fluctuation velocity on the elastic plate is

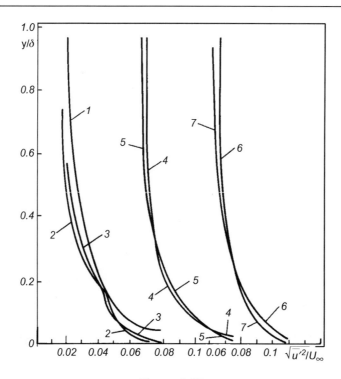

Figure 4.17

Profiles of longitudinal fluctuation velocity on the rigid plate (*1*) and elastic plate without heating (*2, 4, 6*) and with heating (*3, 5, 7*).

closer to the wall compared with that on the rigid plate. This data is similar to that in the laminar boundary layer (see curve *16* in Figure 3.3, Chapter 1, and [305]).

The given experiment series was repeated in the wind tunnel. In Figures 3.3–3.5 and 3.7–3.9, the results of the experiments with plates *11a* (elastic) and *9* (compliant) in an airflow are presented. The experiment conditions are described in Section 3.1. Measurements in air were done in the same section III*a*, as well as in water. The profiles of longitudinal averaged velocities are presented in Figures 3.3, 3.4, and 3.8. The thickness of viscous sublayer of the compliant plate *9* in section III*a* slightly increased, compared with that on the rigid plate. However, the additive constant ΔB increased with heating (see Figure 3.4e). Heating of the compliant plate *9* led to an increase in its elasticity. The elastic properties of the plate in the airflow were not optimal, so heating could not strongly affect the results.

The compliance of elastic plate *11a* increased under heating. Therefore, the efficiency of this plate at heating increased. As is seen from Figure 3.3i, the viscous sublayer thickness did not change, as well as over plate *9*, but the value of δ was larger than that on the rigid plate. The value of ΔB was maximal with heating (see Figure 3.4i).

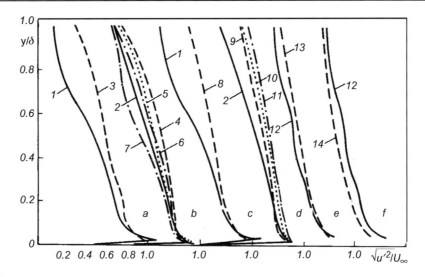

Figure 4.18

The influence of heating the elastic plate on the distribution of fluctuation velocity across the boundary layer in airflow (*a, b, c, d*) and in water flow (*e, f*): *1, 2* — rigid plate, III section, $U_\infty = 10$ m/s and 17 m/s; plate *11a* (see Table 2.4), III section, $U_\infty = 9.9$ m/s (*3*), 17.43 m/s (*4*), 18.23 m/s (*5*); section III*a*: without heating, 19.38 m/s (*6*), with heating, 19.45 m/s (*7*); plate *9*, III section, $U_\infty = 9.765$ m/s (*8*), 15.92 m/s (*9*); section III*a*: without heating, 16.26 m/s (*10*), with heating, 16.14 m/s (*11*); plate *6*, section III*a*, $U_\infty = 0.6$ m/s, without heating (*12*) and with heating during 15 min. (*13*) and 30 min. (*14*).

According to Figure 3.8, plates 9 (curve 22, Figure 3.8e) and 11a (curve 42, Figure 3.8i) have the most pronounced S-shaped velocity profiles (as well as flow rate in the boundary layer). From here it follows that although the efficiency of plate *11a* is better than that of plate *9*, the effect at heating appeared maximal in comparison with the standard in both cases.

The fluctuation profiles on plate *9* with heating were the same as on the rigid plate (see Figure 3.5e, curve *22*). The maximum of the fluctuation velocity on plate *11a* increased and became higher than that on the rigid plate (see Figure 3.5i, curve *42*). Analyzing the flow in the near-wall layer (see Figure 3.9e, curve 22; Figure 3.9i, curve *42*), we can see that in both cases (on both plates) the longitudinal fluctuations increased over the whole layer. The S-shape of the profiles on both plates substantially increased. This indicates the decrease in friction and in shear fluctuation near the wall.

Figure 4.18 presents the fluctuation velocity profiles on the heated elastic plates in water and airflows. The elastic plate *11a* and compliant plate *9* were tested in airflow. In both cases, the maximum of fluctuations decreased and moved closer to the wall, in agreement with data from the visualization of the laminar boundary layer [32, 42]. In contrast, the fluctuation velocities in the buffer layer and in the core became larger than on the rigid plate. It was noted that the lower the flow velocity, the larger the difference. This can be explained by the oscillation of the elastic surface. In the water flow, damping of the surface oscillations by water leads to a decrease in the difference between the rigid and elastic plates (see Figure 4.17).

At heating of elastic plate *11a* (see Figure 4.18b) the level of pulsations of longitudinal speed on the whole boundary layer thickness was decreased in comparison with the corresponding level of fluctuations in the flow on the standard. The fluctuation maximum increased in the near-wall region, its vertical coordinate being decreased. Heating of the compliant plate *9* slightly influenced the distribution of fluctuation velocity across the boundary layer. A similar behavior was observed on the compliant plate *6* in water flow at short heating. An increase in heating duration increases the effectiveness of an elastic plate, and fluctuations of longitudinal velocities decrease on the whole boundary layer thickness.

Figure 3.20c,g shows the spectral densities of longitudinal fluctuation velocity measured in airflow on the heated plates *11a* and *9* in section IIIa.

The heating of an elastic plate increases its compliance and it interacts with a flow better than before heating. As a result, the energy of high-frequency pulsations increased, while it decreased for low frequency. Thus, a maximum and a valley appeared at small wave numbers, indicating the formation of longitudinal vortical systems. The same picture is observed in Figure 4.12 for testing the same plate in water. The change in π-parameters and the between the densities of fluid and plate enhanced the interaction and allowed for the determining of the symptoms of longitudinal vortices generation (see Chapter 1 and Section 4.4).

The compliant plate *9* becomes more elastic when heated, so the flow−plate interaction in airflow worsens. As a result, the spectral properties at heating are practically the same as without heating, although the change in spectrum is similar to that on plate *11a*.

The influence of heating of elastic plates on kinematic characteristics of a boundary layer in air and water flows has been investigated on plates *6, 9* and *11a*. In Figures 4.15−4.17 results for plate *6* in water are shown. The design of plate *6*, which is made of two layers of foam elastomer (FE), is given in Table 2.3. The thickness of all composite plates (see Tables 2.3 and 2.4) is listed in Table 2.8. Thickness of the CCF with sealing rubber film (RF) was 1−1.5 mm for all plates. According to Table 2.8, each FE layer was about 3−3.5 mm, and the thickness of the entire plate *6* was 8−8.5 mm.

Figure 4.18 presents the comparison of kinematic characteristics of plates *9* and *11a* at heating in water and airflows. The design of plate *9* at testing in water was as in Table 2.3, and in air it was as in Table 2.4. Plate 11a had a structure according to Table 2.4. According to the tables, the thickness of the outer layer of plate *9* was 4 mm, the thicknesses of the foam polyurethane (FP) layers were 2.5 and 1.5 mm, and the foam latex (FL) was of the order of 3 mm. The thickness of the entire plate was 12 mm. For the tests in water, plate *9* was made of foam elastomer (FE) in the 8 mm-thick outer layer, then a synthetic CCF and FL sheet, as it was in the first variant of plate *9* (see Figure 2.28). The thickness of plate *11a* was 9 mm. The thickness of its outer layer made of polyurethane was about 4 mm, the thickness of the FL and CCF were the same as for plate *9.*

Section 2.7 describes the results of measuring the static mechanical characteristics of these plates. In Figures 2.51 and 2.52, the mechanical characteristics of plate *4* are represented at various degree of heating. Plate *4* has a thickness of 0.01 m and consists of two FPU (foam polyurethane) layers 4 mm thick pasted over with a rubber film. The heated synthetic fabric was located between these layers. While the mechanical characteristics of plate *4* differ from those of plate *9*, both plates belong to the compliant type of plates. At heating, the mechanical characteristics of plate *4* essentially changed at all values of pressure. The compliance of this plate increased. The data relating to the magnitude of relative deformation depending on loading (see Figure 2.52) were very important. Starting with a certain value of loading, relative deformations begin to decrease. This must be taken into account for regulation of mechanical characteristics of elastomers depending on static pressure of the environment. The results in Figures 2.51 and 2.52 agree with the data in Figures 4.15−4.18.

In Section 2.8 the results of measuring the dynamic mechanical characteristics of elastic plates *10a* and *11a* are described. Plates *10* and *10a* had different structures, but were made of the same materials. In plate *10* a continuous sheet was pasted on an FE base, on which the outer layer of the same material was pasted, but in the form of the longitudinal rectangular strips stuck together. The width of these strips was 5 and 10 mm. The entire thickness of plate was 7−7.5 mm, the thickness of the base was 3 mm. Plate *10a* had its outer layer made of a continuous FE sheet. Between layers the synthetic fabric CCF was placed, like in plate *6,* but the CCF sheet was continuous. Figure 2.54b illustrates the dependence of the damping coefficient on the energy of dynamic shock loading. For plate *11a,* curve *6* designates this dependence without heating and curve *9* with heating; for plate *10a,* the curves are *8* and *10* respectively. Curves *6* and *8* (without heating) and *9* and *10* (with heating) have corresponding equidistant forms of dependences. However, heating of the compliant plate *10a* leads to a reduction of the damping coefficient (rigidity increases), and at the elastic plate *11a* the damping coefficient increases.

The above-presented data indicate the possibility of controlling the mechanical and stabilizing properties of elastic plates by heating. Feasible results depend upon many conditions and require thorough development.

As has been shown, it is possible to control the characteristics of the boundary layer in the flow over elastic surfaces by generating longitudinal vortical disturbances in the boundary layer by means of a special vortex generator or by the special design of elastic plates. Another method of boundary layer control is to change the mechanical characteristics of the plates by controlling the temperature of the elastic material. The design of an elastic plate allows for the control of the pressure and density of liquids filling the cavities of a coating inside the plate (see Figure 2.28) [38, 277, 305]. By changing the pressure inside a coating, it is possible to regulate the rigidity of a coating to counteract the environmental pressure. The density of the liquid filling a coating influences the characteristics of the distribution of fluctuations in the coating caused by the fluctuation field of the boundary layer. This property has been checked in experiments, the results of which are illustrated in Figure 2.56. Water or technical oil was poured over the coating. The obtained results were compared to those when there was air above the plate. The increase in density of the liquid above the elastic plate influenced the phase speed of fluctuations on the plate. The measurements on the elastomer surface have shown that the fluctuations are nonlinear. This means that there is acceleration in the process of oscillation, which leads to increased added mass and, accordingly, to change of parameters of fluctuation in the boundary layer. Figure 2.56 shows schematic representations of photographs of the fluctuation of the elastomer surface caused by the impact of a metal indenter on the surface. When there was air above plate *11a*, curves *33−36* indicate fluctuations of the surface measured by means of two pressure sensors. Curves *37* and *38* characterize fluctuations when there was water above the plate, and *39, 40* for when there was oil. The increase in the density of the liquid essentially influences parameters of fluctuations of the elastomer surface.

The combination of all these methods allows for effective control of the characteristics of flow over an elastic surface. Another effective method is the generation of a set of fluctuations in an elastic coating. All this can allow for the control of the complex wave-guide representing the boundary layer in relation to an elastic plate with a specified structure.

4.6 Investigation of Velocity Fluctuations on Dolphin Skin

Measurements of surface fluctuations on various kinds of elastic plates are described in Section 2.8 (see Figure 2.56). Photographs of oscillograms of the distribution of free fluctuations under various conditions of external loading are given in Figures 2.57−2.61. In Sections 4.1 and 4.2 the results of measurement of spectra of distribution of fluctuations on the surface of elastic plates in a turbulent boundary layer are represented. To define an optimum range of fluctuations and mechanical characteristics of elastic materials it is of interest to consider the results of natural experiments on live dolphins [224, 266].

In the Department of Psychology and Physiology at the University of California experimental proof of active regulation of dynamic interaction of a dolphin skin with a flow has been obtained [224]. Direct research on the adaptation of dolphins to water has been

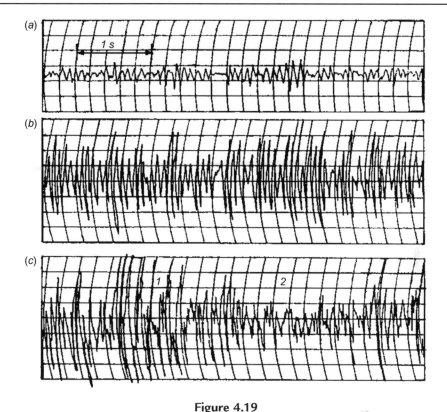

Figure 4.19

Microvibrations of a skin surface of a person (*a*) and a dolphin (*b, c*) [224]: *a* — in air, amplitude 1—5 μm, frequency 11 Hz; *b* — in air, amplitude 5—20 μm, frequency 13 Hz; *c* — in water, operation factors of a fluctuations depend on muscular activity: *1* — active swimming, *2* — movement by inertia.

executed, including the heat regulation of dolphin skin (as the dolphin is a warm-blooded animal). Equipment for the comparison of microvibrations of dolphin and human skin was used. In the air environment dolphin skin is more sensitive than human skin by half the order of the microvibration amplitude. At rest in air, human skin microvibrations are regular, with an 11 Hz frequency and 1—5 mm amplitude. The microvibrations of the skin of a dolphin removed from water and laying on Foamex are also regular, but have a 5—20 μm amplitude and a 13 Hz frequency (see Figure 4.19). In addition, a second periodicity of microfluctuations of a dolphin skin at a frequency of 2.5 Hz was discovered. It was found to correspond to the electrocardiogram and, hence, the frequency of pulsing blood in the skin. Investigation of the microfluctuations of the skin of a slowly swimming dolphin in still water have revealed a change in vibration amplitude and they were less rhythmical than in air in a quiet condition. The authors believe that these changes are caused by the work of the dolphin's muscles while swimming. An interruption (cut-off) of impellent innervations is known to lead to the disappearance of microvibrations on the skin surface.

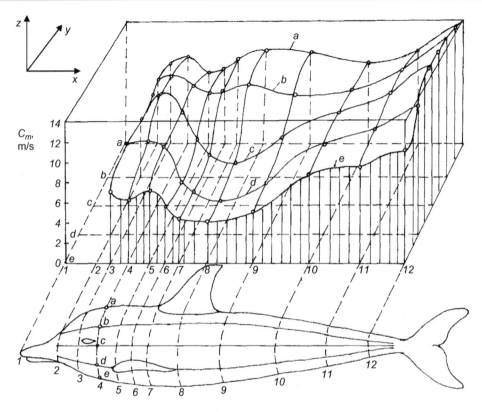

Figure 4.20
Relief of velocity of the forced fluctuations on a dolphin skin surface [266].

This means that microvibrations are related to muscular tone. Thus, a close relationship between microvibrations and the system of epidermis nerves and deeper layers of tissue has been shown experimentally.

In [266] the velocity of oscillation propagation, the dependence of this velocity on the structure of a dolphin's skin and the direction of the oscillation propagation at measurement of forced oscillation by the phase method was defined, as was the character of distribution of the oscillation and its attenuation at propagation along the skin.

The surface oscillations were excited in the range 50−400 Hz on a dolphin's skin (live and dead) and propagated on all sides from the dot activator in the form of traveling waves. Figure 4.20 presents the averaged results of measurements of propagation velocity of oscillations performed on six bottlenose dolphins; x, y are coordinates of a projection of one side of the coating, z is the phase velocity of the oscillation propagation C_m. The velocity is rather high on the head (section 4), decreases about the neck (section 6), increases again from the middle of the body (sections 8 and 9), and then slightly decreases. It grows in the

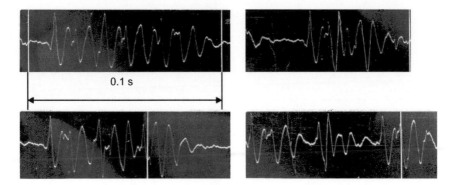

Figure 4.21
Photographs of oscillograms of fluctuations of the surface of a dolphin skin at slow swimming and in motionless position in water [266].

beginning of a tail stem. These results completely correlate with results of measurements of structural aspects of the body and skin coverings of dolphins [25]. At the site of scars or other skin damage, a reflected wave, and also waves of a constant phase and variable amplitude, were formed.

As can be seen in Figure 4.20, $C_m = 4-12$ m/s — that corresponds to the phase speed of the disturbance propagation in the turbulent boundary layer on a rigid plate approximately at this range of speeds. Measurements on artificial elastic surfaces have shown that phase velocity depends on the design and mechanical characteristics of a surface and can essentially differ from the reference data on a rigid surface. $C_m = 25-30$ m/s over tense muscles of a human forearm. At a dolphin's size, C_m does not depend on the direction of oscillation propagation. For a dead dolphin, C_m decreases by 20–25%. Fluctuations on the surface of a dolphin's skin propagate in the form of a traveling wave with a displacement coefficient ≈ 1. All measurements were done in water, and data in Figure 4.20 were measured on a motionless dolphin.

In the further tests, a dolphin was in a coastal 6×30 m enclosed cage. A belt was placed on the dolphin ahead of its back fin. A sensor for vibration was placed on one side under the belt along the direction of motion in such a way that the vibroprobe adjoined to the dolphin skin before the belt. A cable was fastened to the belt on the opposite side of the body.

Sections of free swimming, acceleration and deceleration, including motion by inertia, were fixed. Figure 4.21 shows fluctuations at small speeds of swimming and at rest. Frequency oscillation—pulse chain, single or group, following one after another were recorded with a periodicity of 0.1–0.12 s, a duration of 0.035–0.09 s and a frequency of filling 130–140 Hz.

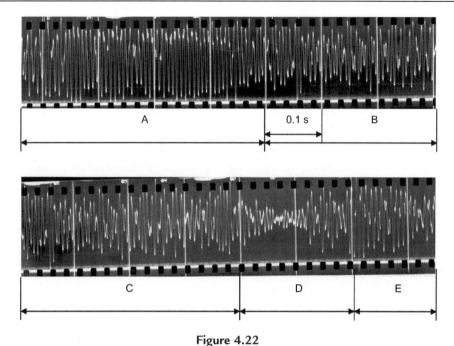

Figure 4.22
Photographs of oscillograms of a dolphin jump through a net: A — uniform swimming,
B — acceleration, C — movement by inertia, D — movement in air, E — entrance into water.
A cycle of a jump: B — E (1s) [266].

Further measurements were made during jumps through a net. The height of a jump was
1.5−2.0 m. The angle A of exit from water was 45−50°; the speed of motion at exit from
water $U \approx 12$ m/s. Figure 4.22 represents oscillograms of fluctuations of the skin during a
jump. Duration of a jump was approximately 1 s. Frequency of fluctuation of the skin is
115−230 Hz. Frequency at acceleration increases, then is constant and decreases after a
jump. Places where the phase varies by 180° during an entry into water were identified.

Wireless equipment has also been used to measure vibrations of the skin at two points and
speed of motion. The measurements were performed in the open sea ($U = 3−4$ m/s) and in
a long hydrodynamic channel, in which speed $U = 7$ m/s was reached. In these experiments
the use of a bending accelerometer and, differentially, a needle sensor for vibrations
(a displacement sensor) was made. At $U = 7$ m/s fluctuations of the skin in the form of
impulses with duration a of 0.1−0.2 s at frequency of filling of 240−270 Hz were
registered.

Based on the performed measurements, some mechanical characteristics of skin coverings
of dolphins were calculated according to the technique described in [27, 28, 285−288]
(see Figure 4.23). The measurement of the oscillating mass of a dolphin's skin [29] was

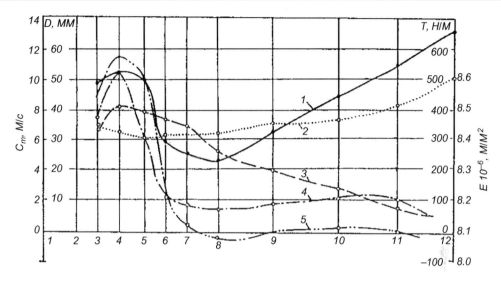

Figure 4.23
Distribution of phase speed of the compelled fluctuations and other mechanical characteristics along a body of a dolphin: *1* — value of C_m; *2* — module of elasticity *E*; *3* — thickness of skin *D*; *4* — static tension *T*; *5* — total tension of skin when moving [226].

used for calculations of tension in the skin coverings. The calculations utilized the formula [27, 288–290, 293, 319, 323]:

$$C_m = \sqrt{(T/M)} \qquad (4.14)$$

Tension was determined along the line *b* (see Figure 4.20) for a dolphin at rest. Curve *5* in Figure 4.23 has hypothetical character. The measured C_m correlates well with results of the measurement of distribution of elasticity and thickness of the skin along the dolphin's body: C_m increases with an increase in elasticity and reduction of skin thickness. The value of *T* changes along the body within the limits of 54–502 N/m, i.e. a change of 9.3 times. According to [29] it is received: $T \approx 5.5$ N/m for small speed of swimming and $T \approx 80.1$ N/m for high speed of swimming, therefore it changes depending on the swimming speed by 14.5 times.

With the exception of C_m, the mechanical characteristics of skin coverings of dolphins shown in Figure 4.23 have a conditional character because *T* and *E* essentially depend on the thickness of the skin covering. Until now, there were no measurements of distributions of fluctuations across the covering depth caused by disturbances of boundary layer, either on live dolphins or using analogues of their skin coatings. The depths to which the decaying oscillations extend define the thickness of skin coverings, according to which real values of *T* and *E* are determined. Calculations of the coefficient of oscillating mass of dolphin skin coverings according to the characteristics of the boundary layer are presented in [39, 41, 43].

It was discovered that there is a pattern of regular transversal micro-roughness on the surface of dolphin skin [25]. During swimming, elastic waves move along the body. Thus, the regular roughness moves along the body. Knowing the frequency of fluctuations of a body and the parameters of the elastic wave's motion [263], and also the speed of disturbances in the boundary layer, it was possible to calculate dimensionless parameters of oscillations in the boundary layer caused by the oscillatory movement of a body of a dolphin [63]. Thus, it has been shown that, during the swimming of a dolphin in a boundary layer the oscillations operate characteristics of a boundary layer, and, hence, its resistance is generated.

Other research [266] has indicated other mechanisms of fluctuations generated in dolphin skin coverings both at rest and while swimming. Such fluctuations arise automatically at traction of the cutaneous muscle of dolphins, which work as a reflex irrespective of the body condition. During swimming the cutaneous muscle and skeletal muscles are strained, which changes the tension of the skin. Therefore, the frequency range and the form of fluctuations on the skin surface change during swimming and depend on the conditions of motion.

Thus, these data confirm what was previous research suggested, that dolphin skin coverings have the ability to actively control hydrodynamic drag [25, 49]. Such active system of fluctuations generated in a boundary layer also confirms the proposed hypothesis about the role of interaction of disturbances in a boundary layer in the control of boundary layer characteristics [25, 33].

However, the physiological mechanism involved in the generation of this form of fluctuations is not understood. It is probable that periodic generation of fluctuations on the surface of the skin is linked to a slight consumption of energy. It is thought that the cause is the physical pattern of liquid flow in the boundary layer. In the viscous sublayer Reynolds stresses periodically lead to bursts from the viscous sublayer to the generation of turbulence and regeneration of the viscous sublayer. The revealed periodic generation of fluctuations in the skin confirms the hypothesis about pre-burst modulation [137]. Such periodic fluctuations in the skin prevent bursts from the viscous sublayer and carry away the stagnated liquid with the big phase speed to the oscillating tail fin where the stagnated liquid is bunched up in a wake behind the tail fin. In [64], on the basis of the results of numerous experiments on the interaction in a boundary layer of various disturbances, the following conclusion was reached. During the movement of hydrobionts, three kinds of disturbing movement are generated in the boundary layer: a) from the action of microfolds on the skin surface; b) from its vibration as viscous-elastic material and; c) due to the work of the tail mover and corresponding muscles.

The above-mentioned research results on the active control of various types of composite elastic materials, which model the dolphin's skin covering, have shown that the most preferable is generation of longitudinal fluctuations in the boundary layer over elastic coatings.

CHAPTER 5

Experimental Investigation of Friction Drag

5.1 Methods of Determining Friction Drag on Plates

The different theoretical methods of calculating the turbulent boundary layer are currently being intensively developed [126]. To calculate the turbulent friction, Karman first used the momentum equation in the integral form [see in 181, 492]:

$$\frac{d\delta^{**}}{dx} + (2 + H)\delta^{**}\frac{dU_\infty}{dxU_\infty} = \frac{C_f}{2} \tag{5.1}$$

where C_f is the local friction coefficient:

$$C_f = \frac{2\tau_w}{\rho U_\infty^2} \tag{5.2}$$

The complete friction coefficient or the mean friction drag coefficient is calculated as:

$$C_F = \frac{2P}{S\rho U_\infty^2} \tag{5.3}$$

$$C_F = \frac{1}{x}\int_0^x C_f dx' = 2\frac{\Theta}{x} \tag{5.4}$$

Much attention has been paid to developing methods of measuring the local friction coefficient [13, 130, 142, 177, 207, 223, 228, 270, 283, 325, 363, 377, 381, 435 and others]. For this purpose, Preston and Stanton pipes, heated wires located in an aperture of a wall, shear stress fluctuation sensors, liquid crystals, tensometric floating elements and other means are used. The shear stress can be determined by three methods: 1) theoretical or experimental formulae; 2) formulae obtained from the velocity profiles expression in the form of power law; 3) formulae based on experimental investigations of the kinematic characteristic of the boundary layer. Formulae obtained from the first method are presented in Table 5.1.

Research on the turbulent boundary layer at small Reynolds numbers is widespread. Coles and Bair [see 153, 514] suggested formulae to calculate the local friction coefficient for $600 < \text{Re}_{**} < 1200$:

$$C_f = \frac{0.3\exp(-1.3H)}{(\lg \text{Re}_{**})^{1.74+0.31H}} \tag{5.4}$$

Table 5.1: Formulae for definition of the factor of friction, received in the first way.

Formula	Formula Number	Reference
$C_f = 0.455 \, (\lg Re)^{-2.58}$	(5.25)	Schlichting [see in 445]
$C_f = (2\lg Re - 0.65)^{-2.3}$	(5.26)	Schlichting [see in 445]
$C_f = 0.026 Re_{**}^{-0.25}$	(5.27)	Blasius [see in 445]
$1/\sqrt{C_f} = 1.77 \ln Re_{**} + 2.62$	(5.28)	Rotta [see in 441]
$C_f = 0.0128 Re_{**}^{-0.25}$	(5.29)	Rotta [see in 441]
$1/\sqrt{C_f} = 1.81 \ln Re_{**} + 2.54$	(5.210)	Squire-Young [see in 441]
$C_f = 0.246 \cdot 10^{-0.678H} Re_{**}^{-0.268}$	(5.211)	Ludweig-Tillmann [see in 441]
$C_f = 2c \, Re_{**}^{-0.17} = 2ce^{-0.391\xi}$, where	(5.212)	Fediaevsky [see in 181]
$\quad c = 0.001[6.55 - 0.0685(\xi - 4.4)$		
$\quad + 0.2506(\xi - 4.4)^2]; \, \xi = \lg Re_{**}$		
$C_f = 0.058\gamma^{1.705} Re_{**}^{-0.268}$, where $\gamma = \lg(8.05/H^{1.818})$	(5.213)	Fernholz [see in 490]
$C_f = 0.01013/(\lg Re_{**} - 1.02) - 0.00075$	(5.214)	Green-Bradshow [see in 490]
$1/\sqrt{C_f} = 1.7 + 4.15 \lg(C_f Re)$	(5.215)	Karman [see in 490]
$C_f = 0.37\lg(Re)^{-2.584}$	(5.216)	Schultz-Grunow [see in 490]
$C_f = [17.08 \lg^2 Re_{**} + 25.11 \lg Re_{**} + 6.012]^{-1}$	(5.217)	Karman-Schehnerr [see in 490]
$C_f = 0.242 C_f/(0.242 + 0.8686\sqrt{C_F})$,	(5.218)	Karman-Schehnerr [see in 490]
$\quad$ where $C_f = (0.242/\lg Re_{**})^2$ or	(5.219)	
$\quad 1/\sqrt{C_F} = 1.24 + 4.13 \lg R_{**}$ or at Re_{**}		
$\quad > 5000 \, H = 1/(1 - 4.8\sqrt{C_F})$		
$C_f = 0.088(\lg Re - 2.3686)/(\lg Re - 1.5)^3$,	(5.220)	Squells-Pain [see in 177]
$\quad$ where $Re_{**} = 0.044 Re/(\lg Re - 1.5)^2$		
$\sqrt{2/C_f} = 5.6 \lg[Re_{**}/(1 - 6.8\sqrt{C_f/2})]$	(5.221)	Clauser [see in 177, 441]
$Re_{**} = 1/6(\xi)^2 + 1/\aleph E[(1 - 2/\aleph\xi)\exp^{\aleph\xi} + 2/\aleph\xi + 1 - 1/6(\aleph\xi)^2 -$	(5.222)	Spalding-Chi [see in 177]
$\quad 1/12 \, (\aleph\xi)^3 - 1/40 \, (\aleph\xi)^4 - 1/180 \, (\aleph\xi)^5]$, where $\aleph = 0.4$;		
$\quad E = 12; \, \xi = \sqrt{2/C_f}$		
$C_f = 0.776/(\lg Re_x - 1.88)^2 + 60/Re_x$ at low Re_x	(5.223)	Granville [see in 214]

$$C_f = 0.023 \, Re_{**}^{-0.24} \tag{5.5}$$

Using the power law for velocity profile:

$$\frac{U}{U_\infty} = \left(\frac{y}{\delta}\right)^{1/n} \tag{5.6}$$

More simple (although less accurate) formulae of the second group can be obtained for calculations of C_f. The ratios of characteristic thicknesses of a boundary layer are obtained on the basis of the Miring equation [490]:

$$\frac{\delta^*}{\delta} = \frac{1}{1+n}$$

$$\frac{\delta^{**}}{\delta} = \frac{n}{(1+n)(2+n)} \tag{5.7}$$

$$H = \frac{\delta^*}{\delta^{**}} = \frac{2+n}{n}$$

Usually $n = 7$ for a turbulent boundary layer at $5 \cdot 10^5 < \text{Re} < 10^7$. In this case, the friction coefficient can be calculated by formulae [445]:

$$C_f = 0.045 \, \text{Re}_\delta^{-0.25}$$
$$C_f = 0.026 \, \text{Re}_{**}^{-0.25} \tag{5.8}$$
$$C_f = 0.074 \, \text{Re}^{-0.2}$$

For an incompressible liquid with arbitrary n we can use the formulae [435]:

$$C_f = 2(\alpha_H^2)^{\frac{1-n}{1+n}} \left(1/Re^{\frac{2}{n+1}} \right)$$
$$C_f = 2(\alpha_H^2)^{\frac{1-n}{1+n}} \left(1/Re_{**} \frac{\delta}{\delta^{**}} \right)^{\frac{2}{n+1}} \tag{5.9}$$
$$C_f = 2[2/(n+3)(n+2)C^n]^{\frac{2}{n+3}} \text{Re}^{-\frac{2}{n+3}}$$

where $\alpha_H^2 = 2\,1\,n$; $n = 1.9(\lg \text{Re} - 3.1)$; $C = \alpha^{\frac{2n}{1-n}} = n + 1.45$; δ^{**}/δ can be taken from (5.7); $\text{Re}_\delta = \dfrac{(1+n)(2+n)}{n} \text{Re}_{**}$. The complete friction is calculated as follows:

$$C_F = \frac{n+3}{n+1} C_f \tag{5.10}$$

when $n = 7$ formula (5.10) transforms into formula (5.8).

The power law of velocity profile can be written in the form [445]:

$$u^+ = 8.74(y^+)^{1/n} \tag{5.11}$$

Hence, on the basis of the relationship [441]:

$$\frac{U}{u_\tau} = \left(\frac{2}{C_f} \right)^{1/2} \tag{5.12}$$

and (5.6) we obtain the equation:

$$C_f = 2 \left(\frac{\nu}{U_\infty 8.47^n \delta} \right)^{\frac{2}{n+1}} \tag{5.13}$$

To the third group belongs the formulae obtained from profiles of averaged and fluctuation components of longitudinal velocity and from the Reynolds stresses. The simplest method is the measurements of the profile of the boundary layer to $y^+ = 5$, when the local shear stress corresponds to Newton formula:

$$\tau_w = \mu \left(\frac{\partial u}{\partial y}\right)_0 \tag{5.14}$$

If the boundary layer thickness is sufficient compared to the sensor size, the measurements can be carried out with high accuracy [142, 435].

When the velocity profile is measured and, correspondingly, characteristic thicknesses of the boundary layer, then we can calculate the distribution of the friction coefficient along the plate, on the basis of the relation $u_\tau = \left(\frac{\tau_w}{\rho}\right)^{1/2}$ and the formula (5.12) from the momentum equation:

$$\frac{d(U_\infty^2 \Theta)}{dx} + U_\infty \delta^* \frac{dU_\infty}{dx} = \frac{\tau_w}{\rho} \tag{5.15}$$

In this way, the formula (5.11) was obtained. The other example is the Squire–Young formula [441]:

$$\frac{\tau_w}{\rho U_\infty^2} = \left[5.89 \lg(4.075 \, \mathrm{Re}_{**})\right]^{-2} \tag{5.16}$$

The local friction coefficients can be determined from the velocity profiles measured according to the Clauser method [142, 177, 441]. Bradshaw, Kline, and others [130, 274, 177] have modified the method.

The value of u_τ and, hence, the value of C_f can be determined by the shape of the linear section of the velocity profile plotted in semi-logarithmic coordinates according to Rotta's method [441], from the expression:

$$u = \frac{u_\tau}{\aleph} (\ln y + const) \tag{5.17}$$

The shear stresses can be determined by maximal values of the fluctuation velocities in the boundary layer [381]:

$$\frac{\sqrt{\overline{u'^2_m}}}{u_\tau} = \{-0.5 \ln[\ln(10^3 C_f)] + 1.36\}^3 \tag{5.18}$$

Maystrello and Montaich suggested calculating the friction coefficient by the measurement of pressure pulsations [177, 468]:

$$\frac{\sqrt{\overline{p'^2}}}{\tau_w} 2.5 + 2.8 \left(1 - e^{-0.1M_\infty^2}\right) \tag{5.19}$$

Table 5.2: The values of C_f calculated by different methods.

Calculation Method (Formula Number)	Section			Section		
	I: $Re_x =$ $7.7 \cdot 10^5$	II: $Re_x =$ $7.8 \cdot 10^5$	III: $Re_x =$ $8.7 \cdot 10^5$	I: $Re_x =$ $1.2 \cdot 10^6$	II: $Re_x =$ $1.2 \cdot 10^6$	III: $Re_x =$ $1.5 \cdot 10^6$
(5.5)	0.0047	0.00468	0.00459	0.00432	0.0042	0.00415
(5.6)	0.00392	0.00391	0.00384	0.00363	0.00353	0.00349
(5.8)	0.00392	0.00381	0.00378	0.0036	0.00347	0.00339
(5.10)	0.00381	0.00371	0.00368	0.0035	0.00338	0.00333
(5.11)	0.00446	0.00432	0.0044	0.00426	0.00411	0.00405
Clauser method	0.004	0.00405	0.0039	0.0037	0.0036	0.0035
(5.37)	0.0038	–	0.0034	0.0034	–	0.0033
(5.38)	0.0052	0.00515	0.00497	0.00402	0.00421	0.0042
(5.40)	0.00405	–	0.0038	0.0036	–	0.0034

The expression (5.19) is well agreed with (5.7). From the equation for the Reynolds stresses obtained by Rotta:

$$-\overline{u'v'} = \frac{\tau_w}{\rho}\left(1 - \frac{\nu}{\aleph y u_\tau}\right), \text{ when } >> \delta_s \tag{5.20}$$

we can conclude that the value of $\overline{u'v'}$ quickly reaches the value of τ_w/ρ, i.e. the shear stress on the wall, at the distance larger than the viscous sublayer thickness. Therefore, the experimental curves shown in Figure 3.13 can be extrapolated in the direction to the wall. We can deduce that the results of measurements near the wall are approximately equal to the shear stress on the wall. Based on this, it is possible to define values of C_f on the wall.

To compare different methods of determining the friction drag in accordance with Table 3.1, values of C_f were determined by measurements on the rigid plate in the wind tunnel (see Table 5.2). The results, obtained by formulae (5.6) and (5.8) and the Clauser method were closest to the experimental data extrapolated from formula (5.40). The values obtained using formulae (5.10) and (5.37) were slightly underestimated. The values obtained with formulae (5.5), (5.11) and (5.38) were essentially overestimated.

The presented comparison showed that some formulae from Table 5.1 are agree well with each other and can be used to calculate C_f. The simplified formula (5.37) is applicable for the initial experimental data processing. The results obtained are also well agreed with the data in [177, 514].

Important conclusions have been made from papers [181, 441, 435]: the value of C_f decreases at increasing values of form-parameter H or decreasing the coefficient n.

Despite the variety of theoretical and experimental methods for defining friction drag on smooth flat plates, direct measurement of friction drag by strain dynamometers is still reliable under new boundary conditions.

5.2 The Complex Apparatus for Experimental Research

The development of numerical methods and computer techniques has promoted the solving of many complex problems in hydromechanics. However, physical experiments continue to be irreplaceable for the analysis of new nonlinear and non-stationary problems, especially when physical representations related to the phenomena considered are not developed. The experimental analogue of computing methods has been utilized at solving new problems.

An experimental complex is presented in Figure 5.1. It consists of the setup *I* (the hydrodynamic channel of low-level turbulence) for physical research of the boundary layer at low Reynolds numbers, and of installations *II-V* (strain dynamometers) for towing tests in a wide range of Reynolds numbers.

The hydrodynamic channel of low-level turbulence is described in Chapter 1. Setup *II* is designed in the form of a thin axisymmetric wing and allows for investigating the same plates in the towing channel (see Figures 2.25, 2.29 and 2.30) as were investigated in installation *I* (see Figures 2.26 and 2.27). In towing tests it is impossible to design a drag balance that would be equally sensitive at various ranges of towing speeds. Therefore, it is necessary either to have several drag balances, or to take into account that one drag balance will produce increased errors at other ranges of towing speeds. The possibility of testing the same surfaces in installations *I* and *II* (see Figure 5.2) allows increasing accuracy of measurements in a wide range of Reynolds numbers.

Installations *III* and *IV* allow for the investigation of the same surfaces as in installations *I* and *II* but in 2D approximation. With these installations it is also possible to investigate various spatial problems of a boundary layer. On installation *III* (see Figure 5.3), a tensometer is placed in a knife-pylon. On installations *IV* and *V* a tensometer is placed in the tail cowls. Engineers and the scientific researchers specializing in a similar direction know similar devices. If readers have details of separate units and features of such devices will interest please address the authors for consultation. It reduces the measurement error caused by the moments from asymmetrical loadings. Installation *V* allows for separate measurements to be made of the loads on the nose, middle and tail parts. Tests on installations *II−V* can be made in both forward and reverse motion.

In designing the experimental complex, unification of the majority of its components was achieved, which allowed regular errors to be reduced, the tests to be unified, and it also increased the interchangeability of units, and so allowed their application for various experiments. Experience shows that it is impossible to create a universal setup on which all problems of a boundary layer can be investigated. However, the given experimental apparatus allowed the researchers to design and carry out a wide range of physical experiments concerning both the 2D boundary layer and other problems of hydromechanics. The experimental complex, when used in conjunction with the devices presented in

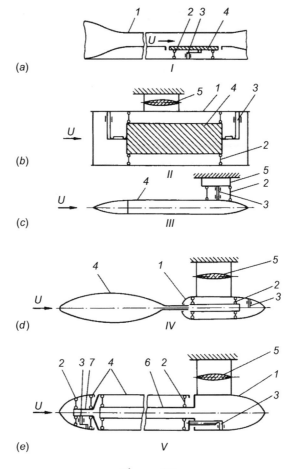

Figure 5.1

Experimental complex for researching the boundary layer. An arrangement of tensometric inserts in the hydrodynamic channel (*a*) and wing (*b*): *1* — hydrodynamic channel and wing, *2* — elastic suspensions, *3* — tensometer, *4* — plate for investigation, *5* — pylon; strain dynamometer schemes in longitudinal streamline cylinders (*c*, *d*, *e*): *1* — cowl, *2—5* — the same as in (*a*) and (*b*), *6—7* — rods; *I—V* — strain dynamometer types.

Chapters 2—4, is an original system, which allowed the development of a new, more economical, technique of experimental research and obtained some qualitatively new results.

The hydrodynamic channel of low-level turbulence is described in detail in Chapters 1 and 2 [14, 285, 286, 305, 310]. The data obtained from the experiments in the channel formed the basis of further investigations, presented below.

The measurement in the channel was carried out by strain dynamometer *I*, located at the end of the working section. Similarly, tensometer *I* was used for the measurements in a wind tunnel (See Chapter 3).

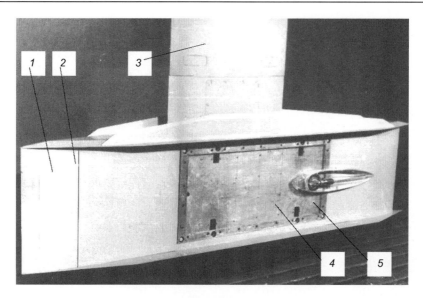

Figure 5.2

Photograph of the axisymmetric wing for the towage of plates (strain dynamometers II): *1* — wing, *2* — wire turbulence promoter, *3* — knife-pylon, *4* — investigated surface, *5* — the sensor for the measurement of elasticity and oscillations (see Figures 2.15 and 2.16).

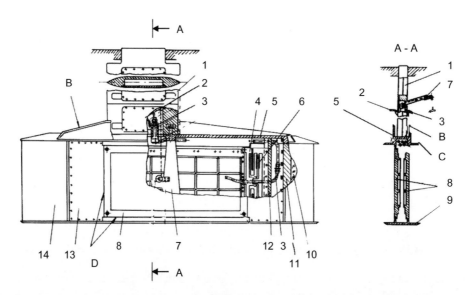

Figure 5.3

Arrangement of strain dynamometers *II* inserts in the wing: *1* — a knife-pylon, electric wires, *3* — tubes for injection of liquids, *4* — tensometer, *5* — the top force washer, *6* — resistive-strain sensor, *7* — calibration device, *8* — panels for investigation plates, *9* — the bottom force washer, *10* — back cowl, *11*, *12* — force rack, *13* — dashboard, *14* — forebody cowl.

To extend the range of Reynolds numbers, the strain dynamometer *II* (see Figure 5.2) was designed in the form of an axially symmetric wing *1*. Due to the small wingspan, it was fitted with end plates. From behind on wing *1,* in the fairing, the sensor of elasticity and displacement *5* is mounted. A wire-tripping device *2* is placed on the forward fairing of the wing.

The force skeleton of the wing (see Figure 5.3) consists of two strong washers *5* and *9* profiled with strengthening ribs and connected by two pairs of supports *11* and *12*. Replaceable fairings are fixed on outer edges of these supports, and symmetric strain sensors on the internal edges. The top force washer *5* has a section of the flange beam. Ribs *C* work as a washer for the prevention flow from the end faces of the short wing.

Ribs *B* provide additional rigidity for the washer and, furthermore, carry out the hydrodynamic function to neutralize the influence of large vortices formed in the corners between the pylon and top washer. Replaceable cowls *10* and *14* are attached to the racks *11* and *12* on the external edges and symmetric drag balances *4* on internal edges. The unified inserts *8* are attached to drag balances from two sides; these are the same removable panels that were established on tensometric dynamometer *I*. The drag balances *4* are closed outside, flush with the model by dashboards *13*.

A system for the injection or suction of liquid or bubbles is provided in the wing design. The design is such (*3*) that it allows the injection of a liquid/gas through the forward slot *D,* between the insert and wing, and suction through the back slot between the insert and wing. There is a slot on the perimeter *D* (see Figure 5.3) of the panel, between its face and the corresponding details of the wing surface. The forward slot is $(1.0-1.2)*10^{-4}$ m thick, the back slot $(1.2-1.8)*10^{-4}$ m, and the lateral slot $(0.8-1.6)*10^{-4}$ m.

The wing model has a built-in calibration device *7*, which provides calibration of the strain beam on the model at any time without the need to dry out the dock and disassemble the model. To the top force washer *5*, a unified knife-pylon *1* is fastened, which is also fastened to the towing carriage of the high-speed towing tank or to the cavitation tunnel. All communications of models *2* and *3* are located inside pylon 1.

The strain dynamometer *III* consists of the pylon fitted with tail fairings (see Figure 5.1). In these investigations the strain dynamometer was used to research the boundary layer on the surface of a cylinder. The cylindrical part of the model consists of several cylinders, which are screwed to each other. This allows for changing the model length over a wide range. The model is described in the following chapter in detail. The calibration device has the same unified principle of action: it allows for calibration during tests without disassembling the setup. If the plate is large, the moments generated at the wing lead to measurement errors. In addition, the strain sensors are in the open "wet" condition. That leads to a drift of indications with changes in the outer temperature.

Figure 5.4

Photograph of models of a body of revolution (strain dynamometers *IV*): *1* — forward cowl, *2* — dashboard, *3* — cylindrical part of the model, *4* — load-carrying case (tail cowl), *5* — short part of the knife-pylon, *6* — long part of the unified knife-pylon, *7* — calibration system, *8* — towing carriage, *9* — towing tank dock.

The strain dynamometer *IV* (see Figure 5.1) consists of the load-carrying case (tail cowl *1* with diameter 0.1 m) fastened to the unified knife-pylon *5*. Unlike strain dynamometers *II* and *III*, the elastic suspension brackets and strain dynamometer are made of separate units. The elastic suspension brackets have a special design and are executed in the form of ring membranes *2*, which are mounted at a sufficient distance from each other. A hollow rod is fixed onto them, having an outside mounting surface. On this rod, various investigated surfaces *4* were established.

In the end, tensometer *3* was placed in a sealed chamber. Such an arrangement was convenient in the operation and allowed for increasing the accuracy of measurements.

Tensometer *IV* is fastened to the load-carrying unit of the towing carriage of the towing tank or cavitation tunnel by means of the unified knife-pylon *5*, consisting of short and long parts (see Figure 5.4). The short part *5* (see Figure 5.4) is fixed on the strain dynamometer, and long part *6* is connected to the towing carriage *8* of the high-speed towing tank *9*.

Both parts of the knife are connected by locks closed by hatches *2*. It allows for the removal of the strain dynamometer *4* and short part of the knife *5*, for adjustment and alignment, without disrupting the alignment or installation of the strain dynamometer in the high-speed towing tank. In the pylon, there are cavities for branch pipes to bring all communications to the strain dynamometer, and corresponding connecting pipes for attaching the devices supplying various liquids to the model. The calibrating system *7* has the same unified principles of operation.

In the load-carrying case *4*, the membranous suspension brackets are connected to a rod, which jets out. The model cylindrical part consisting of three cylinders *3* and forward fairing *1* is mounted on this rod. Cylindrical dashboards *2* cover the junction of the forward fairing and cylinder. Depending on length of the rod, it is possible to mount a certain number of cylinders, *3*. The length of the model can change, and the length of model changes in a similar way in the strain dynamometer *III*.

The most complex and perfect is strain dynamometer *V*; it consists of the same unified units and details as strain dynamometer *IV*. The difference is that diameter of strain dynamometer *V* is increased by 1.75 times. This allowed, in the case *1* (see Figure 5.1), the rigid fixing of the system of hollow rods *6* and *7*. The external pipe *4* leans onto those rods through the system of membranous diaphragms *2*. On this pipe, the investigated surface and nasal cowl are installed. The developed design of strain dynamometer *V* allowed for having three independent surfaces on the model length: in the central cylindrical parts, the nasal and the tail parts. Each of these surfaces had an independent strain dynamometer. It enabled the investigation of the hydrodynamic characteristics of the model simultaneously on the specified parts and on the entire model. The general view of strain dynamometer *V* is shown in Figure 5.5.

Babenko developed the idea of creating of an experimental complex with unified units and principles of work, the design of a drag balance in models of a body of revolution, and division of measurements of drag on separate parts of a body on tensometer *V*. He developed the project of hydrodynamic channel and model *III* (see Figure 5.1a,c). Korobov designed the strain dynamometer and models *II, IV* and *V* (see Figure 5.1b, d, e).

The investigations with strain dynamometers *II, IV* and *V* was done in the high-speed towing tank of the Institute of Hydromechanics of the National Academy of Sciences of Ukraine (see Figure 5.6). The high-speed towing tank has two towing carriages:

- the slow-speed towing carriage with a drive from an infinite cable; its speed of towage is up to 6 m/s;
- the high-speed towing carriage with a drive from a linear electric motor; it has a speed up to 25 m/s, and acceleration up to 7.0 m/s.

The dimensions of the towing tanks are $140.0 \times 4.0 \times 1.8$ m.

All investigations with strain dynamometers *II, IV* and *V* were performed on the high-speed towing carriage by Korobov. Research with strain dynamometers *II* and *IV* (with installation of the short cylindrical part) and processing of the obtained data were executed by Babenko together with Korobov.

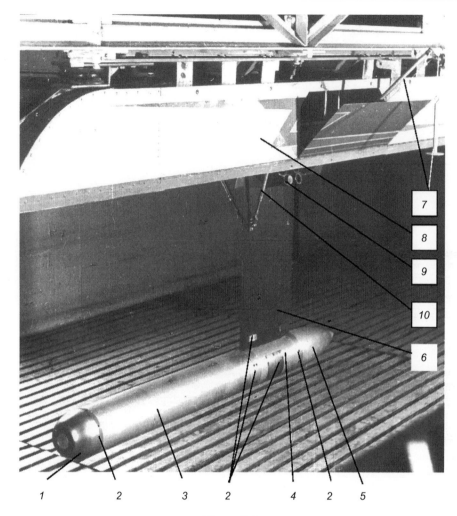

Figure 5.5
The model of a body of revolution (strain dynamometers V): *1* − forward fairing (forward tensometer), *2* − dashboards, *3* − cylindrical part of model, *4* − load-carrying case (tensometer of the central part of model), *5* − back part of the model (back tensometer), *6* − long part of the unified knife-pylon, *7* − calibration system, *8* − towing carriage, *9* − pipelines for supply of an injected liquid, *10* − blocking for increase stability of the model.

5.3 Experimental Investigations of Friction Drag on Elastic Plates

In addition to kinematic characteristics of the boundary layer, friction drag on elastic plates was investigated in the wind tunnel (see Chapter 3) at the same speeds (see Table 3.3). The corresponding Reynolds numbers and values of C_f on rigid and elastic plates are presented in Table 5.3. It becomes evident that calculating C_f with formula (5.3) the value of S was

Figure 5.6
High-speed towing tank of the Institute of Hydromechanics of National Academy of Sciences of Ukraine.

Table 5.3: Friction drag coefficients on the elastic plates in airflow.

Plate Number	10^{-6} Re	$10^3 C_f$	$f_1 = U_\infty/\delta,\ s^{-1}$	$f_2 = u/2\pi,\ s^{-1}$
Rigid Plate	0.99	5.2	404.1	85.4
(Standard)	1.6	4.6	723.1	114.8
5	1.0	5.2	297.6	67.02
	1.66	5.08	499.4	96.8
2a	1.02	5.0	258.2	69.0
	1.6	4.42	429.9	104.4
9	0.95	4.0	410.8	70.4
	1.57	4.1	739.4	109.2
3	0.95	4.58	336.4	69.0
	1.5	3.8	544.9	113.9
10a	0.85	4.95	262.9	69.7
	1.5	4.76	594.2	112
8a	0.95	4.6	339.2	64.9
	1.49	4.2	551.1	112.7
11a	1.03	4.9	406.8	72.3
	1.52	4.51	703.9	120.5
11	0.95	4.68	341.3	67.6
	1.57	4.5	580.5	114

reduced by the area of the rigid fringing of strain-sensor inserts (see Figures 2.25, 2.27, 3.1 and 5.1). This is made in order to compare this data with the results on elastic plates.

The upper limits of energy-carrying frequencies of the turbulent spectrum calculated by formula $f_1 = U_\infty/\delta$ are presented in Table 5.3. Comparing received values C_f for a rigid plate with know experimental data (Figure 5.7, curve 3), it is visible that experiences in a wind tunnel of value C_f are considered overestimated. The values of frequency f_2 at $k = 2\pi f/u = 1$ were calculated from spectra of longitudinal fluctuation velocity (see Figure 3.20). Here f_2 is the average frequency of energy-carrying frequencies. As is shown in Figure 3.20, the region of these frequencies corresponds to $k = 0.34 - 4$. The values of f_1 and f_2 are well agreed with each other if f_1 is calculated by the formula:

$$f_1 = 0.8U_\infty/\delta \tag{5.21}$$

and the upper limit value of f_2- by the formula:

$$f_2 = 4u/2\pi \tag{5.22}$$

where u is the velocity at the point where the spectrum curves were measured, and $0.8U_\infty$ is the phase velocity of disturbances in the turbulent boundary layer.

Note that in accordance with Figure 3.11 and formulae (3.39)−(3.41) minimal friction on elastic plates in the wind tunnel takes place when at small velocities (of Re); the following condition is satisfied:

$$f_3 = 0.48U_\infty/\delta \tag{5.23}$$

and for high U (or Re):

$$f_4 = 0.32U_\infty/\delta \tag{5.24}$$

When analyzing the measurements, the authors considered the correspondence of values $f_2 - f_4$ to values of f in Table 2.19, because the greatest absorption of turbulent fluctuation energy takes place when these frequencies coincide. The dynamic principle (see Section 1.6) is fulfilled better at plates 5 and 2a because the specified frequencies coincide better than at other plates. However, the conditions of coincidence of maxima of fluctuation energy of the boundary layer and damping of mechanical vibrations by the elastic material are not satisfied (see Figures 2.54, 2.55 and 2.62). In addition, these plates have considerable roughness. Other plates showed best results in the airflow when the indicated frequencies were close to each other. The lowest value of C_f is registered for plate 11, at which all conditions are satisfied. Plate 1a showed the anomalous low drag due to susceptibility of fluctuations of the near-wall shear.

In accordance with π-parameters, mechanical characteristics of the elastic plates are non-optimal in airflow. Calculations showed that in water flow the values of π-parameters over the same plates approached optimum, depending on Re numbers. Therefore, measurements of C_f on the same plates were executed in water flow. At low Re numbers, these plates were tested by the strain dynamometer *I* in the range of Reynolds numbers $2.5 \cdot 10^5 - 1.5 \cdot 10^6$ [286], and by the strain dynamometer *II* at high Re numbers, in the range of Reynolds numbers $1.8 \cdot 10^6 - 2.0 \cdot 10^7$ [287, 288].

In the region of Re numbers, where both dynamometers could measure ($1.0 \cdot 10^6 - 2.0 \cdot 10^6$) there is a discrepancy of experimental points caused by the measurement procedures [286, 287]. Interpolation of the experimental curves in this region of Re, as well as correction for a systematic error [286], permit the construction of the unified curve C_f (Re). The results of measurements on the elastic pates in water and airflow are presented in Figure 5.7.

As is shown in [30, 40, 310] the elastic surface influences all the transition stages, including the turbulent regime. As can be seen from Figure 5.7, all elastic plates, depending upon structure, increase the length of transition stages and, hence, the lower the critical Reynolds number grows. Nevertheless, at the stage of development of longitudinal vortices, the transition to turbulence (from lower Re number to higher one) occurs earlier than according to the hypothesis in [40]. This is due to the fact that the structural principle of interaction was not satisfied, with the exception of plate *10a* (curve *13*). An influence onto the turbulent boundary layer at high Re numbers was not observed, except for plate *6* (curve *12*), which satisfied the structural and dynamic principles at Re $= (2-10) \cdot 10^6$.

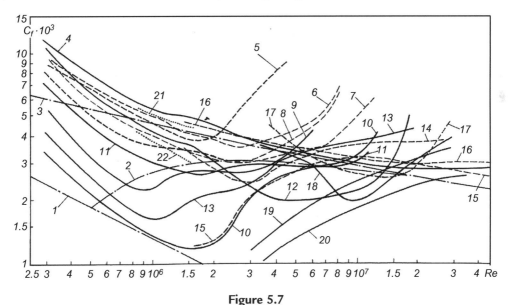

Figure 5.7
Friction drag coefficients on rigid and elastic plates: *1–3* — friction resistance at laminar, transition, and turbulent flows in a boundary layer over a rigid plate [144] (see Table 2.3); water flow: *4* — rigid plate; *5* — plate *11*; *6* — plate *8*; *7* — plate *7*; *8* — plate *1*; *9* — plate *3*; *10* — plate *4*; *11* — plate *9*; *12* — plate *6*; *13* — plate *10a*; *14,15,18–20* — Elastic cylinders (see Section 5.4); *16* — shark skin (plate *11a*); *17* — plate *1* (see Table 2.4); air flow (see Table 3.1): *21* — rigid plate, *22* — plate *11* (see Table 5.3).

All plates showed a substantial increase in drag, starting from a certain value of Re. It indicated that elastomers have insufficient durability of fastening on a metal substrate of an insert. At gaining speeds of towage there was increased discharge near to a surface of a towed elastic plate. This led to deformation of the elastic plates and in some cases an end face of the elastic plate came off. Therefore, practically for all tested elastic plates with mounting speeds of towage the coefficient of resistance increased. The structural analysis of dolphin skin, executed in the first part, has shown there are the special structural formations that cause interference of separation or deformation of the skin covering at fast movements. When this effect was insufficient for the prevention of deformation of the skin covering periodic fluctuations occurred in the skin, resulting in the emission of the braked liquid from near-wall areas of the boundary layer.

The greatest decrease of C_f was obtained on plate *4* in water; however, in the wind tunnel it showed negative effectiveness. This was because in water the effective roughness decreased due to thickening of the boundary layer. The best frequency correspondence, according to measurements in airflow, took place at $\text{Re} = (1-15) \cdot 10^6$. However, due to the increase in boundary layer thickness, the effectiveness of the plate is at higher Re numbers.

Plates *10a* and *6* (curves *13* and *12*) had the worst frequency correspondence compared to that of plate *4*, but their structural correspondence was the best (longitudinal ordered structure of the outer layer of plate).

The majority of the investigated plates were compliant. The elastic plates, according to π-parameters, were recalculated to higher Reynolds numbers (see, for example curves *19* and *20*). However, their effectiveness was limited by the fact that the structural principle of interaction was not satisfied (the outer layer had the transverse structure and generated plane waves in the boundary layer).

The obtained dependences indicate that the influence of an elastic surface on a boundary layer has a selective character. Within the range of towing velocities, the curves have minima at different Re numbers. The obtained results differ substantially from those obtained by other investigators, where the optimum has the opposite sign (see Chapter 1).

It is clear from most of the curves that the elastic plates influence the transitional boundary layer. The physical mechanism of this influence is considered in detail in [40, 305]. However, for the analysis of towing tests it is first necessary to pay attention to the proper fastening of the elastic plates onto metal substrates. At low speeds of towage and separating loads, the elastomer material works more effectively and interacts better with the boundary layer. At increased separating loads, the elastomer is stretched in a volumetric deformation, its parameters change and it ceases to interact with the boundary layer. At further increase of the towing speed, the elastomer undergoes such heave that its roughness becomes invalid. Sometimes it leads to the separation of the elastomer from its substrate and to the occurrence of flutter oscillations either over the entire elastic plate or in the area of the separated section. Thus, drag of the plate essentially increases.

Some curves indicated a substantial influence upon the turbulent boundary layer, the mechanism of which is explained in Chapter 3. To explain the nature of the influence of an elastic surface upon the turbulent boundary layer the authors suggest the hypothesis of frequency interaction. When an elastomer interacts with a flow part of the fluctuation energy is redistributed in the elastomer and boundary layer, which leads to a reduction in the intensity and frequency of bursts from the viscous sublayer.

As an explanation of the selective character of the influence of an elastic surface on a turbulent boundary layer, the hypothesis of frequency interaction is offered. It is known that any elastomer absorbs mechanical fluctuations. The degree of such absorption depends on the frequency of applied loadings (see Chapter 2). Comparing frequencies of the spectral function of turbulent fluctuation energy and spectral function of the energy dissipation coefficient in the elastomer, it is possible to predict the efficiency of the elastomer.

The essence of the suggested approach is demonstrated in Figure 5.8 [285]. The energy spectra (*I–III*) of fluctuation load perceived by the elastic surface at different Re numbers is scaled to the same level. The change of the spectral function of pulsations in the flow on elastic plates in comparison with spectra on a rigid wall confirms the existence of a mechanism of selection and redistribution of flow turbulent energy by a visco-elastic wall. Case I corresponds to the measurement of spectral characteristics at low flow speeds. In these experiments, the upper limit of the energy part of the spectrum does not exceed 100 Hz and cuts off the ascending branch of the absorption coefficient (curve 2, Figure 2.55b). Therefore, the energy of high-frequency fluctuations decreases more strongly than the energy of low-frequency fluctuations.

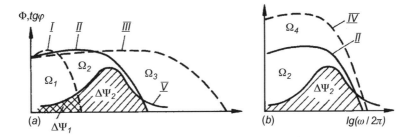

Figure 5.8

Spectra of $\Psi\,(\omega)$ in the boundary layer at different flow conditions *I–III* (*a*) and different levels of perceived energy for one mode of a flow — *IV, V*(*b*); *VI* — frequency law for energy absorption coefficient $tg\varphi(\omega)$ for elastic plate; *I–III* — flow regime at $\mathrm{Re}_I < \mathrm{Re}_{II} < \mathrm{Re}_{III}$ (see Section 5.6); Ω_1, Ω_2, Ω_3, Ω_4 — region of spectral functions; $\Delta\Psi$ — energy dissipated in the elastic plate; $\Delta\Psi_1$, $\Delta\Psi_2$ — regions of fluctuation energy of the boundary layer, absorbed by the plate at different modes of a flow.

With a widening of the frequency range of fluctuation loads (with increase in flow velocity) when the value of $tg\varphi\ (\omega)$ lies within the range of energy-carrying frequencies, i.e. $\omega/2\pi|_{tg\varphi=\max} << U/\delta_T$, the distribution of the spectral function of fluctuations must change substantially. It is corroborated by valleys in the spectra of longitudinal velocity fluctuations near the visco-elastic boundary (see Chapter 3). In this frequency range the extraction of fluctuation energy from the flow is strongest. It corresponds to the case *III* in Figure 5.8.

We can assume that the value of integral effect ξ is in proportion to the part of fluctuation energy dissipated in the wall. Integrating the spectrum of longitudinal fluctuations in the turbulent boundary layer with respect to the whole frequency region $\omega/2\pi|_{tg\varphi=\max} < U/\delta_T$, we can conclude that the maximal hydrodynamic effect takes place when the square $\Delta\Psi_i$, which is cut off by the curve $tg\varphi(\omega)$ from the square Ω_i of the spectrum of fluctuation load on the rigid wall $\Phi(\omega)$, is maximal, i.e.:

$$\xi = \xi_{max}, \text{ when } d\left(\frac{\Delta\psi(\omega)}{\Omega(\omega) - \Delta\psi(\omega)}\right)/d\omega = 0 \qquad (5.25)$$

On the basis of (5.45), we can see from Figure 5.7 that the maximal effect should be in case *II*: $\xi_I < \xi_{II}$, $\xi_{III} < \xi_{II}$. Hence $\xi_{II} = \xi_{max}$.

The waviness of an elastic surface generated by the pressure fluctuations in the boundary layer can be considered equivalent to a certain roughness. The flow velocity being increased, the pressure fluctuations amplify and the amplitude of the elastic boundary increases. If the amplitude exceeds the level of the equivalent roughness, the energy balance will change in the reverse direction. The generation of the fluctuation energy by the roughness will exceed the damping of the fluctuation energy by the elastic surface. Considering the element of the compliant wall under the loading $p'(\text{Re})$, we have:

$$\Delta h = \frac{\sqrt{\langle p'^2 \rangle}}{c} \qquad (5.26)$$

where $c = E'/h$. Using the known condition [381, 445] for the allowable roughness height for a hydraulically smooth surface:

$$\begin{aligned} &\text{Re}_{k(accept.)} = 100. \\ &\text{we have} \quad k_{(accept.)} = \Delta h_{(accept.)} = \Delta y_{(accept.)} = 100\nu/U \end{aligned} \qquad (5.27)$$

Then the minimal allowable rigidity of an elastic plate to hold the hydraulically smooth (static) surface can be calculated as follows:

$$c_{\min} = 0.005\rho U^3\nu^{-1}\text{Re}^{-0.3} \qquad (5.28)$$

Following [453], we can determine the mean value of pressure fluctuation from the expression [172, 401]:

$$\sqrt{\langle p'^2 \rangle} = 0.5\rho U^2 \, \text{Re}^{-0.3} \qquad (5.29)$$

Knowing the rigidity of an elastic plate c_{min} determined from formula (5.46) and the elastomer thickness, it is possible to calculate the theoretical module of elasticity E'_{min}, at which the plate under investigation remains hydraulically smooth. The ratios $E'_{\text{min}}/E'_{\text{meas}}$ were determined for all tested plates when their efficiency dropped to zero. The values of E'_{meas} were taken from Tables 2.19 and 2.20. For all elastic plates, with an error of order 3%, we obtained:

$$E'_{\text{min}}/E'_{\text{meas}} = 1.5 \qquad (5.30)$$

The obtained result can be used as a limit criterion for rigidity when selecting elastic walls intended to reduce turbulent friction. The value of dynamic roughness is also considered in [453, 187].

Optimum mechanical characteristics of an elastic plate under certain conditions of application can appear insufficiently stiff to interfere with the occurrence of a wavy roughness on its surface. According to the hydrobionts approach, the design of such a plate should, for example, provide an arrangement of longitudinal rigid internal walls, which interfere with shear deformations of the elastic plate surface.

5.4 Drag of Cylinders in the Longitudinal Flow

The experiments with towing the elastic plates did not show an appreciable effect at high Reynolds numbers. There are reasons for this. In particular, we failed to achieve perfect smoothness of the plates. There were irregularities on the surface of plates that led to a local pressure gradient and swelling of the surface at high velocities. Some parts of plates separated from the base. Therefore, it was not possible to reach a steady positive effect at a turbulent mode of flow in the boundary layer at high Reynolds numbers.

It is known that the integral characteristics of the boundary layer on a plate and cylinder does not practically differ if the cylinder is long and its radius is less than the thickness of the boundary layer, i.e. $2 \, \delta/d \ll 1$. Thus, the tensiometric cylindrical insert should be located at such distance from the nasal parts that it is in the area of an almost constant and zero pressure gradient. To avoid the above-mentioned disadvantages, the investigations of cylinders coated with elastic coatings were carried out by strain dynamometers *IV* and *V* (see Figure 5.1).

Table 5.4a: Geometric parameters of cylinders according to Figure 5.9.

Test Series and Model Designation	l_{cyl}/d	l_{cyl}/L_i	L_2/d	l_2/d	L_s/d	l_s/d	l_{ef}/d	k_2/d	H_2/d
$A_{1,2,3,4}$	0.84	0.535	1.57	0.73	—	—	—	—	9.09
A_5	0.84	0.328	—	—	2.56	1.72	—	—	—
$B_{1,2,3,4}$	6.45	0.898	7.18	0.73	—	—	5.07	1.39	14.7
B_5	6.45	0.789	—	—	8.17	1.72	5.07	2.38	—
$C_{1,2,3,4}$	11.25	0.838	12.00	0.73	—	—	9.87	1.39	19.5
C_5	11.25	0.865	—	—	13.00	1.72	9.87	2.38	—
D	5.65	0.916	6.07	0.514	—	—	4.92	0.754	10.1

Table 5.4b: Geometric parameters of cylinders according to Figure 5.9.

Test Series and Model Designation	n/d	m/d	a/d	a/c	b/d	b/c	b_1/c	c/d
A, B, C	7.52	2.82	0.40	1.11	3.80	10.56	4.45	0.36
D	4.0	1.83	0.23	1.11	2.17	10.56	4.45	0.206

Table 5.4c: Geometric parameters of cylinders according to Figure 5.9. This table shows the relation of the length of the probationer part of the model of the axisymmetric cylinder (it is shown in figure 5.9 by the vertical shaded lines) to its diameter.

i	1	2	3	4	5
l_i/d	0.55	0.73	1.30	4.50	1.72

These investigations were made with high-speed towing. Previously, experiments were conducted with the purpose of defining the influence of geometric parameters of models on hydrodynamic characteristics: length, diameter of the cylinder and the nasal part shape. The measurement procedure and results are presented in [285–288]. At the beginning, curves of C_F (Re) on the rigid surface were obtained using dynamometers *IV* and *V*. The geometric parameters of the tested cylinders and their arrangement satisfied the recommendations [203, 424, 451]. The details are presented in Table 5.4a, b, c and Figure 5.8. The letters denote the test series (different length of cylinders) and digits 1–5 correspond to the number (shape) of the nose cone. The distance from the cylinder axis to the water surface (h is embedding) and to the bottom (h_M) are presented in Table 5.5.

The drag of the cylinders can be written as follows:

$$C_x = C_{Fcyl} + C_{xw} + C_{xb} \tag{5.31}$$

where C_{Fcyl}, C_{xw} and C_{xb} are the viscous, wave, and base drag.

Table 5.5: The relative embedding models during towing in a high-speed towing tank.

Test Series and Model Designation	h/d	h_M/d	h/L_2	h/L_5	h_M/L_2	h_M/L_5	$10^2 S_{ef}/S$
$A_{1,2,3,4}$	4.5	4.7	2.87	—	3.0	—	12.02
A_5	4.5	4.7	—	1.76	—	1.84	8.49
$B_{1,2,3,4}$	4.5	4.7	0.03	—	0.655	—	2.56
B_5	4.5	4.7	—	0.55	—	0.575	2.35
$C_{1,2,3,4}$	4.5	4.7	0.375	—	0.39	—	1.53
C_5	4.5	4.7	—	0.346	—	0.36	1.45
D	2.44	2.9	0.4	0.48	—	—	2.73

In [285−288] an estimation of the components is given. The greatest components are viscous drag. The bottom drag depended upon the size of the slot in the end of the cylinder *9* (see Figure 5.9, section *I* and *II*) required for the kinematics isolation from the frame of dynamometer *6* and *7* by means of elastic membranes *8*. Depending on the distribution of pressure along the cylinder, the contribution of base drag to the general drag of the model varies. The nose cone shape influenced the pressure distribution along the cylinders and, to a certain extent, the wave drag.

In Figure 5.10, the cylinder drag measurement results are presented, and are dependent on the geometrical parameters. Measurements in series *A*, *B* and *C* are made by the strain dynamometer *IV* (see Figure 5.10) and in series *D* by dynamometer *V*.

In Figure 5.10, letters designate the series of experiments (elongation of the model) and by digits mean number (the shape of nose). The shape of the nasal cowls was developed in the form of a semi-ellipsoid with the ratio of semi-axes (0.5 d/l_i): *1* − 1/1.1; *2* − 1/1.46; *5* − 1/3.44; and also with the nasal needle l_n/d: *3* − 0.58; *4* − 3.78. The diameter of the base of the needle was 0.16 *d* in both cases.

Viscous drag can be presented in the form:

$$C_{Fcyl} = C_{F0} + C_P \qquad (5.32)$$

where C_{F0} is friction drag of an equivalent plate, C_P viscous pressure drag. $L/d > 4$ for series *B*, *C* and *D*, therefore:

$$C_{Fcyl} = (1 + k_{form})C_{F0} \qquad (5.33)$$

where k_{form} is the coefficient of the influence of surface curvature of the axisymmetric model, which depends on elongation of the model.

Base drag C_{xb} is caused by design features of the tensometers (see Figure 5.10) and depends on the pressure coefficient in the area of the end slot. A detailed analysis of

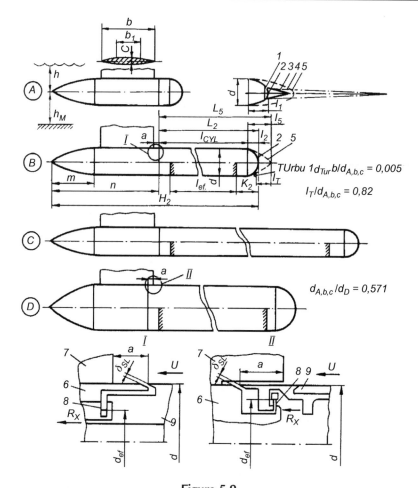

Figure 5.9
Scheme of geometric parameters of cylindrical strain-gauge dynamometers.

definition C_{xb} is performed in [285]. Considering that a comparative experiment was carried out its contribution was approximately identical in each series of experiments.

Wave resistance C_{xw} depends on many parameters: shape of the case (elongation), shape of contours and coefficient of completeness of displacement, Froude number, and the influence of the towing tank walls and relative embedding of models (i.e. relative distance of a model from the bottom), and so on.

In the series of tests A, the wave component was practically absent, as the Froude numbers were greater than 1.4. The coefficients C_P and C_{xb} (in presence of the nose needle) influenced the model drag. The nose needle in series A did not appreciably influence the model drag. The long nose fairing $A5$ essentially reduced C_{xb} as well as the total drag of the

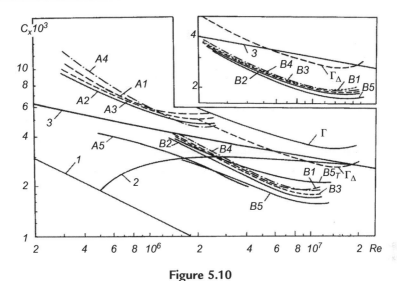

Figure 5.10

Dependence of drag coefficient of the longitudinal streamline cylinders on the Reynolds number. Curves 1−3 are the same as in Figure 5.2. Other designations are in the text.

model. At greater elongation of the entire model in series *B*, this nose fairing did not strongly influence the base drag, and at models in series *C* its influence was insignificant. The curve in series $B5_T$ corresponds to the installation of a turbulizer on the cowl.

However, with increasing elongation of cylinders, wave drag became noticeable on certain modes because relative embedding of models h/L_i decreased and the relative distance to the towing tank bottom h_M/L_i was affected.

The cylinder elongation being increased, the wave drag at certain velocities became significant. In the series *B5*: $Fr_L = 0.706$ at $Re = 1.5 \cdot 10^6$, $Fr_L = 1.412$ at $Re = 3 \cdot 10^6$; in series *C5*: $Fr_L = 0.56$ at $Re = 2.4 \cdot 10^6$, $Fr_L = 1.12$ at $Re = 4.8 \cdot 10^6$; in series *D*: $Fr_L = 0.5$ at $Re = 2.0 \cdot 10^6$, $Fr_L = 1.1$ at $Re = 4.4 \cdot 10^6$.

At low values of Re, Froude numbers were close to the critical value ($Fr_{cr} = 0.5$). This explains the shape of curves.

5.5 Friction Drag of Elastic Cylinders in a Longitudinal Flow

For manufacturing cylindrical surfaces of elastomers, a device making a strip of an elastic material with a specified uniform thickness has been designed. By means of a special gadget these elastic strips were pasted to a cylindrical pipe, which was fixed on tensometer *IV* (see Figures 5.1d and 5.4), variant *B* (see Figure 5.9). Elastic cylindrical surfaces of the type of plate *1* made of monolithic materials FL and various variants of FE (see Table 2.3) have been made. Experiments were performed in the high-speed towing tank (see Figure 5.6). Before and after each series of tests, the tensometer was calibrated. All calibration measurements were obtained in the form of linear dependences of motion of the tensometer on the imposed loading. Figure 5.11 illustrates dependences of resistance force of a cylindrical insert of model at speed of towage changed within the limits of 2−18.5 m/s. Dash-dotted curve and square points mark the measurement on the rigid standard. Curve *1* and round points characterize the elastic plate made of PE of 10 mm thick.

Curve *2* and triangles correspond to the cylinder made of the same material, but its thickness was approximately 3 times less than the thickness of the surface corresponding to curve *1*. The spread of points is approximately identical for all curves, but the spread decreased in tests of elastic surfaces. The efficiency of the elastic surface (curve *1*) was

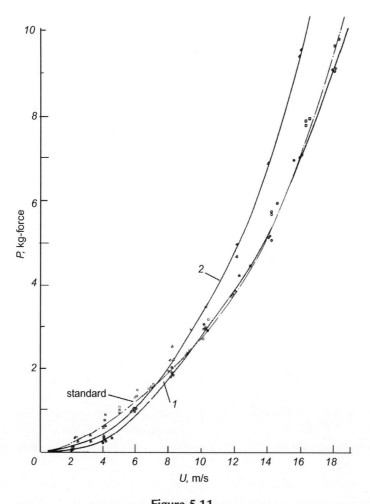

Figure 5.11
Dependence of the short model resistance force on the speed of towage (1 kg-force ≈ 9.8 N).

essentially better than the standard up to a towing speed of 9.5 m/s and then drag of the surface became same as at the standard. The thinner plate at a speed of towage of 7 m/s lost efficiency and further, its resistance was essentially greater than at the standard.

The obtained results are plotted in Figure 5.12 in dimensionless form. Curve *4* corresponds to drag of the reference cylinder made of organic glass (dash-dotted curve in Figure 5.11). Curve *5* corresponds to curve *1*, and curve *6* to curve *2* in Figure 5.12. The same curves *5* and *6* correspond to curves *15* and *14* on Figure 5.7. It appeared that the same material has peak efficiency at different Reynolds numbers, depending on thickness.

At greater thickness of the elastomer, its rigidity becomes less and, consequently, it should be more effective at smaller Reynolds numbers. It was expected that a thinner covering would be more effective at greater Reynolds numbers. However, although a minimum of curve *6* (see Figure 5.12) was shifted to greater Reynolds numbers the coating was not effective because of the poor tensometer fastening on the metal pipe.

Composite elastic coatings (see Table 2.3) were mounted on the long model of tensometer *IV* (see Figures 5.1d and 5.4) and variant *C* (see Figure 5.9 and Table 5.4a, b, c).

As in the first case, the diameter of the model with an elastic coating was the same (0.1 m), but the length of the cylindrical part was almost twice as great. Figure 5.13 represents

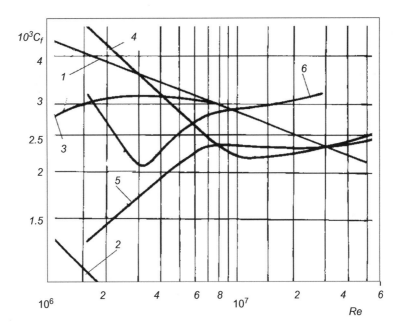

Figure 5.12

Dependence of resistance coefficient on Reynolds number: curves *1–3* — the same as in Figure 5.7; curves *4–6* — see explanation in the text.

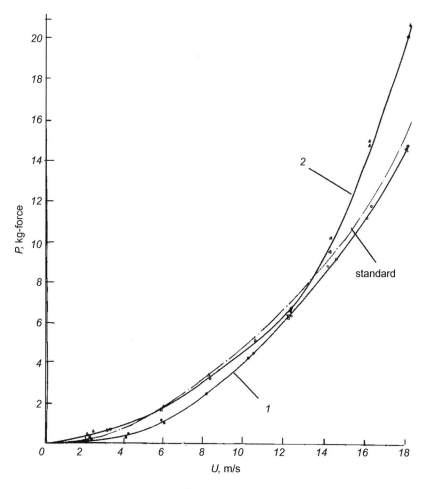

Figure 5.13
Dependence of the long model resistance force on the speed of towage.

results of repeated series of tests of the drag of elastic surfaces in dimensional form. Curve *1* corresponds to the coating made of the monolithic elastic material PE 6 mm thick, and curve *2* to the covering of type *7* in Table 2.3. The dash-dotted curve corresponds to towage of the standard. Figure 5.13 represents the results in dimensionless form. Curve *4* corresponds to resistance of the reference cylinder (standard). Curve *7* corresponds to curve *1* in Figure 5.13 and curve *20* in Figure 5.7, and curve *6* corresponds to the coating of the type on plate *9* in Table 2.3. The same curve in Figure 5.7 corresponds to curve *19*. Curves *5* and *8* in Figure 5.14 correspond to the coatings used in the scheme of Table 2.3 according to plate *8* and *6*. The coatings made of elastic material FL (curve *5*, Figure 5.14) had the glued seam located across the flow at the middle of the cylinder. The surface made

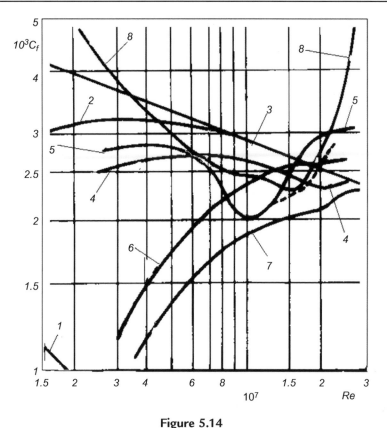

Figure 5.14

Dependence of the resistance coefficient on the Reynolds number: curves *1–3* – the same as in Figure 5.7; curves *4–7* – see explanation in the text.

under the scheme of plate *6* in Table 2.3 appeared insufficiently strong and poorly fixed to the substrate. The same surface in Figure 5.7 is shown by curve *18*.

From the analysis of numerous experiments (only some results are mentioned here) it is possible to draw the following conclusions:

1. The results obtained on tensometer *IV*, variant *B* and *C* (see Figure 5.9, short and long cylinders) have, as a whole, shown steadier results in comparison with experiments on the same plates at towage on tensometer *II*. The efficiency of the investigated materials was better for tests on tensometer *IV*.
2. The best results were achieved on a homogeneous material, which had better resistance to separation loadings at greater speeds of towage.
3. Control of mechanical properties of materials by heating a conductive fabric inside the coating during towage did not give a positive result.

4. This indicates that smoothness of the coating and its durability were critical, because with an increase in towing speeds the boundary layer became thinner and separation loads essentially increased. Therefore, a small degree of roughness becomes a larger feasible roughness. With an increase of speed of towage the coating on the end of the cylinder separated and essentially increased resistance.

5. Tests on the short and long cylindrical tensometers did not allow the reaching of greater Reynolds numbers and/or the defining of efficiency of a coating at the developed turbulent boundary layer.

6. All the tested variants of elastic coatings had a form of dependence of coefficient of resistance, which corresponds to a transitional regime of flow in the boundary layer and is equidistant to curve 2 for the rigid standard. Thus, corresponding critical Reynolds numbers are essentially shifted to greater Reynolds numbers. This means that even imperfect manufacturing techniques of coatings and insufficiently effective designs of coatings can essentially stabilize the boundary layer and reduce friction drag. The investigated types of elastic plates (see Table 2.3) were effective up to $Re = 3 \cdot 107$.

Based on these conclusions, the authors have developed new technology for the manufacturing of coatings and new materials and designs of coatings that better model the structure of high-speed hydrobionts.

All the subsequent experimental investigations were performed on elastic cylindrical inserts by corresponding tensometers made industrially with specially developed techniques and compositions. The material of the coatings was the same, PU; its thickness, length of the cylinder and compounding of material varied. It is known that PU possesses an unlimited variety of prescribed structures. Thus, compositions of PU were used that were able to provide the necessary durability of fastening on the metal substrate and that had the mechanical characteristics to provide the greatest effect under the given test conditions. Special attention was given to the profiling of the metal substrate for the coating and to the fastening of coatings at the transversal joints with the tensometer case. These measures were necessary to prevent separation of the coating from the substrate at greater speeds of towage.

Figure 5.15 represents a photograph of such elastic cylinders intended for towing tests on tensometer *IV*, variant *C* (see Figure 5.9, long cylinders). Cylinders are located in a box on special fixings. There is a small optical distortion of the cylinder shape. The coating color varies according to the structure of the composition. Figure 2.31 represents a sample of such coatings. A photograph of tensometer *V* with an elastic cylindrical insert is shown Figure 5.16. In all subsequent tests a turbulence promoter was established in nasal parts of models. This has allowed the production of a turbulent boundary layer on models at all speeds of towage.

Figure 5.15
Photograph of elastic cylinders intended for towing tests on tensometer *IV* (see Figures 5.1d, and 5.4), variant *C* (see Figure 5.9, long cylinders)

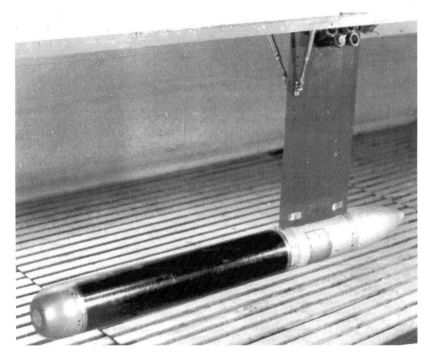

Figure 5.16
Photograph of the tensometer *V* (Figures 5.1e, 5.5, 5.9D) with an elastic coating.

It was necessary to achieve a positive effect at $Re > 5 \cdot 10^6$ in order to get a decrease in drag in the turbulent boundary layer compared with the data in Figures 5.2, 5.12 and 5.14. In accordance with formulae (5.41)–(5.44) and π-parameters (see Table 2.6) the elastomers PU-3 (see Table 2.9) were chosen for testing. The mechanical characteristics of PU are listed in Tables 2.9–2.12, on Figures 2.55 and 2.56 and in Section 2.8. The elastomer PU-3B was tested by strain dynamometer *IV*. The dynamic behavior of elastomer PU-3B differs from that of compliant elastomers FPU-1 and FE-3 (see Tables 2.9, 2.10 and 2.19). According to Table 2.19 and Figure 2.55, the elastomer's natural frequency is over 200 Hz and *tg*φ has a maximum even at higher frequencies. Therefore, according to the method presented in Section 5.3 the elastomer must be effective in the above-indicated region of Reynolds numbers.

The influence of elongation of the elastic cylinder on its efficiency has been investigated. Although the cylinders were not long, the thickness of the boundary layer varied along the cylinder length. The region of energy-carrying frequencies of pressure turbulent fluctuations spectra also must change according to (5.21)–(5.24). Therefore, if the mechanical characteristics of the elastomer are constant along the length, only in a particular place on x will there be the greatest coincidence of the range of frequencies of elastomer fluctuations at the maximal values of a *tg*φ with a similar range of energy-carrying frequencies of the turbulent boundary layer (the dynamic principle of interaction of disturbances). Therefore, two cylinders having different lengths but identical elastic coverings will have different efficiencies. The total efficiency of a short cylinder is higher than the efficiency of a long one because the elastic material will effectively interact with the boundary layer on a greater relative length of the elastic cylinder.

These conclusions agree with [315] and were investigated. Disturbances were introduced on two cylinders of different lengths, *B* and *C* (see Figure 5.8). The elastomer thickness was $h/d = 0.06$; $h/L_2 = 0.00836$ for cylinder *B* and 0.005 for *C*. The obtained curves $C_x(Re)$ are presented in Figure 5.17. It is seen that the effectiveness of the elastic inserts depends on the nose fairing shape. The effectiveness, in accordance with (5.21)–(5.24), was maximal in a certain range of Reynolds numbers.

The efficiency of the short cylinder was higher in a wider range of Reynolds numbers than that of the long cylinder. On short and long reference cylinders the drag curves differed depending on the nose fairing shape. At towage of the elastic cylinders, the short xiphoid tip ($B3_e$, $C3_e$) had no advantage in comparison with the truncated fairing ($B1_e$, $C1_e$); at the same time, the long xiphoid tip on the long cylinder ($C4_e$) was more effective at towage of the long cylinder than the truncated fairing ($C1_e$), and approached the drag law of the most effective fairing $C5_e$. These results differ slightly from known experiments on bodies with xiphoid tips (see Chapter 6). The result obtained for the flow about the long elastic cylinder with the long xiphoid tip is an example of the combined method of drag reduction.

The degree of drag reduction did not exceed 20–30%. Only one condition of interaction of a flow with an elastic surface has been examined. To increase the effectiveness of elastic plates all principles of complex interactions (see Chapter 1) must by satisfied and all π-parameters determined (see Tables 2.5 and 2.6).

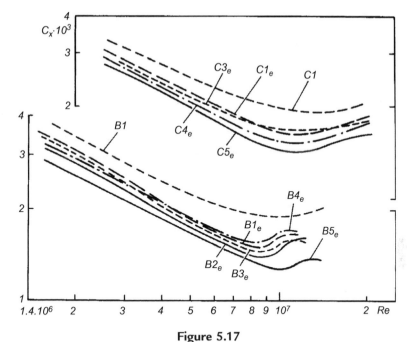

Figure 5.17

Friction drag coefficient of rigid (*B1, C1*) and elastic (*B1$_e$*–*B5$_e$*) cylinders depending on Reynolds numbers. Designations are in Figure 5.16 and Table 5.4a, b, c (*L$_C$* > *L$_B$*).

Table 5.6: Elastomer thickness on cylinders.

Number of the Coated	Elastomer	h/L
1	ПУ-3A	0.01062
2	ПУ-3A	0.00531
3	ПУ-3A	0.00265
4	ПУ-3B	0.01062
5	ПУ-3B	0.00531
6	ПУ-3B	0.00265
7	ПУ-3C	0.01062
8	ПУ-3C	0.00531
9	ПУ-3C	0.00265

The measurements were made by the strain dynamometer *V* to estimate the effect of the elastomer's mechanical parameters upon its frequency properties. The obtained results are presented in Figure 5.17. Elastomer characteristics are listed in Table 5.6. As is seen from Table 2.9 the rigidity of elastomers increases from *A* to *C* as the thickness decreases. The results show that in series *A* and *B* the effectiveness of elastomers practically did not depend on their thickness. Also, the differences in their mechanical characteristics did not influence the efficiency. A difference is revealed for the more rigid elastomers in series *C*, for which efficiency depends on the elastomer rigidity or the frequency range of the loss tangent $tg\varphi$.

All elastomers tested showed drag reduction within a wide range of Reynolds numbers. The reduction did not exceed 35%. In addition to the previous positive results, the positive effect obtained for the tests of elastic coatings on tensometer C indicates the correct choice of mechanical characteristics of the elastomer for manufacturing elastic cylinders. Furthermore, the result was made possible by the high quality of the outer surface of the elastic cylinders and by the correct design of the metal substrate. It provided strong cohesion of the elastomer with the metal substrate. Therefore, higher loadings at rising speeds could not deform or separate the elastomer from the substrate. It was very important to ensure strong fastening of the elastomer the in area that joins the cross-section end faces of the model's metal case. It was necessary to pay special attention to this aspect of the design. The only exception was the cylinder *9* (PU-3*C*), the elastic coating of which separated from the substrate at greater speeds of towage.

The integral effect coefficients are plotted in Figure 5.18b:

$$\xi(x) = (C_{Frigid} - C_{Felastic})C_{Frigid}^{-1} \tag{5.34}$$

where $C_f = 2\tau_\omega/\rho U_\infty^2$, as well as the dependence of the range of energy-carrying fluctuations of the turbulent boundary layer, $U/\delta_{turbulent}$, on the Reynolds number. Indices F_{rigid} and $F_{elastic}$ designate corresponding tests for rigid and elastic cylinders. Thus, the thickness of the boundary layer at $x = l_i + 0.5l_{elasic}$ was calculated as:

$$\delta_T(x) = 0.37x \, \mathrm{Re}_x^{-2} \tag{5.35}$$

The condition of maximal efficiency (5.25) can be simplified. With sufficient accuracy for engineering estimations it is acceptable to assume that the maximum of friction reduction of the given elastic insert will occur at such mode of flow when the top border of the range of energy-carrying frequencies becomes approximately equal (or a little greater) to the frequency corresponding with the a maximum of the elastomer absorption coefficient:

$$U/\delta_T \approx \omega/2\pi|_{tg\varphi(\omega)} = \max \tag{5.36}$$

The registered maximums of effect are correlated with the dynamic properties of the material and the frequency of the fluctuation load.

The revealed effect can be demonstrated on the elastic cylinder *4* (see Table 5.6) made of polyurethane PU-3B. As the velocity increases up to $\mathrm{Re} = 5 \cdot 10^6$ and the range of energy-carrying fluctuation loads widens to 400 Hz at $x_{elastic} = l_{05}$ (curve *15* in Figure 5.18), the Young modulus $E'(\omega)$ and, therefore, the rigidity of the wall increases (curve *5*, Figure 2.55). Hence, the relative level of susceptible energy decreases. Since the loss coefficient of material PU-3B remains constant at increasing ω over the given frequency

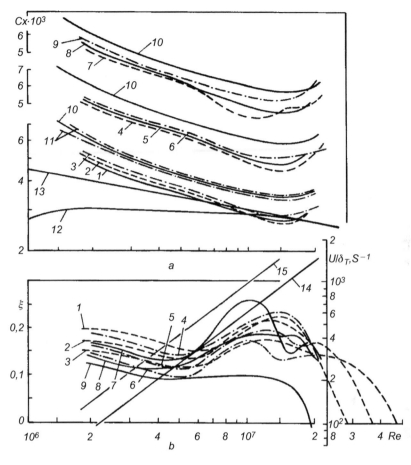

Figure 5.18

Friction drag coefficient (*a*) and effectiveness (*b*) of elastic (*1–9*) and rigid (*10*) cylinders depending on the Reynolds number. Designation of the curves are in Table 5.6. Curve *11* — root-mean-square error in friction coefficient, *12, 13* — friction drag of plate at transition and turbulent flow [446], *14, 15* — dependence of U/δ_T (Re) in the middle and at the end of elastic insert.

range (curve 5, Figure 2.55), the relative part of the turbulent energy dissipated into the elastic wall decreases. Therefore, the integral effect ξ (Re) decreases at Re $\rightarrow 5 \cdot 10^6$.

At frequencies that corresponded to the mode of flow at $5 \cdot 106 < \text{Re} < 1.4 \cdot 107$, the growth rate of $E'(\omega)$ remained same (or fairly decreased), while the coefficient of losses in the material $tg\varphi(\omega)$ sharply increased. Simultaneously, the spectral density of the frequencies corresponding to the maximal coefficient of absorption increased. Thus, the hydrodynamic effect should increase at integration over the whole range of frequencies. That was observed in tests. Thus, the maximum of the effect corresponds in frequency to a maximum of loss coefficient. On the one hand (see Figure 5.18, curves *4* and *15*) ξ(Re) = max at $Re_L \approx 1.4 \cdot 10^7$ and $[U/\delta_T (Re)]_{x=l(0.5)} = 1400\ Hz$; on the other hand (see Figure 2.55, curve *5*), ξ (Re) reaches a maximum at about 800 Hz. As is apparent from the plot $tg\varphi(\omega)$ (see Figure 2.55), the range of energy-carrying part of the spectrum of fluctuation loadings in this case overlaps the bell-shaped frequency dependence of loss coefficient of elastomer PU-3B, in a similar way to case *II* shown in Figure 5.8a. At greater frequencies, function $E'(\omega)$ reaches a constant value, while $tg\varphi(\omega)$ decreases. At a corresponding regime of flow (Re > $1.4 \cdot 10^7$), ξ(Re) also begins to decrease. It corresponds to the scheme in Figure 5.8a, case *III*.

From the experimental results, the conclusion that can be drawn is that the integral effect of elastic inserts made of the same of material is proportional to the geometric parameters characterizing their rigidity and level of turbulent energy absorbed by the material. For example, $h_4 > h_5 > h_6$ for PU-3B on the cylinders 4−6 (see Table 5.6), consequently $(\xi_{max})_4 > (\xi_{max})_5 > (\xi_{max})_6$ (see Figure 5.18). A similar correspondence is observed for modifications A and C.

5.6 Influence of Polymer Additives on the Friction Drag of an Elastic Plate

Investigations on the polymer addition effect upon the friction drag of the cylinders are described in [48, 305] and in Chapter 6. As in other papers, drag reduction is achieved by altering the concentration and consumption of the injected polymer solution. In [48], measurements of longitudinal averaged and fluctuation velocity components in the boundary layer were made by means of a laser anemometer. It registered the thickening of the viscous sublayer of the turbulent boundary layer and the decrease in longitudinal fluctuation velocity and near-wall shear velocity.

These results, by analogy with the data in Chapters 3 and 4, can be explained by the changing of the shape and size of the vortices in the near-wall region, as well as by the change in the cascade process of vortex dissipation in the boundary layer. In [225], it is shown that drag reduction is proportional to the increase in vortex scale near the wall. The spectral characteristics showed, as do the data in Sections 3.4, 3.6 and 4.4, that the energy decreased at low wave numbers and it increased at high wave numbers compared with the rigid standard.

Taking into account the similarity of the results in using the elastic plates and the polymer additions we can expect similarity in the interaction mechanism. The effectiveness of the elastic plates depends on coincidence of the energy-carrying frequencies of the boundary layer with the self-resonant frequency of the elastomer. The polymer solutions in the boundary layer change the level and position of the energy maximum with regard to the wave numbers. This causes the change in the resonance interaction between the flow and surface.

In [305] a hypothesis was advanced that a boundary layer, as a body having some elasticity, has a range of self-resonant frequencies limited by a neutral curve. In accordance with the hypothesis, the energy-carrying frequency region characterizes the region of self-resonant frequency of the turbulent boundary layer. Dissipative properties of the fluid influence the region depending on the flow velocity and a mode of flow in the boundary layer. Therefore, the polymer solution must change the region of natural oscillations of the boundary layer and the condition of resonance interaction with the elastic surface.

To check the stated assumptions, measurements of friction drag on the elastic plates were being made. Polymer solutions were injected into the boundary layer on the model through one nasal slot. Measurements of friction on the rigid cylinder at various Reynolds numbers have been used as reference measurements.

The elastic coating was made of $3 \cdot 10^{-3}$ m thick polyurethane PU-3B ($h/L = 0.00265$ according to Figure 5.9d and Table 5.6) with a density 1250 kg/m^3. The outer diameter of the elastic cylinder was equal to that of the rigid one. The dynamic visco-elastic parameters at $0 < \omega < 320$ s^{-1} are presented in Section 2.8. The static and dynamic Young moduli of the elastomer are $1.6 \cdot 10^3$ kPa and $5 \cdot 10^3$ kPa. The mechanical loss coefficient is 0.55 at frequencies up to 100 s^{-1}, and 0.7 at frequencies up to 300 s^{-1}.

The polymer used was the solution of polyethylene oxide (PEO-WSR-301) with molecular weight $M_\omega = 4 \cdot 16^6$ and mass concentration 10^{-3} (1000 ppm). The polymer solution was injected into the boundary layer through the ring slot $3 \cdot 10^{-4}$ m wide in the nose cowl. The angle between the injection direction and cylinder axis was 20°. The slot edge was at $0.191d$ from the leading critical point. Consumption of the polymer solution was varied by regulation of pressure in the boosting duct by air from a feed tank. Emission was carried out by means of an electromagnetic valve. Measurements on the rigid and elastic cylinders were made in the towing tank (see Figure 5.19) by the strain dynamometer V (see Figures 5.1e, 5.5, 5.9D and 5.16) at towing velocities of 2.0–22.0 m/s. The procedure used for the investigations with polymer solutions is presented in [48, 305].

The series of experiments A and B were made on rigid and complaint surfaces. They differed with the amount of polymer solution injected into the boundary layer. The value of the volume flow rate coefficient C_q is presented in Figure 5.19 and is dependent on the Reynolds number. The coefficient was determined by formula $C_q = Q/US$, where $Q = V/t$ is flow rate of polymer solution, U is the towing velocity, S and L are the slot square and length of the cylinder, V *is* volume of the injected liquid, and t is time of injection.

In comparative experiments involving injection of polymer solutions it is very important to sustain a constant pressure in the injection system. However, there was no possibility of

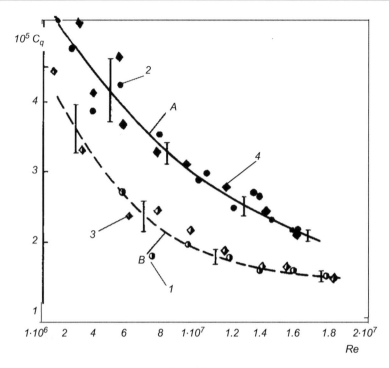

Figure 5.19
Flow rate coefficient of the polymer solution at different Reynolds numbers: A, B flow rates of polymer solution; 1, 2 – rigid and 3, 4 – elastic cylinder. Designation of points as in Figure 5.21.

regulating the pressure with adequate accuracy during experiments. When the values of volume flow rate of a liquid obtained in the experiments were ordered depending on the Reynolds number, a certain spread of these data was revealed. Curves A and B of these flow rates are shown simultaneously for rigid and elastic cylinders. If these average curves are plotted separately for each kind of cylinder, the spread of points is small. Averaging over all the points led to an increase in spread, shown in Figure 5.19 by vertical solid lines. It was found that the higher the speed of towage, the less was the dispersion of points of measurement (error). Therefore, the direct comparison of curves of drag of various cylinders must be done with a measure of reserve, as at each point of the curve flow rates differ depending on the speed of towage.

Despite the stated disadvantage, it is possible to analyze the curves of drag in Figure 5.20. It is possible to define the flow rate of polymer for each Reynolds number in Figure 5.19. In the flow on the elastic cylinder without the injection of polymer solution (curve 6, Figure 5.20), the reduction of drag relative the rigid standard is obtained. The reduction of drag is less than according to Figure 5.18 because in this series of experiments the forward ring slot, which is an additional turbulent trip of the boundary layer, was opened. For the

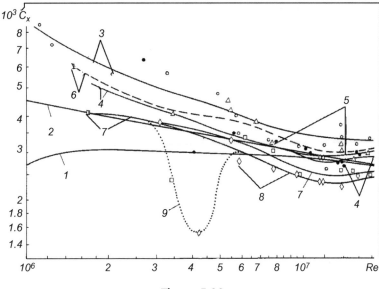

Figure 5.20

Combined method of drag reduction. Coefficient of drag: *1, 2* — a flat rigid plate at a transitional and turbulent boundary layer; *3* — the rigid cylinder (standard); *4, 5* — rigid cylinder at injection of polymer solutions with rates A and B (see Figure 5.19); *6* — elastic cylinder without injection of polymer solutions; *7, 8* — elastic cylinder at injection of polymer solutions with rates A and B. *9* — abnormal drag reduction.

tests of injection of polymer solutions on the rigid cylinder (curves *4* and *5*) drag reduction was almost same as on the elastic cylinder without injection of polymers (curve *6*). Change of the polymer rate at towage of the rigid cylinder did not strongly influence drag reduction (compare curves *4* and *5*).

The injection of the polymer solution on the elastic cylinder (curves *7* and *8*) has an essentially increased effect. At greater values of the polymer rate (curve *8*), the effect increased in comparison with the rigid cylinder (curve *4*) in a wide range of Reynolds numbers.

Abnormal drag reduction (curve *9*) was revealed during the experiments on elastic cylinders (curves *7* and *8*). Unfortunately, these experiments have not been repeated. However, a similar case has been registered during experiments on elastic plates in a wind tunnel (see Chapters 3 and 4). Based on experimental research on live dolphins, a hypothesis has been formulated (see Chapter 1); this is that by the regulation of the skin muscle tension dolphins promote resonant interaction with a flow. This leads to interaction of fluctuations of the boundary layer and natural frequency of fluctuation of the dolphin skin. As a result, the skin coverings vibrate with resonant frequency practically without energy consumption. Fluctuations of this coating occur at a frequency that controls the frequency of bursting

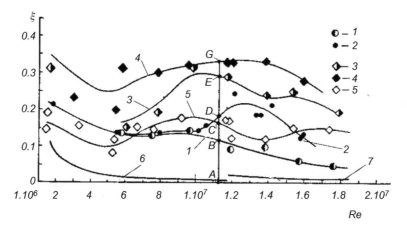

Figure 5.21

Coefficient of drag reduction ξ (Re) versus the Reynolds number: at injection of polymer solutions in the boundary layer on rigid *1, 2* and on elastic *3, 4* surfaces; *5* — without polymer solutions ($Cq = 0$) on elastic surface. The law of PEO rate is denoted by *B* for *1, 3,* and *A* for *2, 4* (see Figure 5.19); *6, 7* — relative root-mean-square error of measurements, $\sigma_{\Delta C_x}/C_x$

from the viscous sublayer. This is also the reason for the abnormally low resistance, which has been discovered in the authors' modeling experiments.

The values of ξ(Re) on the rigid and elastic cylinders in water flow with polymer solution of low concentration are presented in Figure 5.21 (curves 1−5), where ξ(Re) = $(C_{xrig.} - C_{xi})/C_{xrig}$ is the coefficient of relative variation in friction with respect to Reynolds numbers; C_{xrig} is the friction coefficient of the rigid surface; C_{xi} the friction coefficient with modified boundary conditions. Curves 6 and 7 in Figure 5.21 show the relative root-mean-square error of measurements, $\sigma_{\Delta C_x}/C_x$. Curves *1* and *3* correspond to the coefficient of flow rate C_q(Re) according to the law "*B*" (see Figure 5.19), and curves *2* and *4* correspond to the law "*A′′′*". The designation of points in Figure 5.21 are as in Figure 5.19.

As can be seen from the results of the experiments, injection of a small amount $(C_q \approx (2-5) \cdot 10^{-5})$ of low-concentration polymer solution reduces the friction drag. The greater the amount of injected polymer, the more drag reduction. It is very obvious at $Re > 1 \cdot 10^7$, where the coefficients in "*A*" and "*B*" are clearly different:

$$C_{qA} > C_{qB}, \ \xi_2 > \xi_1 \tag{5.37}$$

The indexes at ξ denote the curve number in Figure 5.21. On the compliant surface, the effect of the polymer on the friction drag becomes even clearer:

$$\xi_4 > \xi_3 \text{ and } \xi_4 > \xi_2 \tag{5.38}$$

It should be noted that the effects of the elastic surface and polymer are cumulative. In Figure 5.21, the vertical line marks the segments of the straight line, which defines the efficiency ξ at a given Reynolds number. From Figure 5.21, we have $AC + AB \approx AE$ and $AC + AD \approx AG$.

The combined influence of an elastic surface and polymer solution on a boundary layer has a complex character. It is necessary to carry out special investigations to reveal the mechanisms of flow in the presence of various combined methods of drag reduction.

5.7 Engineering Method for the Selection of Elastic Plates

As it has been shown in Section 1.6, elastic surfaces in the real conditions of a flow are subjected to a complex of loads acting from above and below. Such a complex effect must be analyzed by means of spectral characteristics at the exit of multivariable system.

The interaction of an elastic surface with a flow causing such complex loading can be defined by means of numerical calculations. However, such calculations can be performed only at the corresponding boundary conditions, which are ambiguous because the action of loadings on an elastomer occurs in an interconnected way and is defined by the conditions of motion of the body with the elastic surface. We shall consider some special aspects of the influence of loadings on elastomers.

The elastic surface is influenced by spectral characteristics of fluctuation movements in the form of spectra of velocity pressure fluctuations of a boundary layer. Based on the experimental data of flow on various kinds of elastic plates, it is known that characteristics of the boundary layer on elastic plates differ from the rigid standard at all modes of the boundary layer.

Let us consider the spectral characteristics of longitudinal fluctuation velocity in the boundary layer over rigid and elastic plates. For a laminar flow being exerted over any surface, we can calculate the frequencies of unstable oscillations by the formulae:

$$x = 0.388 \frac{\nu \mathrm{Re}_*^2}{U_\infty} \tag{5.39}$$

$$n = 0.159 \frac{\beta_r \nu}{U_\infty^2} \frac{U^2}{\nu} \tag{5.40}$$

The calculation results and their analysis are presented in [305]. The main features of note are as follows:

1. The region of unstable oscillations depends on the value of U_∞.
2. Elastic plates substantially reduce this region.

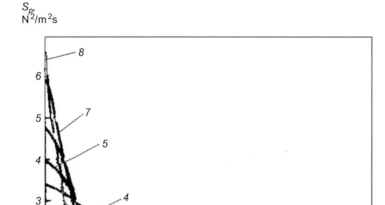

Figure 5.22a
Power-density spectrum of pressure pulsations at a flow of a rigid plate with a velocity of 5 m/s
depending on various values x: *1* − 0.05 м; *2* − 0.1 м; *3* − 0.2 м; *4* − 0.4 м; *5* − 0.6 м;
6 − 0.8 м; *7* − 1.4 м; *8* − 2.0 м.

3. On the plate, in the direction of *x*, at $U_\infty > 1$ m/s, there is a narrow region above which a wide spectrum of unstable oscillations is generated.

Although the laminar segment has small extent along *x*, a wide spectrum of unstable fluctuations arises in this area. In transitional and turbulent boundary layers, the low-frequency part of the spectrum is gradually filled. The elastic surface, with its corresponding selection of design and mechanical characteristics, essentially reduces the spectrum of unstable fluctuations at all stages of transition and in the turbulent boundary layer.

Figure 5.22a, b represents the power-density spectrum of pressure pulsations calculated under the formula [319]:

$$S_p(\omega) = \frac{4b^2\gamma^2\tau^2\alpha U_c}{\pi\left(\alpha^2 U_c^2 + \omega^2\right)} \tag{5.41}$$

where $b = 0.3$; $U_C = 0.8U_\infty$; $\gamma = 4.34$; $\alpha = 2/\delta$; δ is the boundary layer thickness.

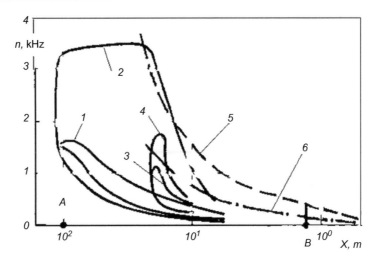

Figure 5.22b

Neutral (*1, 3*) and limiting neutral (*2, 4*) curves of hydrodynamic stability flat rigid (*1, 2*) and elastic (*3, 4*) plates, and power-density spectrum of pressure pulsations of a turbulent boundary layer of a rigid plate at velocity of 5 m/s. Power-density spectrum: *5*—90% of a full range, *6*—75% of a full range. *A* — point of loss of stability, *B* — point of the conditional beginning of a turbulent boundary layer.

Spectra are measured in the flow over a plate at velocity of 5 m/s and various *x*. It is seen that most energy-carrying frequencies are low. With the growth of coordinate *x*, the range of energy-carrying frequencies decreases and the greater part of the energy of pressure pulsations of the boundary layer concentrates in a narrow frequency range.

Figure 5.22b shows areas of unstable fluctuations of a laminar boundary layer in the flow over rigid and elastic plates (see Chapter 1) [31, 32, 305]. When carrying out investigations of hydrodynamic instability the maximum amplitudes of fluctuations are analyzed. The limiting curves bound the range (spectrum) of frequencies, which are determined by visualization of the boundary layer with the tellurium method. In fact, the spectrum of fluctuations will be even more in a laminar boundary layer in the field bounded by the neutral curve. On this plot, dashed and dash-dotted curves illustrate the spectra obtained according to Figure 5.22a, b. Fig. 5.22b shows areas of unstable fluctuations of laminar body layer in the flow over rigid and elastic plates (Chapter 1, Figures 1.39, 1.65) [see 31, 32, and 305]. Curve *5* designates 90% the spectrum of the full range of pressure fluctuations, and curve *6* 75% of the full range. Qualitatively it corresponds to limit (curves *3* and *4*) and normal neutral curves (curves *1* and *2*). It can be seen that the law of dependence of the range of the fluctuations spectrum in laminar and turbulent boundary layers in the flow on a rigid plate with regard to *x* has a similar character.

Maximum ranges of fluctuation frequencies are in the area of stability loss and at the start of development of the turbulent boundary layer. Between these characteristic values of x and further along the plate, the range of fluctuation frequencies is essentially narrowed.

In a wind tunnel with low turbulence intensity, Tetyanko [494] has performed experimental research on spectral characteristics on a rigid plate in laminar, transitional and turbulent boundary layers. Part of his data has been put on plots of spectra at susceptibility of various disturbances by a boundary layer [310].

According to the spectral curves he obtained, the spectrum of low-frequency fluctuations is narrower in a laminar boundary layer, and is not full in the high-frequency area relative that in a turbulent boundary layer. In the process of passage through various stages of development to transitional boundary layer (with growth x), the high-frequency part of the spectrum was more and more filled. In comparison with the spectrum of a turbulent boundary layer, in the low-frequency area the spectrum essentially differs from the turbulent spectrum; there is obviously a symmetric maximum in the spectral curves being expressed, in comparison with the turbulent spectrum, the spectral curve of which has no maxima and tends to 1 at reduction of fluctuation frequency.

The spectral curves obtained in [494] are well agreed with the data in Figure 5.22a, b. According to the model of development of disturbances in a transitional boundary layer (see Chapter 1), in the process of deformation of disturbances in a boundary layer from linear to 3D form, the high-frequency part of the spectrum is more and more filled. The results of experiments on spectral curves at susceptibility of the boundary layer on a rigid plate to 3D disturbances are presented in [310] and have shown the formation of characteristic maxima of spectral curves at low frequencies and valleys in the high-frequency part of the spectrum. A similar picture is obtained in [494]. Figures 4.12 and 4.13 illustrate the spectra on elastic plates in water flow and at susceptibility of 3D disturbances by elastic plates. Here, statistically identical pictures of the formation of maxima are seen at low frequencies. Such data indicate the formation of a narrower band spectrum at a flow on elastic plates and maxima of spectra at low frequencies characterize the presence of longitudinal vortical structures in the boundary layer.

These data and results in Figure 5.22b are the basis for the following hypothesis: it is possible to influence the characteristics of a boundary layer provided that it is possible to reduce essentially areas of instability and the initial section of development of the turbulent boundary layer. Such an assumption is confirmed by the results of physical experiments partially presented in the previous chapters.

Based on this hypothesis, it is possible to assume that dolphins and other high-speed hydrobionts have developed mechanisms for the realization of the revealed laws. Thus, it is possible that dolphins are capable of reducing the range of unstable fluctuations and quickly generate a quasi-turbulent boundary layer with a narrow spectrum of fluctuations by means of their specific skin structure. Regulating the skin's muscle tension, dolphins may adjust fluctuations of the skin to achieve resonance with the narrow spectrum of fluctuations of the boundary layer (see Section 4.6). Along with the formation of the pre-burst modulation mechanism, mentioned previously, this allows dolphins to regulate the resonant work of the skin covering as a complex wave-guide. Fluctuations directed by the skin wave-guide are dissipated or thrown into the flow from the end part of the wave guide (the tail part of the body) due to the work of the fin mover without the wave reflection on the end face of the wave guide.

The same mechanism of organizing a quasi-turbulent boundary layer with a narrow range of low-frequency energy-carrying fluctuations is utilized by high-speed fish. Dolphins use this mechanism by a drastic reduction of the spectrum of neutral frequencies of flat fluctuations and fast passage of other stages of transition with the purpose of formation of quasi-turbulent boundary layer. Swordfish use such mechanism due to the xiphoid tip, which allows the generating of the necessary quasi-turbulent boundary layer along its body. Swordfish skin has the same functionality as high-speed dolphins, including a complex wave-guide. These topics are considered in the following chapter in more detail.

The first task in the engineering method for the selection of elastic plates is to select materials that decrease the amplitude and spectrum of disturbance oscillations as much as possible. Thus, the basis of the first task is the definition of the necessary mechanical characteristics of the materials from which uniform or composite elastic plates with optimum stabilizing properties should be made.

The second task is to develop methods of damping the unstable oscillations of the laminar boundary layer in the region of the widest range of frequencies of unstable oscillations, i.e. in the instability region ($Re_{loss\ stability}$).

The third task is the development of a corresponding design for the elastic plates. To achieve this it is necessary to consider the boundary layer on an elastic plate and a design of an elastic plate from the position of a complex wave-guide, as has been shown in Chapter 1.

As the surface of elastic plates interacts with the field of fluctuation velocities indirectly, and it directly interacts with pressure fluctuations and near-wall shear, we shall further consider the spectrum of turbulent pressure fluctuations. Furthermore, it is justified by the fact that velocity fluctuations correlate with pressure fluctuations:

$$\sqrt{\overline{p'^2}} = 0.5\rho U_\infty^2 Re^{-0.3} \tag{5.42}$$

Calculations of the pressure fluctuation spectrum can be carried out according to [36]. Data in this paper are well agreed with the experimental data in [515, 516].

In the range of x from the expected beginning of the transitional boundary layer (in the area of the start of curves *5* and *6*, see Figure 5.22b) calculations of the range of pressure fluctuation frequencies must be done according to curve *5* in view of 90% of a full range of pulsations of pressure. From $x = 0.8$ m, after point *B* it is possible to consider the range of frequencies of pressure fluctuations on curve *6*.

Therefore, according to the principles presented in Chapter 1 and to the above-presented calculation results, the simplest method of selection of elastomers is as follows. By means of equipment and methods described in Chapter 2 the frequencies of elastomers are determined: the frequency of natural oscillations and frequency range in which the elastomer strongly absorbs mechanical oscillations.

The pressure fluctuation spectrum and the location of regions with the widest range of frequencies of the disturbance motion are then determined. According to this calculation and the results of the investigations performed in Chapters 1−5, the elasomer composite with the natural frequency and $tg\varphi$ range is selected which best corresponds to the calculated pressure fluctuation spectra. This composite is mounted in the place of the body surface with the most corresponded frequencies. Taking into account the absorption of pressure fluctuations by the first composite, the correction is made and then the second composite (next along the body length) is placed with other properties (frequency characteristics), corresponding frequencies of fluctuation loading in the following region. Since the elastic composites are being selected for certain free-stream velocity ranges, the mechanical properties of an elastomer (and, therefore, frequency characteristics) must be controlled according to Chapter 1.

As well as a correspondence in frequency characteristics, it is necessary to reach the energy correspondence. For this purpose, the root-mean-square values of pressure fluctuations are determined using measurements of Reynolds stresses on elastic plates (see Section 3.5) and the formula (5.42). Using the value of a dissipative vortex scale near the wall [130, 232, 233, 445] and its distance from the wall [130, 232] ($y/\delta = 0.15−0.25$), it is possible to calculate the pressure fluctuation energy on the rigid and elastic plates. Comparing the energy that can be absorbed by the elastic composite with the energy of turbulent pressure fluctuations, composites are selected that satisfy maximum correspondence of the above-mentioned energies, i.e. the maximum response of the elastic surface to disturbances generated by the flow are implemented. Only in case of conformity of the frequency range and energy of fluctuations of disturbance movement of the boundary layer and natural elastomer fluctuations will there be the best absorption and redistribution of energy of external disturbances.

Other π-parameters (see Section 2.6) should be taken into account to provide optimal work of the composite elastic coating.

The engineering method for the selection of the elastic plates must be verified by the experiments below (see Sections 5.3 and 5.4). The selected elastic plates must be checked by measuring the friction drag. It is necessary to provide a mechanism for controlling the mechanical parameters of the composite elastic plates for different velocities of flow.

It has been shown above how to define and optimize the mechanical characteristics of the elastomers in view of obligatory sectioning along the body length. Manufacturing a material with the specified change in mechanical characteristics along the length, width and through the thickness of a coating would be optimum. We shall now consider the principles of designing elastic coatings, for which it is necessary to consider the following research findings:

- It is necessary to approach the optimum design of high-speed hydrobiont skin coverings and their specific functioning.
- It is necessary to know the principles of a complex wave-guide.
- It is necessary to apply the results of complex interactions of various disturbances in a boundary layer.
- It is necessary to apply the main principles of the functioning of hydrobionts.

Let us consider these features consistently.

1. The structure of a dolphin's skin covering (see Chapter 1) is such that there is an ordered system of dermal papillae in the outer layer. When swimming velocity increases, the blood pressure in the capillaries of the papillae increases; thus, the shape of the papillae change. At the increase of blood filling, according to the increasing energy of pressure fluctuations of the boundary layer, the fluctuating mass of the skin covering in the outer layer changes. Thus, the parity of energy of shaking force and the caused oscillatory energy of unit fluctuating mass is kept optimal. The temperature of the skin outer layer simultaneously increases and its mechanical characteristics change.

2. The structure of a dolphin skin's outer layers represents the classical sample of a wave-guide. When speed of motion is increased, according to the principles of a complex wave-guide (see Chapter 1) changes to the structure and blood filling of the outer layer automatically change the characteristics of the skin covering as a wave-guide. Indeed, at increased speed of motion the phase velocity of fluctuations in the skin caused by the fluctuations of pressure and the boundary layer shear increases. Hence, it is necessary to change the impedance of such a wave-guide.

3. Experimental investigations of interaction of various disturbances in a boundary layer (see Chapter 1) have shown that there are certain laws of interaction of various disturbances. In particular, introducing flat disturbances into a boundary layer in the area of 3D disturbances in the boundary layer of the plate, there is an intensive interaction of these disturbances leading to accelerated development of stable 3D disturbances. Furthermore, these conclusions are obtained for various types of interaction of disturbances generated in a boundary layer by means of elastic plates. The results confirm that transversal microfolds on the surface of dolphin skin cause

quasi-planar fluctuations in a boundary layer with a certain frequency, which interact with the 3D disturbances generated from below in the boundary layer due to the longitudinal ordered structure of the outer layers of the skin. Interaction of these two kinds of disturbance leads to the formation of steady 3D disturbances in the boundary layer, which develop in a complex wave-guide: "boundary layer—skin coverings."

4. Periodically, at a certain velocity of motion, the skin muscles contract, and that causes such tension in the outer skin covering that intensive fluctuations of the boundary layer cause self-oscillation of the outer layer of the skin (see Section 4.6) and essential reduction of friction drag (see Figure 5.20).

The realization of the above-mentioned features of the functioning of high-speed hydrobionts is implemented in the form of the developed designs of elastic coverings (see Section 2.5). In Chapter 1, and in Figures 2.49, 2.50, 2.54 and 2.63, the structure and mechanical characteristics of dolphin skin coverings are described. In Section 4.6 the results of research on the parameters of fluctuation of dolphin skin coverings at various modes of swimming are represented. Based on dimensionless π-parameters (see Section 2.6) one can define initial data for designing elastic coatings for technical objects. However, each concrete technical device with an external flow or internal flow demands an account of the specific features of the flow structure.

CHAPTER 6

Hydrobionics and the Anatomy of Fast Swimming Hydrobionts

6.1 Interaction of High-Speed Hydrobionts and Flow

The interaction of high-speed hydrobionts with a flow is of interest from the point of view of: 1) the definition of laws of physical interactions with an environment; 2) the search for mechanisms for the economic use of energy; and 3) for the development of new technologies. Therefore, attention has been given to studying the features of water animals, the organism system and its interaction with a flow during motion in water in relation to the reduction of energy consumption. As the movement is in the water environment the power influence of the biosphere on an organism is taken into account in considering these systems. The ways of influencing drag reduction during the motion of a body are shown.

The features of the structure of the outer coverings of hydrobionts have been explored under the action of hydrodynamic and physical fields of the environment, in particular under the influence of the boundary layer structure characteristic for each range of speeds within which each group of hydrobionts swim. Specific structures in the outer coverings developed simultaneously and in an interconnected way during the evolution of hydrobionts to have the effect that the structure of the boundary layer is altered and such a boundary layer is formed that reduces, as much as possible, the hydrodynamic drag and, therefore, the action of the environment on the pain receptors in the skin.

Taking into account the fact that all high-speed hydrobionts swim in the same environment, the mechanisms of laws of interaction with a flow should also be identical, according to the characteristic dimensionless parameters of motion (Reynolds number, Strouhal number, etc.). The adaptations in the structure of a hydrobiont body and skin can be various. The fish are divided into four groups depending on the range of characteristic swimming speeds [297].

Sailfish and swordfish swim in water at the highest speed. They are representative of the fourth high-speed group of fish [296, 308, 309, 397, 399]. This group reaches a speed of 35 m/s (Reynolds number is $1 \cdot 10^8$) and, therefore, it attracts special interest. Ovchinnikov was first to pay attention to the hydrodynamic functions of the gill apparatus [391, 392]. The assumption has also been made that the specific structure of the gill that produces slime improves the flow over the body [147, 148, 294]. Research into the shape of the swordfish body has been undertaken [308, 399, 397, 525, 338, 431, 305, et al.]. The shape of the caudal fins [298, 398] and the heat regulation of the body [299] have been considered. Some hydrodynamic features relating to the form of the body of the swordfish are detailed in [308, 397−399]. A description of the shape and the arrangement of scales on the surface of a young swordfish is given in [19, 392]. A brief description of the skin structure, the channels inside it and the deferent pores is represented in [331], in which [510] is cited. A detailed study of skin coverings of the swordfish in various sections of its body has led to the describing of a number of new structures in its skin [294, 295, 297, 391]. However, these features were considered separately from each other without considering the basic principles of hydrobionics [77, 86]. This has not allowed for a generalized review of their distinctive attributes and hydrodynamic functions. Table 6.1 lists the key investigations on the morphologic, hydrobionic and hydrodynamic features of the swordfish.

6.1.1 Structural Peculiarities of Swordfish Skin Coverings

A comparison of the skin structure of some species of high-speed hydrobiont (tuna, shark, dolphin and swordfish) is given in Figure 6.1.

The total thickness of the skin in the middle part of a body, expressed in shares of length L is about $8.5 \cdot 10^3 L$ for dolphins, and $(0.65−0.95) \cdot 10^3 L$ for fish. The dolphin's skin is an order thicker than that of fish. The digits on the left in Figure 6.1 indicate the absolute dimensions of the skin thickness, and on the right is the ratio to the total thickness of skin (as a percentage). The structure of the skin of the swordfish is more complex than of the other species of fish [296]. Due to evolution, there are specific formations in its skin inherent only to this species. Below, a summary of Koval's research [91] is provided.

Longitudinal folds on the skin surface

There are visible longitudinal folds on the skin surface of the fish, which are formed by non-uniform development of the layer of connective tissue (see Figure 6.2). The folds are located from the end of the gill covers up to the tail fin stalk. The size of the folds varies in longitudinal and transverse directions. On the top and bottom of the body the quantity of folds in 1 cm^2 of surface reaches 7−9. On the centerline, in the middle area, there are much larger folds; their quantity decreases to 3−4 per cm^2 and the height of a single crest reaches 0.7 mm.

Pores on the skin surface

There are pores all over the surface of the skin that produce a slimy substance on a live fish when pressing lightly on the skin. The topography and size of the pores is given in

Table 6.1: Areas of research on swordfish.

Author	Name of Journal	Year	Area of Investigation
Morphology			
Ovchinnikov, V.V.	*Biophysics*, 11, No. 1	1966	Structure of body
Koval, A.P.	*Bionika*, 11	1977	Structure of skin
Koval, A.P.	*Bionika*, 12	1978	Structure of skin and gill covers
Koval, A.P.	*Bionika*, 14	1980	Structure of skin
Koval, A.P.	*Bionika*, 16	1982	Structure of skin
Koval, A.P., Butusov, S.V.	*Bionika*, 24	1990	Structure of tail fin
Koval, A.P., Koshovsky, A.A.	*Bionika*, 25	1992	Structure of heat exchanger
Hydrobionics			
Pershin, S.V., Chernishov, O.B., Kozlov, L.F.	*Bionika*, 10	1976	Hydrobionic parameters of skin
Protasov, V.R., Staroselskaja, A.G.	*Hydrodynamic Features of Fishes*	1978	Theoretical drawing of the structure of body
Pershin, S.V.	*Bionika*, 12	1978	Hydrobionic parameters of body
Babenko, V.V., Yurchenko N.F.	*Biophysics*, 25, No. 2	1980	Hydrobionic parameters of skin
Babenko, V.V., Yurchenko N.F.	*Hydrodynamic Questions in Bionics*	1983	Hydrobionic parameters of skin
Babenko, V.V., Koval, A.P.	*Bionika*, 23	1989	Hydrobionic parameters of skin
Babenko, V.V.	*Bionika*, 25	1992	Hydrobionic parameters of skin and tail fin
Problem of Drag Reduction			
Kozlov, L.F., Leonenko, I.V.	Reports AN Ukraine	1971	Flow about a sphere (experiment)
Kozlov, L.F., Leonenko, I.V.	*Bionika*, 7	1973	Flow about a model of the swordfish (theory, experiment)
Zolotov, S.S., Hodorkovsky,Y.S.	*Bionika*, 7	1973	Flow about a model of the swordfish (theory, experiment)
Aleev, Y.G., Leonenko, I.V.	*Bionika*, 8	1974	Flow about a model of the swordfish (experiment)
Amfilohiev, V.B., Zolotov, S.S., Ivlev, Y.P.	*Bionika*, 14	1980	Flow about a model of the swordfish (theory)
Babenko, V.V., Koval, A.P.	*Bionika*, 16	1982	Flow about a model with injection (experiment)
Voropaev, G.A., Kozlov, L. F., Leonenko I.V.	*Bionika*, 16	1982	Flow about a model (theory)
Babenko, V.V.	*Bionika*, 17	1983	Flow about a model with injection (experiment)
Bandyopadhyay, P.R.	*AIAA* Journal	1989	Drag reduction of the nose body (theory)
Korobov V.I.	*Bionika*, 26	1993	Flow about a model with elastic surfaces and polymer injection (experiment)

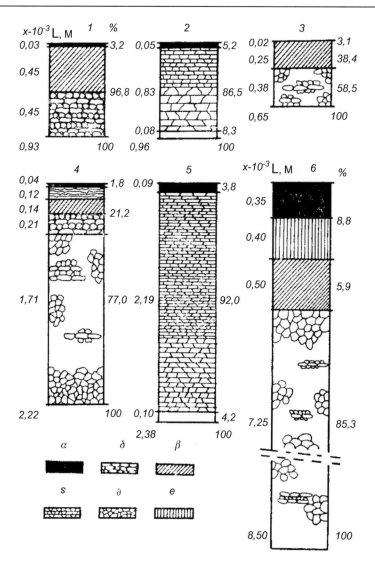

Figure 6.1

Level-by-level structure of histological microscopic sections of the skin of high-speed hydrobionts [309]: 1 — yellow-fin tuna; 2 — mako shark; 3 — swordfish; 4 — spotty tuna; 5 — blue shark; 6 — ordinary dolphin. Layers of skin: a — epidermis; b — fibrous pigmented layer; c — connection-woven layer; d — collagen bunches and fibers layer; e — hypodermic fatty cellulose layer; f — divided papillary layer of a skin of a dolphin.

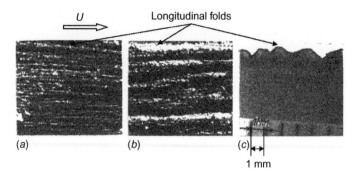

Figure 6.2
Longitudinal folds on the surface of swordfish skin: *a* — fine longitudinal roughness, 3-fold magnification; *b* — longitudinal rugosity, 6-fold magnification; *c* — cross-section of skin in the area of longitudinal rugosity, 6-fold magnification.

Table 6.2: Number and size of pores at various sites of the swordfish body in a sample area 10 mm², L = 2.6 m.

Section of the Body (Percentage of the Fish Length Counted from the Bottom Jaw)												
	30%		40%		55%		70%		80%		90%	
	N	d (mm)	N	d (mm)	N	d (mm)	N	d (mm)	N	d (mm)	N	d (mm)
Dorsal line (back)	6	0.13	13	0.08	13	0.08	12	0.08	9	0.09	6	0.09
Middle line	6	0.1	7	0.1	7	0.1	6	0.1	6	0.09	5	0.09
Ventral line (stomach)	6	0.1	6	0.09	6	0.08	6	0.08	5	0.07	5	0.08

Table 6.2. The density of the pores and their size over the whole body are practically uniform, except for a narrow strip on the top line of the dorsal fin, where their density is approximately double.

System of hypodermic channels

The numerous pores are connected by hypodermic channels, which are located in the dense connective tissue stratum at a depth of about 1 mm from the surface of the body (see Figure 6.3). The average diameter of the channels is 0.5 mm but they can reach up 1 mm in diameter. Smaller channels (0.2–0.3 mm) are located on the head and gill covers. On the lateral surface of the body the channels are angled downwards, perpendicular to the longitudinal folds in the skin. Between the channels there are further channels connecting them to create a uniform interconnected system. On the back top line, the head and gill cover channels are arranged chaotically, without a specific orientation. The channels of the left and right sides of the body are not connected on the top or bottom lines. Thus, there are two independent systems on the two sides of the body.

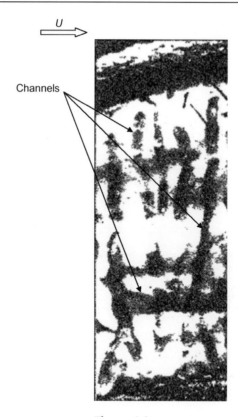

U

Channels

Figure 6.3
System of hypodermic channels on the centerline of the body, 3-fold magnification increase
3 — multiple.

Slots on the body surface

There is a series of unique slots on the swordfish back (see Figure 6.4) in which there are small ducts connected with internal cavities, so-called "ampoules," located in the dermal layer. This series of slots begins at the base of the third beam on the left and right sides of the dorsal fin and ends at the additional dorsal fin located on the tail stem. The slot lengths are different in different locations of the body. The longest slots are found on the dorsal fin. Between large slots there are smaller slots. In a swordfish of length $L = 1.57$ m, the shortest slot is 2 mm long and the longest is 10 mm. With increasing body length, the maximum length of the slots also increases: the length of a slot reaches 18 mm at $L = 2.6$ m.

Behind the swordfish dorsal fin there is a small cutaneous longitudinal fold, which is the vestige of a reduced large dorsal fin. In young swordfish the fold extends up to the second dorsal fin, as in the sailfish [19]. From the first dorsal fin and fold, on both sides, the slots are staggered and slightly shifted to the lateral surfaces of the body. On a swordfish of

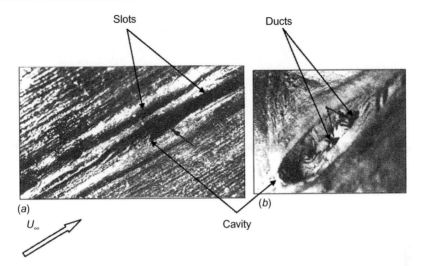

Figure 6.4

Slots (*a*) and excretory channels (*b*) on the dorsal side of the body, 3-fold magnification.

length $L = 1.57$ m 18 slots were found on the left side of the body and 22 slots on the right side. In the dermis the slots become narrower and form a flat bottom and there are one to four small circular apertures—ducts from $0.5-2$ mm in diameter depending on the size of the slot.

Ampoules in swordfish skin

The ducts described above originate from the extensive cavities of ampoules and have a round or oval form with a small thickening at the borders with ampoules in the longitudinal direction (see Figure 6.5).

In most cases, the ducts to two slots come from one ampoule, one is large and the other is small, and they are situated on the opposite sides of the separating fold. The ampoules are distributed in the central part of the back centerline. A fish of length $L = 1.57$ m was found to have 24 ampoules, some of them (14) rather large—up to 7 mm in diameter and 15 mm in length, and 10 ampoules were small, up to 2 mm in diameter and 4 mm in length. Larger fish have larger ampoules. The maximum diameter of the ampoules in a 2.6 m-long swordfish is in the order of 18 mm, with a length of up to 30 mm. Small ampoules are located at the base of the dorsal fin and large ampoules are situated behind the dorsal fin.

The ampoules have a mesh structure (see Figure 6.5). The surface of the ampoules consists of layers of epithelium. The number of layers in a 2.6 m-long fish reached 40 in certain places. The presence of many secretory cells at various stages of development leads to the assumption that there is active and constant secretion of slime inside the ampoules and even in the deferent ducts. Therefore, there is always some slimy substance in the ampoules.

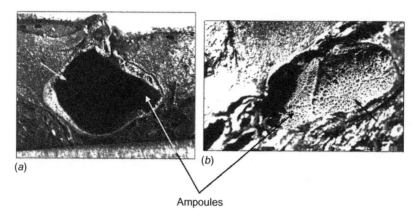

Figure 6.5
Cross-section (*a*) and internal structure (*b*) of an ampoule, 6-fold magnification.

The dense connective tissue of ampoules is well supplied with blood vessels. Under the layer of dense connective tissue (the rigid base of the ampoules) there is friable connective tissue with separate groups of fat calls. At the base of the ampoules are bunches of skeletal muscle cells, which at traction can cause the deformation of the walls of the ampoules and change their volume. This results in the exuding of the slimy substance to the body surface.

Structure of skin in the area of gill slits

In the area of the gill slits and on the internal sides of the gill covers there are specific structures that are similar to the crypts of a stomach mucosa (see Figure 6.6a). These structures are dispersed over the surfaces of the small islets on the gill covers (see Figure 6.6b). In the area of the gill slits the crypt structure is situated in a narrow zone (3−4 cm) on the edge of the gill slot, going a little under the gill cover (see Figure 6.6c). The cellular structure of the crypts does not differ from cellular structure of the skin epidermis. There are plenty of secretory cells located in the top layers and their size can be up to 20 μm in some places. The thickness of the epithelial layer on the connective tissue tubercles is minimal (3−4 layers) but it can be up to 40 layers thick in the center of an islet bordered by tubercles.

Scales on swordfish skin

On the skin coverings of young swordfish there are scales with significant surface relief (see Figure 6.7a). The arrangement of scales and their structure is described in [19, 392]. The body is completely covered with scales, and has from two to five conic spines. The base of the scales (*1*) lay in the dense connective tissue (*2*) of the skin covering, and only the "spines" stick out onto the body surface. The largest spines are located on the head and border of the eye-socket. The scales create several longitudinal rows along the body.

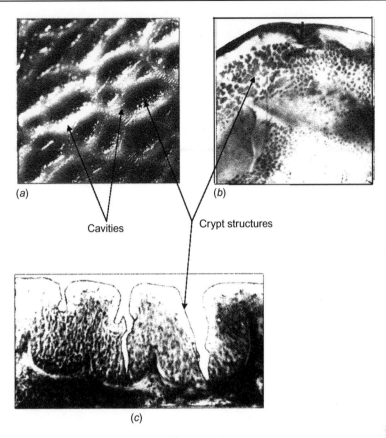

Figure 6.6
Crypt structures in skin in the area of gill slits (*a*), the arrangement of crypt structures on the internal side of gill covers (*b*), 6-fold magnification; a cross-section of secretory apparatus of the crypt structure (*c*), 300-fold magnification.

The most visible are the back and two lateral and ventral rows, between which there are smaller scales covering the whole body. As a fish grows, its scales modify. The scales of a 0.69 m-long swordfish look like small triangles located one over another, similar to tiles.

Rows of scales in the form of bone protuberances are retained on the stomach. The scales on swordfish disappear completely when they reach a length of about 1.5 m. However, the authors' research has shown the existence of scales with conic ledges that penetrate the upper skin layers in the adult swordfish body ($L = 2.6$ m) (see Figure 6.7b).

Structure of the skin covering

The structure of the swordfish skin covering is much more complex than that of other fish due to presence of the structures described above. The epidermis of the swordfish, similar

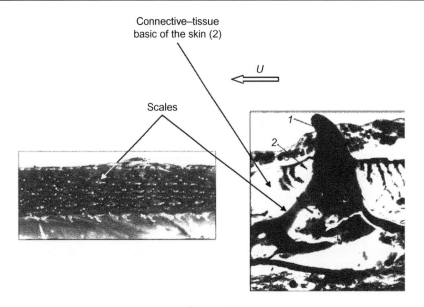

Figure 6.7

Scales of the swordfish: *a* — on the body of 1.2 m-long fish, 3-fold magnification; *b* — on 2.6 m-long fish, 350-fold magnification.

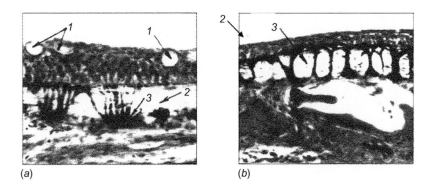

Figure 6.8

A cross-section of skin: *a* — secretory apparatus of epidermis, 250-fold magnification; *b* — fibrous pigmented layer, 250-fold magnification.

to other species of fishes, consists of 8–10 rows of epithelial cells, in which there are numerous secretory cells of oval shape at various stages of development *1* (see Figure 6.8).

The epidermis lies on a thin connective membrane *2*, beneath which is a so-called "fibrous pigmented layer" *3*, revealed earlier in scombroid fish (tunas and related species), which swim in the approximate range $Re = 10^7$ [147]. The structure of the fibrous pigmented layer is similar for all these fish. The thickness of this layer often exceeds the thickness of the

Figure 6.9
A cross-section of hypodermic channel, 300-fold magnification.

epidermis and the membrane lying below it. The second thickest connective layer lies under the fibrous pigmented layer and it is in this that the base of the scales lie.

Below the base of the scales *1*, numerous channels *2* are located, which form an internal surface cover of multilayer epithelium *3* (see Figure 6.9). The thickness of this epithelium varies considerably: it consists of 5–6 rows of epithelial cells stacked in one place and 2–3 in others. Similar to the epidermis of the skin, between the epithelial cells are many cells secreting mucus into the channels.

The channels have branches that run perpendicular to the channel and reach the body surface as pores. The epithelium covering the channels is continous with that on these branches and pores, connecting with the epidermis of the skin. Based on experimental data, a scheme of the structure of swordfish skin has been developed (see Figure 6.10). Although there are essential differences between swordfish skin and that of other hydrobionts, there are also similarities. In particular, the comparison of the skin of dolphins and swordfish has revealed a number of identical structural features [46, 399]:

- The structure of the skin is multi-layered across the thickness.
- There are two layers of thick membrane equidistant to the skin surface.
- There are specific muscular and tendon structures that cause an interfering shift of skin under the action of tangent pressure of the flow.
- The skin is supplied with a specific circulatory system regulating the temperature of the body and mechanical characteristics of the skin coverings.
- The system of the organism controls the structure of the skin coverings in an automatically and interconnected way.

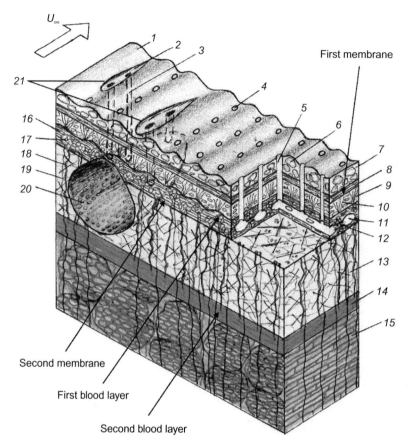

Figure 6.10
Scheme of swordfish skin: *1* — longitudinal folds on the surface of skin; *2* — slots on dorsal side of the body; *3* — excretory channels; *4* — pores; *5* — channels of pores; *6* — secretory cells; *7* — epidermis; *8* — first layer of thin connective membrane; *9* — fibrous pigmented layer; *10* — connecting fibrous tissue; *11* — system of hypodermic channels; *12* — connecting channels; *13* — under-skin layer of fat cellulose; *14* — layer of blood vessels; *15* — locomotor musculature; *16* —second layer of thick connective membrane; *17* — large cells with blood and innervation systems; *18* — blood circulatory system; *19* — innervation system; *20* — ampoule; *21* — cavities.

The equation of motion for an analogue of dolphin and swordfish skin coverings was presented in [66, 67, 72, 73, et al.]. Some specified peculiarities of the skin structure were experimentally investigated in [25, 33−35, 38, 44, 56, 63, 66].

6.1.2 Hydrodynamic Peculiarities of the Skin Structure and Body of the Swordfish

The results of research on the skin structure of the swordfish have shown that there is an essential difference from the skin structure of other high-speed fish. Analysis of the data

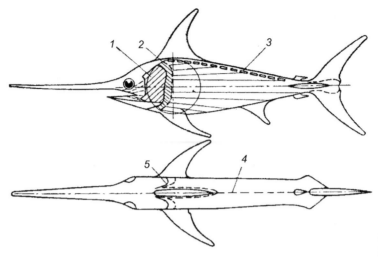

Figure 6.11

Scheme of gills apparatus and ampoules on the swordfish body: *1* — gill cover; *2* — crypt-lice structure generating slim [294]; *3* — ampoules in skin [391]; *4* — slots of ampoules; *5* — slots of a gill.

given in Table 6.1 provides information about some of the hydrodynamic features of the swordfish when swimming. In hydromechanics the resistance of a body consists of form drag, friction drag and ground resistance. Friction drag can make up to 80% of the total resistance of a body. According to the principle of interconnection [77, 86] when analyzing any system of a hydrobiont it is necessary to take into account the influence on this system of other systems of the body. Below are described the hydrobionic peculiarities of the skin structure and body of the swordfish.

Influence of xiphoid tip on drag reduction

The shape of the body and some peculiarities of the skin structure of the swordfish are given in Figure 6.11. Numerous experiments on hydrobionic modeling of the work of systems on the organism relating to hydrobionts were performed. In particular, the role of the xiphoid tip on drag reduction of an axisymmetric body was experimentally investigated [45, 50, 66, 237, 305, et al., see Table 6.1].

The experiments showed that when the model was tested in strong aeration flow, the drag of models with xiphoid tips became less than for the reference model, and drag of the model with a long tip was minimal [305]. Theoretical and experimental results of research of xiphoid tip influence on drag of a body are described in the following sections of this chapter.

Peculiarities of the injection of a liquid through a gill slot

The experimental research has shown that the velocity of a liquid flowing through a gill slot in the area of the back edge should be comparable with the speed of the basic flow in this place [66, 305]. For this, the structure of gill apparatus should be such that the flow speed does not decrease inside the gills. The gill apparatus contains cells generating slime, which reduces drag. Furthermore, the structure of gill apparatus is such that it promotes the expiration of a liquid from the gill slots as a system of longitudinal vortices. Disturbances of the boundary layer having a similar vortical structure are stabilized and promote drag reduction [310].

As can be seen in Figures 6.6a and 6.11, behind the gill slot there is an area with an increased number of slime-generating cells. This promotes leveling of the specified speeds in the area of the back edge of the gill slot, reduces jump of shear stresses and enhances the stability of longitudinal vortical systems generated in the flow over the gill apparatus. These features of the gill apparatus are simulated in [80, 81].

Another important peculiarity is the formation of a longitudinal coherent vortical system in the branchiate device by means of 3D cavities (see Figure 6.6a). Such systems of vortices promote not only stabilization of disturbances in the boundary layer and the elimination of shear stress on the edge of the branchiate slot, but also promote the preliminary preparation of solutions of slime-extending macromolecules in a longitudinal direction. Due to such preparation, polymer solutions become much more effective. A third peculiarity is the problem of interaction of disturbances, which is discussed in more detail later.

In the area where the xiphoid tip connects to the head (see Figure 6.11) there is a curvilinear concave section at which flow Goertler vortices are formed. Goertler vortices interact with longitudinal vortices generated in the branchiate device behind the edge of the branchiate slot. Numerous experiments on the formation of longitudinal vortices behind 3D cavities have been executed [65, 69, 74, 75, 92]. The model experiments on the specified peculiarities of flow in the branchiate device of the swordfish were carried out. In Figure 6.12 visualization of the flow about a longitudinal cylinder is shown. The design of the model allowed the forming of a system of longitudinal vortices inside the slot-hole chamber. At water injection from the slot, visualization showed the presence of such vortical systems (see Figure 6.12). At the injection of a polymer solution, the length of the longitudinal vortical systems essentially increased (see Figure 6.12b).

At injection of polymer solutions in the model, the feeding channel performed in the form of a wavy surface with the certain shape of a wave. This promoted both drag reduction in the channel and the preliminary extension of polymers.

Structure of disturbances in the boundary layer on swordfish skin coverings

The structure of skin coverings of hydrobionts has developed under the influence of disturbing movement of the boundary layer at characteristic speeds of motion. To define the

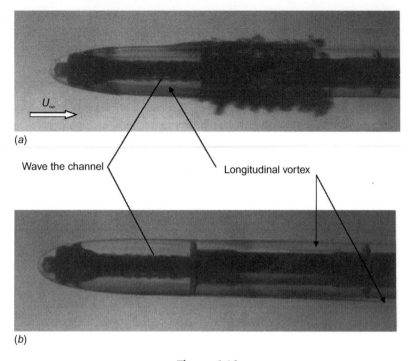

Figure 6.12
Modeling of the work of the gill apparatus. Preparation of vortical structure system before injection of liquid from slit: *a* — injection of water, *b* — injection of polymer solutions.

kind and structure of the disturbing motion, the laws of formation and peculiarities of development of disturbances in the boundary layer on a rigid plate were investigated at various speeds and regimes of flow [305, 310]. The model of development of disturbance motion at various stages of its development is given in Chapter 1 [310].

The peculiarities of the development of the disturbing motion in the boundary layer on various kinds of elastic plates, simulating the structure of skin coverings of hydrobionts, were investigated in [66, 305]. It was revealed that the form and characteristics of disturbances essentially differ from those on a rigid plate. The better the skin covering of hydrobionts is simulated, the stronger the difference is (see Chapter 1).

The comparison of the skin of dolphins, sharks and other high-speed fishes has revealed both essential differences and a number of general trends regarding their structure. It has indicated that the structure of their skin coverings has developed under the action of the same environments [46, 50, 77, 86, 294, 295, 308, 309, 397, 399]. Hydrobionts have produced identical mechanisms of receptivity of the same environmental influences. However, different hydrobionts have different structures relating to these mechanisms. Each scale of a shark is supplied with blood vessels; the scale cover is flexible and the outside is covered with slime. Therefore, their outer cover is like that of the dolphin and it has elastic-damping properties. With increasing swimming speeds the structure of the outer covering changes. Therefore, high-speed fish, such as tuna, are partially covered with scales and partially covered with an elastic-damping layer, like dolphins. The scales are located in the area where the influence of the environment is stronger. The hydrobionts that swim at the highest speeds, the swordfish and sailfish, have a scaleless elastic-damping cover, like the Cetacea. Transversal structures are found in the skin of all high-speed hydrobionts. There are microfolds on the surface of dolphin skin; similar microfolds are formed by the crests of scales on the skin of sharks and other fish; transversal channels are present in swordfish skin (see Figure 6.10, *11*). In addition, there are structural formations in these skin coverings in the longitudinal direction: dermal rollers and longitudinal rows of epidermal papillae in dolphin skin; longitudinal rows of crests of shark scales; longitudinal folds on the surface of swordfish skin (see Figure 6.10, *1*).

To help define the hydrodynamic role of these microstructures, the appropriate calculations were carried out [63]. At the oscillatory motion of a body during swimming, it was found that the specified transversal structures generate wave disturbances into the boundary layer such as Tollmien−Schlichting waves. The calculation results for various hydrobionts and various characteristic speeds of motion were put on the diagram of a neutral curve in coordinates of dimensionless frequency of fluctuation of the specified waves and Reynolds numbers. For all considered hydrobionts, disturbances generated by their skin coverings appeared to be located in the field of stability on the diagram of the neutral curve [63]. Computations were made of the stability of longitudinal vortical disturbances generated in a boundary layer by longitudinal microfolds of skin coverings of the hydrobionts. When a body performs an oscillatory motion, areas of convexity and concavity are formed on its surface. Calculations of the characteristics of these longitudinal vortical systems formed in the boundary layer for all hydrobionts have shown that they are situated in the area of stability on the diagram of the Goertler neutral curve.

Interaction of disturbances in the boundary layer of hydrobionts

The structure of the skin of high-speed hydrobionts is such that it promotes the existence of various types of steady vortical disturbances in the boundary layer. It was first revealed that these disturbances interact with each other. In order to understand the laws of such interactions, numerous experiments on the interaction of various disturbances generated in a boundary layer on a part of its various borders have been undertaken [66, 310]. A particular technique and various pieces equipment were developed for such experiments. Interactions of both identical and different flat and 3D disturbances generated on the boundary layer borders were investigated. These tests were carried out on both a rigid plate and various kinds of elastic plates, simulating skin coverings of various hydrobionts [66]. The following conclusion was drawn: even a uniform monolithic single-layer elastic plate promotes the formation of system of longitudinal vortices in the boundary layer. The interaction of flat and 3D disturbances results in the accelerated formation of steady longitudinal vortices in the boundary layer. The longitudinal vortical systems are most favorable for drag reduction in a wide range of Reynolds numbers.

In the swordfish, due to the specific structure of the gills and the production of a mucous substance from the gill slots, longitudinal vortical systems form in the boundary layer. Stability of these vortices is defined by the longitudinal folds on the surface of the slime-filled skin, which is supplied to the surface through pores (see Figures 6.10 and 6.11 and Table 6.2).

Figure 6.13 illustrates the formation of longitudinal coherent vertical structures (CVSs) in the boundary layer by the skin. Interaction of the boundary layer CVSs and those generated by the skin coverings of high-speed water animals leads to formation of a new kinds of CVSs.

Hydrodynamic significance of the skin covering thickness

Experimental research of the boundary layer on elastic surfaces has revealed that the efficiency of an elastic surface depends on the depth to which the disturbances penetrate from the boundary layer, and it also correlates with the boundary layer thickness. Indeed,

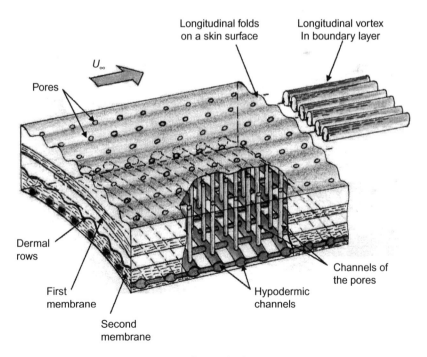

Figure 6.13
Scheme of formation of the longitudinal coherent vortical structures in a boundary layer by swordfish skin.

the depth of distribution of the disturbances in an elastic material depends on the energy of disturbances acting upon the surface of material from the boundary layer. Flow speed, roughness, and curvature of a plate define the energy of disturbances. The boundary layer thickness is defined by the same parameters. Thus, there is a range of optimal values of parameters describing the efficiency of an elastic coating, determined by the boundary layer thickness and the elastic material. In turn, the mechanical characteristics of an elastic material determining its efficiency depend on thickness, structure and design of the material. In engineering, such parameters can have a wide range of values. In nature, this complex parameter has an optimum range of magnitude. Measurements have shown that the following ratio is optimal in the flow over dolphins for the stabilization of the disturbances in the laminar and transitional boundary layers:

$$h_i = (1 - 4)d \tag{6.1}$$

where h_i is the thickness of the oscillating mass of skin, i.e. the depth to which fluctuations of the boundary layer penetrate.

For high-speed hydrobionts the empirical dependencies are given in [63]:

$$h_i = f(x) \tag{6.2}$$

$$h_i/d_i = w(x/L) \tag{6.3}$$

where x is distance along the body of the hydrobiont, L is length of the body, d_i is characteristic thickness of the boundary layer. The empirical formula for $x/L = 0-0.5$ is obtained:

$$\lg(h_i/d_i) = \lg 1.28 - 2.4 \lg(x/L) \tag{6.4}$$

Thus, the uniform law for the considered fast-swimming hydrobionts is obtained. The numerical value of parameter h_i/d_i in the specified range x/L has the meaning:

$$h_i/d_i = 0.6 - 1.28 \tag{6.5}$$

Mechanisms of economical expenditure of a skin-covering slime

As has been discussed above, it is obvious that the slime of on the skin of swordfish has the function of friction drag reduction, working in a similar way to polymer solutions, which have been used in engineering for a long time. The difference is that the slime produced on the surface of the swordfish skin is kept in longitudinal folds due to the formation of longitudinal vortices on the surface. It is known that for a system of longitudinal vortices in transversal section, the flow is washed away from the wall between one pair of longitudinal vortices rotating on a screw line, and is directed to the wall between the next pair of vortices [310]. A stagnant longitudinal area is formed there. In a cross-section of longitudinal vortices there is a large zone with velocity profiles having bends, where friction on the wall is minimal. Thus, the stagnated areas are formed where the slime is not washed away by the flow, which results in an economical expenditure of the mucous substance. Liquid particles move in longitudinal vortices on a screw line. Therefore, in places where the flow goes from the wall, the mucous substance is as if stretched across the flow. The slime gradually diffuses across the boundary layer thickness and, most importantly, is kept in the near-wall area. The tests with polymer solutions have shown that polymers effectively reduce resistance at low concentrations of the solution if they are in the laminar sublayer of a turbulent boundary layer.

The process of formation of slime and its supply to the body surface from cells generating it endures over time. At high-speed swimming and for long periods, the formation of slime only in the surface layer of the skin is not enough for drag reduction. There is an additional system in the skin—channels with deferent ducts where the mucous substance is developed and collected (see Figures 6.4, 610 and 6.13). Calculations show that the area of epithelial channels where the slime is developed is comparable with the area of epidermis producing slime, and it can produce the same quantity of slime. The mucous substance, which is collected in the cavities of the channels, moves to the surface of skin through pores (see Figures 6.10 and 6.13) and is spent more economically than the slime developed in the epidermis. It is connected with the two separate systems of channels of the left and right sides of the body. When the body bends, the apertures of the pores decrease on the concave side and, slime is released on the opposite side. The pores open on each side of the body alternately at a frequency of the body movements. This special arrangement of channels and pores also a plays a role in the minimization of consumption of slime. The channels are situated perpendicularly to the body axis and a have small quantity of anastomosis (cross-connecting vessels) between them. This allows for the mucous substance in channels to be redistributed more economically as slime goes to the surface of skin under the influence of pressure distribution along the body during motion from the oscillations of its tail part.

The pores are distributed reasonably evenly over the body (see Table 6.2), the only exception being on the dorsal fin, where they are almost twice as frequent. The blood vessels in the skin, which influence the thermodynamic function of the produced slime, have an important role. It is known that the degree of dilution of slime in water essentially depends on temperature. The arrangement of channels *11* and *12* in the first layer of blood system and the ampoules *20* between first and second layers of blood system (see Figure 6.10) are especially important for production of slime.

Combined method of drag reduction

One of the major results of hydrobionics research has been the detection of the combined method of drag reduction in nature [25, 33, 66, 305, et al.]. In previous chapters, the results of experimental research on variants of the combined method of drag reduction were presented. In this chapter one variant of the combined method of drag reduction will be considered. However, the physical mechanism of this method has not been fully understood. On the basis of the physical mechanism, the authors consider here a method of interaction of various disturbances in a boundary layer. Results of such an approach have been partially considered in previous chapters.

Methods for the stabilization of vortical disturbances

In the flow in the corner segment between the body and fin, there are large vortical systems increasing drag and lowering the effectiveness of the movement apparatus. The slime developed in the epidermis and in the channels of the skin is not sufficient for stabilizing these disturbances. Behind the vertical fin, in addition to the many pores, the swordfish has a chain of ampoules, where slime is also generated and collected; the slime reaches the surface of the skin through deferent ducts in the area behind the fin. The higher the speed of swimming, the greater the amount of biopolymer is collected in these ampoules. On maneuvering, the strength of vortical systems behind the fin is increased. When the body bends, an additional amount of slime is produced from the ampoules on the bended side. This is how the economical self-regulating consumption of the biopolymer functions

There are elliptical cavities are on the body surface in the area of these slots and ducts, as shown in Figures 6.4 and 6.10. In the skin, in area of the gill slits and on the internal side of the gill covers, specific crypts structures have similar cavities, as illustrated in Figure 6.6a. Experimental investigations of the flow structure in different 3D cavities have been carried out in a hydrodynamic channel with a 1 m length and 0.02×0.2 m cross-section [74, 75]. In Figure 6.14, a photographgraph of the flow in a hemispherical

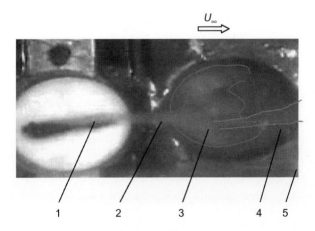

Figure 6.14
Visualization of flow in a hemispherical cavity: *1* — smoke generator; *2* — smoke; *3* — elliptical vortices; *4* — hemispherical cavity, *5* — flat plate. View from above.

cavity is shown. Measurements were carried out in closed-loop water and open wind tunnels. After visualization, measurements of velocities and spectra of the longitudinal fluctuation velocity were carried out with the laser Doppler anemometer and thermo anemometer [65, 69, 92].

Figure 6.15 shows the pattern of flow visualization in water flow in two types of cavities. A pair of longitudinal vortices is formed behind the cavities. Dimensions of the vortices are greater than those formed with the help of skin coverings. Such vortices influence the large vortical systems arising near the vertical fin.

For the stabilization of the large vortex disturbances arising at flow around the lateral fin, the crypt structure, which secretes mucous substance, plays an important role. These structures are located on the internal parts of the gill covers and on the body of the gill covers (see Figures 6.6 and 6.10). At the expense of reducing the resistance of a flow in the area this crypt-structure, the effluent from the gills has the effect that the flow "blows off" and stabilizes the large vortex disturbances arising from the lateral fins. Furthermore, at fast speeds of swimming the hydrobionts are able to press their lateral fins to their body, and so reduce their influence on general resistance.

The morphologic adaptations for the reduction of two components of drag, form drag and friction drag, and also for the reduction of vortex drag caused by the presence of prominent parts (vertical and lateral fins) were considered above. Ground resistance of hydrobionts is absent during active swimming because the tail part of the body and tail fin are a movers, which not only creates thrust, but also creates of a suction force, thus alternately reducing drag in the tail part of the stem.

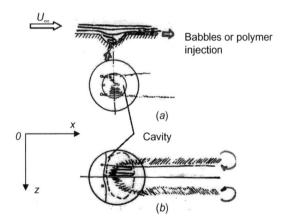

Figure 6.15

Flow visualization on cavities different shapes: a — spherical cavity, diameter $d = 1.5 \cdot 10^{-2}$ m and depth $h = 0.5 \cdot 10^{-2}$ m; b — ellipsoidal cavity with sharp edges, $h = 0.6 \cdot 10^{-2}$ m, lengths of axes are $3 \cdot 10^{-2}$ m and $2.2 \cdot 10^{-2}$ m.

The hydrobionic approach has revealed previously unknown peculiarities of the skin structure of hydrobionts and has suggested new ideas for the problem of drag reduction:

1. The comparative analysis has shown that there are identical adaptations to drag reduction and economical expenditure of energy in the structure of the skin coverings and body of high-speed hydrobionts, and specific formations produced according to characteristic speeds of swimming.
2. New knowledge and understanding has been gained of: the structure and hydrodynamic significance of the xiphoid tip; systems for the production and transportation of slime in swordfish skin; structures indicating a specific role of the slime on the skin.
3. A complex of methods has been revealed for the economic expenditure of slime and optimum arrangement of slime solutions across the boundary layer thickness.
4. On the basis of hydrobionic modeling, some of the peculiarities of the swordfish body and skin have been further checked experimentally.
5. The hydrodynamic significance of microfoliation on the surface of the swordfish skin and thickness of this skin have been analyzed with calculations; also the square of the tail mover of the swordfish has been compared with other species of hydrobiont.
6. On the basis of hydrodynamic analysis of some peculiarities of the structure of the swordfish body, a method of interaction of various disturbances in a boundary layer and combined method of drag reduction have been developed and experimentally investigated.
7. On the basis of the hydrobionic approach, some technical applications of the revealed peculiarities of the structure of skin coverings and body of the swordfish have been developed.

Important areas for future research include the exploration of some of the peculiarities revealed in the interaction of a hydrobiont skin with boundary layer vortical structures, including:

- What is the physics of the combined method mechanism?
- What are skin vibrations and their role?
- What is the boundary layer as a nonlinear wave-guide?
- How is the skin structure of high-speed hydrobionts dependent on the Reynolds number?
- What is interaction of disturbances in a boundary layer?
- What are methods of control of coherent vortical structures in a boundary layer?
- What is the physics of flow over different 3D cavities?

6.2 Experimental Research of Bodies with Xiphoid Tips

The length of a 5–6-year-old swordfish (*Xiphias gladius*) is 1.4–1.7 m, and the length of the xiphoid tip is 35–45% of the body length. The speed of swordfish in extreme situations can reach 130 km/hour. In connection with high swimming speeds, an assumption was

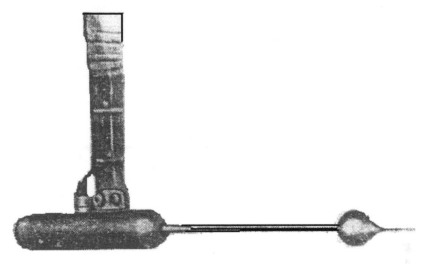

Figure 6.16
Tensometric dynamometer with a sphere [301].

made about the hydrodynamic role of the xiphoid tip. Leonenko was the first to carry out experimental research on the flow over bodies with xiphoid tips [9, 301, 302, 305]. It is better to carry out experiments on canonical bodies, and a sphere was chosen because its drag is standard across many sets of research. Figure 6.16 shows a photograph of a tensometric dynamometer with a sphere with a xiphoid tip mounted.

The sphere has diameter of 70 mm and is made of plastic. The sphere surface area is 154 cm^2, and with the tip is 197 cm^2. A wire ring turbulence promoter was placed on the sphere equator. The xiphoid tip was 140 mm long and 30 mm wide. The thickness at the base is 7 mm. The tip was made of aluminum and screwed onto the sphere by means of stud-bolt. A scheme of the tensometric dynamometer is sketched in Figure 6.17.

Experiments were done in a hydrodynamic tank. The range of working speeds was 2−12 m/s. The sphere was towed at a depth of 0.5 m from the free surface. The interval between runs was 20 minutes; the temperature of water in the tank was 15°C. In tests, the drag of the sphere was measured for the following situations:

- sphere without a turbulence promoter;
- sphere with the turbulizing ring;
- sphere with the xiphoid tip.

Results of the measurements are shown in Figures 6.18 and 6.19. In Figure 6.19 for calculating of the Reynolds number for curve *3*, the diameter of the sphere was used as a characteristic scale, and for curve *4* it was the greatest size equal to the sum of sphere diameter and xiphoid tip length.

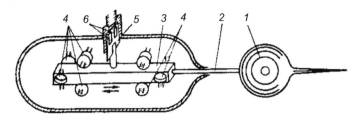

Figure 6.17
Kinematic scheme of tensometric dynamometer [301]: *1* — sphere, *2* — rod, *3* — parallelepiped,
4 — bearings, *5* — measuring spring, *6* — tensometric probes.

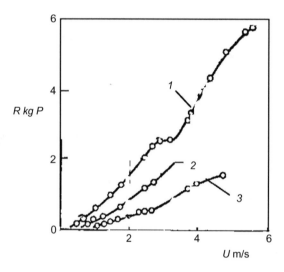

Figure 6.18
Dependence of hydrodynamic drag of the sphere on flow speed: *1* — sphere, *2* — sphere with the
turbulizing ring, *3* — sphere with the xiphoid tip [301].

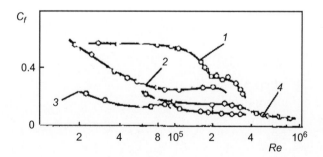

Figure 6.19
Dependence of the sphere drag coefficient on Reynolds number: *1* — sphere, *2* — sphere with the
turbulizing ring, *3, 4* — sphere with the xiphoid tip [301].

Like a tripping ring, the xiphoid tip on the sphere essentially reduces hydrodynamic drag (by 2.0–2.5 times). In both cases, the effect indicates a shift in the separation point of the boundary layer in the stern-part of the sphere. For well-streamlined bodies (a body of revolution with large elongation), the basic component of hydrodynamic drag at motion under water is friction drag (80–90%). The results obtained have led to the assumption that a xiphoid tip will also influence the drag of a body of a revolution model. Experimental research on such a model was undertaken (see Figure 6.20). The model of a body of revolution was fixed on the same tensometric dynamometer and towed in the same tank.

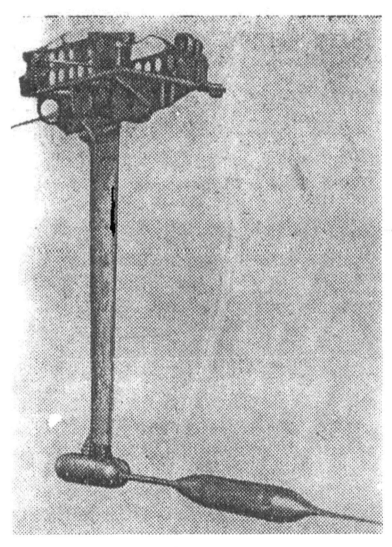

Figure 6.20
Installation for towing a model with a xiphoid tip [302].

The body of revolution was made of aluminum and had the following parameters: length without a tip, 72.8 cm; length with the tip, 109.5 cm; diameter of the body of revolution, 10 cm; the wetted surface without a tip, 2019 cm^2; the wetted surface with the tip, 2129 cm^2; diameter of the turbulizing wire, 0.2 cm; length of the turbulizing wire, 15.9 cm; distance of the turbulence promoter from the leading tip of the model, 2.5 cm. The model consisted of a cylinder with an ogive-shaped nose and stern parts. The cylindrical part was 27.3 cm long. The nose outline of the model had the shape of semi-elliptical body of revolution with the half-axis ration 1: 4. The stern outline was made according to coordinates in [302, 305].

Before towing tests, the pressure distribution along the model was measured. For this purpose, drainage apertures, connected by means of tubing with a manometer, were made in the lateral surface of the model. There were 16 drainage apertures in the model without a tip, and 20 apertures in the model with a tip.

The experiments were carried out in the open hydrodynamic pipe, the working section of which has been made in the shape of a channel with the dimensions $0.5 \times 0.6 \times 1.5$ m, at a speed of 2.5 m/s. Non-uniformity of the velocity field in the working section is not more than 2%, turbulence intensity of the stream not more than 3%. The results of the measurement of pressure are presented by a standard image in the form of the dimensionless pressure coefficient:

$$\bar{p} = 2(p_i - p_{st})/\rho U^2 \tag{6.6}$$

where p_i is pressure at the point of measurement, p_{st} static pressure, U flow speed. On the test data, the distribution of pressure along the model without and with the tip is constructed (see Figure 6.21).

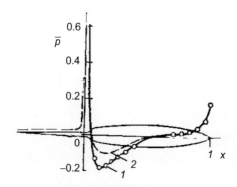

Figure 6.21
Influence of xiphoid tip on pressure distribution along rotations body: *1* — without a tip, *2* — with a tip [302].

The results showed that presence of a tip insignificantly changes the distribution of pressure for the given shape of the body of revolution. However, for the first time it was possible to find an essential difference of the distribution of pressure along a model with xiphoid tip. The distribution of pressure on the models with a tip became uniform, maximum pressure at the beginning of the model decreased twice and gradients of pressure along the model decreased. This is important for live organisms. The results of towing tests have been plotted in Figure 6.22. The model was towed at a range of speeds 4.6—10.4 m/s. The xiphoid tip had the same shape as same that of a swordfish. The detailed data of the towing tests are listed in the tables in [302, 305]. Curve *1* corresponds to tests without a tip, but with the wire ring turbulizer. There were no tests without the turbulence promoter. As the surface of the tip was smooth and no turbulence promoter was mounted, a natural transition occurred in the boundary layer on the model with the tip, depending on the speed of towage. For comparison of the results of towage of the model with the xiphoid tip (curve *3*) with data of towage without a tip (curve *1*), own drag of turbulence promoter defined from the formula in [305] was subtracted from the data of the model drag:

$$R_{turb} = 0.62 \cdot \rho(U^2/2) \cdot d \cdot l \tag{6.7}$$

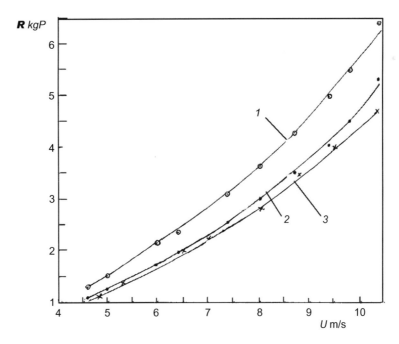

Figure 6.22

Dependence of drag of the body of revolution on the speed of towage: *1* — without a tip and with turbulizing ring, *2* — without a tip but with own resistance of the turbulence promoter subtracted, *3* — with the tip.

Figure 6.23
Photograph of various shapes of models and nose parts [9].

where ρ is density of water at temperature during tests, d and l are diameter and length of the wire turbulence promoter, respectively. Comparison of curves *2* and *3* shows that the installation of the xiphoid tip led to drag reduction on the model by approximately $7-8\%$, despite the increased wetted surface of the model with the tip by approximately 5.5%. In addition, the efficiency increases at greater Reynolds numbers, when the boundary layer on a model with a tip is turbulized.

Leonenko carried out further tests with models that had nose parts of various shapes, and he also researched the pressure distribution on models that, unlike the previously described model (see Figure 6.20), matched the shape of the swordfish body [9]. A theoretical draft of a 1.5 m-long swordfish served as a prototype for construction. Figure 6.23 represents a photograph of the nose parts of the model, and Figure 6.24 provides a theoretical scheme of the swordfish shape with drainage apertures. The model was made according to this theoretical drawing, its length with the attached tip was 1.05 m. The greatest height of the body (without the back fin) was 0.15 m, and its thickness 0.1 m.

The biggest cross-sectional area of the model was 0.0118 m^2, and the wetted surface 0.314 m^2. Drainage apertures were made with a diameter of 2.5 mm and located along the lateral (10 apertures) and bottom (12 apertures) sides of the model. The drainage apertures were connected with a battery of pressure pipes by means of rubber pipes, as usual. The high-speed pressure was measured by means of Pitot tube. Calibration of zero was done by immersing the model and Pitot tube into a pendant bath with water. The water level was the same as the free surface in the flow. The pressure distribution along the model was measured in the same hydrodynamic pipe at flow a speed of 2.5 m/s. Figure 6.25 shows these data.

In comparison with previous measurements on cylindrical models (see Figure 6.21) the influence of the xiphoid tip on pressure distribution along the body is essentially stronger.

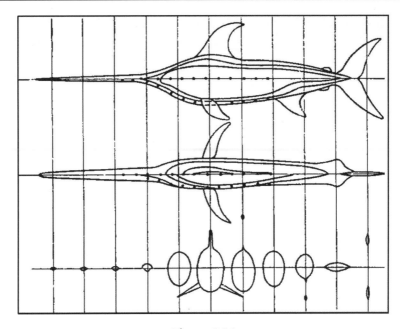

Figure 6.24
Theoretical drawing of the swordfish and the arrangement of drainage apertures on the model [9].

The same conclusions are valid for this model: the xiphoid tip leads to an essential reduction of maximum pressure in the leading part of the model, smoother distribution of pressure along the body and a reduction of pressure gradients. The maximum pressure shifts into the leading part of the tip. Its absolute value is less, as well as the wetted surface area on the tip. The magnitude of pressure along the model with the tip is less. In the opinion of the authors of the experiments all this results in earlier turbulization of the boundary layer and reduction of total friction drag of the body, which is confirmed by the corresponding measurements represented in Figure 6.7.

This series of measurements was done in 1969. At that time, the principles of hydrobionics and, consequently, the mechanism and functions of a xiphoid tip had not been formulated. The authors concluded that the effects they revealed did not explain the high swimming speed of the swordfish. Unfortunately, the technique used for the experiments was not correctly developed, as the functions of the tip were not known. As a result of this research, the authors stated that the xiphoid tip accelerates turbulization of the boundary layer. It was not considered that the top surface of the swordfish tip was essentially rough and worked as a turbulence promoter. Therefore, in experiments on the sphere, the length of the tip was twice the diameter of sphere, although the authors noted that length of the swordfish tip is 35−45% of body length. Therefore, in such experiments it would be expedient to use either

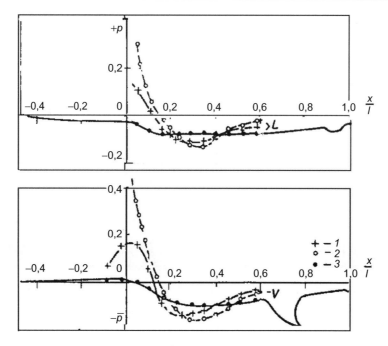

Figure 6.25

Distribution of pressure along the model of swordfish: *L* — measurement along lateral meridional sections, *V* — measurement along the bottom vertical section, *1* — with xiphoid tip, *2* — without a tip, *3* — arrangement of drainage apertures [9].

a rough tip on the sphere, or a turbulizing ring. Furthermore, it is important to correctly specify the value of the Reynolds numbers corresponding to the swimming speeds of the swordfish. At such Reynolds numbers a boundary layer is always turbulent. The same disadvantages relate to the experiments on models of a body of revolution. In these tests, the tips were not rough and no turbulence promoter was applied on the tips. Other requirements of hydrodynamic modeling were also not met. It is a fact that the swordfish swims at high speeds with a constantly open mouth, so there are round near-wall jets over the most of its body going from gill slits. However, the role of the xiphoid tip under such conditions of motion remains unexplored. Despite these disadvantages, Leonenko managed to obtain new data on the peculiarities of the flow over a body with the xiphoid tip.

In 1969—1970, theoretical and experimental research on the flow on a body of revolution with a nose needle was undertaken at the Faculty of Hydromechanics at Leningrad Ship-Building Institute [525]. The experimental research was done in a wind tunnel with body of revolution models having a total length with a tip of 1.8 m, and without a tip of 1.26 m, with a maximum diameter of 0.304 m. The length of the tip was 0.54 m, which is 30% of the length of the body with the tip and 43% of the length of the body itself.

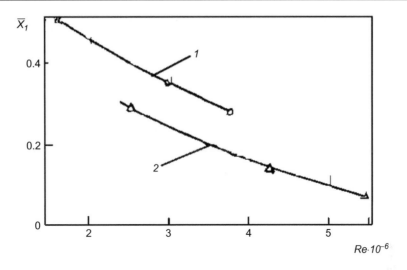

Figure 6.26
Dependence of the length of the laminar section of the boundary layer on the Reynolds number:
1 — without a nose needle, *2* — with the nose needle 0.3 L_2 long [525].

The results of the measurements performed by means of a hot-wire air flowmeter are drawn in Figures 6.26 and 6.27.

The measurements have shown that the dimensionless length of the laminar section on the model with the needle is less than that in the absence of a needle, in the whole range of Reynolds numbers investigated. The distribution of momentum thickness on both models at a flow speed of 20 m/s ($Re = 2.4 \cdot 10^6$ for the model with needle and $Re = 1.7 \cdot 10^6$ without a needle) have shown that the momentum thickness on the model with the tip is higher, beginning from $\bar{x} = 2$.

Viscous drag of the models was determined by the method of impulses from measurements of the high-speed pressure in the stream and aerodynamic wake behind the body. Results of the measurements are shown in Figure 6.28 for the whole range of measurements at $Re = (1.7-5.4)10^6$. Experimental points for both models are taken at identical flow speeds, but the obtained points are spread in Figure 6.13 due to different Reynolds numbers based on the model length. In the investigated range, the spread of viscous drag coefficients of the model with the needle appeared greater than for the model without a needle. This indicates that in the given range of Reynolds numbers viscous drag is essentially influenced by the length of the laminar section of the boundary layer (see Figure 6.26). From Figure 6.28 it follows that at $Re > 5 \cdot 10^6$, a tendency to reduction of C_x on the model with the needle is observed, while C_x increases on the model without a needle and could become greater than on the model with the needle.

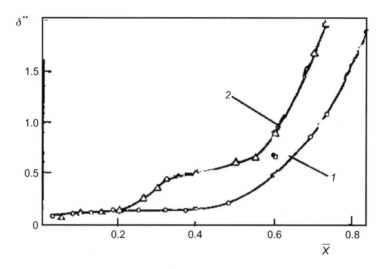

Figure 6.27
Momentum thickness along the body: *1* − without a nose needle ($Re = 1.7 \cdot 10^6$), *2* − with 0.3 L_2 long nose needle ($Re = 2.4 \cdot 10^6$) [525].

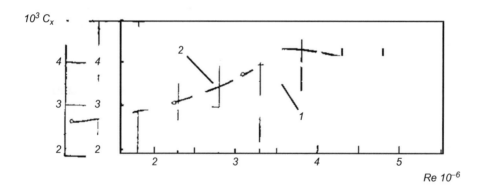

Figure 6.28
Dependence of viscous drag coefficient of the model on the Reynolds number: *1* − without nose needle, *2* − with nose needle 0.3 L_2 [525].

Despite the interesting results obtained in [525], it suffers from the same disadvantages as in works [9, 301, 302, 305].

6.3 Theoretical Research on Bodies with Xiphoid Tips

Originally it was supposed in theoretical calculations that body shape does not essentially influence the distribution of tangential stress along the body, and a mixed scheme of flow

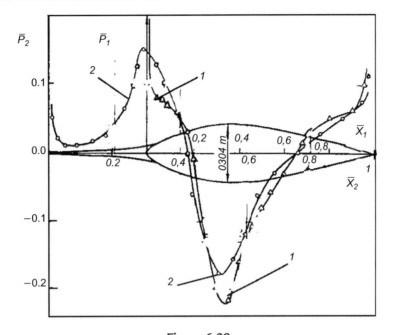

Figure 6.29

Distribution of the pressure coefficient along the model length: *1* — without a tip, $L_1 = 1.26$ m,
$x_1 = x/L_1$; *2* — with the xiphoid tip, $L_2 = 1.8$ m, $x_2 = x/L_2$ [525].

in a boundary layer was accepted. It was considered that the development of a boundary layer occurs naturally from the laminar mode of flow to the turbulent. It was also supposed that the shift of maximum tangential stress, due to presence of the xiphoid tip, into the tip area, which is small, leads to a reduction of total drag of the body. According to these assumptions, the pressure distribution on the body of revolution with and without a tip (see Figure 6.29) was calculated in [525], in the assumption of potential flow. The shapes of the model of a swordfish and xiphoid tip on which the experimental research of the boundary layer parameters (see previous section) have been performed are also shown. The pressure distribution obtained theoretically agrees qualitatively with similar results in Figure 6.25 in the previous section, according to the experimental data from Leonenko [9].

For the bodies of revolution in Figure 6.29 [525] there is a simplified calculation of friction resistance based on the assumption that the model friction drag is similar to the drag of an equivalent flat plate, the width of which changes along its length under the law *r (x)* for the corresponding model. The length of the laminar section was assumed to be according to Figure 6.26 in the previous section. Such a simplified calculation has shown that up to $Re = 3.2 \cdot 10^6$ the model without a tip has smaller drag, and at higher Reynolds numbers the drag coefficient of the model with a tip becomes less.

Kozlov and Leonenko [302] have made calculations of the drag of a body with a tip using the same initial formula as Zolotov and Hodorkovsky [525], who have chosen the method of calculation in view of the laminar section of the boundary layer. It is known that the swordfish tip is rough and the speed of its motion high [302] and so in their computations they have assumed that the law of distribution of tangential stress along the body corresponds to the turbulent mode of flow on the body.

The integral drag of the body is calculated at turbulent distribution of tangential stress along the body. Theoretical dependences of efficiency of drag reduction are obtained at presence of the tip depending on the dimensionless length and dimensionless area of the tip surface. For all parameters considered in the calculations, the tip reduces drag at the turbulent mode of flow in the boundary layer in comparison with absence of a tip. These data agree with experiments by Leonenko described in the previous section (see Figure 6.22). However, as in Leonenko's experiments, the peculiarities of the motion of swordfish are not considered in the offered scheme of calculation.

Amphilohiev et al. [17] in their calculations used the same assumptions as Zolotov and Hodorkovsky [525], that there are laminar sections in the flow over the body, a point of transition, and a turbulent section with no separation. In the calculations for two shapes of the body of revolution shown in Figure 6.30, it was accepted: 1_w — dirigible shape without a tip, 2_w — with a long cylindrical insert and ellipsoidal outlines without a tip. Three sizes of tip were analyzed for each shape of the model: S — short, M — middle and L — long.

All parameters necessary for calculation of a dimensionless kind can be found in [17]. The calculation was done on a computer using the Van Driest method, which is convenient for calculating long bodies and has no restrictions in the area of the joint of a tip and basic

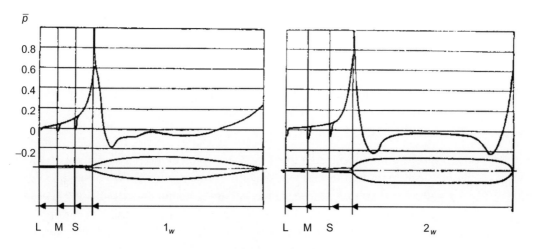

Figure 6.30
Distribution of pressure along bodies [17].

body related to inflections of the surface. Unfortunately, the calculation results for the distribution of pressure on bodies of revolution do not coincide with both the simplified calculation by the same authors in an earlier work (see Figure 6.29) and experimental data in Figures 6.21 and 6.25.

During calculations of drag coefficients of bodies of revolution shown in Figure 6.30, theoretical dependencies of the drag coefficient and relative change of friction drag were obtained. Figure 6.31 illustrates the calculations of drag coefficients of models in view of the change of Reynolds number of the moistened surface square at the location of the tip. Curves without a tip are located between laminar and turbulent laws of friction for a flat smooth plate, approach the curve for transitional boundary layer at $Re = 5 \cdot 10^6$, and are practically in the area of turbulent friction for a smooth flat plate at $Re = 7 \cdot 10^7$.

These reference curves indicate that there is an error in calculation, because it is known from the experimental results that curves of friction for a body of revolution are always situated above laws for a flat plate. Furthermore, transition on a plate begins at $Re = 5 \cdot 10^5$ while, according to the calculations, transition on the reference body of revolution begins at a higher order of Reynolds numbers. The authors do not explain what causes the delay of transition in the flow over the reference body of revolution. Curves for the bodies of revolution with tips lie in the area of laws of turbulent friction for a flat plate, which also confirms an error in the calculations. Considering that the calculations had a comparative character, interesting results are obtained indicating that for the assumptions made, the drag of a body of revolution without a tip is smaller than that of a body with a tip, up to

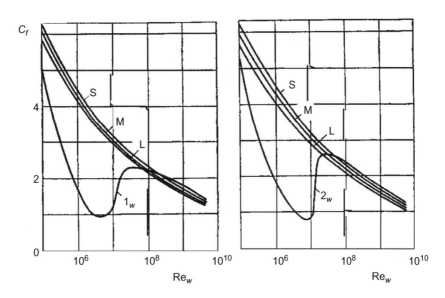

Figure 6.31
Dependence of the coefficient of a model's resistance on the Reynolds number [17].

$Re = 7 \cdot 10^7$. At further increase of the Reynolds number, tips cause drag reduction in comparison with a standard body of revolution. The longer the tip, the stronger the effect.

Voropaev [505] has also calculated the potential flow over bodies of revolution with tips. However, he performed numerical modeling with the use of the finite-difference method of calculation at presence of inflection points in the contour. With the development of the algorithm of numerical calculations, Voropaev paid great attention to the peculiarities of the joining of the xiphoid tip with the body of revolution of spherical shape or of an ellipsoid of revolution shape. Further, the algorithm allowed for the calculation of corresponding characteristics and for other shapes of a body of revolution.

The ellipsoid had the ratio of axes 1:4. The parameters varied in calculations: length of the tip was 0.5, 0.25, 0.1; fillet radius 1, 0.5, 0.3; nose rounding was 0.025 ellipsoid length, the slope of the tip 5°, flow speed 10, 20 and 30 m/s. The parameters for the sphere were: length of the tip 1, 0.5, 0.3; and fillet radius 1, 0.7, 0.35 the sphere radius. At the specified values of parameters, distributions of velocity and pressure along the bodies were calculated. Figure 6.32 represents the distribution of pressure in the flow over the ellipsoid of revolution. The abscissa axis corresponds to dimensionless coordinate S along the body contour. Curve *1* corresponds to a long tip with fillet radius equal to 0.5 longitudinal axes of the ellipse, curve *2* to a twice reduced length of the tip, and curve *3* to the same tip as curve *1* but at fillet radius equal to longitudinal length of the ellipsoid. Distribution of pressure corresponding to curve *3* is optimum, i.e. when both length of the tip and fillet radius are maximum in this calculation. The maximum length of the tip in the calculations approximately corresponds to the relative size of the swordfish tip. Thus, the calculations have confirmed the basic positive results for a tip in the experimental data in Figure 6.10 in the previous section: the peak of pressure decreases at the area where the tip and ellipsoid join, gradients of pressure along the length decrease and efficiency of the long tip is higher than that of the short tip.

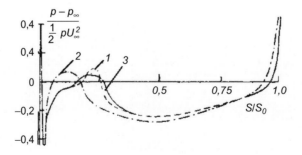

Figure 6.32
Distribution of pressure along the ellipsoid of revolution with a tip [505].

The calculations showed that the use of a tip reduces resistance by 8−10% for bodies of small elongation (i.e. a sphere), for an ellipsoid of moderate elongation by 6−8%, and for bodies of large elongation by 4−5%. These results agree with the findings of experiments by Leonenko discussed in the previous section and lead to the conclusion that the greatest efficiency improvement made by the application of a tip is on a body of revolution of small elongation. According to the calculations a short tip appeared more effective for bodies of moderate and large elongation. It is also revealed that a greater fillet radius of a tip with body is preferable. Thus, for the coupled calculation of potential flow and turbulent boundary layer it is possible to choose optimum lengths of tip and fillet radius for a continuous flow with the least tangential stresses on the entire body.

The problem of flow on a curvilinear surface has been investigated for more than 50 years. With reference to the problem of the influence of a tip on body resistance, Voropaev investigated the role of curvature in the field of the join between body and tip from the aspect of a potential flow. However, it is known that longitudinal vortical systems of the Goertler type always form on curvilinear concave surfaces. The role of these vortical systems has not been investigated. Bandyopadhyay [103] also investigated the role of curvature in the area of join between tip and body on the distribution of pressure along the body. His work consisted of two parts. In the first part he gave a detailed review of the known theoretical results in comparison with corresponding experiments on the flow over a flat plate having convex or concave sections. Initially, variations of an arrangement of a convex section behind, before or in the middle of a flat surface were considered. The length of the curved section was also analyzed in a dimensional form. The analysis showed that in all cases the convex surface leads to a reduction of the drag coefficient, which slightly increases behind the convex section. Also systematized was work in which similar combinations of a concave section behind a flat surface, or other combinations of sections of the surface, were investigated. In this case the drag coefficient increased in area of the concave section. These conclusions agree with the known peculiarities of the flow in a confuser and diffuser and correspond to the Bernoulli equation. When the length of a convex and concave section was small the results were interesting. In these cases, the laws of change of drag coefficient along length were similar: the drag coefficient decreased in the area of curviness, and then increased. Such an increase was small behind the convex section and significant at the end of the concave section. But further along the drag coefficient changed slightly. Based on this analysis it is possible to conclude that convexity favorably influences reduction of the coefficient of friction and concavity leads to an increase in the coefficient of friction. It was thought likely that vortical systems of Taylor−Goertler type develop in the concave section.

Based on the detailed analysis of the features of the flow for convex and concave sites on a flat surface, the problem of flow over a cylindrical body of revolution with a cylindrical nose and a cone-shaped tip is examined. Using a program of numerical calculations on

FORTRAN for inviscid subsonic flow over axisymmetric bodies, Bandyopadhyay [103] made calculations for these two types of tips. Figure 6.33 shows the results of these calculations for the cylindrical tip, and Figure 6.34 for the cone tip. Bandyopadhyay [103] analyzed the results as follows: Figure 6.33 shows a straightforward application of Bushnell's [138] concept to a cylindrical body. The following salient features of the nose body are to be noted: 1) the purpose of the small-diameter forebody is to grow a turbulent boundary layer that can "absorb" the viscous-drag reducing feature of the convex curvature that follows. The dimensions of the forebody, its diameter and length, should be small to keep its drag low; 2) the unavoidable concave region between the forebody and the convex region should be short, a) to avoid growth of Taylor–Goertler-like roll cells, b) to reduce

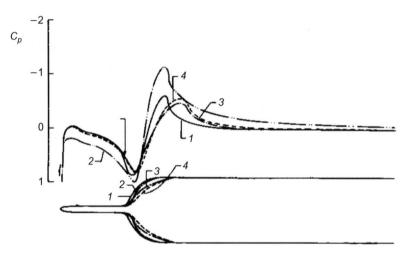

Figure 6.33
Computed surface-pressure distributions in single-stage nose geometries with straight cylindrical fore bodies. Computed separation location in all four geometries is indicated, $M = 0.5$ [103].

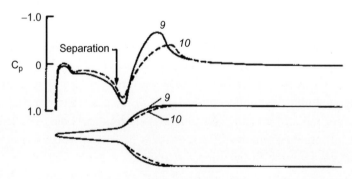

Figure 6.34
Computed surface-pressure distributions in single-stage nose geometries with mildly flared fore bodies. Computed separation location in both geometries is indicated, $M = 0.5$ [103].

integrated turbulence amplification, and c) to utilize the asymmetric effect of a convex–concave surface curvature; 3) the surface length of the convex region in terms of δ should be as large as possible, and δ/R on the convex surface should be >0.05 (analysis of available convex curvature data shows that an asymptotic state is reached for $\delta/R > 0.05$, and relaxation is slower with increase in $\Delta s/\delta$ [99]; 4) as per the concept, the levels of wall shear stress at the shoulder will now be lower than the equilibrium levels, and a viscous drag reduction can be expected due to the relaxation of the flow over the main cylinder.

The above relates to a single-stage nose body. Its surface pressure distribution is also shown in Figure 6.33. All nose-body computations have been performed at $M_{ref} = 0.5$. At this reference speed, compressibility effects are present, but shock waves do not appear at the nose–cylinder junction. The relevant Reynolds number based on the main cylinder diameter is $U_e\, d/\nu$ is $5 \cdot 10^6$. The inviscid flow pressure distributions were computed using the code developed by Keller and South [264]. The boundary-layer calculations were carried out using these inviscid pressure distributions. There are two potentially troublesome regions of pressure gradient: one is the strong adverse pressure gradient near the forebody/concave junction; the other is the strong acceleration followed by an adverse pressure gradient region at the main cylinder shoulder. Calculations showed that in the worst case the former leads to separation and the latter to the formation of shock waves and/or separation. The location of separation (zero wall shears) is marked in Figures 6.33 and 6.34. A parametric study was then conducted to essentially control the pressure gradient in these two regions.

In the present work, the surface tangents are matched at the curvature junctions. The nose-body dimensions are nondimensional by the main cylinder diameter d. The parameters studied for the nose body shown in Figure 6.31 are forebody diameter and length, ratio of concave to convex radii of curvature, and concave/convex match point location (defined by the turning angle φ), which determines the lengths of the curved regions. Figure 6.32 shows the changes in pressure distribution due to changes in the parameters of the curvature, keeping the forebody unchanged. However, none of these geometries could prevent separation near the flat/concave junction without a passive bleed.

Changes in the length of the forebody did not prevent separation either. To alleviate the adverse pressure gradients, the forebody was slightly flared (5°). The pressure distributions of the nose bodies with this additional parameter are shown in Figure 6.33. Although the flare reduced the adverse pressure gradient, it was still not adequate to prevent separation. There are several ways of alleviating this boundary-layer separation. One is by the so-called inverse design approach of Zedan and Dalton [523], for example. According to this approach, a desirable pressure distribution, i.e. likely to prevent separation, should first be chosen from experience. In Figures 6.33 and 6.34 such a distribution could be obtained by smoothing the dip in the neighborhood of separation. Then the body geometry that would create such a distribution could be computed back. Boundary-layer computations can then be carried out to check if separation has indeed been prevented. However, this approach has not been taken here, because it is likely to substantially redefine the geometry in a way that separation will be prevented, but the essential geometric features like δ/R and $\Delta s/\delta_i$ of convex curvature and other viscous drag reduction concepts, which are the main thrust of the present work, are likely to get obscured in the process.

Regular calculations of the influence of curvature radius of the joining surface on pressure distribution along a body with a tip, shown in Figures 6.33 and 6.34, agree well with similar theoretical calculations presented in Figures 6.29 and 6.30, and with experimental data in Figures 6.21 and 6.25 of the Section 6.2. Bandyopadhyay [103] conducted computations for bodies moving in air. But as his calculations are for small Mach numbers, the results in Figure 6.34 (curve *10*) are especially important—they are closest to the experimental data for the swordfish body model, the nose tip of which has a large joining radius to its body and a shape similar to that used in the calculations.

In much research work the features of the shape of nose parts of various bodies on their hydrodynamic characteristics have been investigated. For example Davis [158] investigated the distribution of pressure on elliptic nose parts of profiles and cylinders. Theoretical research on the influence of xiphoid tips located on bodies of revolution on characteristics of long bodies have shown that tips change some characteristics of bodies depending on some parameters. What was important was the relative length of tip and radius of joining with the body. It has been revealed that the pressure distribution along a body essentially changes with the presence of a tip. Tips lead to drag reduction at certain Reynolds numbers. Certain problems have been determined, but these have not been sufficiently studied yet.

The experimental and theoretical research on characteristics of long bodies of revolution considered here has shown that there has been no focused and interconnected study of this problem. Approaches from technical perspectives have not solved the problem in sight. At the same time, the hydrobionic approach has allowed researchers to define a circle of problems, which will help focus research on the problem of the flow over long bodies with tips [90, 91, 93]. The authors note, for example, that the shape of the swordfish body, and that of other xiphoid fish, has a particular joining radius developed as a result of evolution. It allows for the defining of the dimensions of these radii for theoretical and experimental research on models. Analysis of the considered results has shown that it is necessary to conduct experimental and theoretical research based on the hydrobionic approach.

6.4 Combined Method of Drag Reduction of a Body with a Xiphoid Tip and Injection of Polymer Solutions

The search for new ways of energy saving and methods of drag reduction is a task for scientific research. Babenko [34, 38, 44, 45] offers a new way of reducing drag by controlling CVSs with the so-called "combined method." The methods of CVS control he developed [62, 68, 70, 71, 78] have influence only on the coherent vortical structures, not on the entire flow. The combined method has been found in all high-speed hydrobionts [25, 45, 67, 305]. The combined method was first tested in a situation in which CVSs in the boundary layer are simultaneously influenced with the help of an elastic coating and injections of polymer solutions in the boundary layer [34, 66]. Other combined methods of drag reduction were investigated as well [38, 45, 66, 91, 305].

A method of turbulent boundary layer control with help of high-molecular polymer additives of low concentration, which imparts non-Newtonian properties to liquids, is an effective way of influencing hydrodynamic processes in a boundary layer. Such additives reduce the turbulence intensity and increase the laminar sublayer thickness, which results in significant friction drag reduction of the surface. The effect of drag reduction is shown when polymeric additives are present in the buffer zone between the viscous sublayer and turbulent core.

There are several methods of injecting a polymer solution into a turbulent boundary layer. One is the use of a soluble coating on the objects. Being gradually dissolved it gives non-Newtonian properties to the flow in the boundary layer. Porous surfaces are also used for the injection of polymer solutions into a turbulent boundary layer. Test results for a porous surface with uniform injection of a polymer solution onto the entire surface of both sides of a plate are presented in [268]. The design of a porous surface is described in the patent by Babenko [34].

The most common way to inject polymeric additives into a boundary layer is the injection of a polymer solution through a slot. There are a number of projects devoted to the problem of injecting polymer solutions through a slot. Much of the research in this field has been published in the Proceedings of Seawater Drag Reduction Conference [365, 400]. The method of injecting different liquids through several slots has also been used [78, 80, 81, 84, 85, 268, 365]. Recently the combined method of drag reduction has involved the increasing development of the slots [162, 458]. The "combined method" is used by swordfish, in which the simultaneous use of several methods of drag reduction is achieved with the help of:

- slime—a solution of biopolymers;
- the xiphoid tip;
- the formation of CVSs as a system of longitudinal vortices in the boundary layer;
- the control of the elastic skin coverings.

In the present work, the combined method, which includes the first three elements from the above list, was experimentally investigated. The parameters of the first two methods vary and the parameters of the third method are constant—the scale of longitudinal vortices in the transversal direction is identical in all cases. The intensity of the vortices changes depending on the injection speed of polymer solutions.

6.4.1 Experimental Equipment and Technique of Measurements

In order to investigate the influence of polymer solutions on the characteristics of a boundary layer a model of a body of revolution was designed. Figure 6.35 provides a scheme of this model made of Plexiglas®. The model tail part consists of cone *1*, case *2* and fairing *3*, and has the shape of a symmetric airfoil. Cases *2* are fastened to strain beam *5* with two screws *4* established inside knife *6*. The cylindrical part of the model *7*, which has a diameter of 0.04 m and consists of two parts, is screwed to case *2*. One or both sections can be installed. When two parts are mounted, the model length is 0.415 m. The nose part of the model *8* is screwed in front on the cylindrical section. One axial and eight inclined apertures are drilled in its forward face part. The conic fairing *9* and nose fairing *10* are mounted in front on the cylindrical part so as it is possible to make one or two slots of various width. The parameters of the forward slot are: width—0.8 mm, diameter—29 mm. The nose fairing *10* was designed in various shapes and, like the tail fairing, was made of Plexiglas® or metal (brass). Three tubes of 10 mm in diameter are established in the knife and are tightly connected to apertures inside the model. The design of the model allows the supply of a homogeneous liquid, or liquids with different densities, into the model. Figure 6.35 shows a scheme of the injection of liquids with different densities into the model. The design allows for the mixing of these liquids inside the model feeding into a slot or supplying different liquids into each slot.

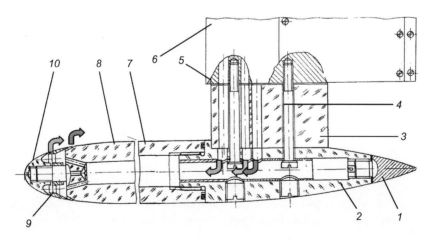

Figure 6.35

Scheme of the model for injection of polymer solutions into the boundary layer: *1* — tail fairing, *2* — tail part of the model, *3* — fairing, *4* — screw, *5* — strain beam, *6* — knife, *7* — cylindrical part of the model, *8* — nose part of the model, *9* — conic fairing, *10* — nose fairing.

Figure 6.35 corresponds to tests when the ogive nose fairing is mounted. Figure 6.36 shows the scheme of two types of xiphoid nose fairings—short and long tips.

The xiphoid tips were made of Plexiglas® and brass with the forward part at 9° and 16°. In addition, the nose part of the tip was smoothly matched to its rear part, the contours of which were the same in all tips. Tripping devices were mounted on all the tips. The turbulence promoter on the ogive tip (OT) had an outer diameter of 20 mm and was placed at distance of 6 mm from its leading edge. These measurements for the short xiphoid tip (SXT) were 11 mm diameter and 25 mm distance, and on the long xiphoid tip (LXT) they were 7.5 mm and 31 mm respectively. All turbulence promoters had a square cross-section with a 1.5 mm side and the internal part matched the nose surface. When analyzing the data obtained, it was necessary to bear in mind that the own drag of the turbulence promoters was not defined and was not subtracted from the total drag of the model.

The designs of all nasal fairings and the forward cylindrical part allowed a liquid to be injected into the boundary layer in which the system of longitudinal vortices was generated. The lengths of the ogive and the two xiphoid tips were 0.02, 0.05 and 0.08 m, and the wetted surface area of the models was 465, 468 and 471 cm², respectively The range of Reynolds numbers was $Re = 3 \cdot 10^5 - 1.5 \cdot 10^6$.

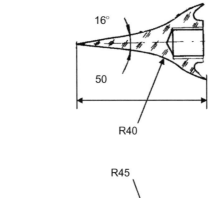

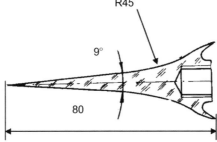

Figure 6.36
Two types of nose xiphoid fairing, short and long tips.

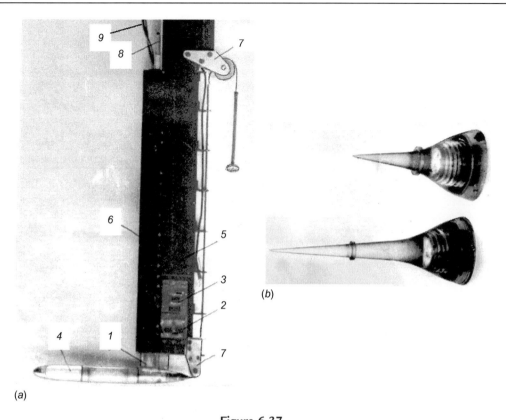

Figure 6.37
Photograph of a body of revolution (*a*) and nose xiphoid tips of the model (*b*). In *a*: *1* — fairing; *2* — strain beam; *3* — resistor element (sensor); *4* — model; *5* — knife; *6* — forward fairing of the knife; *7* — calibrating device; *8* — pipe for supplying polymer solutions, *9* — electric wires.

Figure 6.37 represents the model in working position and includes a photograph of the nose fairings. Model *4* was fastened to the strain beam *2* with fairing *1*; the strain sensor (tensometer) *3* was pasted on the strain beam *2*. The wetted surface area of fairing *1* was 65 cm^2. The strain beam was fixed in the knife *5* having forward *6* and tail fairings (not shown in the photograph). Calibration equipment *7* was placed in the tail fairing, allowing calibration before and after the experiments without removal of the model. It consisted of the removable bottom roller, fixed top roller, and Capron® rope installed on the tail part of model passing through these rollers and fixed on a loading platform. Tubes *8* for injection of a liquid through nose slots in the model and electric wires connecting strain beam with strain amplifier were established in the nose fairing of knife *5*. Electric wires *9* of direct current inside the model could be connected to the metal nose and tail fairings. The design of knife *5* allowed the measuring the drag of model *4* with fairing *1*, the area of which was much less than area the of the model.

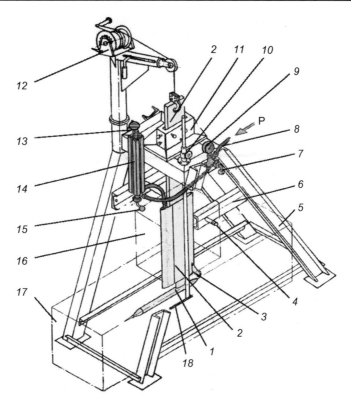

Figure 6.38

Experimental installation for research of polymer solutions: *1* — model, *2* — knife, *3* — calibration device, *4* — limit stops, *5* — frame, *6* — persistent beam, *7* — electro-crane for air submission under pressure, *8* — highway of compressed air, *9* — manometer, *10* — microscrew for Pitot gauge movement, *11* — suspension bracket for knife, *12* — crane beam, *13* — filler plug, *14* — feeding container, *15* — electro-crane of a solution submission in a model, *16* — a working site pit, *17* — water tunnel working part, *18* — Pitot pipe.

Knowing the drag coefficient of the profile with the fairing at various Reynolds numbers, it was possible to define the fairing drag and to subtract it from the total resistance of the model with the fairing.

The experiments were carried out in a closed hydrodynamic tunnel [262]. The working section was 1.8 m long, with a 0.4×0.4 m cross-sectional area, and was made of glass, as illustrated in Figure 6.38. The maximum flow speed was 3 m/s. There was a pit in the cover of the working site. To avoid the influence of vibrations of the hydrodynamic pipe case on measurement, Babenko fastened the model to a frame made of heavy channels, as shown in Figure 6.38. In the beginning the model was placed in the working section through the pit, and was then fixed on the frame with the use of a crane beam. To avoid the transfer of vibrations through the basement, the frame was established on rubber sheets to absorb the vibrations and, in addition, was fastened to the walls of the laboratory with two channels.

Model *1* was fastened to knife *2*, which was fixed in the unit of its suspension bracket *11*. The knife was fixed in its middle part by limit stop blocks *4*, established in a stop beam *6* of frame *5*. These limit stop blocks were necessary to eliminate the model vibrations caused by the flow. The knife was placed in pit *16*, and the model was in the working section *17* of the hydropipe. With the help of crane beam *12*, the model was aligned with the working section axis.

The technique of the experiments consisted of the following. The model drag at various speeds of flow was determined by means of the strain beam. The container *14* was filled with water and the filler plug *13* was closed. The crane *7* supplying pressure through the pipeline *8* was opened, thus, the pressure in the system was controlled by means of manometer *9*. Then, the crane *15* supplying pressure in the feeding container *14* was opened, and an electro seconds counter switched on simultaneously. After releasing a fixed volume of liquid from the container, the crane *15* was closed. The electro seconds counter defined the duration of the release of the liquid.

Measurement of the model drag was similarly measured for the injection of other liquids through nose slots and with two kinds of xiphoid tips installed. Before each test with polymeric solutions, the feeding container *14* was washed out with water. The closed hydrodynamic pipe was filled with pure water after each series of experiments so that the internal surface was cleared. Calibrations of the strain beam were made at the beginning and the end of each series of measurements, in order to check both the drag balance system and for the absence of an influence of polymer solutions on the characteristics of the basic flow.

A result of the injection of a liquid into motionless water was observed. Based on the calibration lines, indications of the drag balance were used to plot the model drag force against the speed of towage. Drag force R of the model is defined by the known formula:

$$R = C_f \cdot q \cdot S \tag{6.8}$$

where C_f is the model resistance coefficient, $q = 0.5 \cdot \rho \cdot U^2$ dynamic pressure, S area of the model surface wetted by the flow.

The data obtained were plotted as the dependence of C_f on Re. Considering the specific structure of the model, the total drag contains the drag of fairing resistance *3* (see Figure 6.35) and drag of the turbulence promoter (see Figure 6.37b). Korobov [292] has described in detail how to identify these components. Because the experiments were not comparative, these additional components of drag were not subtracted from the total resistance to obtain only the drag of the cylindrical part of the model. The technique of Kozlov [305] can be used to estimate the trip drag. The fairing drag contributes 14% of the total resistance. The efficiency of the polymer solutions increases up to 14% if these values are subtracted from the data obtained at injection of polymer solutions. When the direct current was fed to the nose fairing *10* and end fairing *1* of the model (see Figure 6.35), the influence of electrostatic field on the expansion of polymer macromolecules was defined and, hence, their efficiency.

6.4.2 Drag of the Model with No Injection of a Polymer Solution

Figure 6.39 presents the findings of the conducted experiments. Curves *1, 2* and *3* designate the standard dependencies of drag of a longitudinal flat plate on Reynolds number at laminar, transitional and turbulent boundary layers. Curve *4* represents the corresponding dependence of the model OT with the turbulence promoter and with closed nose slots, curve *5* the same but without turbulence promoter, and curve *6* with turbulence promoter and opened nose slots. Dashed lines give the data for the model with SXT: *7*—without turbulence promoter and with open slots at strongly aerated flow, *8*—the same, but in the pure flow, *9*—the same as *8* but with turbulence promoter. Dash-dotted lines show results of test of the model with LXT: *10*—without turbulence promoter and with open slots, *11*—the same but with turbulence promoter.

The model with OT in the further tests will be called the "standard" model. Its resistance without the turbulence promoter and with closed slots indicates that the shape and roughness of its surface are such that its drag corresponds to the transitional boundary layer on a plate at corresponding Reynolds numbers (curve *4*). Curve *5* testifies that the turbulence promoter on standard model is insufficiently effective. However, the combination turbulence promoter with opened nasal cracks provides a turbulent boundary layer on the standard (curve *6*).

It is known that a xiphoid tip is effective at a turbulent regime of flow; therefore, no tests were conducted with no tripping and with closed slots. The drag of models with xiphoid tips with the turbulence promoter on the standard model was not sufficiently effective. However, the combination of the turbulence promoter with open nose slots provided a turbulent regime of flow in the boundary layer greater than at the standard. Thus, the intensity of the growth of resistance slows down with an increase in xiphoid tip length. Based on these results, it is possible to state that xiphoid tips under the given conditions of the tests can reduce or increase drag in comparison with the standard.

The longer the tip, the greater the positive influence on the drag reduction. There are possibly optimum sizes of a xiphoid tip for drag reduction. The wetted surface of a larger

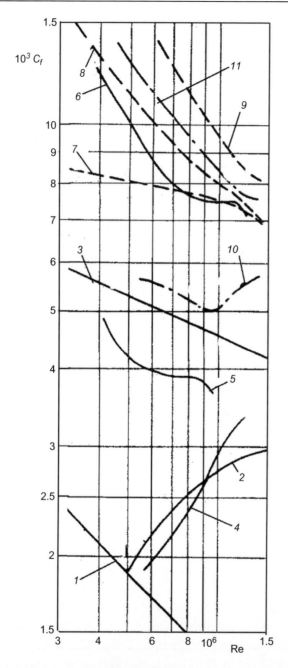

Figure 6.39

Dependence of factor of resistance on a Reynolds number at the different form nasal parts of model: *1, 2, 3* — standard dependences of resistance of the longitudinal streamline flat plate at laminar, transitive and turbulent boundary layers. Resistance of a model with OT: *4* — without the turbulence promoter and with closed nasal slots; *5* — with the turbulence promoter and closed

(Continued)

tip has a greater power to neutralize its positive effect. It is also likely that the shape of a tip cross-section also has an effect on the complex influence on drag reduction. In these experiments, the turbulence promoter was mounted on the xiphoid tips in such a way that the effective lengths of the model with various tips did not strongly differ at defined Reynolds numbers. The effective length of the model was measured from the beginning of the turbulence promoter. Considering the roughness of the swordfish tip, it was considered expedient to check the efficiency of the results obtained by placing the turbulence promoter at the beginning of the xiphoid tip.

Comparing curves *8* and *10*, it can be seen that the longer xiphoid tip, the more effective it is. Comparing curves *7* and *8*, it is clear that the efficiency of the tip essentially increases in strongly aerated streams. A comparison of curves *6* and *7* shows that the model with the less effective short tip in the aerated stream has less resistance than the reference model in the fairly aerated stream.

6.4.3 Drag of the Model with the Ogive Tip and at an Injection of Polymer Solutions

Figure 6.40 shows hydrodynamic drag of the OT model at different flow rates and concentrations of liquids injected through the nose slots. Only one forward slot was used in all the tests, as the second slot was located close to the first and so at the injection of only one kind of liquid its efficiency was insignificant.

The dotted lines mark results of the first measurement series on the reference model. Curve *4* represents the dependence of the model drag on the Reynolds number when a turbulence promoter was mounted on the nose part and slots for injection were open. There was no injection through the slots. Curve *5* shows the drag of this model when water was injected through open nose slots. The average water flow rate through the slots was $Q_m = 57.6$ cm^3/s. The average water flow rates through the slots in each series of measurements are presented in Table 6.3.

To find the dependence C_f on u_{sl}/U, the values of the momentum coefficient C_μ were determined:

$$C_\mu = 2Qu_{sl}/\rho U^2 \cdot S = 2\ Cq \cdot \rho_c \cdot u_{sl}/\rho\ U \tag{6.9}$$

◄
  slots; *6* — with the turbulence promoter and open nasal slots. Resistance of a model with SXT (shaped lines): *7* — without the turbulence promoter and with open nasal slots; at a flow over by a strongly aerated flow; *8* — without the turbulence promoter, with open slots; *9* — with the turbulence promoter and open slots. Resistance of a model with LXT (dot-dashed lines): *10* — without turbulence promoter with open slots; *11* — with the turbulence promoter and open slots.

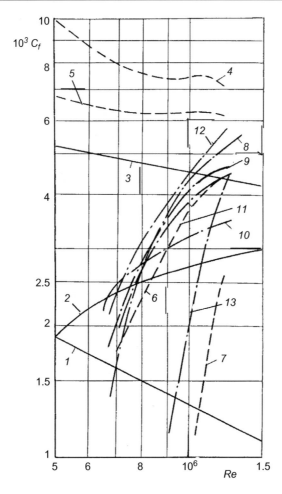

Figure 6.40

Effect of injection of PEO on the drag of the model with OT: *1, 2, 3* — same as in Figure 6.39; *4* — with turbulence promoter, open slots and without injection; injection of water at average flow rates: $5 - Q_m = 57.6$ cm^3/s, *8*—71 cm^3/s; injection of PEO with concentration and average flow rate: *6*—06—0.1%, 84 cm^3/s, *7*—0.1%, 100 cm^3/s, *9*—0.05%, 54 cm^3/s, *10*—0.1%, 59 cm^3/s, *11*—0.1%, 49 cm^3/s, *12*—0.15%, 54 cm^3/s, *13*—0.1%, 80 cm^3/s.

Table 6.3: Parameters of injection of water.

N (m/s)	U (m/s)	u_{sl} (m/s)	u_{sl}/U	Q (cm^3/s)	$Cq \cdot 10^3$	$C_\mu \cdot 10^3$
1	1.19	1.61	1.35	55.7	1.07	2.9
2	1.29	1.69	1.31	58.5	1.01	2.65
3	1.82	1.66	0.92	57.4	0.71	1.31
4	2.52	1.67	0.67	57.6	0.51	0.68

where ρ and ρ_c are densities of the injected liquid and flow, accordingly. Considering that in tests $\rho_c \approx \rho$, we obtain:

$$C_\mu = 2C_q u_{sl}/U \qquad (6.10)$$

It is important to note that the pressure in the injection system of was not controlled to provide the flow rates in each series of experiences to be such that $u_{sl}/U \sim 1$. The same pressure was in the system, while the speed of flow in the hydrodynamic pipe varied. Air was pumped up in a special bulb, and after its reducer was open air flowed into the system of boosting (position *8* in Figure 6.38 shows the path of compressed air). Further experiments were also conducted using the above technique. Pressure in the first series of experiments was the order of 2 atmospheres, and 3 atmospheres at increased flow rate. Therefore, in each series of tests with injection of a liquid, the velocity and amount of injected liquid were approximately identical, but the specified ratio changed. To obtain optimal results on drag reduction, it was necessary that the ratio $u_{sl}/U \approx 1$ was maintained. If $u_{sl}/U > 1$, the positive effect is defined not by friction drag reduction in the boundary layer but by the impulse describing the tractive force. Therefore, in the series of tests *3* in Table 6.3 this ratio comes closer to 1 and other results are not optimal for friction drag reduction in the boundary layer. Thus, values of Q conform to the average value of the whole series of experiments in Table 6.3. All the results obtained should be analyzed based on these considerations and the corresponding tables. For example, value of C_f in curve *4* at $Re = 0.9 \cdot 10^6$ is optimal; other values on the right and left from this curve are not optimal. Therefore, it is necessary to consider the corresponding values listed in the tables providing comparison of other curves. Comparison of curves *4* and *5* shows that drag of the model decreases at the injection of water.

Curve *6* represents the drag at injection of Polyox solution with a concentration 1000 ppm through the nose slots. The average flow rate of Polyox solution in each series of measurements is given in Table 6.4. The total average flow rate of the liquid for curve *6* was $Q_m = 84$ cm^3/s. The average values of flow speed were approximately same in most measurements, as well as in the previous experiments (see Table 6.3).

Pressure in the injection system was approximately the same as in the previous experiences. However, due to drag reduction, the flow rates Q and speed u_{sl} of polymers increased.

Table 6.4: Parameters of injection of polymer solutions.

N (m/s)	U (m/s)	u_{sl} (m/s)	u_{sl}/U	Q (cm^3/s)	$Cq \cdot 10^3$	$C_\mu \cdot 10^3$
1	1.04	2.16	2.08	74.7	1.61	6.7
2	2.02	2.68	1.33	92.6	1.03	2.74
3	2.48	2.38	0.96	82.0	0.74	1.42
4	2.90	2.38	0.82	82.0	0.63	1.03

It is known that polymers change the flow speed in the injection system, and this should be considered in calculations for particular technological devices and for the planning of similar experiments. On the basis of these data, it is possible to estimate the efficiency of polymers for drag reduction. It is evident that injection speed from the slot increased by approximately 40%. The optimal ratio specified above is satisfied in this series of experiments at the flow speed of 2.48 m/s (series 3, Table 6.4). In the previous experiments (see Table 6.3), the optimum ratio took place at a flow speed of 1.66 m/s (series 3, Table 6.3). Comparing curves 5 and 6, it can be seen that increasing the flow rate for the test involving injection of the polymer solution into the boundary layer of the model, drag essentially decreases in comparison with the injection of water.

Curve 7 represents the drag under the same conditions of the experiment corresponding to curve 6. Thus, the flow rate of Polyox solution (see Table 6.5) increased. The total average flow rate of the liquid for curve 7 was $Q_m = 100$ cm^3/s.

The flow rate would increase even more and be equal to $Q_m = 110$ cm^3/s if it were not for the last series of tests where the flow rate sharply decreased. Bearing in mind this correction in the comparison of curves 6 and 7, it can be seen that the model drag essentially decreased at the same concentration of polymer solution, but increased flow rate by approximately 50%. To check the reliability of the results obtained, hydrodynamic drag of the reference model was measured again. Thus, flow rates and concentration of polymer solutions changed. The results of these measurements are presented in Figure 6.40 by dot-dashed lines. Curve 8 represents the model drag for water injected through nose slots with the average flow rate $Q_m = 71$ cm^3/s, curve 9 for Polyox solution of concentration 500 ppm injected and the average flow rate $Q_m = 54$ cm^3/s (see Table 6.6), curve 10 for the injection

Table 6.5: Parameters of the injection of mixed polymer solutions.

N (m/s)	U (m/s)	u_{sl} (m/s)	u_{sl}/U	Q (cm^3/s)	$Cq \cdot 10^3$	$C_\mu \cdot 10^3$
1	1.00	3.16	3.16	109.0	2.24	14.0
2	2.06	3.16	1.54	109.0	1.19	3.7
3	2.48	3.22	1.30	111.0	1.00	2.6
4	3.02	2.06	0.68	71.0	0.53	0.72

Table 6.6: Parameters of the injection of mixed polymer solutions.

N (m/s)	U (m/s)	u_{sl} (m/s)	u_{sl}/U	Q (cm^3/s)	$Cq \cdot 10^3$	$C_\mu \cdot 10^3$
1	1.4	1.57	1.12	54.0	0.86	1.94
2	2.08	1.58	0.76	53.9	0.58	0.88
3	2.4	1.57	0.65	54.1	0.5	0.66
4	3.0	1.56	0.53	54.0	0.4	0.425

of preconditioned Polyox solution with $C = 1000$ ppm and $Q_m = 59$ cm^3/s (see Table 6.7), curve *11* for injection of the same concentration of Polyox solution prepared 7 days before the experiment with $Q_m = 49$ cm^3/s (see Table 6.8), *12* with $C = 1500$ ppm and $Q_m = 54$ cm^3/s (see Table 6.9), *13* for injection of a synthetic glue solution with $C = 1000$ ppm and $Q_m = 80$ cm^3/s. The corresponding average flow rates for each flow speed are listed in Table 6.10.

As a rule, polymer solutions are preconditioned in a long mechanical mixer with a rotating rubber screw. In the process of mixing, the Polyox powder undergoes a phase of swelling

Table 6.7:

N (m/s)	U (m/s)	u_{sl} (m/s)	u_{sl}/U	Q (cm^3/s)	$Cq \cdot 10^3$	$C_\mu \cdot 10^3$
1	0.98	1.72	1.76	59.1	1.35	4.75
2	2.2	1.71	0.78	58.9	0.6	0.93
3	2.62	1.7	0.65	58.8	0.5	0.665
4	3.0	1.65	0.55	59.0	0.44	0.485

Table 6.8:

N (m/s)	U (m/s)	u_{sl} (m/s)	u_{sl}/U	Q (cm^3/s)	$Cq \cdot 10^3$	$C_\mu \cdot 10^3$
1	1.0	1.41	0.71	48.0	1.08	1.54
2	2.24	1.43	0.64	48.0	0.48	0.61
3	2.68	1.42	0.53	51.0	0.43	0.45
4	3.04	1.43	0.47	48.8	0.36	0.34

Table 6.9:

N (m/s)	U (m/s)	u_{sl} (m/s)	u_{sl}/U	Q (cm^3/s)	$Cq \cdot 10^3$	$C_\mu \cdot 10^3$
1	1.0	1.56	1.56	54.1	1.21	3.8
2	2.0	1.56	0.78	54.1	0.6	1.02
3	2.5	1.56	0.63	53.9	0.485	0.61
4	3.0	1.58	0.53	54.0	0.4	0.425

Table 6.10: Parameters of injection of synthetic glue solution.

N (m/s)	U (m/s)	u_{sl} (m/s)	u_{sl}/U	Q (cm^3/s)	$Cq \cdot 10^3$	$C_\mu \cdot 10^3$
1	1.08	2.14	1.98	74.0	1.54	6.15
2	2.05	2.44	1.19	84.0	0.92	2.20
3	2.48	2.52	1.02	87.0	0.78	1.60
4	2.98	2.20	0.74	76.0	0.57	0.85

when water molecules join Polyox molecules. During further mixing, Polyox uniformly distributes in the liquid and its molecules gradually elongate. The preconditioning processes for Polyox are investigated in a number of papers [365, 411]. Babenko [78] developed a device for preconditioning polymer solutions. If Polyox is well preconditioned for application its efficiency increases. If the Polyox is not used immediately after preparation its molecules begin to fold into balls, thus, its efficiency decreases. To confirm this, a series of experiments were performed. The data are presented by curves *10* and *11*. The results show that a fresh solution (curve *10*) gave the best results in comparison with the Polyox solutions prepared a week before use.

Curve *9* ($C = 500$ ppm and $Q_m = 54$ cm^3/s) illustrates the injection of a polymer solution of low concentration and the same flow rate as for the injection of water (curve *5*, $Q_m = 57.6$ cm^3/s). It can be seen that the injection of the low-concentration polymer solution is essentially more effective than water. For a higher concentration of the polymer solution (curves *10* and *11*) its efficiency increases (compare to a curve *9*). However, a further increase in concentration at the same flow rate reduced the effect (curve *12*). The increase in concentration and flow rate (curves *6, 7* and *13*) gave maximum efficiency for the given test conditions.

Curves *8, 6* and *13* characterize tests with approximately identical flow rates of the injected liquid. The drag reduction was essentially higher for the injection of polymer solutions into the boundary layer than for the injection of water. However, about the same effect was obtained for the injection of water (curve *8*) as for the injection of polymer with a smaller flow rate (curve *12*). In relation to this, the assumption was made that because of the small size of the model and great flow rates of injection of liquids through the nose slots, that what was at work was the creation of a hydro jet thrust. To check this assumption, in each series of measurements on the model in the mooring mode (i.e. in still water) a liquid was injected through the nose slots. It was found that the model design created propulsion due to injection of solutions through the nose slots. Thus, a double mechanism was in effect for the injection of polymers, the creation of propulsion and friction drag reduction in the boundary layer due to specific properties of polymers.

Considering the complexity of preparation of Polyox solutions and their high cost, tests were conducted with its analogue in the form of synthetic glue (curve *13*), the efficiency of which was comparable with that of Polyox.

At injection of Polyox solutions measurements in a mooring mode were not taken. In order to define the role of polymer solutions in drag reduction, it is enough to subtract the corresponding data in a mooring mode for synthetic glue or water.

The mechanism of generation of thrust by means of a liquid injected through nose slots is rather attractive. However, it has the essential disadvantage that a jet in the nose part of a body essentially increases speed near the body surface, which leads to an essential increase in the friction drag. Therefore, such a scheme has not been utilized in engineering. However, the results obtained have shown that this technique of creating a hydro jet thrust by means of polymer solutions means that the friction essentially decreases and, thus, the disadvantage of such a scheme is eliminated. Low values of model drag have been obtained in tests. The derived friction coefficient was essentially less than in curve *1* (see Figure 6.40) and consequently are not shown on the graphs. It is possible to assume that the areas of curves located below curve *1* characterize the hydro jet thrust created in such a way. However at increased speeds (i.e. Reynolds number), as can be seen in all the plots, the contribution of thrust to the drag reduction decreases. To maintain the thrust at a certain level at $Re > 1 \cdot 10^6$, it is necessary to apply some additional means, for example, increasing the impulse of the jet or more ballast for the jet, i.e. an increase in the concentration of the polymer solution.

Injection of polymers or glue solutions produces a better effect than water injection, especially at small Reynolds numbers. The greatest reduction of model drag was achieved with the injection of a polymer or synthetic glue solution (curves *7* and *13*) with a concentration of 1000 ppm and a large flow rate (100 cm^3/s and 80 cm^3/s respectively).

At identical low flow rates (curves *9−12*) and different concentrations, the best result was achieved with the polymer solution at a concentration of 1000 ppm (curve *10*). The model drag became about the same as the drag of the model without tripping and with closed slots (curve *4*, Figure 6.39). An increase in the liquid flow rate through the slots gave a greater effect than the optimal concentration of the solution.

Injection of polymer solutions through nose slots resulted in a new law for the model drag in comparison with water injection.

6.4.4 Drag of the Model with Xiphoid Tips and Injection of Polymer Solutions

Similar research was carried out with two kinds of xiphoid tips installed on the nose part of the model. The data from these experiments are presented in Figure 6.41. The dashed lines represent the model drag with S XT and dot-dash lines show the drag with LXT. As on the reference model, mooring tests were carried out to determine the influence of the liquid injection on thrust. It was found that the injection of water through the nose slots on the model for both xiphoid tips did not influence the reduction of hydrodynamic drag. However, the thrust in mooring mode was found to be greater than for the reference model.

The following conclusions can be drawn from the comparison of the measurements on the models with xiphoid tips with similar results for the reference model: In spite of the fact that the flow rates of injected polymer solutions were essentially less on the models with xiphoid tips than on the standard, their resistance appeared smaller. Only for the injection

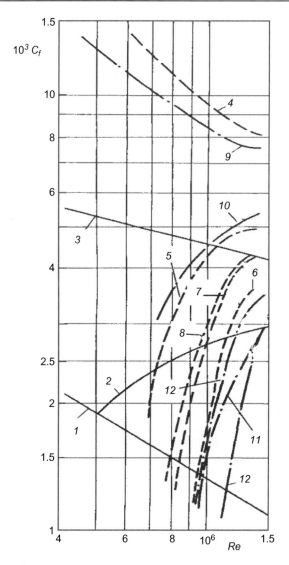

Figure 6.41
Effect of injection of PEO on drag of the model with SXT (dashed lines) and LXT (dot-dashed
lines): *1, 2, 3* — same as in Figure L.39; *4, 9* — with turbulence promoter, open slots and
without injection of a liquid; *5, 10* — at a water injection with the average charge at
$Q_m = 51$ cm^3/s; injection of PEO concentration and average flow rate: *6*—0.05%, 47 cm^3/s,
7—0.1%, 34 cm^3/s, *8*—0.15%, 28 cm^3/s, *11*—0.05%, 50 cm^3/s, *12*—0.1%, 45 cm^3/s,
13—0.15%, 38 cm^3/s.

of the synthetic glue solution was the drag of the standard comparable with the drag of the models with xiphoid tips.

The character of dependencies of the drag on Reynolds number is almost identical for all the models. The model with the long tip had the least drag, and the best result was seen for the polymer solution concentration of 1000 ppm, as well as for the standard. The only exception was the model with SXT, where the best result was obtained for the polymer solution at a concentration of 500 ppm.

6.4.5 Friction Drag for the Injection of Polymer Solutions into a Boundary Layer

Considering that injection of water gave the same effect at any shape of tip, it is possible to compare the obtained data by subtracting the thrust arising at the injection of a liquid through nose slots in mooring tests. Such comparison has qualitative character, because the thrust was not measured at mooring tests with polymer solutions injected. To balance the absence of such measurements, the values of model drag coefficient located below curve *1* on all graphics were not considered in the calculations. The magnitude of force determined at the injection of the solution of synthetic glue during mooring tests was subtracted from data in Figures 6.40 and 6.41. It obviously reduced the efficiency for the injection of polymer solutions on the models with SXT and LXT, but has allowed the defining of the basic differences compared with the reference tip. Mooring tests do not allow the assessment of a reliable value of friction drag because the wall jet is injected into a motionless liquid. It is a case of a submerged wall jet. Kinematic characteristics and structure of such a jet differ from the corresponding parameters of the wall jet injected into a boundary layer. However, the data in Figure 6.42 allows the analysis of qualitative characteristics of the problem considered.

Analyzing all the data of the experimental research of injecting a liquid through nose slots, it is evident that after the ratio $u_{sl}/U \approx 1$ is satisfied, on all the figures the curves become equidistant to the reference curves in the absence of a liquid injection. This indicates that the influence of the thrust of the nose jet, at the parameters of the injected jet, becomes minor in comparison with the influence of wall jets upon the friction drag. The assumption has been made that, at the specified ratio, the difference between drag coefficients on the reference curve and at the injection of polymer solutions is constant in the whole range of Reynolds numbers investigated. So it is possible to draw curves of drag polymer injection equidistant to the reference curve for the range of Reynolds numbers. Below is the analysis of the data with drag at mooring tests subtracted.

Figure 6.42 illustrates the computed data for polymers injected on the model with OT. It is interesting to compare these results with the data in Figure 6.40. It appears (see Figure 6.42, curve *5*) that injection of water does not influence the drag reduction at small Re, and

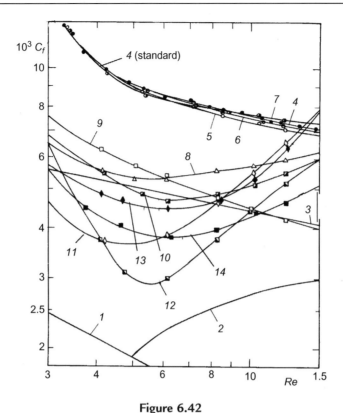

Figure 6.42
The effect of a PEO injection on the drag of the model with OT: *1, 2, 3* − same as in Figure 6.39; *4* − with turbulence promoter, opened slots and without injection (standard); injection of water at average flow rate: *5* − $Q_m = 57.6$ cm³/s; *6*−71 cm³/s; *7*−73 cm³/s; PEO injection with concentration and average flow rate: *8*−0.05%, 54 cm³/s; *9*−0.1%, 59 cm³/s; *10*−0.1%, 49 cm³/s; *11*−0.15%, 54 cm³/s; *12*−0.1%, 80 cm³/s; *13*−0.1%, 84 cm³/s; *14*−0.1%, 100 cm³/s.

slightly reduces drag at greater Re. Precisely same conclusions were made for increased water flow rate, as shown in Figure 6.40 (curve *8*) and Figure 6.42 (curves *6* and *7*). The data shown in Figure 6.42 have led to a number of additional conclusions. Practically all curves have extrema obtained at $u_{sl}/U \approx 1$. The shape of curves to the left of the extremum is equidistant to a certain extent to the reference curve *4*. At increasing Re the efficiency of polymer solution injection on the model with OT decreases. The only exception was curve *9* in Figure 6.42 (and Figure 6.40, curve *10*), which characterizes the influence of fresh polymer solution injection. With growing Re its efficiency increases. An increased concentration (Figure 6.42, curve *11*, and accordingly Figure 6.40, curve *12*) and, in some cases, polymer flow rate (Figure 6.42, curve *13*, and accordingly Figure 6.40, curve *6*) show that the drag increased in comparison with the standard at Re = $1.4 \cdot 10^6$. The effect was

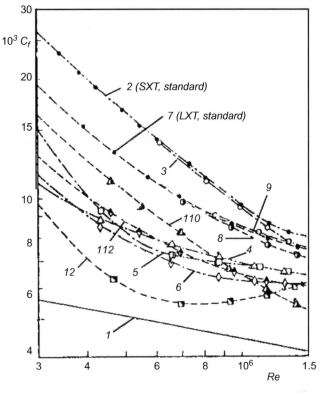

Figure 6.43

The effect of a PEO injection on the drag of the model with SXT and LXT: *1* — same as *3* in Figure 6.39; model with SXT: *2* — with turbulence promoter, opened slots and without injection (standard); *3* — injection of water at average flow rate $Q_m = 51$ cm^3/s; PEO injection with concentration and average flow rate: *4*—0.05%, 47 cm^3/s; *5*—0.1%, 34 cm^3/s; *6*—0.15%, 28 cm^3/s; model with LXT: *7* — with turbulence promoter, opened slots and without injection (standard); injection of water at average flow rate: *8* — $Q_m = 51$ cm^3/s, *9*—70 cm^3/s; PEO injection with concentration and average flow rate: *10*—0.05%, 50 cm^3/s; *11*—0.1%, 45 cm^3/s; *12*—0.15%, 38 cm^3/s.

observed for the whole range of Re numbers at injection of a polymer solution with a high concentration and flow rate (Figure 6.42, curve *13*, and accordingly Figure 6.40, curve *7*). In the range of Re = $(3.5-9.2) \cdot 10^6$ the solution of synthetic glue had a maximum efficiency and competed with the polymer solution practically in the whole range of Re (Figure 6.42, curve *12*, and accordingly Figure 6.40, curve *13*).

Figure 6.43 represents the same computations for polymers injected on the model with SXT and LXT. Similar to the model with OT, the data in Figure 6.43 leads to some additional conclusions, in comparison with Figure 6.41. The injection through a slot of water did not affect the model drag. Unlike OT, the majority of curves have no extremum and maximum efficiency is observed in a wide range of Reynolds numbers. In the whole range of Reynolds numbers considered, the shape of curves is equidistant, to a certain extent, to the reference curves *2* and *7*. At greater Re the efficiency of polymer solution injection on the model with SXT and LXT also decreases, but remains positive at any Re. Thus, the efficiency of xiphoid tips is considerably higher in spite of the fact that Q_m is essentially less than on the model with OT. Therefore, efficiency has twice increased at small Re, and even more at greater Re. With OT, when the efficiency decreases with growing Re, and even disappears in many cases, utilizing xiphoid tips makes it possible to maintain high efficiency at any Re. The only exception is the model with OT, curve *9* (0.1%, 59 cm^3/s) in Figure 6.42, when efficiency is kept at any Reynolds number. Similar results were achieved in tests with LXT, curve *10* (0.05%, 50 cm^3/s), but there the polymer solution concentration was 2 times less.

The efficiency of a polymer solution injection increases with the increase of xiphoid tip and depends less on Q_m (see Figure 6.43, curves *5* and *11*). Tests with the model with SXT and LXT have shown that injection of a polymer solution is most effective for SXT and LXT at $C = 0.15\%$ (curves *6* and *12*), but since Re = $1.2 \cdot 10^6$ the maximal effect is reached at $C = 0.05\%$ (Figure 6.43, curve *10*). Unlike tests with OT, the difference in speeds of flow and injected liquid did not have such a strong effect as the efficiency of a polymer solution injection, which was displaced in the area of higher Re.

The experimental research presented here examined the efficiency of one kind of combined method of drag reduction, and also the hydrodynamic value of the xiphoid tip of the swordfish. Results of the research have shown that using the xiphoid tip, injections of polymer solutions into a boundary layer and forming longitudinal vortical systems in the boundary layer is an effective combined method of drag reduction. Unfortunately, it was not possible in these tests to apply tips of sufficient length with tripping in the start of the tips. But this research has shown that the drag reduction increases with increasing tip length.

It was also not possible to create optimal sized longitudinal vortices for each flow speed. It is obvious that in order to increase the efficiency of the combined method of drag reduction it is necessary to form proper longitudinal vortices in a boundary layer system for the given conditions of the experiments. It is known that the size of longitudinal vortical systems in a turbulent boundary layer near the wall of a rigid flat plate depends on the flow speed. Babenko [66] has experimentally found that the size of longitudinal vortical systems change in the flow on an elastic surface. The same results are obtained from research on longitudinal coherent vortical structures in a boundary layer in the presence of a polymer solution [361]. These factors are necessary for controlling the sizes of such vortical structures. Maximum efficiency of the present combined method is obtained on the model with LXT at $C = 0.05\%$ and $Q_m = 50$ cm^3/s (curve *10*). The use of solutions of cheap synthetic glue is also promising.

Use of the present combined method has revealed high efficiency with all kinds of tips for the formation of thrust at the injection of solutions through nose slots at Re $\leq 10^6$. To make the method more efficient it is necessary to use fresh polymer solutions and to reduce diffusion of the polymer solutions across the boundary layer thickness. Babenko [78, 81] has developed devices for the effective preparation of polymer solutions. Babenko and Shkvar [84] and Moore [364, 365] have conducted research on the reduction of polymer solution diffusion by means of multilayered injection of different liquids across the boundary layer thickness. The results obtained have revealed the functional significance of the structure of some systems of the swordfish. It is known that the gills of the swordfish have strong muscles, regulating the size of the gill slots, the flow of a liquid through the gill apparatus and, probably, the size of the morphological structure inside gills. Such control is necessary not only to control the drag at various speeds of swimming but also for effective maneuvering during swimming. Tests have shown that increased flow rate and increased concentrations of polymer solutions drastically increase thrust at Re $\leq 10^6$. Regulating the size of the gill slot and thrust created by the tail fin, it is possible to control the additional nose thrust.

Essential future research will involve investigating the physical mechanism of the combined method of drag reduction that the swordfish effectively applies in a combination with elastic skin coverings and heating of the liquid inside the gills.

6.5 Physical Mechanism of the Influence of Xiphoid Tip on Drag Reduction

The findings of various authors on the hydrodynamic significance of the xiphoid tip for drag reduction are represented above. Below, the main conclusions from this research are listed:

* For a body of revolution of cylindrical, spindle- or other shape the presence of a xiphoid tip of different lengths can have an insignificant or significant effect, to the order of 8%, on the reduction or increase in resistance.
* The measurements have shown that, with the presence of a xiphoid tip, the distribution of pressure along a body essentially changes in the area of the tip, gradients of pressure decrease.
* A xiphoid tip mounted on a blunt body (a sphere) reduces drag by 2−3 times.
* Water injection through ring slots in the nose part of a cylindrical body of revolution with a xiphoid tip, where there is a maximum in the pressure distribution, creates additional thrust.

- If polymer solutions or other similar liquids are injected through this crack, depending on certain parameters, there is an essential additional draft in the nose parts, which remains significant in the absence of a xiphoid tip, but at smaller Reynolds numbers.
- The presence of a xiphoid tip has a double function: at smaller Re numbers there is essential nasal thrust, and at increasing Re the additional thrust disappears but steady drag reduction of the order of 20−30% is maintained under given test conditions.

The last conclusion is drawn from the data of tests with injection of polymer solutions when only the component of drag reduction was considered. At the injection of polymer solutions we shall define for each Re according to values of efficiency of drag reduction the parity expressed as a percentage:

$$\Delta C_f\% = 100 \cdot (C_f - C_{f\,\text{inj}})/C_f \qquad (6.11)$$

where C_f is the coefficient of drag without injection and $C_{f\,\text{inj}}$, with injection. In Figure 6.44 the results of efficiency are given at injection of polymer solutions on the model with OT, and in Figure 6.45 with SXT and LXT [67, 72].

According to Figure 6.44, in a cylinder with OT ΔC_f is a function of concentration C. The best results relate to curve *5*, whereas with $C = 0.15\%$ (curve *10*) with increase in Re,

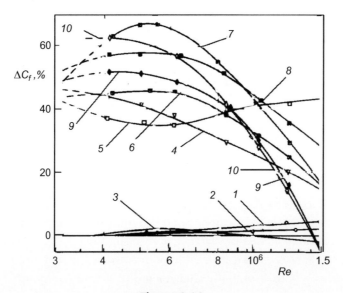

Figure 6.44

Efficiency of injection of polymer solutions of model with OT: *1, 2, 3* − injection of water at average charges $Q_m = 57.6$; 71; 73 cm³/s; injection of PEO concentration and average charge: *4*−0.05%, 54 cm³/s; *5*−0.1%, 59 cm³/s; *6*−0.1%, 49 cm³/s; *7*−0.1%, 100 cm³/s; *8*−0.1%, 80 cm³/s; *9*−0.1%, 84 cm³/s; *10*−0.15%, 54 cm³/s.

injection increases C_f compared to the standard. At greater Re efficiency declines as well with a solution of synthetic glue (curve *9*). The difference is dependent not only on average charge Q but on the age of the solution when its concentration C is the same (curves *2* and *3*). Preparation of polyethylene oxide (POE) solution just prior to use is preferable to preparing it 7 days prior to the experiments (curve *6*).

In a cylinder with SXT, the change in ΔC_f was virtually independent of C, whereas with LST it was a function of C. The decrease in C_f for all nose sections is most significant with $C = 0.1\%$, although best results were obtained with the cylinder with LST starting at $Re = 1.4 \cdot 10^6$ at $C = 0.05\%$ (curve *7*).

In spite of the lowest Q, the best results were obtained for a cylinder with SST and in some cases for one with LST. With a rise in Re ΔC_f decreased. It was not possible to carry out studies with $Re > 1.5 \cdot 10^6$. This decline occurred because of change in u_{sl}/U ratio. With lower Re, the ratio is optimal and commensurate with *1*. With increase in Re, u_{sl}/U became non-optimal as u_{sl} at changing U in these experiences practically did not vary.

Comparison of results of efficiency for the model with OT and with xiphoid tips has led to the following basic conclusions. The efficiency of the decrease in resistance of friction for the model with OT appeared up to numbers $Re = 1.4 \cdot 10^6$. With the growth of Re,

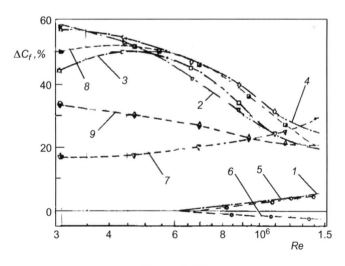

Figure 6.45
Efficiency of injection of polymer solutions of model with SXT and LXT. Model with SXT: *1* — injection of water at average charge $Q_m = 51$ cm³/s; injection of PEO concentration and average charge: *2*—0.05%, 47 cm³/s; *3*—0.1%, 34 cm³/s; *4*—0.15%, 28 cm³/s. Model with LXT: *5, 6* — injection of water at average charge $Q_m = 51$ cm³/s and 70 cm³/s; injection of PEO concentration and average charge: *7*—0.05%, 50 cm³/s; *8*—0.1%, 45 cm³/s; *9*—0.15%, 38 cm³/s.

efficiency of the model with OT essentially decreases, and, in some cases, resistance increases. Peak efficiency was in a range of numbers $Re = 4 \cdot 10^5 - 10^6$.

Curves of efficiency for the model with xiphoid tips had no expressed extrema in comparison with the model with OT. Thus, high efficiency was observed in the whole range of investigated Re, which was kept practically constant, unlike the model with OT, and at $Re \geq 1.5 \cdot 10^6$. The advantage of the model with xiphoid tips was that up to $Re = 10^6$ a high nasal draft was formed.

Until now the results of theoretical and experimental research has not resulted in a hypothesis of a physical model of influence of the xiphoid tip on drag reduction. It is known that there are various components of the common body resistance moving in a liquid. The greatest contribution to the common resistance under certain conditions of movement gives resistance of the form. Therefore, at movement of surface vessels the greatest effect allows for the optimum body form, for example, for the reduction of wave resistance. In addition, for moving bodies under water, the great value is the resistance of friction. In all cases, however, in the beginning it is necessary to try to define the optimum form of a body that gives the greatest effect.

Based on the peculiarities of movement of high-speed hydrobionts and on the basis of the test results given in the previous sections, the authors offer a hypothesis for the physical mechanism of influence of the xiphoid tip on drag reduction.

It is possible to distinguish two basic ways of movement for hydrobionts: anguilloid and scombroid. The first is characteristic of low-speed animals and it differs in the big constant along the body amplitude of the propulsive wave. The second is used by high-speed hydrobionts, involving a propulsive wave of low and variable amplitude and high frequency, and the movement of the body is provided with oscillation of only the tail fin. Logvinovich [345, 347], Lighthill [340] and Kozlov [306] have offered a hydrodynamic theory of the scombroid mode of motion, and Logvinovich [345] and Lighthill [340] have given a hydrodynamic theory of anguilloid motion.

However, the theories regarding the movement of water animals has not covered certain peculiarities until now, in particular, the work of the branchiate device, the shutters of which are mobile and control the charge of the fluid injected through the branchiate crack. In this section, the hydrodynamic approach to the research on the mechanism of a flow in relation to water animals from the aspect of laws of separation flow is offered. On the basis of the results of experimental investigations, the role of the branchiate device for the scombroid way of movement has been analyzed.

Figure 6.46 shows photographs of visualization by means of the tellurium method of a flow of the cylinders located across a stream. In Figure 6.47 images are given of a flow of the

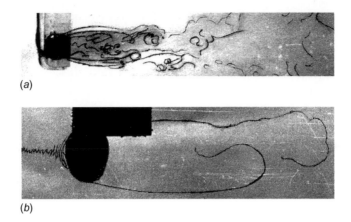

(a)

(b)

Figure 6.46
Flow of cylinders in a diameter of 6 mm (*a*) and 15 mm (*b*), located across a stream.

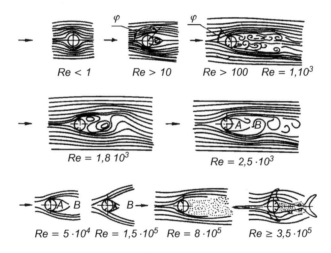

Figure 6.47
Flow around an endless cylinder at different Reynolds numbers.

infinite cylinder located across a stream visualized by means of the tellurium method [305] up to values $Re = 2.5 \cdot 10^3$ and at $Re > 5 \cdot 10^4$ on Dryden's measurements [144, 145].

Reynolds numbers are calculated on the diameter of the cylinder d. At $Re < 1$ flow of the cylinder was continuous (unseparated) and at $Re > 10$ separation of the laminar flow occurred leading to the formation of a zone of weak vortices in the fodder part of the cylinder. Strokes inside the cylinder and corner φ designate zones of flow separation.

Further increases in Re led to the formation of two symmetrical vortices, which with the increase of Re grew in size, and at Re > 100 began to separate, forming a Kármán street. Visualization by means of tellurium allowed the defining of a flow of the cylinder not only depending on Re but also on d. In Figures 6.46 and 6.47 flow lines around a cylinder with $d = 6$ mm are shown at Re $= 1 \cdot 10^3$ and with $d = 15$ mm at Re $= 1.8 \cdot 10^3$ and $2.5 \cdot 10^3$.

At Re $= 1.8 \cdot 10^3$ a Kármán street was formed at once behind the cylinder, and with the increase in Reynolds number was displaced—being removed from area B and forming zone A, near to the cylinder, free from Kármán vortices and filled by retrogressive flow. At smaller diameter d, at Re $= 1 \cdot 10^3$ the picture of a flow was similar to the picture at Re $= 2.5 \cdot 10^3$. The difference was in the vortex sizes and their intensity. These results were coordinated with measurements of dependence of the coefficient of ground pressure from the diameter of the streamline cylinder [440].

The increase in Reynolds number leads to the growth of zone A, which then decreases while the zone B constantly increases. At Re $> 2 \cdot 10^5$ zone A practically disappears and the Kármán street collapses and behind the cylinder the area of a turbulent flow (see Re $= 8 \cdot 10^5$) begins. In the range Re $= 10^3 - 10^5$ laminar separation of flow takes place at $\varphi = 80 - 85°$. Transition of a laminar regime of flow to turbulent in a boundary layer occurs behind a zone of separation. However, if at Re > 10 the area of transition was behind a point of connection of separation flow lines, at an increase in Re it moves against a stream. In the range Re $= 2 \cdot 10^5 - 3.5 \cdot 10^6$ flow separation remains laminar, but transition follows immediately after the separation. Then, the boundary layer joins the cylinder and repeated separation of the turbulent boundary layer occurs at $\varphi = 110°$, so on the surface of the cylinder between these corners φ a "bubble" forms of increasing ground pressure (lowering under pressure behind the cylinder) and reducing coefficient of resistance C_f.

At Re $> 3.5 \cdot 10^6$ the "bubble" disappears; area B essentially decreases because the transition of the boundary layer occurs before a separation zone. As well as separation flow in the boundary layer, it becomes turbulent. At Re $= 10$ and 10^6 boundary layer separation occurs at the same φ, however the coefficients of resistance in both cases differ by two orders [427] owing to the distinction of coefficients of ground pressure [440]. It is known, that coefficient C_f is minimal at the cylinder at Re $> 5 \cdot 10^6$.

Pershin [396] investigated the hydrodynamic characteristics of various kinds of water animals. In relation to the principle of receptoring regulation [36] it is possible to consider the evolution of water animals from the point of view of flows separation. The energetic capabilities of protozoans did not enable them to overcome a boundary of Reynolds numbers calculated on the length of the body Re $= 1$ [396] and, therefore, the form of their body consists either of spherical or cylindrical elements. Plankton moves at Re $= 10^0 - 10^2$. From Figures 6.46 and 6.47 [440] and Britter et al. [134] it is shown that behind a cylindrical body under pressure the degree of turbulence increases. Increased loadings according to [36] have led to the lengthening of a body. The results of experiments in which vortex separation has been eliminated by means of a thin partition along the longitudinal axis of a flow in the cross-section of the cylinder have shown an essential reduction of ground pressure and a decrease in C_f [440].

If calculations are made with Re based the diameter of the cylinder in Figure 6.47, and on Re defined by length, Re at which various kinds of water animals move are reached. Fish move at $Re_L = 10^3-10^7$ and certain kinds at $Re_L \approx 10^8$. Comparison of lines of a current at cross-section flow of the cylinder (see Figure 6.47) and forms of fish [431] has shown that the area borrowed at these numbers Re_L by cross-section section of the cylinder, the braked liquid before it and zones *A* and *B* is similar to the form of cross-section section of a body of fish along a longitudinal axis of a body. For an illustration, at $Re > 3.5 \cdot 10^6$ in Figure 6.47 contours of a body a swordfish are represented in the corresponding scale. Apparently, nonstationary flows of the considered cylinder and presence of Kármán Street for poorly streamlined bodies have caused, in the evolution of water animals, the occurrence of curved-oscillatory movement of a body or its tail part in conformity with of their speeds, movements and Re numbers. It has resulted either in the undulating or scombroid way of movement. Research [431, et al.] has shown that cross-sections of fins and bodies along a stream correspond to the best modern aerodynamic structures [350, 427]. Babenko [94] has developed a new structure of a wing by analogy with the shape of some birds.

It is known that at cavitations and supercavitations in the flow of a body (for nasal and obstacle of various form) separation arises inside of which the moving body is located. Thus, resistance becomes minimal and is defined only by the area cavitator. This principle of movement corresponds to the present hypothesis regarding how water animals have evolved for the reduction of resistance by the placement of a body in the field of separation for the nasal part of their body.

For high-speed water animals—swordfish, sailfish, marlin and spearfish—Reynolds numbers at their navigation are $Re_L \approx 10^8$. Figure 6.48 provides photographs of fish of this kind. In Figure 6.49 the contours of a swordfish body [392] and marlin [431] are shown. Considering the given hypothesis, the body can be presented conditionally in the form of a cylinder of small scope located across a stream with rounded-off end-faces (in Figure 6.49 its cross-section is designated by a circle). As was specified above, the nasal and tail parts of a body of a fish borrow stagnant areas concerning this cylinder. Re calculated on the diameter of such a cylinder for a swordfish has been calculated as $1.2 \cdot 10^7$.

In [305] it is shown that a xiphoid tip on a sphere is similar to a turbulator and reduced the coefficient C_f of a sphere at $Re = 10^6$ by 2.5 times and reached sizes commensurable with C_f aerodynamic profiles. It is also shown that a xiphoid tip in a combination with a steeply pitched head permits turbulization of a body's boundary layer and a shift in separation point to $\varphi = 110°$ (Figures 6.47 and 6.49) and is essential to drag reduction.

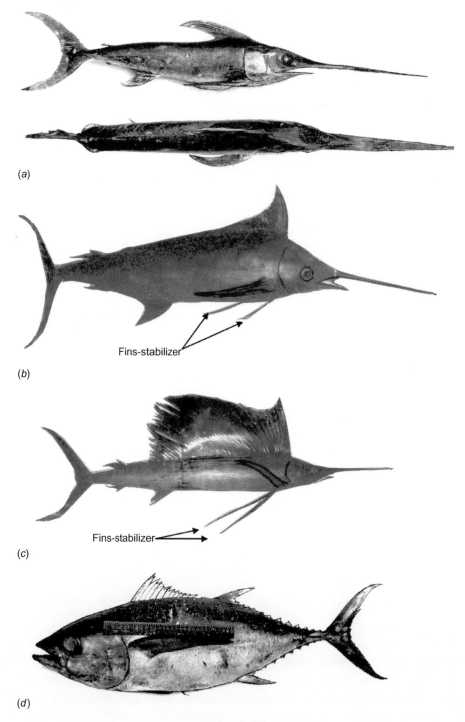

(a)

(b)

Fins-stabilizer

(c)

Fins-stabilizer

(d)

Figure 6.48
Photographs of swordfish (*a*), marlin (*b*), sailfish (*c*) and tuna (*d*).

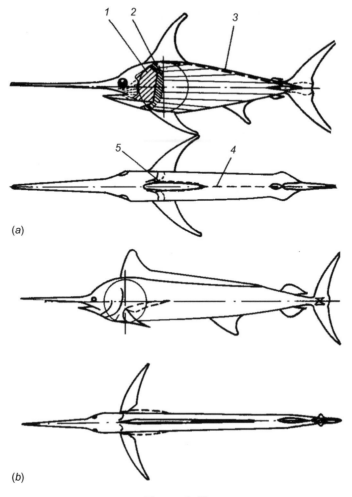

Figure 6.49

Scheme of body forms of swordfish (*a*) and marlin (*b*): *1* — gill cover; *2* — crypt-like structure for the generation of slime [294]; *3* — ampoules in skin [391]; *4* — crack of ampoules; *5* — crack of a gill.

The prevention of separation in the converging part of the body of the fish is performed by a special structure—the branchiate device [294, 396] and external covers [63]. The other hydrodynamic function of the branchiate device includes the regulation, through a branchiate crack, of the injection of flow, depending on the body curvature at its fluctuation. At high speeds of movement the tail part of the body oscillates and then flat wall jet injected from gills prevent separation from the butt-end parts of the "cylinder" (see Figure 6.46), reducing the role of butt-end effects at a flow of "cylinder." It also has a known function as ending washers for the reduction of inductive resistance.

Techniques relating to calculation of boundary layers on flaps of a wing widely apply to calculation of blow away efficiency, the so-called coefficient of movement quantity of a blown jet C_μ [145, 253]. This coefficient also characterizes the optimum charge of a jet and defines an effective parity of jet speed on an output from a crack and of flow speed on a wing. For revealing dependence ΔC_f from u_{sl}/U the obtained data have been certainly depending on the coefficient of quantity of movement C_μ (see Figure 6.50), appearing as:

$$C_\mu = 2Q \cdot u_{sl}/\rho U^2 \cdot S_0 = 2\,C_q \cdot \rho_{sl} \cdot u_{sl}/\rho\,U \tag{6.12}$$

where S_0 is the area of the cylinder washed by the injected jet; ρ and ρ_{sl} the density of an injected jet and incident flow respectively; c_q the coefficient of the charge. Considering that according to experience $\rho \approx \rho_{sl}$, expression (6.12) looks like:

$$C_\mu = 2\,C_q \cdot u_{sl}/U \tag{6.13}$$

It is known [350, 427] that a blown jet adjoins to the surface of a flap (the Coandă effect), leading to the growth of local speeds and bringing in a boundary layer and additional kinetic energy, neutralizing a positive gradient of pressure on the flap and providing its continuous (unseparated) flow. The coefficient of elevating force of a wing intensively increases at $C_\mu = 0.01-0.2$, and at $C_\mu = 0.5$ the effect of blowing of a jet essentially decreases [253, 427]. For maintenance of C_μ an optimum crack has before it a zone of maximal discharge on the flap to provide sufficient hashing of a jet with a boundary layer.

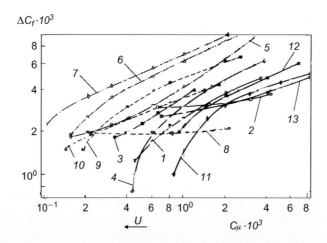

Figure 6.50

Coefficient of cylinder drag as a function of C_μ. Efficiency of the injection of polymer solutions of model with OT: *1* — C = 0.05%, Q = 54 cm³/s; *2*—0.1, 59; *3*—0.1, 49; *4*—0.15, 54; *11*—0.1, 80 (synthetic glue); *12*—0.1, 84; *13*—0.1,100. For SXT: *5*—0.05, 47; *6*—0.1, 34; *7*—0.15, 28. For LXT: *8*—0.05, 50; *9*—0.1, 45; *10*—0.15, 38.

A similar phenomenon is found in water animals. For example, according to tests in aero-hydrodynamic installations [305, 525] the branchiate crack of the swordfish settles down before the maximum midsection (see Figure 6.49) in the field of the maximal under pressure. As the branchiate device is intended for the reduction of C_f instead of for the influence on lift force the range of optimum values of C_μ should be defined.

In Figure 6.50 dependences of efficiency of drag reduction ΔC_f from the factor of quantity of movement C_μ are presented; thus, the size ΔC_f is defined not in a relative way, as in Figures 6.44 and 6.45, but in an absolute one:

$$\Delta C_f = C_f - C_{f\ inj} \tag{6.14}$$

Continuous lines designate the results of measurements for the model with OT, stroke-dashed lines for model with SXT, and shaped lines with LXT. It is possible to see that the greatest gain in C_f is made at smaller U, at $u_{sl}/U \approx 1$ in a range $C_\mu = 1-10$. Sizes ΔC_f decrease at $C_\mu = 0.1-1$, when $u_{sl}/U < 1$.

In comparison with the other conclusions made based on the analysis of data shown in the previous figures, those in Figure 6.50, have shown that:

- Xiphoid tips provide efficiency at smaller values of coefficient C_μ in comparison with OT. Thus, values of coefficient C_μ become closer to known optimal sizes at injection wall jets on flaps.
- At identical values of C_μ absolute efficiency ΔC_f in comparison with relative efficiency (in percentage) appeared essentially greater than for the model with OT in the whole range of the measured values of coefficient C_μ.

In Figure 6.48 only four types of high-speed fish are shown. For the analysis of these photographs it is important to bear in mind that they are made from dead water animals. Therefore, the form of their bodies does not fully correspond to that of live animals. Furthermore, it is necessary to consider scale of these photographs. The swordfish had a general length of 1.6 m, the marlin ≈ 2.5 m, the sailfish ≈ 2 m and the tuna ≈ 0.85 m.

Of course water animals move at various speeds. The core adaptations developed during evolution are shown at cruising speeds, and it is at these speeds that the animals can move for a long time. Under extreme conditions they can also reach their greatest possible speeds for short periods of time. In addition, there are modes of slow movement. The analysis of the developed adaptations discussed is focused on movement at cruising speeds. It is obvious that the forms of the fish shown in the photographs correspond to the hypothesis offered: that high-speed fishes have developed such adaptations that during movement their body settles down in the field of a trace of a badly streamlined body. Thus, the body settles down in the separation zone (negative pressure) due to the additional draft realized, and the surface of the body is not flowed round by a high-speed stream, and the border of section of this stream and the braked trace. The resistance of friction at a flow of such a stream will be less than at a flow the basic stream. From the work of the fin propulsor there is undersuck force, which also gives additional draft. Thus, a system of receptors is used to control the optimum speed of movement. If speed increases in relation to the optimum, the body leaves the area of the braked stream and sets up conditions of a flow when resistance to movement essentially increases. In addition, for a reduction of speed of movement, the separation area for a badly streamlined body becomes isolated on a body. Thus, according to a Kármán street, fish increase the amplitude and reduce the frequency of fluctuation of the tail mover to get into a mode of movement with the least losses. Tuna swim at speeds that correspond to our modeling experiment with the model with OT. Effective use is made of nasal draft and drag reduction at injection of polymer solutions up to Reynolds numbers corresponding in our modeling experiment Re = $1.5 \cdot 10^6$.

Xiphias species swim at much greater speeds, but the mode of their movement also corresponds to the conditions of flow shown in Figure 6.47. For these fish, a flat top of the surface of the head from the tip prior to the start of the vertical fin is characteristic, being a continuation of the tip. It is clearly visible on Figure 6.48a (top view). Therefore, it is necessary to consider the length of the tip prior to the start of the vertical fin. Relating to this, in one experiment parities of a body and xiphoid tip have not been executed, and all known results of experiments did not allow the maximal effect from the application of a xiphoid tip.

The body a swordfish has a rectangular form that allows increasing the moment of application of force from muscles to fin mover. The branchiate cover is flat and the line of the edge is a straight line; therefore, the injection jet influences simultaneously the body and the fins. The top surface of the tip is rough and creates turbulization in a boundary layer. Therefore, in our experiments on tips a turbulence promoter was established in the form of a ring. Turbulization of a boundary layer shifts the area of distribution of tangents exertion along a body in the area of a tip. Simultaneously, it confirms the hypothesis put forward that the application of a tip on a sphere reduces resistance by 2.5 times. The bottom surface of the tip of a swordfish is smooth, and at greater speeds of the movement this smooth surface can play a role as a gliding surface.

For these fish it is characteristic that at greater speeds of movement their lateral fins nestle against the body, which essentially reduces the general resistance of the body. In the bottom part of the body are three long thin fins, which appear to leave from a trace separation, according to the scheme in Figure 6.47. At greater speeds of movement, they also function as gliding surfaces. So, at higher speeds the body moves with minimal resistance, and moves steadily due to the stabilization of the body by means of the gliding surfaces of nasal tip and these three thin stabilizers, which are located in the forward part of the body. This also confirms the hypothesis.

By regulating the width of a crack simultaneously from two sides of a body it is possible to change the speed of injection and thus efficiency of drag reduction and sizes of nasal drafts. By regulating the width of a crack it is possible to receive the necessary moment for maneuvering in a horizontal plane of movement. By slightly changing the corner of attack of the hypoid tip it is possible to receive the necessary moment for maneuvering in a vertical plane of movement. Three thin stabilizers at the bottom of a body (see Figure 6.48b,c) play the same role and simultaneously can cause damping of the size of the moment arising on a tip from a change of a corner of attack. The form of a body of a swordfish is such that the center of gravity and the center of pressure of a body are located the close. All this allows the swordfish to regulate its speed of movement and to maneuver during movement.

At greater speeds of movement the vertical fin of a sailfish (see Figure 6.48C) develops the shape of a sail, which also reduces resistance. However, the role of this sail has not been sufficiently investigated. Babenko has suggested using a similar form of a sail for planning a body in water [420, 422].

6.6 Kinematic Characteristics of the Model Flow at the Injection of Polymer Solution through a Ring Slot

Despite the revealed optimum values of C, Q and C_μ, understanding of the data and the mechanism of a half-bounded jet interaction with a boundary layer formed earlier necessitated the measurement of the average and longitudinal profiles of fluctuation velocities, which was done on an ellipsoid of rotation with the use of a laser Doppler anemometer (LDA) [48, 246, 247]. The dimensions of the ellipsoid were two times smaller in comparison with the model. Thus, the boundary layer thickness was also smaller, which made it possible to carry out the experiments at smaller values of C and Q. The description of the experimental facility and the conditions of the experiments are presented in [246] and its scheme is shown in Figure 6.51.

The research was conducted in a tank with the dimensions $4.6 \times 1.5 \times 1.5$ m, in which a submerged jet was created flowing with a constant speed $U_0 = 3.33$ from a conic nozzle with the angle 15°, diameter of the outlet $d_0 = 1.24$ cm. An ellipsoid with axes $x_0 = 20$ cm and $y_0 = 2$ cm was inserted in the jet at a distance of 8 cm from the nozzle outlet. The ellipsoid surface area was $S = 98$ cm^2. There was a 4 cm-long cylindrical segment in the middle part. The major ellipsoid axis was aligned with the symmetry axis of the jet. The ellipsoid was fastened onto the tail part via a vertical fairing of an elliptic cross-section inside of which was a channel for feeding a liquid. A liquid was injected through a 0.08 cm-wide circular slot cut through, 22.5° to the major axis of the ellipsoid at a distance of 2 cm from the nose parts of the ellipsoid. The slot area was 0.21 cm^2. A liquid was

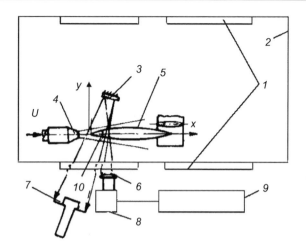

Figure 6.51
Scheme of the experimental facility for research on the kinematic characteristics of an injected near-wall jet: *1* — glass windows; *2* — tank; *3* — mirror; *4* — conic caps for formation of a submerged jet; *5* — ellipsoids of revolution; *6* — focusing lens; *7* — photograph receiver; *8* — beam splitter; *9* — laser; *10* — slot.

Table 6.11: Parameters of injection of water and polymers solutions.

N	U (m/s)	u_{sl} (m/s)	u_{sl}/U	Q (cm^3/s)	$Cq \cdot 10^3$	$C_\mu \cdot 10^3$	Injected Fluids
1	3.33	0.228	0.068	4.8	1.47	0.2	Water
2	3.33	0.238	0.072	5.0	1.53	0.22	POE, C = 0.005%
3	3.33	0.62	0.186	13.0	3.98	1.48	Water
4	3.33	0.714	0.214	15.0	4.6	1.97	POE, C = 0.01%

injected through the slot by means a free-flow device enabling the regulation of the flow rate through the slot within the range 1–15 cm^3/s. The coordinate device made it possible to move the trial volume of the LDA of dimensions $60 \times 60 \times 500$ μm along x and y, while not breaking the optical scheme adjustment. The kinematic characteristics of the boundary layer were measured in cross-sections, the first of which was at 0.4 cm upstream, and sections 2–9 were at 1.0, 2.5, 3.5, 5.5, 7.5, 9.5, 11.5, and 14.0 cm downstream from the nose part of the ellipsoid. Coefficients c_q and C_μ (see Table 6.11) showed that at small flow rates on the ellipsoid c_q is comparable with the flow rate coefficients on the model at speed U_∞ up to 2 m/s. As $U = $ const, at growing Q the magnitude of c_q on the ellipsoid becomes greater than on the model. At the same time, C_μ on the ellipsoid and the model are comparable only at the flow rate on the ellipsoid $Q = 13–15$ cm^3/s. Therefore, it appears that the data obtained on the ellipsoid can be used to explain the physical nature of the integral characteristics on the model only at large flow rates.

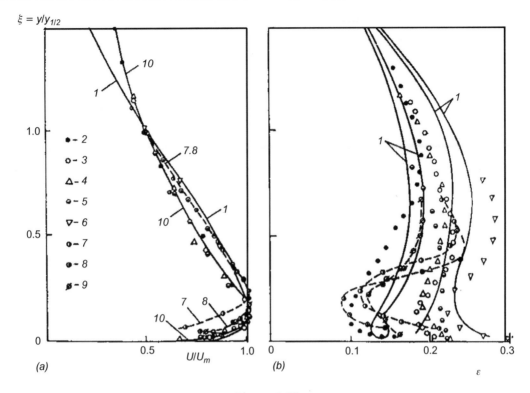

Figure 6.52

Profiles of average (*a*) and longitudinal fluctuation (*b*) velocities on the ellipsoid of rotation: *a*: *1* — measurement without injection for a plane plate and cylinder placed across a flow [517]; *2–6* — for the corresponding sections on the ellipsoid; *10* — generalized curve for sections *2–5*. Measurements with injection: water in third section at $Q = 13$ cm^3/s (*7*); in the forth section (*8*); polymer solution at $Q = 15$ cm^3/s (*9*). *b*: *1* (on the left) — measurement without injection for a plane wall flow [517]; *1* — (on the right) — for cylinder placed across a flow. Other designations as in *a*.

The kinematic characteristics of the boundary layer and data on the interaction of the submerged jet with the ellipsoid placed in its core with a half-submerged jet injected on its surface are shown in Figure 6.52. The following designations are used in the figures:

$\xi = y/y_{1/2}$— dimensionless coordinate;
y — distance from a the model surface;
$y_{1/2}$— coordinate of the point;

where $U = 1/2U_m$, U is current flow speed in km/s, and U_m is the maximum speed, $\varepsilon = \langle u^2 \rangle^{1/2}/U_m$ is intensity of turbulence, and $\langle u^2 \rangle^{1/2}$ is the root-mean-square longitudinal fluctuation velocity.

The analysis of dimensional velocity profiles on the ellipsoid with the opened slot but without jet injection shows that the profile in section 1 has the shape typical for a submerged jet flowing from a nozzle [247]. Because of the decelerating influence of the ellipsoid on the velocity of the submerged jet it has the value $U/U_0 = 0.96$ at the axis of symmetry. In sections 2−7 the velocity profile resembles the shape typical for a half-submerged jet [517]. In section 3 (at 0.5 cm behind the slot) the area of maximum velocity is a little stretched along y. In sections 8−9 the velocity profile comes to be characteristic for a uniform flow about a body (as on the model). These results in dimensionless form are shown in Figure 6.52, where there are data for sections 2−6 only. In comparison with the velocity profiles in a jet flow over a plate or cylinder located across the flow [517], the velocity profiles in sections 2−5 differ from the Gaussian distribution:

$$U/U_m = \exp(-0.693\,\xi^2) \tag{6.15}$$

Apparently this is due to a negative pressure gradient in the nose part of the ellipsoid. Beginning from section 6, the velocity profile is described by expression (6.15) with an allowable error. It agrees well with the shape ellipsoid, which is matched with its cylindrical part at this section. The maximum velocity in the jet near the ellipsoid were found for section 2 at $\xi_m = 0.133$; 3 −0.143; 4−04−0.146; 5−05−0.152; and 6−06−0.166. In accordance with the data in [517], $\xi_m = 0.16$ for a plane wall.

The water injection changes the velocity structure only in sections 3 and 4, just behind the slot. A difference in profiles caused by the variation of the water flow rate through the slot is also noticeable. At greater water flow rates, ξ_m in section 3 (see Figure 6.52, curve 7) was 0.245, and in section 4 (see Figure 6.52, curve 8) it was 0,184. At lower water flow rates, velocity profiles in sections 3 and 4 correspond to Figure 6.52, curve 8. The injected half-bounded jet reduces the relative velocity in the boundary layer, which leads to an increase in ξ_m; thus, the boundary layer thickness is increased, and the tangents stresses on the wall decrease. At the injection of the polymer solutions with $Q = 5$ cm³/s and $C = 0.005\%$ the velocity profiles up to ξ_m correspond to curve 10 on Figure 6.52, and at $\xi_m < \xi < 1$ to curve 8 on Figure 6.52 (the experimental points are not shown in view of the large saturation of the graph). There, ξ_m was 0.16. The velocity profile at the polymer solution injection with $Q = 15$ cm³/s and $C = 0.01\%$ corresponded everywhere to curve 8 on Figure 6.52, and ξ_m in section 3 was 0.21, decreasing to section 7 up to $\xi_m = 0.18$. The increase of the polymer solution concentration from 0.005% to 0.015% at $Q = 5$ cm³/s resulted in an insignificant increase of δ and reduction of τ, and the increase of Q from 5 cm³/s to 15 cm³/s at $C = 0.01\%$ caused an increase in δ and τ.

The measurements have shown that the injection of polymer solutions in comparison with injection of water has not resulted in essential change of the velocity profile near the ellipsoid surface, while the difference became appreciable in the interval $\xi_m < \xi < 1$ and indicated an increase in the flow rate of the submerged jet bounded by the ellipsoid and the undisturbed flow. Further, the influence of polymer solutions on the boundary layer is kept up to section 7.

Figure 6.52b shows average values of longitudinal fluctuation velocity for sections 2−6. In the first section there is a distribution of ε typical for a submerged jet with the magnitude of 0.06 in the jet core, and 0.3 at $y/y_0 = 1$. In sections 2−6 the dependencies of ε on ξ become similar to those in [517] for a jet attached to a wall. The difference consists of the following: the shape of velocity profiles on the ellipsoid was gradually deformed downstream, approaching the profile of a wall-attached jet with a Gaussian distribution that is typical for both a plane wall and a cylinder placed across a flow. At the same time, moving away from the ellipsoid tip, the shape of fluctuation velocity profiles gradually transformed from that typical for a plane wall to the shape typical for profiles on a curvilinear wall. ε near the wall and on the entire ξ grew simultaneously. Such a transformation indicated the transition of the boundary layer flow from the laminar to a turbulent regime.

The injection of water at $Q = 4.8$ cm^3/s resulted in increased ξ_m near the wall and reduced ε in the region $\xi < 0.3$. These distinctions were observed in sections 3 and 4, as well as in measurements of the velocity profile. Increased flow rate up to $Q = 13$ cm^3/s (dashed curves in Figure 6.52b) resulted in decreased ε in comparison with the standard in a greater range of ξ, up to $\xi = 0.4$. Moreover, maximum values of fluctuation velocities near the wall decreased, and these effects were kept up to section 7. The increase of ξ_m occurred qualitatively similar to that described in [11].

The injection of a polymer solution at $Q = 5$ cm^3/s and C = 0.005% much reduced the value ε in comparison with the standard at a wider range of ξ, from 0 to 1, and up to $\xi = 0.15$ and further in comparison with water near the wall in the interval $0.3 < \xi < 1$. Increased concentrations of up to 0.015% at the same flow rate resulted in keeping the effect on greater distance along the ellipsoid, up to section 8. If the flow rate grew simultaneously, ε essentially decreased near $\xi = 0.2$ just behind the slot, and in the other sections along the ellipsoid length the effect fairly decreased in comparison with the growth of C, which agrees with measurements of the velocity profiles.

Water injection promotes the reduction of tangent stresses on the wall, but as soon as its influence on the average and fluctuation velocity fields rapidly weakens behind the slot, the integral characteristics of the boundary layer are almost the same as those for the reference. The polymer solutions reduce transversal and longitudinal fluctuation velocities, and also tangent stresses [215, 326], which results in a longer area of the influence of polymer solutions upon the ellipsoid boundary layer.

The conducted experiments on the axisymmetric submerged jet about an ellipsoid are self-valued. Another problem was also investigated: the influence of the near-wall jet injected on the ellipsoid upon submerged jet characteristics. The variations of parameters C and Q at polymer solution injections were insufficient to deduce laws of influence on the submerged jet. The data obtained were nondimensionalized by the parameters of the submerged jet at the nozzle outlet. The speed at the nozzle outlet u_c and thickness S of the incoming jet on the ellipsoid surface were considered. According to our experiments, $S \approx 0.6$ y_0, where y_0 is the nozzle radius. S was defined from the velocity profiles in the first measurements section near the ellipsoid. Thus, dependencies of the dimensionless coordinate $y_{1/2}$ and U_m on the dimensionless coordinate x were defined. The axes were situated as shown in Figure 6.51. The results of the obtained dimensionless parameters and comparisons to the data of other authors are represented Figure 6.53. Because the shape of the nose part of the ellipsoid has a big radius of curvature the deduced laws of increase in the distance of the maximal values of the average speeds along the ellipsoid are similar to data of other authors for the flow on a flat plate (see Figure 6.53a). It was unexpected to obtain such data for the variation of maximal value of the average velocity along the ellipsoid (see Figure 6.53b). The reduction of maximum speed along the ellipsoid corresponds to the laws typical for a near-wall jet on the cross-flow cylinder. For these experiments it is likely that losses of energy of the submerged jet increased due to the influence of the ellipsoid shape. In the experiments of other authors for the near-wall jet on a flat plate or cylinder, energy losses in the jet were caused only by losses due to friction along the boundary.

In our experiments the boundary layer on the ellipsoid developed naturally. Therefore, in the area of measurement of the flow parameters, the transition to turbulence occurred only in last sections along the ellipsoid in which measurements were taken. It is known that polymer solutions are effective at turbulent mode of flow when the macromolecules of polymers are stretched. Therefore, the efficiency of polymer solutions injected on the ellipsoid differed insignificantly from the injection of water. Nevertheless, the influence of polymer solutions on fluctuation velocities in a cross-section of the submerged jet was also revealed in this case. Based on the above-mentioned results, it is possible to draw the following conclusions: at a flow over the ellipsoid the influence of polymer solutions on the kinematic characteristics of the boundary layer extended through the entire thickness. It resulted, in particular, in the shape of profiles of average and fluctuation components of velocity (see Figure 6.52) becoming the same as in the flow over a plate. Furthermore, the flow rate of the submerged jet determined by the average velocity profile increased at injection of a plane jet on the ellipsoid due to those polymer solutions extending their influence on the characteristics of the boundary layer, not only along the length but also across the boundary layer thickness.

These results for the flow about an ellipsoid have provided a better understanding of the mechanism of injection of polymer solutions on the model. It is obvious that xiphoid tips change the pressure distribution along the model; therefore, the diffusion of a polymer solution across the boundary layer thickness decreases. The variation of the flow rate of the

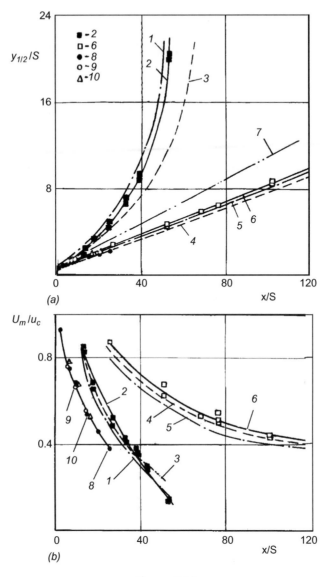

Figure 6.53

Dependence of thickness (*a*) and the maximal speed (*b*) fit wall jets from longitudinal coordinate. *a*: laws on a curvilinear wall: *1* — Fekete [182], *2* — Wilson, Goldstein [517], *3* — Newman [379]; laws on a flat wall: *4* — Kruka, Eskinezi [321], *5* — Meier et al. [321], *6* — Wilson, Goldstein [517], *7* — flat free jet, Nakaguchi [375]; measurements on an ellipsoid: *8* — without injection, *9* — injection of water with $Q = 13$ cm^3/s, *10* — injection of a polymer solution with $Q = 15$ cm^3/s and $C = 0.01\%$. *b*: laws on a curvilinear wall: *1, 2, 3* — the same as for laws on a flat wall: *4* — Wilson [518], others, same designations as for *a*.

submerged jet flowing about the ellipsoid indicates that losses in the flow have decreased. If these results are applied to the results of the flow on the model, the occurrence of additional nose thrust on the models at increased flow rates of polymer solutions becomes clearer, as does the increase in efficiency of the xiphoid tips. On the ellipsoid, polymer solutions increase the range of influence upon the flow in comparison with injections of water. The same effects are also observed in the flow on the model.

6.7 Method and Apparatus for the Optimum Injection of Liquids in a Boundary Layer

The hydrobionic approach (see Section 1.8) has helped to solve a number of new problems, and in order to do this numerous experimental and theoretical research projects have been conducted (see Sections 1.5−1.7 and 1.9−1.11). In the future it will be necessary to develop new engineering solutions, which at least in part will be found by exploring and utilizing the optimum solutions found in nature and developed through natural evolution. Below, the authors consider four patents that have implemented the knowledge gained through the hydrobionic research discussed.

6.7.1 Method for Reducing the Dissipation Rate of Fluid Injected into a Boundary Layer [81]

Technical field

This invention relates to a method and apparatus for the injection of high molecular-weight materials such as polymers into the boundary layer of a fluid flow, and it has been shown to reduce skin friction drag significantly for both vessels moving relative to water and for pipeline applications. The large polymer molecules interact with the turbulent activity in the near-wall region, absorbing energy and reducing the frequency of burst (high-energy fluid moving away from the wall) and sweep (low-energy fluid replacing the high-energy fluid in the near-wall region) cycles. The reduced burst frequency results in less energy dissipation from the wall and can result in skin friction drag reduction of up to 80%. Experiments have shown that the efficacy of polymer molecules for drag reduction is closely related to their molecular weight, their location in the boundary layer and the degree to which they have been stretched, or "conditioned."

Background

In the past, polymer mixture ejectors have been simple slots that ejected a fluid mixture/ polymer solution at an angle to the wall. To attain high drag reduction for a reasonable distance downstream with this ejection approach, large quantities and high concentrations of polymers must be ejected in order to flood the entire boundary area, creating a "polymer ocean" effect. The high polymer consumption rates of these systems have made them impractical for many drag reduction applications. To be useful for practical applications a more efficient method for ejecting polymer mixtures for drag reduction needed to be devised.

The invention

This invention enables the efficient ejection of fluid mixtures/solutions into the near-wall region of a boundary layer of a fluid flow. The first objective of this ejector is the conditioning of the polymer prior to ejection, so that drag reduction occurs almost immediately following ejection. The second objective of the invention is to release the polymer only into the boundary layer region, where it can provide the greatest drag reduction. The third objective of the invention is to retain the polymer in the near-wall region of the boundary layer, the most effective region for drag reduction, for as long as possible.

This ejector system preconditions the polymer mixture/solution for improved drag reduction performance using a unique arrangement of flow area restrictions, as well as by employing dimples, grooves and elastomeric materials. The dimples, grooves and flow area restrictions are sized relative to one another and to the Reynolds number of the flow for optimal polymer molecule conditioning (lengthening, unwinding, or stretching) so as to provide optimal drag reduction after ejection into the fluid flow. In addition, this ejector uses a new approach to structuring the flow in order to reduce migration/dissipation of the polymer away from the near-wall region. This is achieved by a unique system of slots, each having a carefully designed surface curvature and surface features that establish a duct-like system of longitudinal (i.e. in the direction of the flow) Goertler vortices [189, 428, 445, et al.]. Goertler vortices are formed by the centrifugal effect of a fluid flow that is given angular velocity by a concave surface. The duct-like system of Goertler vortices formed by this invention mimic the spacing of naturally occurring quasi-longitudinal vortex pairs in the boundary layer, but are paired in the opposite oz1entation. The pairing of naturally occurring quasi-longitudinal vortex pairs is such that they migrate from the wall and are believed to contribute to the development of bursts and sweeps, which account for a large portion of hydrodynamic drag. The vortices created by this ejector are such that the pressure differentials they create cause the vortices to remain near the wall. This advantageously causes the polymer that has been ejected into the boundary layer to remain in the near-wall region.

Brief description of the drawings

This invention can be more fully understood from the detailed description given below and the accompanying drawings (see Figures 6.54–6.59), which are given by way of illustration only and are not limited to the invention discussed, e.g. Figure 6.54 depicts Goertler vortices forming due to centrifugal forces caused by drag on a concave surface.

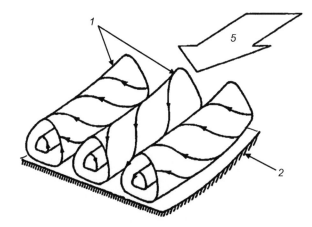

Figure 6.54

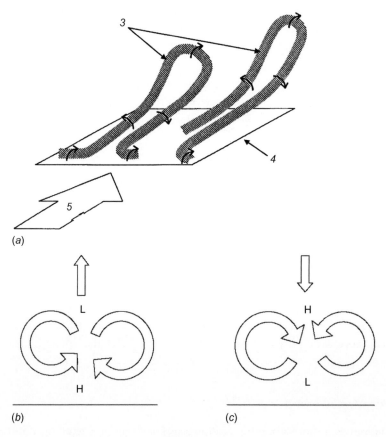

Figure 6.55

a — in isometric view, naturally generated quasi-longitudinal vortex pairs; *b* — a cross-sectional view of naturally occurring vortex pairs; *c* — a cross-sectional view of longitudinal Goertler vortex pairs formed by the vortex duct ejector of the invention considered.

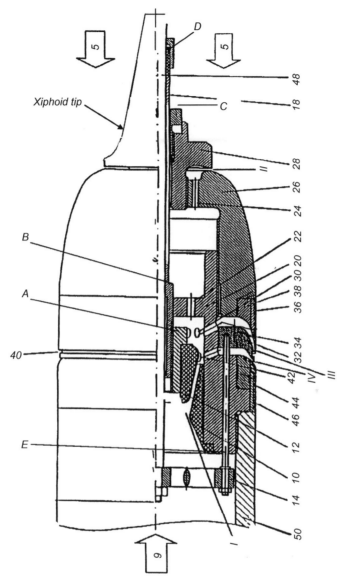

Figure 6.56
A side view of the vortex ejector of the invention, with the lower portion being a cross-sectional view, which shows the inner components of the vortex duct ejector.

Best mode of carrying out the invention

This system achieves a more effective polymer mixture/solution ejection by releasing the water—polymer mixture/solution into the near-wall region of the boundary layer and by adjusting the mixture/solution flow characteristics so that it remains in the near-wall region.

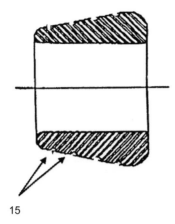

15

Figure 6.57
In a cross-sectional view, a cone component of the ejector shown in Figure 6.56.

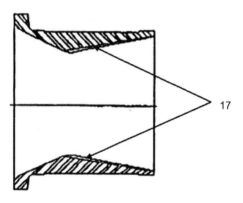

17

Figure 6.58
In cross-sectional view, a diffuser component of the ejector shown in Figure 6.56.

By producing a mixture/solution with flow characteristics that adhere it to the wall the ejector extends the polymer residence-time in the near-wall region before it is diffused into the surrounding water, and, thus, reduces the polymer consumption of such a drag reduction system.

Goertler vortices are formed by the centrifugal effect of a fluid flow that is given angular velocity by a concave surface. Figure 6.54 depicts naturally occurring Goertler vortices forming, *1* due to centrifugal forces caused by drag on a concave surface, *2* due to a fluid flow indicated by the arrow *5*. The surface features of this ejector create Goertler vortices that mimic the spacing of naturally occurring quasi-longitudinal vortex pairs in the boundary layer, but they are paired in the opposite orientation. The pairing of natural

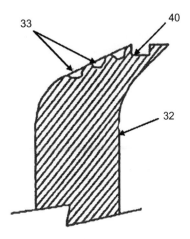

Figure 6.59
In cross-sectional view, a portion of an ejector ring.

quasi-longitudinal vortex pairs is such that they migrate from the wall and are believed to contribute to the development of bursts and sweeps that account for a large portion of hydrodynamic drag.

Figure 6.55a depicts an isometric view of quasi-longitudinal vortex pairs. It is generally accepted that flow over a stationary surface creates transverse structures, which become distorted into hairpin-shaped vortices *3* near the wall *4*. The arrow *5* indicates the flow direction. The quasi-streamwise "legs" of each hairpin-shaped vortex produce a pressure differential normal to the wall that makes the vortex pair migrates away from the surface. Figure 6.55b is a transverse cross-sectional schema of a vortex pair inducing a pressure differential that will move it away from the wall. The "H" represents a local higher-pressure region, and the "L" represents a local lower-pressure region. In contrast to these naturally occurring vortex pairs, the Goertler vortex pairs generated by this invention are paired and spaced so that the pressure differential they create causes them to hug the surface. Figure 6.55c is a cross-sectional view of a vortex pair that creates a pressure differential that drives the vortex pair in a direction towards the wall, thereby causing the vortex pair to hug the wall. Because the vortices of such a pair remain near the wall, they keep the polymer that has been ejected in the near-wall region, thereby reducing the occurrence of bursts and sweeps. Hereafter, the ejector of this invention will be referred to as the "vortex duct ejector" because of its innovative use of vortical structures to control polymer mixture/ solution dissipation.

Figure 6.56 illustrates the vortex duct ejector. The polymer mixture/solution *9* flows into the ejector from the left, moving toward slot "I." The boundary layer to be injected with the polymer mixture/solution envelops the vortex duct ejector and the direction of flow of liquid *5* is from right to left, just as if the ejector were on a body moving to the right in a stationary medium. Polymer mixture/solution I is ejected from the slots II, III, and IV into the boundary layer of the ejector body. Optimal solution concentrations and volume flow rates are determined as required for each application.

The polymer mixture/solution flowing into the ejector from the left is directed toward slot I by diffuser *10* and cone *12*. The interaction between one or more vanes (not labeled) attached to the framework *14* reduces the irregularity of the flow. As the flow passes through slot I, dimples in cone *12* and longitudinal slots in diffuser *10* create quasi-longitudinal vortices.

Figure 6.57 is a cross-sectional view of the cone *12*, illustrating the dimples *15* in cone *12*. Figure 6.58 is a cross-sectional view of the diffuser *10*, illustrating the longitudinal slots *17* in diffuser *10*. Interaction of vortices created by the dimples *15* and slots *17* promotes further mixing and stretching of the polymer molecules in the mixture/solution. The width of slot I can be adjusted, or varied, by sliding the central tube *18* with attached cone *12* longitudinally. The materials and features of the diffuser *10* and cone *12* can also be changed or modified to alter the vortical structures. The throttled and conditioned flow then passes out of slot I and through a system of passageways in framework *20*. The size of the passageways in framework *20* governs the shape of the dimples on cone *12* according to condition:

$$0.25\, d_{\text{passageways}20} \leq d_{\text{dimples}12} \leq 0.25\, d_{\text{passageways}20} \tag{6.16}$$

where $d_{\text{passageways}20}$ is the diameter of the passageways in framework *20* and $d_{\text{dimples}12}$ is the diameter of the dimples in cone *12*. The depth (h) of the dimples is given by equation:

$$h_{\text{dimples}12} = 0.25\, d_{\text{dimples}12} \tag{6.17}$$

where h is the depth of the dimples in cone *12*, and d is as defined above. In addition, the grooves in diffuser *10* are defined by equations:

$$B_{\text{grooves}10} = d_{\text{dimples}12} \tag{6.18}$$

$$W_{\text{grooves}10} = h_{\text{grooves}10} = 0.25\, d_{\text{dimples}12} \tag{6.19}$$

where $B_{\text{grooves }10}$ is the center-to-center distance between the grooves in the diffuser *10*, $W_{\text{grooves }10}$ is the width of each groove in the diffuser *10*, and $h_{\text{grooves }10}$ is the depth of each groove in diffuser *10*.

Vortex formation can be enhanced by fabricating the cone (*12*) from an elastomeric material with characteristics what satisfy the equation:

$$(E/\rho)^{1/2} = 0.5\, U_\infty \tag{6.20}$$

where E is the modulus of elasticity, ρ is the density, and U_{∞} is the velocity of the exterior flow. For additional vortex enhancement, one may use an anisotropic elastomeric material characterized by the following condition:

$$2 \leq E_{\text{long}}/E_{\text{inverse}} \leq 5 \tag{6.21}$$

where E_{long} is the longitudinal modulus of elasticity and E_{inverse} is the transverse modulus of elasticity.

The system of passageways in framework *20* can be divided into four groups. The first group *22* passes the solution in the longitudinal direction through a second group of passageways *24* in the fairing *26*, having a diameter one-half that of the dimples in cone *12*, and out into the flow path through slot II. Slot II is a laminar region ejector, and it is intended to thicken and condition the boundary layer upstream of the slots III and IV. The concave shape of the forward surface of the slot formed by stopper *28* creates longitudinal Goertler vortices and the shape formed by fairing *26* (see Figure 6.56), which provides a Coandă surface. The surfaces of slot II are parallel at the aperture. As the flow from slot II enters the boundary layer it is characterized by longitudinal Goertler vortical structures immediately adjacent to the attached flow coming off the downstream Coandă surface. These longitudinal Goertler vortices condition the flow upstream of slot III. Slot II acts to thicken and condition the boundary layer, reducing the disturbances caused by the ejected flow at slots III and IV.

Another group of passageways *30* passes the mixture/solution obliquely through the framework *20*, the fairing *26*, and rings *32, 34, 36* and *38* to exit from slot III. The curvature of the upstream surface of slot III is concave in order to produce a system of longitudinal Goertler vortices; these vortices are then amplified by dimples on an elastic downstream surface of ring *32*. Figure 6.59 illustrates, in cross-sectional view, a portion of such a ring *32*. The dimensions and pitch of the dimples *33* in ring *32* are given by:

$$\lambda_{\text{dimples } 33} = d_{\text{dimples } 33} = [(7.9 \cdot 10^5)/\text{Re}_{\text{x}}] + (3.56 \cdot 10^{-5})(\text{Re}_{\text{x}}) + 1.71 \tag{6.22}$$

and:

$$h_{\text{dimples } 33} \leq 0.5 \, d_{\text{dimples } 33} \tag{6.23}$$

where $\lambda_{\text{dimples } 33}$, $d_{\text{dimples } 33}$ and $h_{\text{dimples } 33}$ are the pitch, diameter and depth, respectively, in wall units y^*, of the dimples in ring *32*, and Re_{x} is the Reynolds number of the water flow immediately downstream of slot IV. As is well known, wall units are a non-dimensional measurement of distance from a wall. They can be expressed as a length dimension using the following equation:

$$y = (y^* v)/\mu \tag{6.24}$$

where y is a dimensioned length, v is the cinematic viscosity of the fluid and μ is the friction velocity of the fluid.

Fabricating ring *32* from elastomeric material can further enhance the Goertler vortices forming in slot III. If an elastic material is chosen, its characteristics should satisfy equation (6.20), above. For additional enhancement effects, one may use an anisotropic elastomeric material, characterized by condition (6.21), above.

When ring *32* is located in a more upstream position than that illustrated in Figure 6.56, such that its transverse groove is located beneath the edge of ring *36*, the transverse groove *40* creates a stationary transverse vortex within transverse groove *40*. The low pressure created by this transverse vortex draws the flow ejected from slot III, including the longitudinal Goertler vortices, against the wall and stabilizes the flow ejected from slot III. When ring *32* is located further from ring *36*, the transverse groove generates a series of transverse vortex rings, which escape and migrate downstream with the flow. The frequency at which these transverse vortices are released can be controlled by periodic motion of rings *32* and *34* (i.e. by oscillating central rod *48*, which indirectly supports ring *34* via framework *14*), or by changing the elastic characteristics of the material of ring *32*. The dimensions of the transverse groove are given by:

$$w_{\text{inverse } 40} = h_{\text{inverse } 40} = 0.5 \, d_{\text{dimples } 33} \tag{6.25}$$

where $w_{\text{inverse } 40}$ is the width and $h_{\text{inverse } 40}$ is the depth, respectively, of the transverse groove *40*.

The last group of "desperate passageways," *42* in framework *20* passes the polymer mixture/solution obliquely into the space between adjustable rings *32, 34, 44* and *46* and out into the flow stream through slot IV. As with slot III, the curvature of the upstream surface of slot IV creates a system of longitudinal Goertler vortices that are amplified by the dimples in rings *44* and *46*. These Goertler vortices interact with the vortices coming from slot III to form longitudinal wave-guides, which act to retain the polymer solution near the wall. The dimensions and spacing of the dimples in rings *44* and *46* are governed by the same equations as the dimples in rings *32* and *34*.

The width of slots I, III and IV can be either adjusted or oscillated by sliding cone *12* and/ or the rings *32* and *34* longitudinally. Cone *12* is articulated on the end of central tube *18*, and the central rod *48* via fasteners to framework *14* articulates rings *32* and *34*. By adjusting the slot widths one can vary the ejection velocity of the mixture/solution. The most effective drag reduction usually occurs when the ejection velocity is between 5% and 10% of the free stream velocity. The ejector body *50* and slot widths should be adjusted to provide a mixture/solution flow velocity in this range for the desired mixture/solution flow rate. An entirely different slot structure can be achieved by removing rings *32* and *34* and replacing rings *44* and *46* with rings featuring longitudinal slots. The longitudinal slots are positioned at an approximate multiple of the spacing of the naturally occurring quasi-longitudinal vortex pairs and create high-powered longitudinal vortices.

Industrial applications

Of course, the ejector of this invention is not limited to the set-up illustrated. Numerous variations of the ducted vortex ejector are possible. For example, rings *32, 34, 44* and *46* may be replaced with rings that have different material and structural characteristics. Rather, the scope of the invention is defined in the following claims and legal equivalents. Various modifications will occur to specialists in this field from reading the description given above, and all such modifications are intended to be within the spirit of this invention.

According to the above-stated patent specification, an operating prototype of mixer and injector has been designed. The Cortana Corporation paid for its design and manufacture. The project was discussed at length with the President of the Cortana Corporation, KJ Moore, who paid for the research and registration of all patents in relation to it. Figure 6.60 shows the model of the device for injecting liquids with different densities into a boundary layer.

Figure 6.61 illustrates the details of the injector, which is denoted as for Figure 6.56: *A* in Figure 6.56 designates the plug, on which a cone *12* is fixed. The tube *18* is fixed on tube *B*, which is screwed in plug *A*. The rod *48* is located inside of tube *18* and is fixed on the other side in frame *14*. The tube *18* is located inside of the plug *C* and its position is fixed

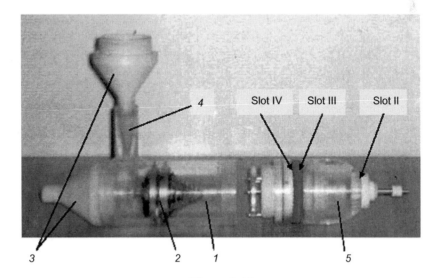

Figure 6.60
Photograph of the model of a mixer and injector: *1* — cylinder, *2* — mixer, *3* — cone, *4* — small cylinder, *5* — injector.

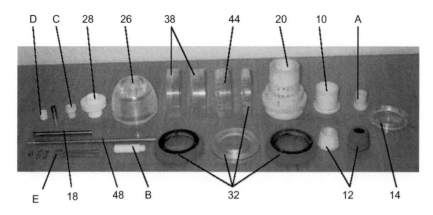

Figure 6.61
Photograph of the injector details: *10* — diffuser, *12* — cone, *14* — frame, *18* — central tube, *20* — frame, *26* — fairing, *28* — stopper, *32* — ring, *38* — ring, *44* — ring, *48* — central rod.

relative core *48* by means of captive nut *D*. The frame *14* is fixed by three studs *E*, which are connected to the central ring *34*. On this ring, the ring *32* is fixed.

Such system allows the independent movement of both the central ring 34 and cone *12*. To do this the captive nut *D* should be unscrewed and those details moved by means of core *48* or tube *18*. This system can be rearranged to be automatically controlled from inside the injector. If a xiphoid fairing is mounted outside then it is possible to regulate the width of each slot independently during motion. Furthermore, it is possible to make the slot width oscillate with the use of a specified program.

The injector operation is based on the method of boundary layer receptivity to various disturbances. This method consists of creating a resonant interaction of the perturbations in the flow with disturbances generated by solutions injected through a slot. The chamber under the slot is designed to be flexible and controllable for this purpose.

Increasing the efficiency of polymeric solution injection is addressed by the simultaneous use of several methods. The first is that two chambers are installed under the slot. The first chamber promotes reduction of the speed discontinuity on the slot outlet. The second consists of installing longitudinal vortex generators on the internal surface of the second chamber. Thus, a polymer solution flows through a slot that does not have smooth rigid walls, but rather a system of longitudinal vortices on one side. The third method is that the first chamber under the slot generates a system of transversal vortices in the boundary layer, and the second chamber creates a system of longitudinal vortices. The fourth method is that the chambers under slot are controlled and work in the foreground of the monitoring system and automatic control. The fifth is that the chambers under slot are fabricated with a coating of elastic material. The sixth is that the system of longitudinal slots for polymer solution injection is installed in front of each transversal slot.

The flow consists of a complex of vortical disturbances, the systems of longitudinal vortices and transversal (toroidal) ring will approach the system of longitudinal vortices originating from the slot II.

The main purpose of the injector is to decrease the diffusion of polymers across the boundary layer thickness, thereby decreasing the consumption of polymers. The injector should also be used to precondition the polymer solution with the purpose of increasing its efficiency, including further dissolving of the polymer.

As has been shown by the research of Pogrebnjak [411] that the efficiency of polymeric solutions depends very much on preconditioning of polymers. The molecules of the polymer initially should be stretched before injecting through a slot. If this is done the molecules of the polymer begin reducing friction in a boundary layer immediately after injection from the slot. However, the effectiveness of polymers is very sensitive to stretching. If a molecule is stretched moderately it continues to expand after injecting through the slot. If a molecule of polymer is stretched strongly it becomes resilient, like a string, and does not damp disturbances of the boundary layer. All this reduces the polymer solution's effectiveness. The molecules of the polymer are expanded in a turbulent boundary layer under action of the Reynolds stresses. However, the stresses are small in a laminar boundary layer and cannot stretch a polymer molecule. The authors' research has shown that in a transient boundary layer, when the longitudinal vortical structures are formed, the stresses are capable of stretching a polymer molecule. Therefore, three means of stretching the polymer molecules are used in the device discussed:

- width of the gap between cone *10* and diffuser *12*;
- formation of longitudinal vortices with the use of holes of various configurations;
- formation of longitudinal vortices with use of longitudinal irregularities on the surface or inside details *10* and *12*.

Various combinations of all these three factors achieve the objective. Thus, both holes and longitudinal elements on a surface and inside of details *10* and *12* can be utilized. In addition, it is important to simultaneously regulate the gap between these details.

The molecules preconditioned in such a way remain stretched if the distance from the diffuser to the slot outlet is small. Therefore, the solution injected from slot II will have some molecules of polymers that begin to roll up, in comparison with a solution from slots III and IV. The same system for stretching of molecules of polymers is on the surface of slots III and IV. Other mechanisms for stretching molecules of polymer are given in the area of slot II. Thus, it is possible to alter the preconditioning of polymer molecules in each slot. Polymer solutions with different kinds of preconditioning will move through these three slots into the boundary layer. The result is a vertical three-layer system of the boundary layer. Theoretical calculations of the multilayered injection into a boundary layer have been developed by Shkvar [84].

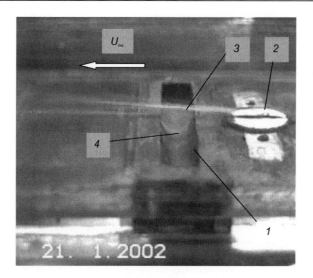

Figure 6.62
Visualization of flow in the semi-cylindrical cavity: *1–21–2D* cavity, *2* — smoke generator,
3 — smoke jet, *4* — large vortex.

In the previous chapters, various methods of formation of longitudinal disturbances in a boundary layer have been considered. In this patent various types of deepening are applied to the formation of coherent vortical disturbances. Various kinds of vortical chambers are also used for the preconditioning the solutions injected through slots in the design. Numerous experiments of such types of flow have been conducted. The fundamental characteristics of the flow have been elucidated and various methods of control have been explored for such flows. This has enabled the development of optimum methods of injection of liquids into a boundary layer. In Figure 6.62 pictures of visualization in 2D and 3D cavities (see Figures 6.14, 6.15 and 6.63) are shown.

The average and fluctuating characteristics of the boundary layer velocity fields on a plate that has local non-uniformity in the form of a transversal semi-cylindrical cavity were experimentally investigated. The research was carried out in an open-circuit aerodynamic tube with a ⌀102 mm working section. The axial section of this working part contained the plate with a ⌀18.5 mm cavity 800 mm long. The experiment conditions were as follows: $U_\infty = 1.11–20.1$ m/s; $\mathrm{Re}_x = 3 \cdot 10^4 – 8 \cdot 10^5$; $\delta = (2–11) \cdot 10^{-3}$ m; $\delta^* = (0.2–2) \cdot 10^{-3}$ m; $\theta = (0.15–1) \cdot 10^{-3}$ m. The flow was visualized by the smoke from combusted oil and the velocity field was measured by a DISA thermo anemometer.

A large vortex *4* appears in the cavity *1* and occupies the entire cavity. Parts of this vortex are washed away periodically by blowing-up flow (see Figure 6.63) and move away downstream. The periodicity of such low-frequency emissions is not constant and depends

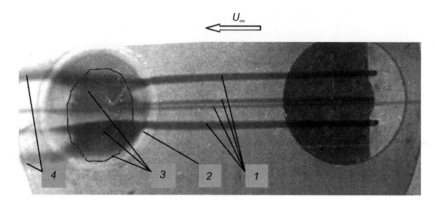

Figure 6.63

Formation of typical spindle-shaped vortices in the hole and two longitudinal vortices: *1* — four dyed jets, *2*—3D hole, *3* — spindle-shaped vortices, *4* — two longitudinal vortices.

on the flow speed. After emission (bursting) a new vortical system forms in the cavity, which is also washed away after certain period. In this way the cycle of "origin—growth—partial bursting" of the large-scale vortical systems in the cavity is formed.

The development and renewal of vortical structures in the cavity leads to changes of the velocity fields in the boundary layer on a plate. The gas flow is accelerated and unstable in certain places, not only above and behind the cavity, but also in front of it. These zones, the kind, quantity and arrangement of coherent vortical structures in the cavity also changes depending on the flow velocity.

It has been discovered that such local non-uniformity of the surface considerably influences the velocity profiles, boundary layer thickness, displacement thickness, momentum thickness, shape factor, surface friction coefficients, and the statistical characteristics of the velocity fluctuation field.

Figure 6.14 is a photograph of the visualization pattern of airflow over a 3D hole taken in the same facility where the 2D hole was investigated (see Figure 6.62). Much research on 3D holes in water flow has been conducted. Figure 6.15 depicts the visualization patterns of the flow over various 3D deepenings. Four dyed jets visualized the water flow. This has given a better understanding of the character of vortical flow in 3D deepenings. The flows in 2D, 3D deepenings and in a vortex chamber were filmed with a high-speed camera and a video camera. Figure 6.63 is a photograph of water flow in a 3D hole. Formation of typical spindle-shaped vortices in the hole and two longitudinal vortices behind the hole can be seen.

Figure 6.64 shows visualization of flow in a vortical chamber. Rational control of gas or liquid mixing and energy dissipation processes in vortex chambers is impossible without a detailed study of differently scaled coherent vortical structures. According to

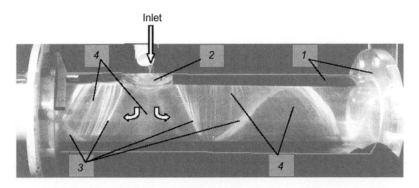

Figure 6.64
Photograph of coherent vortical systems in the vortex chamber: *1* — vortex chamber, *2* — skewed inlet nozzle, *3* — large-scale vortical systems, *4* — small-scale vortical systems similar to Goertler vortices.

various conditions, these structures are capable of decreasing or increasing and stabilizing mass- and heat transfer and blasting processes in the cylinders of internal-combustion engines, magnetohydrodynamic (MHD) generators, power-generating equipment, chemical technology, and so on.

Experimental investigations were conducted in an aerodynamic facility that consisted of a vortex chamber. A complex of devices was used for flow visualization: optical disc, fast-response photographic control and measuring, pneumatic sensors, and thermo- and laser Doppler anemometer devices for spectrum analysis of speed and pressure fluctuations in the flows. The vortex chamber included a cylindrical tube with transparent wall, with internal diameter $d_0 = 102$ mm and length $l_0 = 635$ mm, with the cylinder end fixed in various positions. The relative depth of the dead zone of the cylinder (between the middle of the air-inlet nozzle and the dead end) was varied in the range $L/d_0 = 0.1-4.4$. Parameters of flow in the nozzle of the vortex chamber were as follows (see Figure 6.64): slope of the flow to the internal surface of cylindrical tube $\gamma = 67°$, angle of nozzle relative the cylindrical tube generatrix $\alpha = 0°$; $\mathrm{Re} = 4.7 \cdot 10^5$; $L/d_0 = 4.4$.

The air was supplied to the internal tube cave through the changeable single-jet nozzles or the series of inlet windows with different tangential and axial angles. The Reynolds number based on the nozzle parameters varied in the range $\mathrm{Re} = 2 \cdot 10^5 - 8 \cdot 10^5$. The mean velocity and fluctuation velocity profiles in vortex chamber were established. The experimental data have shown that an interconnected system of large-scale coherent vortical structures was formed along the cylindrical chamber, including the dead-end zone. Visualization and the use of a high-speed video camera allowed the process of formation of secondary vortical systems on the internal surface of the chamber and shear boundaries of the axial large-scale vortex to be registered and described. New experimental data were obtained for the

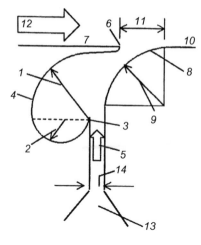

Figure 6.65

statistical characteristics of fluctuation velocity fields in the main parts of wall vortical jets, which were formed by the inlet nozzle of the vortex chamber. With the distance from the curved wall there was a redistribution from large-scale systems to small-scale vortical ones. It was revealed that the large-scale vortical systems stretch along the concave wall, while the small-scale vortical systems stretch along in a transversal direction. With the distance from the nozzle the wall jet of the semi-bounded flow is saturated with the large wave-number vortices owing to the break-up of their large-scale structures. The angular arrows indicate the flow direction in the vortex chamber.

The principles of organization of multi-layered injection of various solutions into a boundary layer considered in the above patent (Section 6.7.1) are developed in other patents.

6.7.2 Method and Apparatus for Increasing the Effectiveness and Efficiency of Multiple Boundary Layer Control Techniques [364]

Figure 6.65 is a schema of a basic ejector element of the ejection system. It includes a nozzle *13*, which preconditions the additive, a vortex chamber *4* on the upstream side of the element, which has form and scale defined by two radii, *1* and *2*, a knife-edge *3* where the chamber *4* and ejection stream intersect *5*, either the knife-edge or surface having a radius of curvature *6* sufficiently large to have the effect of a Coandă surface at the location where the chamber *4* and outer wall intersect *7*, a Coandă surface *8* of radius *9* on the downstream edge of the ejector, which connects to the outer wall *10*, and an aperture *11*, through which the ejected stream *5* joins the established boundary layer flow *12*. In situations where there are constraints on the geometry of the ejector, the Coandă surface *8* may have a compound radius vice of a fixed radius. At the inlet of the ejector is a nozzle *13* or other device that produces a convergent flow into the ejector stream *5*. The ejection stream has a width that equals h_1. The purpose of the nozzle is to establish a laminar contraction flow sufficient to uncoil, align and stretch the additive molecules such that they are in the condition necessary to be effective. The flow through the ejector will be laminar, since the ejection velocity should be about 10% of the free stream velocity. The ejector should be sized to accommodate mass flow rates of $10Q_s$.

The mass flow rate may vary by a factor of about two upwards or downwards, and will depend upon the length and character (e.g. roughness and visco-elastic properties) of the wall being treated, the free stream velocity, the type and concentration of the additive and the level of drag reduction desired. The range of these parameters for the most commonly used additives will result in laminar flow through the ejector. The velocity of the ejected fluid stream is bound at the low end by the value sufficient to keep the ejected stream attached to the Coandă surface *8*. It is bound at the upper end by the velocity of the near-wall boundary layer flow that is displaced by the ejected stream. By not exceeding that velocity, the two flows can merge without producing a significant increase in the local level of turbulence. The desired mass flow rate and ejection velocity determine the nozzle or slot width h_1. By using a properly configured nozzle, the additive will be effective immediately upon merging with the boundary layer and, thus, begin to affect the level of turbulence, which is the principal mechanism of diffusion in a turbulent boundary layer. The concentration of the additive is, of course, greatest at the point of ejection. Thus, reduction in the level of turbulence at this location is critical to diffusion control and maximizing the effect of the volume of additive ejected. As described in the literature, a nozzle with a length in the order of 10 mm and an angle of about $10-45°$ between the nozzle walls has been shown to be adequate to precondition the additive at mean flow rates through the nozzle of about 1 m/s for polymers such as Polyox WSR-301.

As described below, the Coandă surface and vortex chamber function as a unit. The purpose of the Coandă surface is to keep the ejected stream attached to the downstream external wall. When polymer additive is ejected, the value of the radius of the Coandă surface *9* should be about $4h_1$. Eliminating the separation region at the downstream edge of the ejector avoids the unsteadiness introduced by such separation, which is inherent in traditional slot designs.

The velocity component of the ejected stream normal to the boundary layer is reduced to near-zero because of the Coandă surface and low mass flow rate. By eliminating boundary layer "blow-off" the increase in pressure drag and the rapid diffusion of ejected additive associated with that phenomena are avoided.

The purpose of the vortex chamber, located on the upstream side of the ejector, is to reduce or eliminate sources of vorticity that otherwise would contribute to the disruption of the established boundary layer, thereby increasing local drag and enhancing the rate of additive diffusion across the boundary layer. The shape of the chamber is defined by two radii, *1* and *2*. The center point for *1* is the tip of knife-edge *3* and the value of *1* is approximately $4h_1$. The center of *2* is midway along a line extending from knife-edge *3* to the opposite wall of the chamber. When *2* is one half the length of *1*, the two curves will provide a continuous surface. While this 2:1 ratio need not be precise, variations from this ratio will require a short wall segment to avoid any discontinuity or inflection point in the profile of the chamber. The top of the chamber is formed with a tangent connecting knife-edge *6* to the surface formed by *1*. As mentioned above, the knife-edge at *6* can be replaced by a small curved surface to facilitate manufacture and increase the strength of the wall. If the curvature is sufficient to keep the flow attached until it merges with the free stream flow, there will be no degradation in ejection performance. For the parameters associated with full-scale marine applications, the radius of that curve should be about $0.5\,h_1$ with its center on the outer wall, such that the dimension of the opening *11* to the established boundary layer *12* is about $3\,h_1$.

The presence of an internal vortex chamber on the upstream wall modifies the behavior of the flow relative to the flow in a curved channel and eliminates the vortices otherwise introduced by the curvature of the upstream wall. Neither Dean-type nor Goertler-type vortices are formed. The motion of the ejected stream induces circulation in the vortex chamber. For a properly formed and scaled chamber, a stable vortex is established within the chamber. The boundary layer on the upstream boundary of the ejected stream does not continue to develop. Rather, vorticity, which is produced by the internal wall upstream of the chamber, is dissipated by the vortex entrained in the chamber. The velocity profile of the ejected stream is modified relative to established channel flow such that the flow along the upstream edge of the internal stream is slowed less than without the vortex chamber, thereby producing a more stable layer of ejected fluid as it merges with the near-wall region of the established boundary layer. Hence, the unsteadiness introduced into the boundary layer at the upstream edge of the ejected stream is reduced.

Without the vortex chamber, the curvature necessary to form the Coandă surface could result in the production of Goertler-type (over a concave wall) or Dean-type (in a curved pipe) vortices. Hence, the net effect of the Coandă surface on the ejection process is improved because the vorticity at the upstream edge of the internal channel is dissipated by the vortex chamber. In addition, preconditioning of the additive by the contraction flow through the nozzle initiates the drag-reducing effect of the additive. Specifically that effect includes the dissipation of small-scale vorticity. These separate mechanisms work together to improve the behavior of the ejected stream as it merges with the established boundary layer.

Combining the improved behavior of the ejected stream with the preconditioning of the additive during the ejection process results in a more rapid suppression of turbulence and, hence, a reduction in diffusion of the concentrated additive. In this invention, diffusion of the concentrated additive is further reduced by preconditioning the flow just upstream of the principal ejector. Several techniques can be employed. For unsteady or complex turbulent boundary layer flows, the present invention includes a separate additive ejector that is configured to eject a low concentration of additive and is located just upstream of a principal ejector. The concentration can be of order 10 ppm by weight, since the intent is not to be effective far downstream but only immediately upstream of and directly where the concentration of the material from a principal ejector is greatest (i.e. where the penalty for diffusion is the greatest). Thus, for the cost of a modest amount of additive, much greater quantities of additive from the principal ejector will remain in the near-wall region.

For relatively steady flows, more simple ejectors configured as transverse grooves, properly scaled to produce a stable entrained vortex or vortex system, are positioned upstream of the principal ejector. A stable and entrained vortex system will dissipate small-scale vorticity produced at the wall and interrupt the development of the upstream boundary layer. Groove profiles that produce stable-entrained vortices specifically for after-body separation control have been published in the literature. In addition to properly shaping the groove, the present invention introduces small amounts of additive in order to contribute further to the stabilization of the entrained vortex.

Three profiles of upstream groove configuration are given in Figure 6.66. Figure 6.66a is a schema of a transverse groove cross-section of elliptical form with a major axis *15*, a minor half-axis *16*, and a depth *17*, relative to the outer wall. This shape when scaled properly (*15 > 17*), can be more tolerant of low levels of unsteadiness in the boundary layer than a rectangular shape.

Figure 6.66b is a schema of a cross-section of a rectangular groove of width *18* and depth *17* (where *17 ≈ 18*), which can be fed an additive through a nozzle *19* with a Coandă surface *20* at the bottom downstream edge of the groove. For this configuration, the addition of small amounts of additive will increase the stability of the entrained vortex. Figure 6.66c is a schema of a cross-section of a similar rectangular groove that can be fed an additive through a nozzle *21* at the top upstream edge of the groove.

For this configuration, the additive expenditure rate will be slightly greater than for the configuration in Figure 6.66b, but the additive will suppress small-scale vorticity in the near-wall region of the boundary layer as well as stabilize the entrained vortex. In all cases, the external flow *12* is from left to right.

In addition to these techniques, it is also possible to precondition the flow upstream by employing other drag reduction techniques just upstream of a principal ejector. These techniques include, but are not limited to, riblets, drag-reducing coatings of various types

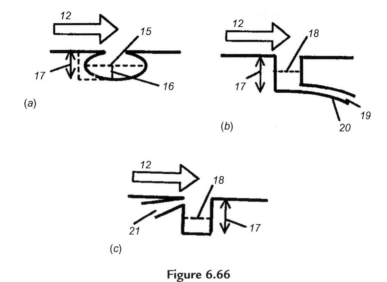

Figure 6.66

and boundary layer suction. As described in the literature, each has its advantages relative to the characteristics of the upstream flow.

Since this design of ejector is much less disruptive than prior designs it is possible to stratify different additives using tandem ejectors. Ejection of fluids of differing viscosities through multiple ejectors permits the establishment of a favorable viscosity gradient in the near-wall region, thereby enhancing system performance. For example, the ejection of fluid from a similar but smaller ejector located immediately downstream of a principal ejector and scaled for a value of $Q_s \approx 1$ will act to displace the additive from the upstream ejector away from the wall and into the region where it is effective in reducing the level of turbulence. For gas microbubbles, this also reduces the potential of the bubbles to act on the wall as roughness elements during ejection. For both gas and concentrated solutions of polymer, it can provide a favorable rather than unfavorable viscosity gradient at the wall. The ejected fluid can be the solvent alone, for example water, or a dilute solution of the additive such that the viscosity is the same or less (such as for heated water) than the ambient solvent. When just water or no additive is used in the downstream ejector, the requirement for a nozzle can be relaxed. Since the flow rate through the downstream ejector is reduced to $Q_s \approx 1$ the ejection velocity should be about 5% of the free stream velocity. This is about half of the ejection velocity of the larger upstream ejector. To accommodate the lower ejection velocity, the ratio of the diameter of the Coandă surface to the slot width should be increased over that of the upstream ejector to a value of 6 to 8 times the downstream slot width *26*, to inhibit the development of local separation on that surface. The size of opening *11* should remain about three slot widths *26*; hence, the segment between the curved wall of the chamber and edge *6* must be extended in comparison with the upstream ejector.

Figure 6.67 is a cross-section schema of one configuration of a triple ejector system. In this configuration, an elliptical transverse groove *22* is located upstream of the first principal ejector unit *23*. Additive may be fed into the elliptical groove in the same manner as displayed in Figure 6.67b, c. Depending on the character of the upstream flow, additional grooves *25* or, instead of the grooves, a small ejector scaled to eject 5 to 10 Q_s of additive at a concentration in the order of 10 ppm can be positioned to suppress the level of turbulence at the first principal ejector. "Sacrificing" this small amount of additive will reduce the level of turbulence and, thus, the amount of diffusion at the first principal ejector.

The fluid, f_4, from the first principal ejector may be a mixture of gas microbubbles which, according to the literature, including work by Merkle and Deutsch [160, 162, 190, 429, 430], can be effective within 300 viscous units of the wall, i.e. further from the wall than most polymers are effective. Deutsch also reports that the microbubble layer seems to act to screen the near-wall layer from the larger structures in the outer regions of the boundary layer. Thus, multiple tandem ejectors *23* and *4* can be used to position microbubbles of different scales and polymers of different molecular weights and configurations at the stratum where they are effective. Downstream of the principal ejectors (*23* and *4*) is a smaller ejector *26*, having a slot of width h_2 (*27*) that is scaled to the mass flow rate of the fluid, f_3, which is ejected from this downstream ejector. When only the solvent is intended to be ejected from the downstream ejector, for example for establishing a favorable viscosity gradient, the requirement for a nozzle or similar device to produce a convergent laminar flow can be relaxed. However, nozzles of various configurations are often used to produce the desired scale microbubbles; hence, specific nozzle designs are likely to be

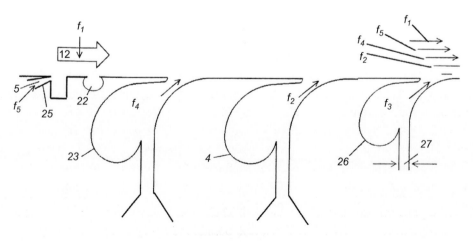

Figure 6.67

required for microbubble ejection as well as for preconditioning the polymer prior to ejection.

Thus, in addition to altering the rheological characteristics of the near-wall fluid, multiple ejectors can be employed to stratify additives that are known to be effective in specific strata of the boundary layer. For example, some additives, such as microbubbles of a particular scale, are considered to be effective further from the wall than are polymers. In Figure 6.67 a set of three tandem ejectors each scaled for the desired mass flow rate, could provide a three-tiered strata of water (low viscosity), f_3, under a concentrated solution of polymer, f_2, over which microbubbles, f_4, are ejected.

Similarly, multiple layers of appropriately scaled bubbles or multiple layers of different types of polymer can be ejected from tandem ejectors. Over these fluids flows the additive from the upstream grooves or "sacrificial slot," f_5, and the free stream fluid, f_1.

In the past, shipbuilders concluded that high concentrations and high flow rates of additive from a single ejector system are more efficient than the same amount of additive being ejected from multiple ejection sites distributed along the length of a hull. The increase in local skin friction produced by traditional ejectors and boundary layer blow-off, which led to an increase in pressure drag, were contributors to these phenomena. By avoiding those effects, the present invention makes it possible to employ sets of ejectors at multiple locations along a vehicle or propulsor, and, thereby, optimize the distribution of additive as a function of shape and length of the wall (vehicle). Thus, very long walls may be treated without a significant loss in efficiency.

Ejectors also can be configured to energize the near-wall flow to avoid separation during changes in the free stream angle of incidence, since the ejector is adaptable to local changes in flow conditions. The post-ejection processes include treatment of the wall to reduce diffusion of the additive downstream of the ejector, treatment of the outer flow to reduce diffusion of the additive both along the wall and around any protrusions, and the downstream ejection of either different additives or a different concentration of additive to achieve a more efficient additive expenditure rate.

The patent below applies the principles considered above in Section 6.7.1 for the organization of longitudinal vortical structures, for example in slot I, III and IV, as one of methods of drag reduction in external flows.

6.7.3 The Underwater Apparatus with Fin Propulsive Device [421]

Lin, Howard and Selby [342] experimentally investigated the influence of longitudinal grooves in a plate on flow separation control. Their results have shown that it is possible to reduce the separation area by 60%. Thus, it is possible to apply this principle by using a profiled surface on a fin propulsive device (Patent US 3773011 B63, 1973).

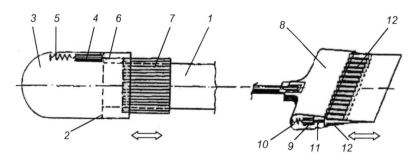

Figure 6.68

In this patent, the design of an underwater apparatus with fin propulsive device (see Figure 6.68) is presented. The apparatus consists of the case *1* in the forward part, on which the disk *2* is fixed, on which the head part *3* of the device is established. Inside the head part the solenoid *4*, a spring *5*, and the rod *6* are mounted. The rod is connected with the cylinder *7*, the external surface of which is designed in the form of large riblets.

The apparatus is put in motion by the fin propulsive device *8*. The solenoid *9*, spring *10* and rod *11* are mounted in *8*. The back part of the fin propulsive device *12* is connected to the rod *11* and has ribletted surfaces on both sides.

For various modes of movement, but especially for maneuvering, in order to eliminate the boundary layer separation the case *1* and back part of the fin propulsive device *12* are advanced a certain distance by means of the specified electrical system. Thus, the riblets on these surfaces promote reduction of the separation. For an increase of the overall performance of the apparatus, solutions of polymers or microbubbles can be injected onto the ribletted surface.

The principles of organization of longitudinal fluctuations of frame *14* for controlling the width of slots *1*, *3* and *4* are considered above in Section 6.7.1 and are applied as a method of drag reduction in external flows in this patent

The principles of organization of longitudinal vortical structures were applied, as well, to internal flows for increasing the performance of a fluid converter. Figure 6.69 shows a photograph of the system of longitudinal deepenings on the internal surface details of a fluid converter.

6.7.4 The Underwater Apparatus [167]

In the US patent "Aerodynamic element with low aerodynamic resistance" (4,434,957 B64C), the surface has transversal elements forming vortices. Solenoids *2* and *3* with the core *4* are mounted in the nose parts *1* of the design of the underwater apparatus

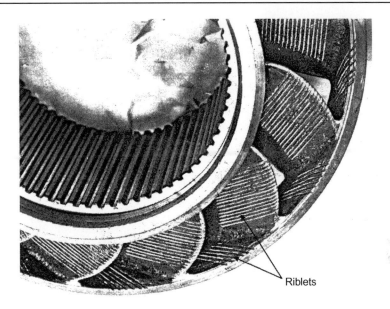

Figure 6.69

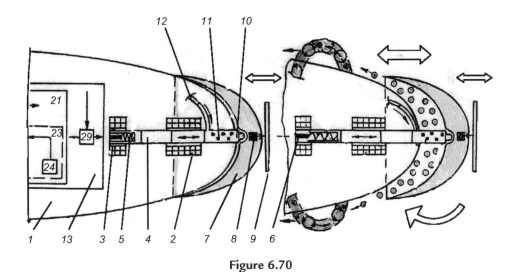

Figure 6.70

(see Figure 6.70). The spring *5* placed inside core *4* is connected to the finger *6*. On the opposite side, the rod *4* is connected with a hinge to fairing *7* inside of which an electric motor of rotary movement *8* is placed; ring *9* is placed on its axis. On that side the case of the rod is made in the form of a hollow permeable cylinder *10* with a various apertures *11* on its surface. The cylinder *10* is connected to a flexible tube *12* for feeding various solutions, for example, polymers, microbubbles, their mixtures, and so on. The microprocessor *13* is for controlling the motion of the specified elements.

After the electric motor *8* is turned on the ring *9* starts rotating, forming a system of rotating vortical rings. When a voltage is removed from the electromagnet *3* the spring *5* pushes out core *4* and fairing *7* a specified distance from the nose parts of the case *1*. In this case, the solenoid *3* works as a locking device. When a voltage is fed to the electromagnet *2,* core *4* is drawn in. Controlling electromagnets *2* and *3* make the cowl *7* oscillate. The block *13* regulates the amplitude and frequency of the oscillation of fairing *7* and the parameters of rotation of ring *9*. Oscillatory movement of ring *9* under certain conditions is provided in the same way.

The specified types of motion allow vortical rings to be generated from the fairing surface on the case of the apparatus. The rings also rotate around the longitudinal axis of the case. This provides steady movement of the rings along the entire surface of the case. Vortical rings during motion along the case utilize the energy of the boundary and the friction drag into the rolling resistance. When various resistance-reducing solutions are injected from the slot formed by the fairing *7* and case *1*, the stability of the motion and the efficiency of the vortical rings essentially increase.

The fairing can be turned in two opposite directions. At such turns, the cowl is tightly pressed to the case on the one side, and the slot on the other side increases. This causes the moment on the nose part of the case and improves its maneuvering. The same is provided by mutual motion of the fairing *7* and rings *9*.

6.8 Modeling of Disturbance Development in the Flow Behind a Ledge

In various kinds of shear flows there are areas with high concentrations of shear stresses (see Sections 1.1 and 1.7, and Figure 1.48). Examples of such areas include the place near an aperture edge at an outflowing flooded jet or the place near to a slot edge through which liquid solutions are injected in the boundary layer; similarly, in places near to edges of various ledges or deepenings on a body in flow. Areas of separated flow form behind back edges of various types of ledges, which lead to cavitation about the body with growth of flow speed. Pressure jump occurs on this edge, leading to the appearance of disturbances behind the edge on the border of the separation area and basic flow. The second type of area with high concentrations of shear stresses is the zone of attachment of flow to the wall.

The integral effects have been investigated arising in cavitation flows, for example spectral characteristics of oscillating axisymmetric cavities, and characteristics of plane flows behind a ledge on a plate and on rotating bodies. The influence of the boundary layer thickness on the structure of the near-wall flow behind a 2D ledge was also investigated.

These problems and other problems of cavitation flows have been considered in the proceedings of conferences [84, 85].

In [133] the CVSs formed behind cavitators executed in the shape of a disk, sphere, hemispheres or ogive are experimentally investigated. Using high-speed filming, the disturbances in the form of waves were identified, the parameters of which were put on the diagram of stability of Tollmien—Schlichting waves. Gromov and Korotkin [85] measured the low-frequency quasi-linear fluctuations near to the cavity surface.

Brennen [133] took photographs of the vortical structures developing on a cavity surface at subsequent stages of development of the disturbing movement, after the formation of disturbance waves. Hoyt [236] measured CVSs on the border of a submerged jet using the technique by Brennen.

It is of interest to investigate in modeling experiments the laws of development of wave instability of the disturbances arising at formation of a cavity identified in [133]: various issues of turbulent flow about a ledge, and, on the basis of the susceptibility method, to develop combined methods of stabilization of flows in areas of high shear stress.

Figure 6.71a is a schema of the cavitation flow about a ledge under the plane approach [133]. Brennen used this schema in his theoretical calculations of hydrodynamic stability of the flat waves identified by means of high-speed filming. Examples of these photographs are shown in Figure 1.9. The measurements are presented in the form of a dependences of the dimensionless wavelength D/λ and dimensionless distance X_1/D on Reynolds number

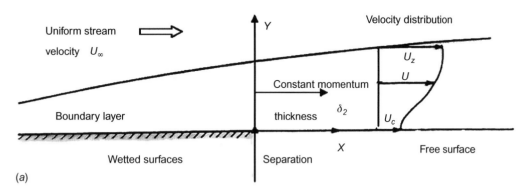

Figure 6.71
Schemes of a ledge flow: a — settlement scheme by Brennen [133] of the cavity ledge flows; b — modeling of current behind a ledge for experimental research on membranous surfaces [32, 305]; c — scheme of formation of boundary layers on a border of section of two environments, d — scheme of flow stabilization in an area of cavity closing by means of an elastic surface; e — scheme of energy recycling by means of various stabilizers of a cavity.

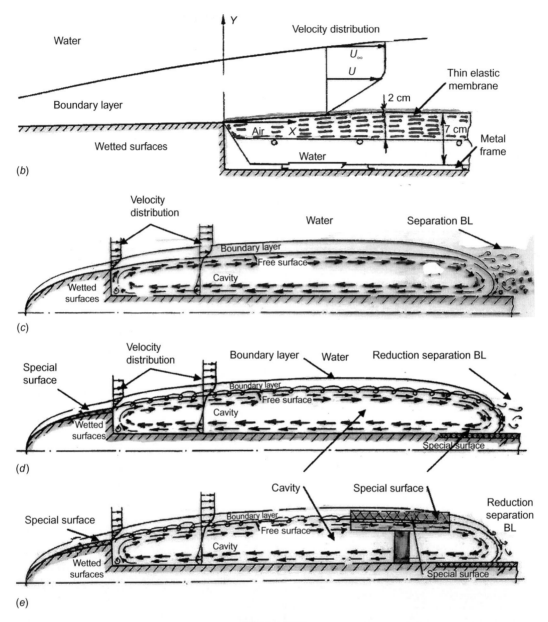

(b)

(c)

(d)

(e)

Figure 6.71
(Continued)

$U_T D/\nu \cdot 10^{-5}$ based on diameter D of the bodies investigated in a cavitation tunnel. Wavelengths on the cavity surface have been measured in flows about three kinds of sphere and ogive tip. It was revealed that the wavelength λ and X_1/D decrease with growth of Reynolds number.

According to Figure 6.71a, Brennen calculated the hydrodynamic stability and performed a qualitative comparison with data from experimental research. The comparison was done by means of dependences of dimensionless wave number $\alpha_r = 2\pi/\lambda$ and dimensionless frequency $\gamma = 2\pi f \cdot \delta_2/U_\infty$ on Reynolds number based on the momentum thickness δ_2. The shapes of cavitators were analyzed. Frequency characteristics of the disturbing movement essentially differed from the neutral curve plotted in corresponding coordinates for Blasius profile, established numerically by Lin. Experimental values of dimensionless frequency at the cavitation flow about 3D tips were essentially greater than Lin's neutral curve. Thus, in the zone of stability loss the frequency in the cavitation flow were 2.2 times higher than in the flow about a 2D plate. At $\mathrm{Re}_{\delta_2} = 300$ and 400 the difference increased to 3 and 3.45 times. These results are caused by the specific boundary layer in the cavitation flow about 3D bodies. It is important to note that the theoretical calculations by Bremen were based on certain assumptions. Figure 6.71b is a schema of the experimental facility used for modeling cavitation flows about a ledge. Experimental research is led to 2D statement—in the flow over a plate. A steam-gas medium inside the cavity can be treated as a medium with high compliancy, and a layer of high-speed border of the liquid adjacent to the steam-gas medium can be considered as a thin elastic film. Experimental research on the hydrodynamic stability of such a representation was carried out in a low-turbulence hydrodynamic tunnel by means of the tellurium method [31]. The design of the hydrodynamic facility and the technique used for the measurements is described in Section 1.5 [31, 305]. Various kinds of bottom were mounted in the 3 m-long working section of the facility (see Figure 1.20). For modeling of the flow behind a ledge the bottom was made in the form of a metal frame fitted outside with a thin polyvinyl film (see Section 1.5, Figure 1.22). It was possible to regulate the height of the ledge under the film, its elasticity and tension. Water was under the film during the experiments (series B1−B15, B21−B42), or air (series B16−B20), or heated water (series BT43, BT44).

Hydrodynamic stability was experimentally investigated by the classical method [305]. All parameters of the disturbing movement were investigated, and the findings are presented in the form of various laws, including in the form of neutral curves [31, 305]. In Sections 1.5−1.7 some results of the research on hydrodynamic stability are presented for various conditions of flow. Section 1.5 contains information on the techniques used in this research and some of the results. In particular, neutral curves are represented at the flow over rigid and elastic plates in Section 1.6 (see Figure 1.39) and Section 1.10 (see Figure 1.65) respectively.

Table 6.12 shows the mechanical characteristics of elastic plates in the present experiments. According to Figure 6.72a, the area of instability essentially increased, and Reynolds number of the stability loss decreased. It was revealed that the area of instability depends on the thickness of the water layer h under the membrane. The area of instability was less at smaller thickness h. At the same time, dimensionless wave numbers and phase speeds

Table 6.12: Mechanical characteristics of elastic plates.

Numbers of Experiences	$E \cdot 10^{-4}$ (N/m^2)	T (N/m)	ρ_M (kg/m^2)	$t_1 \cdot 10^4$ (m)	$h \cdot 10^2$ (m)	$U_\infty \cdot 10^2$ (m/s)
B1−B5					7	11.0
B6−B10	1.92	79.0	950	1	1	11.0
B11−B15	1.55	36.0	950	1	1	9.1
B16−B20	1.15	17.5	950	1	2(air) + 5 (water)	10.5
B21−B25	1.35	25.0	950	1	7	10.5
B26−B34	1.04	15.0	950	1	1	10.4
B27−B35					1	7.8
B36−B39					5	11.0
B40					1	10.1
B41−B42						9.1
BT43−BT44						11.0

depend to a lesser extent on h, although the laws are the same (see Figure 6.72b). Thus, the maximum values of these quantities appeared to be higher than those at the flow over a rigid wall and are shifted to greater Reynolds numbers.

In the series B1−B5, approximately at the same values of Reynolds numbers as in Brennen's experiments, frequencies of unstable fluctuations increased by 1.4, 2.1 and 1.7 times respectively. An ambiguity was revealed in the dependence of maximum values of the considered quantities on Reynolds number. Increased phase speed and graphs of amplitudes of cross-section velocities of the disturbing movement against the fluctuation frequency indicate a deteriorated stability. In comparison with the rigid standard the maximum amplitudes of cross-section speed of the disturbing movement increased and the coefficient of attenuation decreased.

In series B16−B20 (see Figure 6.71b) an air layer 2 cm thick was under the membrane and a 5 cm water layer was under the air layer. A reduction of tension in the membrane and the presence of air led to an increase in stability and a reduction in the range of unstable oscillations. In this series of experiments the frequencies of unstable fluctuations increased accordingly by 1.2, 1.3, and 1.5 times in comparison with the flow over a rigid plate [31, 305].

The so-called limit neutral curves [305] have also been established. In these experiments the entire range of unstable fluctuations were identified by means of the tellurium method, has been discussed. All these points are bounded by the enveloping curve, which has been called a limiting neutral curve. It appears that this curve bounds the range of fluctuations, which exceeds the range of fluctuations limited by the usual neutral curve 2−3 times [305].

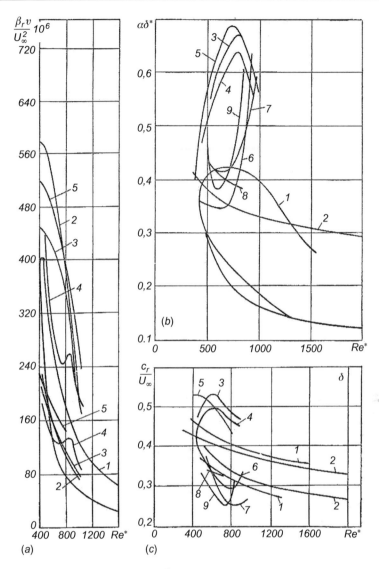

Figure 6.72

Neutral curves in coordinates of dimensionless frequency (*a*), wave number (*b*), and phase speed (*c*) at a longitudinal flow membranous surfaces: *a* — measurements on rigid (curve *1*) and membranous surfaces at *h* = 7 cm (curve *2*, series of experiences B1−B5), *h* = 1 cm (curve *3*, series of experiences B6−B10), *h* = 2 cm (curve *4*, series of experiences B16−B20), *h* = 1 cm (curve *5*, series of experiences B11−B20); *b*, *c* — measurements Babenko [31] (curve *1*) and Schubauer and Scramsted (curve *2* on a rigid plate); measurement on membranous surfaces [31]: *3* — experiences B1−B5, *4* − B6−B10, *5* − B16−B20, *6* − B26−B34, *7* − B36−B42, *8* − BT43−BT44, *9* − B21−B25.

The range of usual neutral fluctuations at flows over membranous surfaces exceeds the range of such fluctuations at a flow over a rigid plate 1.5 times. If the neutral curves in Figure 6.72 are interpolated for membranous surfaces to the left on Reynold numbers, then the critical Reynold number of the stability loss is essentially less than for the neutral curve for a rigid plate. These data coincide with results of measurements by Brennen, in which experimental points of unstable fluctuations are also essentially shifted to the left from the stability loss Reynolds number for a rigid plate.

The above-mentioned results have shown that the parameters of the disturbing movement essentially depend on the basic flow speed, size of the ledge, tension of the elastic film simulating the velocity of the cavitation flow and the properties of the liquid above and under the ledge. As in Brennen's experiments [133] the frequency of unstable fluctuations has essentially increased in comparison with classical data for the flow over a rigid plate. Other characteristics of the disturbing movement have changed as well, which leads to accelerated development of CVSs on the border of the gas—liquid shear layer. The development of CVSs behind a cavitator in the field of high concentration of shear stresses on the cavity border actually occurs in the same sequence as in Figure 1.42 (see Section 1.7), which is a model of development of disturbances in a boundary layer [310].

The flow about a cavitator will now be considered from the point of view of the susceptibility problem (see Section 1.11). Movement of the liquid environment behind a cavitator causes a corresponding movement of gas in the cavity, as shown by arrows in Figure 6.72c. Movement of the gas-steam medium near the border with the liquid occurs owing to the known laws of movement of a liquid near a moving wall. Thus, two boundary layers exist at the liquid—steam border. One is located in the liquid; it arises on the cavitator surface, as shown in Brennen's scheme (see Figure 6.72a). The second boundary layer arises near the interface in the gas-steam medium on the moving liquid border. The boundary layer in the gas-vapor environment arises according to the circulating movement of gas in the cavity caused according to Tulin's closure condition as shown by arrows in Figure 6.72c. These two boundary layers interact at the boundary between two media.

Fluctuations in the liquid boundary layer on the liquid—gas border arise behind the cavitator edge owing to the pressure jump and cause fluctuations in the boundary layer in the gas-steam medium. The parameters of the disturbing movement in the liquid and gaseous environments are different. Therefore, the interaction of fluctuations in two media leads to the accelerated development of disturbances along the boundary of two media.

In the area of attachment of the cavity to the body there is complex pattern of the interaction of disturbances moving along the interface of the two media and those arise in the attachment area. The disturbances generated at closing of the cavity on the wall cause disturbances in the flow. These disturbances also interact with each other.

In [84, 310] the interaction of various disturbances in a boundary layer has been experimentally investigated. It is convenient to present real disturbances possessing a complex form as combinations of simpler waves or vortical motions of various kinds, frequencies and amplitudes. Thus, the concept of susceptibility includes the process of interaction of various disturbances in a boundary layer and the result of such interaction.

The findings of research on the laws of a boundary layer flow at each stage of transition under the influence of a finite plane or vortical disturbance have led to two basic conclusions. First, the process of interaction of disturbances in a boundary layer can be presented as a superposition of the "frozen," so-called natural structures of disturbing motion at each stage of transition (at "absence" of disturbances) with structures of the introduced disturbances. Second, interaction of the natural structures of disturbing motion in a boundary layer with disturbances introduced from outside has a resonant character depending on the kind, energy and peak-frequency-wave characteristics of these structures. These conclusions concern not only the transitional boundary layer, but also the turbulent boundary layer and other types of disturbance interactions.

The principle of susceptibility states that for interaction of disturbances it is necessary to create vortical structures similar to those existing in the flow, as shown in Figure 1.70 (see Section 1.11) and Figure 6.13 (see Section 6.1). Yet another principle of susceptibility is that it is necessary to create certain artificial vortical structures that could easily be damped by similar structures. On the basis of these principles it is necessary to organize certain CVSs on a cavitator, as shown in Figure 6.72d. Such vortical structures can be created, for example, by means of special elastic surfaces (see Section 2.5, Figures 2.28, 2.30 and 2.31). CVSs generated by this surface will be transported downstream along the boundary between two media. In the area of the cavity reattachment it is necessary to apply the CVS control methods developed (see Section 1.13) for the effective stabilization of the artificial disturbances introduced on the cavitator. For this purpose an elastic surface, having a structure similar to the cavitator surface, can be placed in the reattachment area. However, geometric and other parameters of these special surfaces will differ because, depending on the cavity length, the parameters of the disturbing motion introduced on the cavitator will change during its development along the flow.

Such methods of interaction of CVSs in a boundary layer work in nature—for high-speed water animals. This method of stabilization of the disturbances developing on the boundary of two environments reduces the deformation of the cavity shape, stabilizing the disturbances on the cavity surface and the disturbances arising at reattachment of the cavity to the body.

Two problems, along with others, are important at cavitation flows:

• It is necessary to stabilize fluctuations of the cavity relative to the body and, hence, to stabilize the motion of the body itself.

- At great speeds of motion, at which a cavitation flow arises, greater energy is spent, which is also necessary for further maintenance of the cavitation mode of flow. Therefore, it is important to investigate methods of recycling of the spent energy.

Figure 6.72d represents a method of stabilization of disturbances introduced on a cavitator and damped in the reattachment area, which promotes an increase in the cavity length. In the reattachment area the flow splits—partially reversed motion of the liquid upstream and partial burst of the liquid–gas mixture downstream is observed. With a proper design for the forward part of the elastic coating located in the reattachment area, there is partial recycling of the energy spent from the area of the reversed flow. The second half of the elastic coating established in the reattachment area has another structure and promotes the stabilization of flow in the area of the cavity collapse. It ensures the reduction of losses of the gas-steam medium from the cavity.

Utilization of the susceptibility method allows greater recycling of the spent energy to take place in comparison with the method considered above. Therefore, the means to do this are now given. In the first method, the nose cavitator has a design of elastic coating such that it allows not only the generation of CVSs on its surface but also causes swirling of the flow in a helix. Then helical CVS form on the cavity surface, and as a result the development of the disturbances on the cavity surface is better stabilized, the cavity shape is stabilized and a gyroscopic effect appears. The body motion at the cavitation flow is better stabilized. In this case the elastic surface located in the back part of the body (see Figure 6.72e) has a completely different structure to the structure of the coating considered in the previous method. This new design of a coating stabilizes the oblique CVSs going from the cavity surface, utilizes the energy and increases the gyroscopic moment formed on the cavitator.

A variant of this method consists of an elastic coating being applied to a special stabilizer located equidistant to the body surface in addition to the back part of the body (see Figure 6.72e). Such a stabilizer is placed ahead of the reattachment area. In this case, the stabilizer additionally stabilizes CVSs. In addition to CVSs it is stabilized on the elastic surface located on a streamline body, which increases efficiency of CVS stabilization.

The stabilizer can have a cylindrical shape or consist of discrete parts (skis) that glide on the internal surface of the cavity. The stabilizer can change the angle of attack and distance from the body. The stabilizer surface can be attached in a fixed way or with a hinge, so that it is possible to maintain the optimum angle of attack for gliding. The design of the elastic surface on the stabilizer is similar to that on the body surface. It increases efficiency of energy recycling and the magnitude of the gyroscopic moment.

It is possible to apply other designs for the formation of the gyroscopic moment (see Figure 6.72e). For example, a cavitator can be fixed with a hinge on the body and rotate along the longitudinal axis of the body under the action of the gyroscopic moment arising on the cavitator. A stabilizer or a part of its surface can be made to rotate along the longitudinal axis the body.

It was shown in some cavitation experiments with dilute polymer solutions [133] that injection of polymer solutions on the surface of a spherical cavitator increases the separation angle of flow at cavitation. At the same time, an injection of polymer solutions on the surface of a cylindrical cavitator does not influence the flow separation angle. Polymer solutions can be injected through the slots in the cavitator surface or in the area of its face edge, and also on the stabilizer's surface. Instead of a polymer solution or simultaneously with it, microbubbles can be injected, as shown in Figures 2.39 and 2.40 (see Section 2.5). It is possible to apply a multilayered scheme of injection of liquid solutions, as shown in Section 6.7. In order to utilize the specified methods of stabilization of CVSs and a body at cavitation, a new shape of a gliding surface can be used, similar to fish scales (see Figure 2.41).

For the development of theoretical methods of modeling the problem was considered and it was thought that another scheme for a cavitation flow could be used. In this case the cavitation flow behind a ledge can be presented in the form of a reversed jet. The papers [237, 238] experimentally studied a submerged jet when a liquid flows from an aperture into air (see Figure 1.10). Flow behind a ledge can be treated as a reversed jet when air is located inside and a liquid jet moves outside.

6.9 Basic Conclusions

The findings have led to the conclusion that the basic problem at injection of polymer−water solutions into a boundary layer is the revealing of the physical mechanism of the influence of polymer molecules on CVSs in the boundary layer, and the methods of reduction of polymer solution diffusion.

At the present time much experimental research is being undertaken with the purpose of revealing the physical mechanism of the interactions of a polymer solution with a boundary layer. However, few research projects so far have focused on studying the interaction mechanism of polymer solutions with CVSs, and more attention should be paid to this problem.

A number of theoretical models of a polymer solution's influence on the characteristics of a boundary layer have been created. Those projects have mainly investigated the variation in the diffusion and kinematic viscosity in a boundary layer.

Under action of shear stresses in a boundary layer polymer molecules expand into a string, which chemically interacts with molecules of water. A grid of molecules with a dominant longitudinal orientation, or so-called associations, is formed. This grid promotes the interaction of associations with longitudinal vortical structures of a boundary layer. The associations damp the influence of pressure pulsations. As a result, longitudinal vortical structures in a boundary layer merge and the ratio of fluctuations of velocity components and pressure changes, as well as the Reynolds stresses and turbulence anisotropy. The associations play a very important role in mutual damping of the fluctuation components.

The influence of polymer solutions on a boundary layer is maximal if the optimum concentration of polymer solution is maintained in the area of the boundary between the viscous sublayer and buffer zone. However, until now this has not been achieved because polymer solutions diffuse rapidly in a boundary layer. To solve this problem, the method of multilayered injection of various liquids across the boundary layer thickness [78, 79, 81, 84] has been put forward and theoretically proved. Experimental research has confirmed the capabilities of such anapproach [365, 366, 400]. Optimization of a polymer injection in a boundary layer consists of a distributed injection through a surface [34]. In the present research, a new method of increasing efficiency is considered, which consists of the utilization of a sword-shaped tip.

The physical mechanism consists in two factors: first, it redistributes pressure and tangential stress along the body and, second, it artificially turbulizes the boundary layer on the body up to the place of polymer injection. Thus, in the boundary layer the extended strings act mainly as associations. The relative length of the tip and radius of joining with the swordfish body are very important. It is seen in photographs that the swordfish radius of joining is very large. Therefore, the best results in experiments were obtained when the tip was longer and, thus, the radius of joining greater.

This given hypothesis regarding the flow about the swordfish body from the location of flow separation is promising. Not only does the specific shape of the swordfish body confirm the correctness of this approach, but also the shape of its head, which, as is apparent from photographs, has a solid frontal part and solid lateral parts in the form of strong developed gills. Together with the xiphoid tip, such a head makes it possible to for the animal to withstand strong frontal pressure at high speeds of swimming. A hypothesis about specific shape of the swordfish body and the sailfish body has been set out in Section 6.5. The body shape can be considered as a transversal cylinder of small elongation (head), in the wake of which the animal body is situated. This arrangement decreases the drag of such a cylinder due to the elimination of the separation zone. The flapping fin mover sheds the separation zone behind the cylinder into the wake in the process of oscillation. Other mechanisms of drag reduction are considered mainly behind the gill slit, i.e. on the body in the separation zone of the flow about the head. Since the flow speed is great, consequently, pressure gradients are great, so soft body tissues cannot take such loads in the region of the head. Therefore, the authors assert that all the conclusions are relevant only for the segment behind the separation zone, i.e. behind the gill slit. At the same time, it appears from the structure that the head part is solid and capable of taking the loads. In addition the rough xipoid tip plays an important role in redistributing the pressure, because maxima of pressure gradients occur on the head part. The tip also causes turbulization of the boundary layer, which is capable of stretching polymer macromolecules (e.g., slime) and so achieve greatest effectiveness.

One direction for research on the optimization of polymer injection is the search for methods of preconditioning of polymer solutions (i.e. the elongation of strings) before injection in a boundary layer. This has been worked on by Pogrebnyak [411] and Babenko [78, 81]. Another direction is the search for methods of reducing diffusion of polymers in a boundary layer. In relation to this, uniform injection of polymer solutions through a porous surface [38] and through longitudinal slots [76, 78 and 79] has been developed, similar to the process in swordfish skin. A new method of multilayered injection of various liquids [78, 79, 81, 84, 85] has also been explored. This method relates to the hypothesis of its application by sharks. The idea for this method arose during numerous discussions with K.J. Moore [365].

The findings of this research have shown that the combined methods of influence on the characteristics of a boundary layer are promising. In this research the combined methods of drag reduction in the form of utilization of xiphoid tips and injection of polymer solutions have been investigated. On a small model, polymer solutions were injected in the form of a system of longitudinal vortices of a constant size. Another combined method consists in the simultaneous use of an elastic surface and injection of a polymer solution. Preliminary measurements were also taken in the latter case at the presence a xiphoid tips. The combined methods are rather promising and so it is important to develop research in this direction. It is also necessary to carry out the appropriate hydrobionic research in field.

The following combined methods of drag reduction have been investigated:

1. Elastic surfaces in combination leading to the formation in a boundary layer of longitudinal vortical structures and 2D disturbances. Thus, the generation of disturbances has been investigated at its placement on an elastic surface and inside its rigid structure for the formation of vortices. Disturbances have been generated by means of a special design of elastic surface, and also by heating and regulation by pressure of separate elements of an elastic surface. In addition, joint influence on resistance of an elastic surface and its curvature have been investigated.
2. A combination of elastic surfaces, injections in a boundary layer of polymer solutions and formations of longitudinal disturbances.
3. A combination of the form of nasal contours of a rigid surface, in particular, various xiphoid tips, and injection in a boundary layer of polymer solutions. In a boundary layer longitudinal vortical disturbances were formed.
4. As above, but at formation around of model of a static electric field.

5. Injection in a boundary layer of a rigid model of polymer solutions from two slots at the simultaneous formation of two longitudinal vortical systems.

6. A combination of elastic surfaces, injections in a boundary layer of polymer solutions, formation in a boundary layer of longitudinal vortical systems and application of nasal contours as xiphoid tips.

7. Various methods of injecting microbubbles of gas in a boundary layer with the help of thin slots, current-carrying wires, strips or diaphragms placed lengthways or across a stream.

8. Injection in a boundary layer through three slots of different solutions in a combination to create three systems of longitudinal vortical structures.

Some variations to the combined methods of drag reduction, according to the above mentioned list, have also been explored.

6.9.1 Flow of Elastic Surfaces

On the basis of the hydrobionic approach (see Section 1.8) a variety of elastic surfaces has been developed and made (see Section 2.5). Along with homogeneous monolithic surfaces multilayered anisotropic elastic surfaces, which are modeled on the structure of the skin of high-speed hydrobionts, have been investigated. Studies have involved the placement of various kinds of vortex generators inside and outside of elastic plates. In some cases, current-carrying heating elements were placed inside and outside of elastic plates. In other cases, inside the plates were longitudinal and cross canals containing a liquid, the pressure of which could be adjusted statically or dynamically. Such elastic surfaces made it possible to combine two or more methods of CVS control (see Section 1.13). This provided the facility to regulate the characteristics of the boundary layer via the regulation of the mechanical characteristics of elastic surfaces and the formation in set combinations of various kinds of CVSs. The inclusion of current-carrying elements on the surface of plates allowed the creation of longitudinal layers of microbubbles in a boundary layer.

The characteristics of the boundary layer of such surfaces have been investigated in various hydrodynamic installations and in a wind tunnel. Some results of these experiments are given in Chapters 3–6. Practically all investigated elastomers have a reduced hydrodynamic resistance up to 35% in a wide range of Reynolds numbers.

6.9.2 Injection of Polymer Solutions at a Flow of Elastic Surfaces Generating Longitudinal Vortical Structures

In Section 2.5 the design of an elastic plate intended for use with the specified combined method of drag reduction is given. The external layer of the elastic plate is made from permeable elastic material. Internal canals in the covering are filled with a polymer solution with a concentration of 2000 ppm. According to the distribution of pressure along a streamline body the insignificant part of a polymeric solution will act on an external layer of an elastomer. The designed composite elastic covering promotes the formation in a boundary layer of longitudinal CVSs. Such a design models the integuments of a swordfish (see Section 6.1).

A similar combined method can be carried out at the injection of polymer solutions through a nasal slot in a boundary layer of an elastic surface. Such experiments have been discussed in Section 5.6 and Chapter 6. The integrated efficiency of such a method is 40–60%.

6.9.3 Influence of Nasal Parts of a Body on the Efficiency of Polymer Solution Injection into a Boundary Layer

The results of research on the influence of nasal contours on the characteristics of a boundary layer of models, in particular at the injection of polymer solutions, have been discussed in this chapter. Research of a flow of a model with various kinds of tips has shown that the model with least resistance has an ogive tip. However, a xiphoid tip resistance is decreased to a greater extent than with an ogive tip. From the results, the greatest efficiency was seen for the model with a long xiphoid tip. Xiphoid tips have efficiency in a wider range of Reynolds numbers. At certain concentrations and specific charges of polymers, for the models with xiphoid tips formation nasal drafts is fixed.

The influence of various forms of nasal contour has been investigated, including on longitudinal and streamline cylinders of large sizes. The length of the cylinders changed over a wide range, and the diameter of cylinders was 0.1 m and 0.175 m.

6.9.4 Influence of a Static Electrical Field on the Characteristics of a Boundary Layer

The model (see Section 6.4) was made of an organic glass and had a replaceable tail and nasal cowls. These cowls were made from organic glass and brass. The cowls made from brass incorporated wires that were connected to a source of direct current. The resulting static electrical field around the model has the characteristic that under certain conditions it influences the efficiency of the injection of polymer solutions.

It is known that molecules of water can make various liquid crystal structures, so-called clusters, which essentially differ among themselves: clusters of seawater differ from clusters of freshwater, for example. Therefore, the efficiency of polymer solutions in seawater decreases. It is also known also that optimal clusters of water for an organism appear in defrosting water. Figure 6.73 shows a photograph of clusters. It is possible to assume that these clusters will be the most effective at the dissolution of polymers in water. Effective polymer solutions can be utilized in defrosting water or in the presence of an electromagnetic field, the power lines of which are directed along a streamline body.

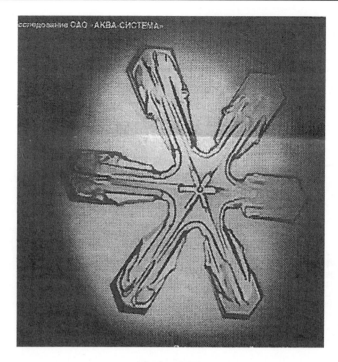

Figure 6.73
Liquid crystals of water after it has defrosted.

6.9.5 Injection of Polymer Solutions from Two Slots

The design of the model in Section 6.4 allowed the nasal part of model (see Figure 6.35) to be continuous, and was without slots or had two slots formed in the nasal parts. Thus, the design of the tips made it possible to inject, from the first slot, a continuous veil of near-wall jets with polymer solutions. From the second slot the near-wall jet containing a system of longitudinal vortices was generated. There was an opportunity to generate a system of longitudinal vortices both at the injection of solutions and from the first slot. The design of the nasal parts allowed for the preliminary stretching of polymer molecules, which increased the efficiency of the polymer solutions. The formation of one or two systems of longitudinal vortices in the injected near-wall jets raised the stability of the near-wall jets and reduced the degree of polymer diffusion through the thickness of a boundary layer.

6.9.6 Injection of Polymer Solutions from One Nasal Slot for a Flow of the Cylinder with an Elastic Surface

The results of the experiments performed are given in Section 5.5. The models had an external rigid surface or various kinds of elastic surfaces. The length of the first model, and the form of the nasal parts for both models varied. It was possible to form various sizes of longitudinal CVSs in injected near-wall jets with polymer solutions. The conclusion based on the results was that the combined methods provide increasing efficiency of drag reduction. The results obtained depend on many parameters, and are listed above for this method of drag reduction. Therefore, it is necessary to investigate the influence of each parameter on drag reduction separately and independently, and then to investigate the influence of each consecutive and additional parameter.

6.9.7 Various Methods of Injection into a Boundary Layer of Gas Microbubbles

In Section 6.4 (see Figure 6.39) the measurements of resistance of a model with short xiphoid tip are presented. In a strongly aerated stream resistance of same model essentially decreased. The effect of drag reduction is kept at all Reynolds numbers.

In [335] the analysis of various methods of microbubble injection is provided. In Section 2.5 various ways of injecting microbubbles into a boundary layer are given.

6.9.8 Injection of Polymer Solutions Through Three Slots

Drag reduction is theoretically investigated in the presence of a gradient of viscosity across a boundary layer. Such an approach is achieved by means of the model shown in Figure 6.35. Three cylindrical channels enter into model. Through the back channel inside the model electric wires join a trailer tip to a forward tip. Polymer solutions are transported inside of a model through this channel and two forward channels in a knife. In this set up, nasal slots can inject two polymer solutions independently in various concentrations. Thus, it is possible to provide a gradient of viscosity across a boundary layer.

The idea to develop a system of multilayered injections of various liquids in a boundary layer arose during discussions about scientific problems with K.J. Moore, the President of the Cortana Corporation, and, in particular, the role of shark slots. In Section 6.7 the device developed for injection into a boundary layer of various solutions, switching to microbubbles, through three or any number of cracks, is described and the method substantiated. Shkwar developed the theory of multilayered near-wall jets of various solutions injected in a boundary layer (see Chapter 7).

The application of three-layered or multilayered injection results not only in the formation of a gradient of viscosity across a boundary layer, but also in the reduction of diffusion of the solutions. As indicated in Section 6.9.4, it has also been established that the diffusion of polymer solutions decreases in an electrostatic field. In Section 6.9.5 this is solved by the injection through the first slot of polymer solutions of high concentration, and through the second slot of solutions of low concentration. Maintenance of the corresponding sizes of longitudinal vortices results in an essential reduction of diffusion of a solution injected from the second slot.

In Section 6.7 a similar idea according to research in [364] is set out. Thus, through the first crack water is injected, through the second a polymer solution is injected and through the third a microbubble water solution is injected. The diffusion of microbubbles is interfered with by the layer of polymer solutions located above it, which in turn has its diffusion interfered with by the layer of injected water. This combined method has made it possible to reduce resistance by 60%.

CHAPTER 7

Mathematical Modeling of the Turbulent Boundary Layer with Injection of Polymer Additives

7.1 Introduction

High levels of environmental pollution are common nowadays, affecting the atmosphere, water and other sources. These are usually located near big cities, and include nuclear power stations, chemical plants and underwater storage of chemical weapons (which have been on the bottom of large water reservoirs since World War II). These environmental problems have stimulated interest in the use of propagation of admixtures of low concentrations injected locally into either a gaseous or fluid medium by means of forced convective and diffusion transfers [407] inherent in these media. The multicomponent and multiphase flows are essential parts of a wide range of operating processes that take place in engines of various designs and functions, as well as in other industrial equipment [160, 202, 335, 368, 387]. For example, in aviation turbojet engines used under conditions of a torrid climate, the problem of rise in traction during a take-off is solved through the injection of water into the air inlet, resulting in the formation of an air-and-water mixture. More complicated processes occur in the combustion chambers, where a fuel blend is sprayed in a compressed air jet. On burning, this fuel blend reacts chemically and energywise with the main flow, whereupon the flow represents a complex mixture of gaseous and finely dispersed combustion products. At the same time, the injection of either high-molecular polymers or surface-active substances of low concentration into the fluid flow is an efficient method of influence on both the flow pattern and the processes of turbulence formation in the boundary layer [383, 403, 404, 405, 425, 426, 449, 450, 457, 486].

An alternative method of influence on the wall flow, which is similar to the foregoing one in the principle of feasible constructional implementation but not in the mechanism of influence, consists of injection of a suspension of gas microbubbles into the fluid flow [175, 335].

The development of these approaches for the control of the formation of a wall flow consists of the combination of these methods, i.e. the simultaneous injection of both the polymers and the microbubble suspension into the boundary layer [175, 190, 335, 403, 457]. This idea is based on the well-known fact of hydrobionics that an intensive oxygen exchange occurs in the gills of fast-moving fish with water passing through them. This process is accompanied by both a rise in temperature of the water that has passed through the gills and the saturation of water with a weak solution of mucus. The mechanism of influence of this mucus on the flow is, in many ways, similar to the principle of influence of polymer additives. Since the discharge water is ejected from the gill cavities in a tangential direction to the flow, i.e. into the domain of maximum shear stresses, it is assumed that the basic modification of the flow properties, which happens in the gills, is the formation of the gradient, normal to the surface, of viscous properties of the wall flow. This modification reduces the turbulent friction drag at high velocities. The possibility of extending this mechanism of control over the formation of the turbulent flow, which occurs in nature, to technical objects, providing an improvement in their hydrodynamic characteristics, is of considerable interest. Thus, the need for both optimization of the equipment that performs the injection of such substances (gas suspensions) into wall flows and for the reduction in consumption of the injected additives stimulates the use of local injection near a surface, in some section of the boundary layer or in several sections along the direction of its development, so that the concentration of the active admixture near the wall were kept at the optimum level. The simulation of propagation of the injected substance with consideration for the convective and diffusion properties of both the carrying flow and the substances transferred by it, along with the consideration for the viscosity gradient normal to the surface is a topical problem. When considering a local injection of a system of phases, determination of the optimum relative positions of the injectors along the vertical extent of the cross-section, from which the substances are injected simultaneously, is important. Also crucial are the optimization of both the geometrical extents of the injectors and the consumption of different phases along with the processes of interpenetration of the phases and the combined action of them on the main flow parameters.

Shkvar has developed the mathematical modeling of injection of various liquids with different densities and other characteristics into the turbulent boundary layer through one or several slots [76, 84, 369, 370].

7.2 Statement of Problem

The aim of the present research is to build a mathematical model of the turbulent boundary layer, into which the components of high-molecular-weight polymers are injected with the purpose of shear friction reduction. An advantage of the given statement is the account of heterogeneity of concentration distribution of the injected substance across the thickness of shear flow in that cross-section of the boundary layer where the polymeric component is injected. The marked phenomena assume:

- the effect of the distance (height) from the surface to the ejector of a polymer solution;
- the effect of the ejector nozzle width;
- the possibility of the presence of several ejectors of various thickness located on different heights from the surface and its effect;
- modeling of diffusion of the polymer additive injected into a stream across the boundary layer thickness as it is convected by the average flow.

Thus, given a distribution of velocity exterior in relation to a boundary layer of flow, and given initial distributions of velocity and concentration of the molecular-weight component of the polymer, it is necessary to calculate fields of velocity and concentration of polymer additives in the flow area of interest.

The following traditional simplifying assumptions have been adopted in the mathematical model developed:

1. The flow has a precisely expressed dominant direction of development, which justifies application to the considered task the boundary layer theory and equations, which are obtained because of this theory.
2. The concentration of polymer in the solution, which is injected from an ejector, is low enough to allow the fluid to be considered as Newtonian.
3. The surface is supposed flat and hydrodynamically smooth.
4. The parameters of the polymer solution injected into a boundary layer do not change the monotone distribution of the average longitudinal velocity component and do not modify the static pressure p on a normal y to the surface. This assumption does not influence the correctness of the boundary layer assumptions $p\,(y) = \text{const}, \frac{\partial p}{\partial y} = 0$.

The assumptions mentioned above make it possible to simplify as much as possible the problem of boundary layer modeling, so as to focus attention on the multiphase simulation of the polymeric components.

7.3 Brief Analysis of Known Results for Turbulent Flows in the Presence of Solutions of High-Molecular-Weight Polymers

Outcomes obtained in various papers [114, 193, 221, 227, 238, 275, 284, 332, 341, 351, 352, 383, 394, 404, 405, 425, 426, 449, 450, 486, 499, 509] on flows of polymer solutions can be generalized as follows.

7.3.1 Physical Aspects of Influence of the Polymeric Components on the Structure of a Boundary Layer Flow

1. The presence of low concentrations ($10^{-3}-10^{-1}$ kg/m^3) of some polymers (for example, guar, polyethylenoxid WSR-301, collagen etc.) essentially reduces shear stresses [114, 193, 221, 227, 238, 275, 284, 332, 341, 351, 352, 383, 394, 404, 405, 425, 426, 449, 450, 486, 499, 509] in shear flows (known as the Thoms effect).

2. The effect of drag reduction does not depend on the absorption properties of a surface and, as a result, on the amount of the polymer additive molecules settled on the wall, but rather is determined by properties of the molecules in the shear flow [449].

3. Only those polymers with molecules that have the ability to unfold and stretch along the direction of streamlines have the property to reduce drag [404, 405, 425, 426, 449, 450]. The same effect can be achieved in the presence of suspensions of solid oblong shaped particles, such as asbestos and paper filaments, and nylon threads in a stream.

4. The efficiency of a polymeric additive ejection is determined by its chemical and mechanical properties, which is why its molecular weight and the flow temperature can be considered as factors that characterize the polymer properties in a shear flow. At the same time, as noted in [486], "the molecular aspects of turbulent drag reduction mean that hydrodynamic efficiency of a polymer is mainly determined, not only by its molecular weight, but rather by conditions of the micromolecular skein and its sizes, which are a result of conformation of the molecular 'chain' depending on the external conditions and medium composition."

5. The reaching of maximum effectiveness of the ability of polymeric additives to reduce the drag is mainly determined, by the property of shear flow to expand the skeins of polymer molecules. So, the increasing of shear deformations and longitudinal stretching is the favorable factor for the lining up of molecules along a stream [449, 486]. This peculiarity explains the greatest effectiveness of polymeric additives near a surface that is in the domain of great transversal (normal) velocity gradients. Analysis of experimental data [351] shows that "... the effect of turbulent friction drag reduction can be recognized only when shear stresses reach some threshold magnitude," which does not depend on the polymer concentration in accordance with the results of one group of researchers. On the other hand, such dependence has been revealed at small concentrations of polyethylene ($c < 5.10^{-5}$ g/cm^3) [22].

6. The concentration of the polymer additive is an essential factor influencing the flow in a turbulent boundary layer, and an amplification of the Thoms effect at an increased concentration happens only up to its threshold value [114, 193, 221, 227, 238, 275, 284, 332, 341, 351, 352, 383, 394, 404, 405, 425, 426, 449, 450, 486, 499, 509]. It is also necessary to note that the influence of the same concentration of different polymeric additives is different.

7. With regard to the application aspect, polymer solutions are characterized by the property of aging during storing, which can be intensified by the influence of external factors and leads to a decrease in the drag reduction ability [405]. Moreover, polymers demonstrate the feature of shear degradation during use, which is amplified at increasing shear stresses [383]. This property means that drag reduction happens only a short distance from the ejector of the polymer additive.

8. Under the influence of polymer additives a turbulent shear flow with a dominant direction gains such rheological properties as visco-elasticity and anisotropy. These properties are a result of the appearance of a system of stretched polymer molecules aligned with the flow. Such a latticed structure of the moving liquid influences the process of turbulence generation, intensively suppressing normal turbulent disturbances and, to a much smaller degree, the longitudinal turbulent pulsations. These physical circumstances lead to anisotropy of the turbulent transport, which can be characterized by the ratio of the normal component of turbulent viscosity $\mu_{\tau x}$ to the longitudinal one $\mu_{\tau x}$ [383, 425, 449]. These effects are dominant in the mechanism of friction reduction by means of polymer additives.

7.3.2 Influence of the Components of a Polymer on the Profile Longitudinal Component Velocity

The presence of polymeric components in a flow leads to a deformation of the profile of the longitudinal component of velocity of the averaged motion in the near-wall area of the shear layer. In connection with the conventional law-of-the-wall, it is convenient to describe the character of this deformation scaling the longitudinal component of velocity u and transversal coordinate y with the friction velocity and the near-wall scale of length, correspondingly ($u^+ = \frac{u}{v_*}$, $y^+ = \frac{y v_*}{v}$), and presenting the function $u^+ = f(y^+)$ in semilogarithmic coordinates.

1. The viscous sublayer thickness in the viscous units $\delta^+ = \frac{\delta v_*}{v}$ increased proportionally to the coefficient of anisotropy of turbulent viscosity $A = \frac{\mu_{\tau y}}{\mu_{\tau x}}$ [425], which, in turn, is proportional (up to certain threshold value) to the polymer component concentration.

2. The logarithmic part of the epy velocity profile shifts in the direction of higher velocity proportionally to the polymer concentration. This influence can be accounted for by the constant C in the logarithmic law $u^+ = \frac{1}{k} \ln y^+ + C$ ($C = 4.9-5.5$; $k = 0.4$ for the flow over a smooth, flat surface without a pressure gradient, without polymeric components), which is increased by some value $\Delta u^+ > 0$, depending on the properties and concentration of the polymer and the shift properties of the flow as well [351, 425, 449 and 486]. Examples of empirical functions simulating the shift of the logarithmic part of velocity Δu^+ as a function of properties of a shear flow and a polymer can be found, for example, in [352, 425]. Thus, the presence of polymer additives can be modeled by

changing the fullness of the longitudinal component of the velocity profile. As for Karman's factor, unfortunately it is not possible to formulate a general point of view about the constant k. Some experiments indicate the invariance of the conventional value $k = 0.4$, while the slope of the logarithmic part increases in other research [351, 425, 449, 486], which indicates a decrease of k as the polymer concentration grows. Nevertheless, for limit values of concentration at which the condition of saturation is reached, limiting logarithmic distributions of velocity were revealed, which are characterized by the following pairs of "constants" offered by different authors: ($C = 17$, $k = 0.0855$; $C = 18.2$, $k = 0.0885$; $C = 20.2$, $k = 0.077$). These rather small values of the Karman's constant support the assumption that the Karman's constant is not a constant, being a function of the polymer concentration in a wide range of concentrations.

3. In calculations the above-mentioned effect of the viscous sublayer thickening is higher and the shift of the logarithmic part of the velocity profile is taken into account by the modification of van Drist's damping factor; therefore, the formula for mixing length l takes the following form $l = ky\left[1 - \exp(-y^{+}F/26)\right]$, where F is a parameter characterizing the damping properties of the viscous sublayer under the influence of polymer additives in the shear flow. Examples of similar empirical relations can be found, for example, in [284, 351, 352, 509].

The above-mentioned data on the structure and peculiarities of the development of flow in a boundary layer with polymeric components make it possible to construct a semi-empirical model of the flow considered.

7.4 Governing Equations

In connection with the above-formulated assumption about the validity of the concept of a boundary layer with reference to the investigated class of flows, let us consider the following system of the non-dimensional equations of a boundary layer:

$$\frac{\partial \overline{u}}{\partial x} + \frac{\partial \overline{v}}{\partial y} + \frac{1}{u_H}\frac{\partial u_H}{\partial x} = 0; \tag{7.1}$$

$$\overline{u}\frac{\partial \overline{u}}{\partial x} + \overline{v}\frac{\partial \overline{u}}{\partial y} + \overline{u}^2\frac{1}{u_H}\frac{\partial u_H}{\partial y} = -\frac{\partial \overline{p}}{\partial x} + \frac{\partial \overline{\tau}}{\partial y} \tag{7.2}$$

$$\overline{u}\frac{\partial \overline{c}}{\partial x} + \overline{v}\frac{\partial \overline{c}}{\partial y} = \frac{\partial}{\partial y}\left(\overline{D}_{eff}\frac{\partial \overline{c}}{\partial y}\right) \tag{7.3}$$

where (7.1) is the continuity equation; (7.2) is the equation of impulse for longitudinal component of velocity u; (7.3) describes transport of the polymer component. Non-dimensional coordinates x and y were constructed on the basis of typical linear size of the calculation domain L. The velocity of outer flow u_H is supposed to be a known function of longitudinal coordinate x, the non-dimension pressure is determined by the outer velocity in correspondence with Bernoulli's equation $\frac{dp}{dx} = -\frac{1}{u_H}\frac{du_H}{dx}$. The dimensionless shear stress in correspondence with the assumption about flow of a medium is simulated by the Boussinesk's formula:

$$\overline{\tau} = \overline{\nu}_{eff}\frac{\partial\overline{u}}{\partial\overline{y}} \tag{7.4}$$

here $\overline{\nu}_{eff} = \frac{(\nu + \nu_t)}{u_H L}$ is the dimensionless kinematic factor of effective viscosity.
Accordingly, the concentration of polymer c in the equation (7.4) is scaled by the value c_o in the initial cross section, $\overline{c} = c/c_o$, and effective diffusivity is simulated by the expression $\overline{D}_{eff} = \frac{(\nu/Sc + \nu_t/Sc_t)}{u_H L}$, where Sc and Sc_t are molecular and turbulent Schmidt numbers, which are determined as ratios of kinematic factors of molecular and turbulent viscosities (ν and ν_t) to the corresponding diffusivities (D and D_t), that is $Sc = \nu/D$ $Sc_t = \nu_t/D_t$. The value Sc is determined mainly by physicochemical properties of the polymer and solvent (water), whereas the value Sc_t, being a criterion of similarity, establishes a measure of analogy of the turbulent diffusion of the solvent relative the turbulent diffusion of the additive transported by this solvent. As a first approximation in the present research the turbulent Schmidt number Sc_t was constant.

The system (7.1−7.4) is solved under the following boundary conditions:

$$\overline{y} = 0,\ \overline{u} = 0,\ \overline{v} = 0,\ \frac{\partial\overline{c}}{\partial\overline{y}} = 0 \text{ on the surface} \tag{7.5}$$

$$\overline{y} - \infty,\ \overline{u} - u_H(\overline{x}),\ \frac{\partial\overline{c}}{\partial\overline{y}} \to 0 \text{ on the outer boundary of boundary layer} \tag{7.6}$$

$$\overline{x} = \overline{x}_o,\ \overline{c}_o = \varphi(\overline{y}) \text{ in the initial calculated cross-section } (x = x_o) \tag{7.7}$$

The function $\overline{u} = f(\overline{y})$ sets the initial velocity profile; it can be prescribed analytically or, for example, taken from experimental data. The function $\overline{c}_o = \varphi(\overline{y})$ is determined by the known concentration c_o of the polymer solution injected into the stream through the ejector, and also by geometric parameters of the ejector, namely, height h of its the lower edge over the surface and width of the slot s. That function is $\overline{c} = \begin{cases} 0 & npu & 0 \leq \overline{y} < \overline{h} \\ 1 & npu & \overline{h} \leq \overline{y} < \overline{h} + \overline{s} \\ 0 & npu & \overline{y} \geq \overline{h} + \overline{s}. \end{cases}$

7.5 Calculation Method

A two-step implicit noniterative method is used for solving the equations (7.1−7.4) with the specified boundary conditions (7.5−7.7). The equations are solved on a rectangular grid,

Table 7.1

Equation	Terms of the Equation (8)				
	Variable φ	A_x	A_y	F	S_φ
(7.2)	u	$\bar{u}$	$\bar{v}$	$\bar{\nu}_{eff}$	$(1-\bar{u}^2)\dfrac{1}{u_H}\dfrac{du_H}{d\bar{x}}$
(7.3)	c	$\bar{u}$	$\bar{v}$	$\bar{D}_{eff}$	0

irregular in both directions. The number of nodes in the x (i_{max}) direction has been found from the condition that the first calculated step is associated with two thicknesses of the boundary layer in the initial profile, and further steps were increased under the law of geometrical progression $\Delta x^{i+1} = \Delta x^i q_x$ with the ratio $q_x = 1.05-1.1$. In the direction normal to the surface, the grid is also clustered under the law of geometrical progression. The first step from the wall is determined by the condition $\Delta y_1^+ = 0.8$, and the last step is limited by $\Delta \bar{y}_{j_{max}} = 0.05\delta$. Thus, the grid is created in correspondence with the flow modeled. Usually, the number of nodes in the streamwise direction is equal to $i_{max} = 40-70$ nodes, and the normal direction is covered by $j_{max} = 70-140$ nodes.

For convenience of using the calculation method, equations (7.2) and (7.3) are represented in the generalized form:

$$A_x \frac{\partial \varphi}{\partial x} + A_y \frac{\partial \varphi}{\partial y} = S_\varphi + \frac{\partial}{\partial y}\left(F\frac{\partial \varphi}{\partial y}\right) \tag{7.8}$$

where $\varphi = \{u,c\}$ is a generalized variable; A_x A_y are factors of convective transport along x and y respectively; F is diffusion factor; S_φ is a source term. The parameters of (7.8) are shown in Table 7.1.

The flow characteristics in section x^{i+1} are calculated on the basis of the known values in section x^i through two stages. At each of these stages the equations (7.2) and (7.3) are solved in the following finite-difference form corresponded to the six-node stencil:

$$A_{xj}\frac{\bar{\varphi}_j^{i+1}-\bar{\varphi}_j^i}{\Delta \bar{x}^i} + \frac{A_{yj}}{2}\left(\frac{\bar{\varphi}_{j+1}^{i+1}-\bar{\varphi}_j^{i+1}}{\Delta \bar{y}_j} + \frac{\bar{\varphi}_j^{i+1}-\bar{\varphi}_{j-1}^{i+1}}{\Delta \bar{y}_{j-1}}\right)$$

$$= S_{\varphi j} + \frac{2}{\Delta \bar{y}_{av}}\left(F_{j+1/2}\frac{\bar{\varphi}_{j+1}^{i+1}-\bar{\varphi}_j^{i+1}}{\Delta \bar{y}_j} - F_{j-1/2}\frac{\bar{\varphi}_j^{i+1}-\bar{\varphi}_{j-1}^{i+1}}{\Delta \bar{y}_{j-1}}\right) \tag{7.9}$$

where the indices i, j determine the node in which the value of the grid variable equation (7.9) is determined; $\Delta \bar{x}^i = \bar{x}^{i+1} - \bar{x}^i$, $\Delta \bar{y}_j = \bar{y}_{j+1} - \bar{y}_j$, $\Delta \bar{y}_{av} = 0.5(\bar{y}_{j+1} - \bar{y}_{j-1})$ are current values of the grid steps; $F_{j+1/2} = 0.5(F_j - F_{j+1})$, $F_{j-1/2} = 0.5(F_{j-1} - F_j)$ is the diffusion factor in half-integer points between nodes along y.

The equation (7.9) can be presented in a different form:

$$W\overline{\varphi}_{j-1}^{i+1} + P\overline{\varphi}_j^{i+1} + E\overline{\varphi}_{j+1}^{i+1} = R \tag{7.10}$$

This equation connects the values of φ^{i+1} in three adjacent nodes along the y axis. The factors W, P, E and R are functions of parameters of grid parameters and variables. The sequential application of this equation to all nodes of the grid in the calculation section $x^{i+1} = const$ allows for the building of a system of $j_{max} - 2$ linear algebraic equations with a three-diagonal matrix, which is solved by the Thomas algorithm [369, 370].

At the first stage of calculation the values of grid variables defined by the convective factors A_x A_y; diffusive factor F and source term S_φ are taken from the previous calculation section along x, which has the number i, that is:

$$A_x = \overline{u}_j^i\ A_y = \overline{v}_j^i\ F = \overline{v}_{eff}^i\ S_\varphi = \left(1 - \overline{u}_j^{i2}\right)\frac{1}{u_H}\frac{du_H}{dx} \quad \text{for Equation (7.2)} \tag{7.11}$$

$$A_x = \overline{u}_j^i\ A_y = \overline{v}_j^i\ F = \overline{D}_{eff}^i\ S_\varphi = 0 \quad \text{for Equation (7.3)} \tag{7.12}$$

At the second stage the values known from the first stage are used for validating the equations parameters, namely:

$$A_x = \tilde{\overline{u}}_j^{i+1}\ A_y = \tilde{\overline{v}}_j^{i+1}\ F = \tilde{\overline{v}}_{eff}^{i+1}\ S_\varphi = (1 - \tilde{\overline{u}}_j^{i+1^2})\frac{1}{u_H}\frac{du_H}{dx} \quad \text{for Equation (7.2)} \tag{7.13}$$

$$A_x = \tilde{\overline{u}}_j^{i+1}\ A_y = \tilde{\overline{v}}_j^{i+1}\ F = \tilde{\overline{D}}_{eff}^{i+1}\ S_\varphi = 0, \quad \text{for Equation (7.3)} \tag{7.14}$$

Here "$\sim$" designates outcomes of the first stage (phase) of the calculation. If the second stage (phase) results are marked with "$\approx$," the final values of variables in the section x^{i+1} are found from the formula $\overline{\varphi}_j^{i+1} = 0.5(\tilde{\overline{\varphi}}_j^{i+1} + \tilde{\overline{\varphi}}_j^{i+1})$. The next stage of calculation is to determine the shear stress on the surface that has been realized by approximation of (7.4) in the vicinity of the wall utilizing the first and second nodes:

$$\overline{\tau}_w^{i+1} = 0.5(V_{eff_1}^{1+1} + V_{eff_2}^{1+1})\frac{\overline{u}_2^{i+1}}{\Delta\overline{y}_1} \tag{7.15}$$

Local values of the friction drag coefficient c_f and friction velocity v_* are determined by the formulae $c_f = 2\overline{\tau}_w^{i+1}$, $v_* = u_H\sqrt{\overline{\tau}_w^{i+1}}$. Then the displacement thickness is $\delta^* = \int_0^\delta (1 - \overline{u})dy$ and the momentum thickness $\delta^{**} = \int_0^\delta \overline{u}(1 - \overline{u})dy$.

The continuity equation (7.1) is approximated on the four-node stencil. The solution to this equation is sought for after integrating Equation (7.8) at $\overline{\varphi} = \overline{u}$, which makes it possible to calculate the vertical velocity component $\overline{v}$ in the section x^{i+1} using the horizontal component $\overline{u}$ in two neighboring sections along x. After that, the calculation advances to the next layer along the x coordinate and the procedure is repeated until the final cross-section of the calculation domain is reached.

Thus, the calculation method is of the "predictor-corrector" type [393, 410]. In correspondence with its structure it ensures the second order of accuracy in both directions, x and y. The finite-difference representation of the generalized equations in the form (7.9) with the way of modeling its terms described above has a significant advantage in comparison with the frequently used Crank–Nikolson scheme. There is no need to iterate on nonlinearity of the equation (7.2) due to the structure of the applied two-stage calculating procedure, which provides the second order of accuracy independently on operating conditions. In particular, this advantage is exhibited at calculations of flows with a significant pressure gradient and, first of all, when the adverse pressure gradient initiates separation or pre-separation conditions of the flow development, where the convergence of the iterative process is essentially worsened.

7.6 Turbulence Model

In the present research, with the purpose of modeling turbulence the algebraic model of turbulence offered by the Ukrainian researcher Movchan has been applied as the basic tool. This model has been modified by the authors of this book for modeling a series of peculiarities of near-wall flows, such as roughness of the surface, near-wall jet flow, three-dimensionality, heat- and scalar-additives transport, etc. [369, 370]. The choice of an algebraic model is stipulated by the circumstances, and this class of turbulence models, being relatively simple, approximates well the properties of the mean turbulent flow in the vicinity of a wall. As for more complicated differential models of turbulence, it is important to note that they have greater capabilities for modeling the transport of components of the Reynolds stresses tensor and also other components of turbulent motion. However, in the vicinity of a surface, the reliability of their approximations is less satisfactory. The physical effect of the influence of polymeric components on a turbulent flow development is exhibited near a surface, which indicates the necessity for a significant volume of research with the purpose of finding physically justified and mathematically correct approximations of right-side terms of the corresponded differential equations. While this is considered to be an achievable and realistic aim, simpler and adopted algebraic approaches have been used for modeling of turbulence at the present stage.

The model of turbulent viscosity applied in the present work has the following view:

$$\mu_t = \chi \rho \delta^* u_H \gamma \tanh\left(\frac{\ell \sqrt{\overline{\tau}}}{\chi \Delta}\right) \qquad (7.16)$$

$$\ell = ky_1 \tanh \frac{\sinh^2[\chi_1 y_1^+]\tanh[\sinh^2(\chi_2 y_1^+)]}{ky_1^+ \sqrt{\bar{\tau}}} \tag{7.17}$$

where $\chi = 0.0168 - 0.0215$, $\chi_1 = 0.068 - 0.072$, $\chi_2 = 0.223$, $k = f(c, \ldots)$, are the model coefficients; l is length scale of the near-wall turbulent motion, $\bar{\tau} = \tau(y)/\tau_w$ is dimensionless shear stress in the vicinity of the surface, which is determined in an association with the sign of the pressure gradient $\frac{dp}{dx}$ by relations: $\bar{\tau} = 1 + \frac{dp}{dx}y$, at $\frac{dp}{dx} \geq 0$; $\bar{\tau} = 1/(1 - \frac{dp}{dx}y)$, at $\frac{dp}{dx} < 0$; $\Delta = \delta^* u_H^+$ is Rotta–Clauser length parameter; y_1 is the coordinate y biased concerning the wall with the purpose of registration of the influence the surface roughness and polymer components, determined by the formula $y_1 = y_1^+ \nu/v_*$.

The value y_1^+ in 'wall law' coordinates is determined by a relation $y_1^+ = \begin{cases} 0 & if \quad s \leq 0 \\ s & if \quad s > 0 \end{cases}$,

where $s = y^+ + \Delta y_{rhn}^+ - \Delta y_{pol}^+$; Δy_{rhn}^+ is a parameter accounting for the influence of the surface roughness; Δy_{pol}^+ is a parameter introduced here with the purpose of modeling of the influence of polymeric components. The introduction of the shift function $\Delta y_{rhn}^+ > 0$ allows reflecting within the framework of the turbulence model, the effect of the surface roughness is known from experimental research. This effect involves the downward displacement of the logarithmic part of the velocity profile in the semilogarithmic coordinates, relative to its location for the flow on a smooth surface, by some magnitude Δu^+. This magnitude Δu^+ is a function of parameters of the surface roughness. The structure of this function and its connection with Δy_{rhn}^+ is described in [369, 370]. Similarly, the shift function $\Delta y_{pol}^+ > 0$ takes into account the displacement of the logarithmic part of the velocity profile in the opposite direction, which arises under the influence of the polymer components (see Section 7.3.2). Similarly to the influence function of roughness, within the framework of the given model there is uniqueness and universal dependence between Δy_{pol}^+ and the corresponded shift of the velocity profile Δu^+. However, the analysis and generalization of experimental data show the impossibility of establishing a universal and unequivocal dependence between Δu^+ and the polymeric component parameters. Therefore, for each case considered it is necessary to use the known experimental information on the influence of concentration and other properties of the polymer on the turbulence and parameters of the boundary layer modeled. The empirical dependencies of such are found in [351, 425] and other research on various polymers. This information was also used as the first approximation within the framework of the present model of turbulence.

The following hypothesis is adopted for registration of non-uniformity of the polymer concentration through the thickness of the flow: the shift function Δy_{pol}^+, unlike the function of roughness Δy_{rhn}^+, is not constant in the section $x^{i+1} = const$. It is determined in each node y_j^{i+1} by the value of the local polymeric additive concentration c_j^{i+1}, which is known from the solution of (7.3), that is $\Delta y_{pol}^+ = f(c, y, \ldots)$.

With the purpose of checking the assumptions used as the basis of the constructed
turbulence model and the calculation method, the results of their testing will now be
considered.

7.7 Calculations Results and Discussion

The first stage of testing the model is checking the computed profiles of the longitudinal
velocity component of the averaged turbulent flow. With this purpose, the following
transformed form of (7.4) was integrated numerically:

$$\frac{\partial u^+}{\partial y^+} = \frac{1}{1 + \nu_t/\nu} \tag{7.18}$$

The integration was fulfilled along the y coordinate inside the near-wall zone of the
boundary layer thickness. Equations (7.16) and (7.17) were applied for turbulent viscosity
modeling. Figures 7.1–7.3 compare the computed longitudinal velocity component
$u^+ = f(y^+)$ (lines) with experimental data (points) for flows of homogeneous solutions of
different polymers with various concentrations. In the process of modeling of the turbulent
viscosity, the type of polymer and level of its concentration were accounted for in
agreement with known empirical information. Figure 7.1 corresponds to the experimental
data of Khabakhpasheva [221], Figure 7.2 to experiments by Ivanuta and Chekalova [248,
249] and Figure 7.3 integrates various experimental data of different authors for various

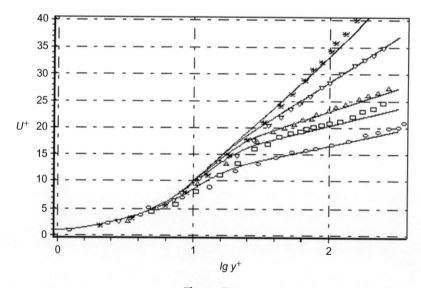

Figure 7.1
Comparison of the predicted velocity profiles and the experimental data of Habakhpasheva [221].

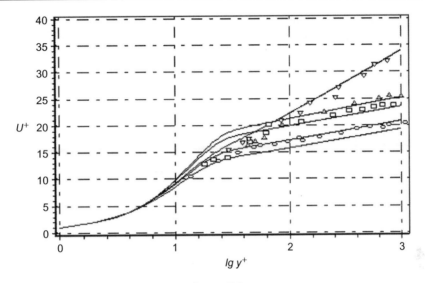

Figure 7.2
Comparison of the predicted velocity profiles and the experimental data of Ivanuta and Chekalova [248] (polyethylene oxide WSR-301, $c = 2 \cdot 10^{-6} - 50 \cdot 10^{-6}$ g/cm^3).

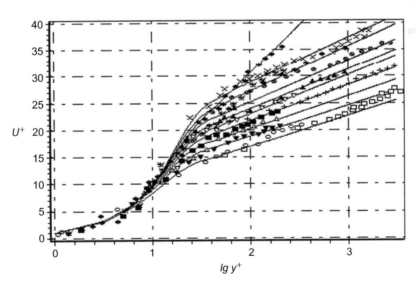

Figure 7.3
Comparison of the predicted velocity profiles and the experimental data of different authors for different types of polymer additives with a wide range of concentrations ($c = 0 - 7.4 \cdot 10^{-4}$ g/cm^3).

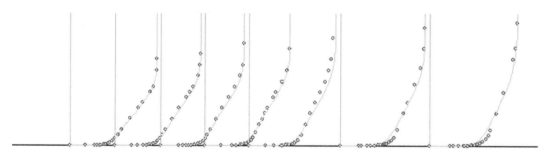

Figure 7.4
Prediction of the flow development (unfavorable pressure gradient): blue lines are velocity profiles; red circles are experimental data of Bradshaw & Ferris [129] (Stanford Conference, 1968, id. 2400).

polymers in a broad range of concentrations. All velocity profiles are represented in semilogarithmic coordinates.

It follows from the represented comparisons that the calculated distributions of velocity agree satisfactorily with the experimental data of various authors. As can be seen in Figures 7.1–7.3, the effect of thickening of the viscous sublayer is correctly simulated in calculations at increased concentration of polymer in the solution, as well as the transformation of the linear dependence $u^+ = y^+$ in the viscous sublayer into the logarithmic law $u^+ = \frac{1}{k}\ln y^+ + C$. Furthermore, the numerical experiment has shown that the dependence of Karman's coefficient on the concentration of polymer $k = f(c)$ is not universal and, as well as the dependence $\Delta y_{pol}^+ = f(c)$, is determined by the polymer properties.

The following series of comparisons show the results of modeling of the external flow developing along a surface in the absence or presence of the polymeric additive solution injected into the boundary layer through a nozzle (injector). The corresponding computer code has been written with the programming language Object Pascal in the IDE Delphi-6 with a wide use of graphical tools.

Figure 7.4 illustrates the calculations of airflow with an unfavorable pressure gradient, investigated experimentally by Bradshaw, et al. [129] and selected at the Stanford Conference in 1968 as canonical data (id. 2400) [154]. The red points show the experimental data and blue lines represent the computed velocity profiles. Figure 7.5 represents a comparison of the computed and experimental profiles of velocity of the flow considered, presented in semilogarithmic coordinates. The correspondence of the calculated results with the data of the experiments which can be deemed satisfactory in both Figures 7.4 and 7.5. Figure 7.6 illustrates the process of convection, diffusion and mutual penetration of the system of different phases of polymeric components at injection

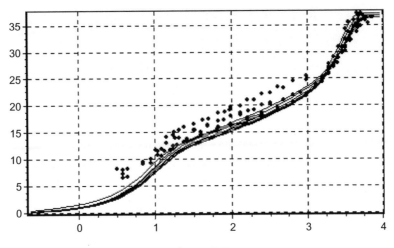

Figure 7.5
Comparison of the predicted (lines) and experimental (points) velocity profiles in semi-logarithmic coordinates under conditions of the experimental data of Bradshaw & Ferris [129] (Stanford Conference 1968, id. 2400).

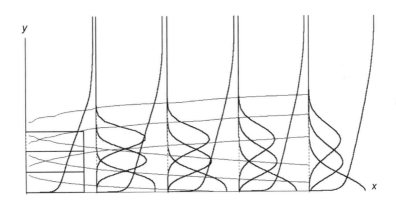

Figure 7.6
Prediction of flow development: monotonic lines are velocity profiles, dome-shaped lines are profiles of polymer concentrations.

immediately into the boundary layer in different places along the boundary thickness (thicknesses of the nozzles were equal each other).

7.8 Conclusions

1. The semi-empirical approach based on an algebraic model of the turbulent viscosity for accounting for the influence of a system of polymeric additives injected into fluid flow has been elaborated.

2. The method of calculation of the turbulent boundary layer has been developed to describe the flow in which the local injection of a polymer solution has been carried out.

3. The average velocity profiles of flow of homogeneous solutions of polymers have been computed.

4. The calculations of development of flows along a surface have been carried out in the presence of injection of a polymer solution for various heights of the ejector over the surface, and for various thickness of the ejector nozzle.

5. The results obtained indicate the advantage of injecting a polymer solution from a thin ejector immediately near a surface. The best values of geometric sizes and regime parameters should be determined for each case investigated.

6. These results describe qualitatively and correctly the processes in the controlled boundary layer, which makes it possible to develop some possible directions of further research in this field.

7.9 Probable Directions for Further Developments

1. Determining an optimum arrangement of an ejector along x and distances between several sequentially installed ejectors in view of convection, diffusion and modification of properties of the polymer component.

2. Determining an optimum angle of injection concerning a wall.

3. Investigation of the possibility of installation of several ejectors along a surface and analysis of their common work at various concentrations of polymer solutions, and also at various dispositions. Choice of an optimum configuration.

4. Theoretical calculation of injection of polymer solution through a porous surface.

5. Modeling of the combined influence of polymeric additives and roughness of the surface on drag reduction.

6. Improvement of empirically determined dependences of the turbulence model, accounting for the influence of the pressure gradient on the effect of the drag reduction by using polymeric additives, analysis of the mutual influence.

7. Modeling of the combined action of tangential blowing, wall-jet flow and polymeric additives.

8. Modeling of the influence of polymeric components on the Reynolds stresses tensor components. An analysis of physical effects of the mechanism of drag reduction by polymeric injection. On this basis, improvement of the approaches used in the present work and principles of semi-empirical modeling of the near-wall shear flows.

References

[1] Abramov LI, Halipha BA. Velocity profile of a turbulent boundary layer on the flat plate streamlined by the flow with a high degree of turbulence. Proc Moscow Energy Inst 1972;part 127:49–54.

[2] Ali A-M, Guljaev VI. Distribution of periodic waves in transversely to directed layers of the composite environment consisting of three homogeneous materials. Appl Mech 1982;18(3):87–91.

[3] Aerodynamic Drag Reduction Technologies. Editor, P. Thiede. Berlin Heidelberg: Springer-Verlag; 2001. p. 390.

[4] AGARD Report 827 High Speed Body Motion in Water. Copyright AGARD. 1998.

[5] Agarkov GB, Babenko VV, Fereneth ZI. About innervations of a skin and skin dolphin musculature in connection with a hypothesis of stabilization of flow in a boundary layer. In: Problems of Bionics, Moscow: Nauka; 1973. pp. 478–483.

[6] Ahenbah JD. Fluctuations and waves in the directional-reinforced composites. In: Mechanics of composite materials, vol. 2. Moscow: World; 1978. pp. 354–400.

[7] Airapetov AB. About one approach to calculation of a turbulent boundary layer on a surface with a pliable covering. Eng Phys J 1982;40(4):657–63.

[8] Akilli H, Sahin B, Rockwell D. Control of vortex breakdown by a transversely oriented wire. Phys Fluids 2001;13(2):452–63.

[9] Aleev YG, Leonenko IV. Hydrodynamic value of a swordfish rostrum. Bionika 1974;8:21–3.

[10] Reza A, Carpenter PW. Effect of wall compliance on streak-like flow structures. In: Proc. of the 2nd Intern. Symp. on Seawater Drag Reduction. ISSDR 2005. May 23–26, 2005. Korea:Busan; 2005. pp. 489–496.

[11] Alimpiev AI, Mamonov VN, Mironov BP. Power spectra of speed pulsations in a turbulent boundary layer on a no tight plate. PMTF 1973;3:115–9.

[12] Ambartsumjan SA. Difference-modular theory of elasticity. Moscow: Phys. Math. State Publishing House; 1961. p. 384.

[13] Amphilohiev VB, Droblenkov VV. Stability of a boundary layer on an elastic plate under various boundary conditions. In: Proc. of the Scientific and technical society. Resistance to movement and seaworthiness of ships. Leningrad: Shipbuilding; 1967, Part. 89:16–21.

[14] Amphilohiev VB. Turbulent flow with elastic borders. Bionics 1969;3:46–53.

[15] Amphilohiev VB, Droblenkov VV, Savorohina AC. Growth of small disturbances in a boundary layer on an elastic surface. App. Mech Tech Phys 1972;(2)part 127:137–139.

[16] Amphilohiev VB, Ivlev YuP. Pulsations of speed in a laminar boundary layer of models with an elastic surface. Hydromechanics 1973;23:61–63.

[17] Amphilohiev VB, Zolotov SS, Ivlev YP. Estimation of a friction resistance reduction of rotation body's at use extended nasal extremities. Bionika 1980;14:22–6.

[18] Anderson DA, Tannehill JC, Pletcher RH. Computational fluid mechanics and heat transfer. New York: Hemisphere Publishing Corporation; 1984 (translated into Russian, Moscow, Mir, 1990, v.1: 792 p. v.2: p. 384).

[19] Arata GFJ. A note on the flying behavior of certain liquids. Nautilus 1954;69:1–3.

[20] Ash RI, Bushnell DM. Compliant wall turbulent skin-friction reduction. AIAA Paper No 833; 1975;833: 1–33.

[21] Avdeev AN, Plessky VP. Distribution of Raleigh's waves along a periodic rough surface of an isotropic body. Acoustic J 1982;28(part 3):289–93.

[22] Avilov GM, Tartakovsky BD. About an opportunity of expansion of temperature-frequency area of efficiency three-layer vibro-absorbing designs. Acoustic J 1982;28(part 2):279–80.

[23] Babenko VV, Morosov DA. Some physical laws in action when dolphins dive. In: Mechanisms of movement and orientation of animals. Kiev:Naukova Dumka 1968. pp. 49–57.

[24] Babenko VV, Gnitethskj NA, Koslov LF. Preliminary results of elastic properties investigation of living dolphin's skin. Bionika 1969;3:12–9.

[25] Babenko VV, Surkina RM. Some hydrodynamic peculiarities of dolphins swimming. Bionika 1969;3:19–26.

[26] Babenko VV, Gnitethskj NA, Koslov LF. Preliminary results of investigation of temperature distribution on the surface of dolphins' bodies. Bionika 1970;4:83–8.

[27] Babenko VV. Principal characteristics of flexible coatings and similarity criteria. Bionika 1971;5:73–6.

[28] Babenko VV. Some mechanical characteristics of dolphins' skin covers. Bionika 1971;5:76–81.

[29] Babenko VV, Surkina RM. Definition of parameter of varying weight of skin covers of some sea animals. Bionika 1971;5:94–8.

[30] Babenko VV, Gnitethsky NA, Koslov LF. Hydrodynamic tunnel of low turbulence, apparatus and method of investigation of laminar boundary layers hydrodynamic stability. Bionika 1972;6:84–90.

[31] Babenko VV, Koslov LF. Experimental investigation of hydrodynamic stability on rigid and elastic damping surfaces. J Hydraulic Res 1972;10(4):383–408.

[32] Babenko VV, Koslov LF. Experimental investigation of laminar boundary layer hydrodynamic stability on elastic surfaces in water flow. Bionika 1972;6:22–4.

[33] Babenko VV, Koslov LF, Pershin SV. Damping coating. Patent of Ukraine 413286; 1974.

[34] Babenko VV, Koslov LF, Korobov VI. Damping coating. Patent of Ukraine 483538; 1975.

[35] Babenko VV. Fin propulsive device. Patent of Ukraine 529104; 1976.

[36] Babenko VV, Nikishova OD. Some hydrodynamic regularity of the skin covers structure of the marine animals. Bionika 1976;10:27–33.

[37] Babenko VV, Koslov LF, Kajan VP. Waving fin device for movement. Patent of Ukraine No 484129; 1976.

[38] Babenko VV, Korobov VI, Koslov LF. Regulating damping coating. Patent of Ukraine No 597866; 1978.

[39] Babenko VV. Investigation of the elasticity of living dolphin's skin. Bionika 1979;13:43–52.

[40] Babenko VV. On interaction of flow with elastic surface. Mechanisms of turbulent flows. Moscow: Nauka 1980;292–301.

[41] Babenko VV. On oscillate mass of integuments of dolphin's skins. Bionika 1980;14:57–64.

[42] Babenko VV, Voropaev GA, Kozlov LF, Korobov VI. Some kinematics characteristics of boundary layer in streaming of elastic plate. Eng Phys J 1980;38(6):1049–55.

[43] Babenko VV. One method of experimental investigation in hydrobionic. Bionika 1981;15:88–98.

[44] Babenko VV, Yurchenko NF. Damping coating. Patent of Ukraine No 80267; 1981.

[45] Babenko VV, Koval AP. About hydrodynamic functions of the gill apparatus of the swordfish. Bionika 1982;16:11–15.

[46] Babenko VV, Koslov LF, Pershin SV, Sokolov VE, Tomilin AG. Self-regulation of skin damping in cetaceans during active swimming. Bionika 1982;16:3–10.

[47] Babenko VV, Ivanov VP, Yurchenko NF. Measuring with laser anemometer of boundary layer receptivity to plane and three-dimensional disturbances. Autometria 1982;3:91–6.

[48] Babenko VV, Kozlov LF, Ivanov VP, Korobov VI. Experimental investigation of boundary layer interaction with injected half-limited jet. Hydromechanika 1983;47:28–34.

[49] Babenko VV, Sokolov VE, Koslov LF, Pershin SV, Tomilin AG, Chernishov OB. Property of a cetacean skin to actively regulate hydrodynamic drag to navigate by means of control of local intersection of integuments of skin with a flowing round stream. Moscow: State committee of the USSR on affairs of invention and discovery. Discovery 1983;265:4. X1.1982. Publ.15.5.1983. Bull. N17:Pub. 15.5.1983.

[50] Babenko VV. A flow around mechanism in aquatic animals. Bionika 1983;17:39–45.

[51] Babenko VV, Ivanov VP. Investigation of boundary layer structure in beginning of turbulence. Bionika 1984;18:28–35.

[52] Babenko VV. Investigation of wall region structure of turbulent boundary layer. Near-wall turbulent flows. Novosibirsk: Inst. of Heat physics SO AN SSSR; 1984;5–12

[53] Babenko VV, Yurchenko NF, Koslov LF. The control of the three-dimensional disturbances development in the transitional boundary layer. In: Proc second IUTAM symposium on laminar-turbulent transition. Novosibirsk, Novosibirsk, July 24−27, 1984. Berlin: Springer−Verlag;1985:329–335, or 509–513.

[54] Babenko VV, Kozlov LF, Dovgy SA, et al. The influence of the outflow generated vortex structures on the boundary layer characteristics. In: Proc. second IUTAM symposium on laminar-turbulent transition. Novosibirsk: Berlin: Springer-Verlag; 1985. pp. 509−513.

[55] Babenko VV, Yurchenko NF. Interaction of disturbances in the boundary layer. Mech Fluids and Gases 1985;5:68−74.

[56] Babenko VV, Kanarskj MV. Method for determination of dynamic viscous-elastic characteristic. Patent of Ukraine No 1183864; 1985.

[57] Babenko VV, Kozlov LF, Yurchenko NF. Generation of the given type of disturbances in boundary layers. In: Third Int. Symp. on Stratified Flows. Pasadena, USA; 1987. p. 27 (exact date unspecified due to visa restrictions).

[58] Babenko VV. The problem of turbulence development in boundary layers. scientific and methodological seminar on ship hydrodynamics, 17th Session, Varna, 17−22 October 1988. Proc., vol. 1; 1988. pp. 13.1−13.3.

[59] Babenko VV. The problem of boundary layer receptivity to various disturbances. Bionika 1988;22:15−23.

[60] Babenko VV, Yurchenko NF. Development of organized vortex structures in boundary layers over rigid and elastic plates. Turbulence '89: Organized structures and turbulence in fluid mechanics. Grenoble, France, September 18−21; 1989, p. 63.

[61] Babenko VV. Problem of interaction between different disturbances in a boundary layer. Colloquium on Goertler Vortex Flows "Euromech 261," June 10–13, 1990, Nantes, France, p. 70.

[62] Babenko VV. Methods of influence on coherent structures of a boundary layer. Research workshop "Ordered and turbulent patterns in Taylor-Couette flow," May 22–24, 1991, Columbus, USA.

[63] Babenko VV. On interaction of hydrobionts body with flow. Bionika 1992;25:1−11.

[64] Babenko VV. Boundary layer as inhomogeneous nonsymmetrical waveguide. Bionika 1992;25:40−5.

[65] Babenko VV, Musienko VP, Korobov VI, Ptukha YA. Experimental investigation of spherical groove influence on the intensification of heat and mass transfer in the boundary layer. Euromech Colloquium 327, Effects of Organised Vortex Motion on Heat and Mass Transfer. August 25–27, 1994, Institute of Hydromechanics, National Academy of Sciences of Ukraine, Kiev. Book of Abstracts, 1994: pp. 23−24.

[66] Babenko VV, Kanarsky MV, Korobov VI. Boundary layer on elastic plates. Kiev: Naukova dumka; 1993. p. 263.

[67] Babenko VV. Hydrobionic Principles of drag reduction. In: Second engineering aero-hydroelasticity conference (EAHE) Pilsen, Czech Republic, v II; 1994. pp. 277−293.

[68] Babenko VV. Methods for control of coherent vortical structures in a boundary layer. Review meeting on "Bio-locomotion and rotational flow over compliant surfaces," Baltimore, USA: Johns Hopkins Univ.; 1995.

[69] Babenko VV. Development of three-dimensional disturbances over concave elastic surface and with help of spherical grooves. In: Proceedings of the 10th international couette-taylor workshop. Paris, France; 1997. pp. 1−12.

[70] Babenko VV. Method of influence on coherent vortices structures of a boundary layer. In: Proceedings of the international symposium on seawater drag reduction, Newport, Rhode Island, USA; 1998. pp. 113−120.

[71] Babenko VV, Yaremchuk AA. On biological foundations of dolphin's control of hydrodynamic resistance reduction. In: Proceedings of the international symposium on seawater drag reduction, Newport, Rhode Island, USA, 1998, pp. 451−452.

[72] Babenko VV. Hydrobionic principles of drag reduction. In: Proc. of the Intern. Symp. on Seawater Drag Reduction, Newport, Rhode Island, USA. 1998. pp. 453−455.

[73] Babenko VV. Hydrobionic principles of drag reduction. AGARD FDP Workshop on "High Speed Body Motion in Water," Ukraine: Kiev 3; 1998. pp. 1−14.

[74] Babenko VV, Musienko VP, Korobov VI, Pjadishus A. On selection of the geometrical parameters of groove for generation disturbances in the boundary layer. Bionika 1998;27−28. pp. 42−47.

[75] Babenko VV, Korobov VI, Musienko VP. Formation of vortex structure on curvilinear surfaces and semi-spherical cavities. In: 11th International couette-taylor workshop. July 20—23, Bremen, Germany; 1999. p. 103.

[76] Babenko VV, Korobov VI, Shkvar EA. Hydrobionic principles of polymer injection optimisation on an elastic surface. In: 11th European drag reduction working meeting, Prague, Czech Republic; 1999. p. 55.

[77] Babenko VV. Hydrobionic Principles of Drag Reduction. App. Hydromechanics 2000;2(74): 2: pp. 3—17.

[78] Babenko VV. Method and apparatus for mixing high molecular weight materials with liquids. United States Patent 6,200,014 B1; 2001.

[79] Babenko VV. Control of the coherent vortical structures of a boundary layer. Aerodynamic Drag Reduction Technologies, Proceedings of the CEAS/Drag Net European Drag Reduction Conference, 19—21 June 2000, Potsdam, Germany. Springer-Verlag Berlin Heidelberg; 2001. pp. 341—350.

[80] Babenko VV. Method for reducing dissipation rate of fluid ejected into boundary layer. Patent US 6,349,734; 2002.

[81] Babenko VV. Method and apparatus for increasing the effectiveness and efficiency of multiple boundary layer control techniques. Patent US 6,357,374; 2002.

[82] Babenko VV. Method for reducing dissipation rate of fluid ejected into boundary layer. United States Patent No US 6,357,464; 2002.

[83] Babenko VV, Voropaev GA. Department of boundary layer control and hydrobionics. In: Institute of the hydromechanics. Kiev: Academy of Science NASU; 2002. pp. 33—52.

[84] Babenko VV, Shkvar EA. Combined method of drag reduction. Cheboksary, Russia: International Summer Scientific School "High speed hydrodynamics;" 2002; 321—326.

[85] Babenko VV. Combined method stabilization of body in shear flow. In: Proceedings of the Fifth International Symposium on Cavitation (Cav2003), Osaka, Japan, November 1—4; 2003.

[86] Babenko VV, Carpenter PW. Dolphin hydrodynamics. In: Carpenter PW, Timothy JP, editors. Proceedings of the IUTAM symposium on flow past highly compliant boundaries and in collapsible tubes. 2001. Dordrecht/Boston/London: Kluwer Academic Publishers; 2003. pp. 293—323.

[87] Babenko VV. Hydrobionic principles of drag reduction. Int J Fluid Mech Res 2003;30(2):125—46 (Translated from Prykladna Gidromechanika, 2000, (74), No 2: pp. 3—17).

[88] Babenko VV, Kusnetsov AI, Kusnetsov An I, Moros VV. Technique of realization towing tests in experimental tank through two models gliding vessel. App. Hydromechanics 2003;5(77): No 4: pp. 5—11.

[89] Babenko VV, Moros VV, Kusnetsov AI, Kusnetsov An I. Experimental research into the wing drag of a model of a three-case vessel gliding. In: Collection of the proceedings of the Nikolaev shipbuilding institute, Ukraine, Nikolayev; 2004; N4 (397). pp. 11—17.

[90] Babenko VV. Interaction quickly floating hydrobionts with flow. In: First international conference on bionics, Hanover, Germany; 2004. pp. 153—159.

[91] Babenko VV, Koval AP. Structure and hydrodynamic peculiarities of swordfish skin. App. Hydromechanics 2004;6(78): No 2: pp. 3—19.

[92] Babenko VV, Voskoboinick AV, Voskoboinick VA, Turick VN. Velocity Profiles in a boundary layer above a plate with deepenings. Acoustic Bull 2004;7(3):3—12.

[93] Babenko VV. Interaction of quickly swimming hydrobionts with flow. In: Proceedings of the second international symposium on seawater drag reduction (ISSDR) 2005, Busan, Korea; May 23–26, 2005. pp. 579—592.

[94] Babij VV, Babenko VV, Moros VV. Profile of a steering complex of a ship. Patent of Ukraine, N 52328; 2002.

[95] Bainbridge R. The speed of swimming of fish as related to size and to the frequency and amplitude of tail beat. J Exp Biol 1958;5(1):17—33.

[96] Bace S, Ivanov AV, Fernholz HH, Neemann K, Kachanov YS. Receptivity of boundary layer to three-dimensional disturbances. European J Mech D/Fluids 2002;21(1):29—48.

[97] Banah L Ya, Perminov MD, Petrov VD, Sinev AV. Methods of calculation of matrixes of rigidity and damping for complex spatial systems. Vibration isolation machines and vibroprotection of the person — the operator, Nauka, Moscow, 1973; pp. 67—81.

[98] Bandyopadhay PR, Gad-el-Hak M. Rotating gas—liquid flows in finite cylinders: sensitivity of standing vortices to end effects; 1985.

[99] Bandyopadhyay PR. Review of mean flow in turbulent boundary layers disturbed to alter skin friction. Trans ASME J Fluid Eng 1986;108:127−40.

[100] Bandyopadhyay PR. Drag reduction outer-layer devices in rough wall turbulent boundary layers. Exp Fluids 1986;4:247−56.

[101] Bandyopadhyay PR. Rough-wall turbulent boundary layers in the transition regime. J Fluid Mech 1987;188(23):1−266.

[102] Bandyopadhyay PR. Resonant flow in small cavities submerged in a boundary layer. Proc R Soc, London, A 1988;420:219−45.

[103] Bandyopadhyay PR. Viscous drag reduction of a nose body. AIAA J 1989;27(23):274−82.

[104] Bandyopadhyay PR, Rieser JE, Ash RL. Helical-Perturbation device for cylinder—Wing vortex generators. AIAA J 1992;30(4):988−92.

[105] Bandyopadhyay PR, Gad-el-Hak. M. Reynolds number effects in wall-bounded turbulent flows. App. Mech Rev 1994;47(28):139.

[106] Bandyopadhyay PR, Henoch C, Hrubes JD, Semenov BN, Amirov AI, Kulik VM, et al. Experiments on the effects of aging on compliant coating drag reduction. Phys Fluids 2005;17:8, 085104-1-9.

[107] Bartenev GM, Zelenev Yu V. Physics and mechanics of polymers. Moscow: Higher school; 1983;391.

[108] Basin AM, Korotkin AI, Kozlov LF. Control of a boundary layer of a ship. Leningrad: Shipbuilding; 1968.491.

[109] Bechert DW, Bruse M, Hage W, Meyer R. Fluid mechanics of biological surfaces and their technological application. Naturwissenschaften 2000;87:157−71.

[110] Bechert DW, Meyer R, Hage W. Drag reduction of Airfoils with miniflaps. AIAA Paper. 2315; 2000. p. 29.

[111] Bendat J, Pirsol A. Application of the correlation and spectral analysis. Moscow: World; 1983;312.

[112] Benjamin TB. The threefold classification of unstable disturbances in flexible surfaces bounding inviscid flows. J Fluid Mech 1963;16:436−50.

[113] Berlin AA, Shutov FA. Foam-polymers on the basis of reactive oligomers. Moscow: Chemistry; 1978;296.

[114] Berne K, Skrivener O. Diminution of a resistance and structure of turbulence in the diluted solutions of polymers. In: G.R. Hough (ed.), Proceeding of a symposium on "viscous flow drag reduction," Dallas; 1979.

[115] The collection of articles. Bionics 1998;27−8.

[116] Bippes H. Experimentelle Untersuchungen des laminar turbulent umschlages an einer parallel angestrom den konkaven wand. Heidelb. Akad. Wiss., Math. Naturwissenschaften 1972;3:103−11.

[117] Blackwelder RF, Haritonidis JH. Scaling of the bursting frequency in turbulent boundary layers. J Fluid Mech 1983;132:87−103.

[118] Blackwelder RF. Analogies between transitional and turbulent boundary layers. Physof Fluids 1983;26(10): 2807−16.

[119] Blackwelder RF, Gad-el-Hak M. Method and apparatus for controlling bound vortecs in the vicinity of lifting surfaces. US Patent No 4,697,769; 1987.

[120] Blackwelder RF, Roon IB. The effects on longitudinal roughness elements upon the turbulent boundary layer. AIAA Paper 88-0 134; 1988. pp. 1−6.

[121] Blackwelder R. F., M. Gad-el-Hak. 1990. Method and apparatus for reducing turbulent skin friction. United States Patent No 4,932,612.

[122] Blick EF. Theories of compliant coating drag reductions. Phys Fluids 1977;20(10, pt. 2):132−6.

[123] Boboshko VA, Rindja NV, Shmedro JA. Characteristics of a turbulent boundary layer on the concave surface of turn on 90°. Proc Acad Sci USSR. Mech Liquid and Gas 1982;5:155−8.

[124] Bojko AV, et al. Occurrence of turbulence in near-wall flows. Novosihirsk: Science; 1999;327.

[125] Bojko AV. Receptivity of a flat plate boundary layer to a free stream axial vortex. European J Mech: Fluids 2002;21(3):325−40.

[126] Book of Abstracts of the 10th European Drag Reduction Working Meeting. 1997. Berlin. p. 49.

[127] Borshevsky JT, Rudin SN. Control of a turbulent boundary layer. Kiev: Higher school; 1978;319.

[128] Bradshaw P. The effect of wind-tunnel screens on nominally two-dimensional boundary layers. J Fluid Mech 1965;22(4):679–88.

[129] Bradshaw P, Ferriss DH, Alwell ND. Calculation of boundary layer development using the turbulent energy equation. J Fluid Mech 1967;28:593–616.

[130] Bradshaw P. Introduction in turbulence and its measurement. Moscow: World; 1974;278.

[131] Brandritt E, Wotts KS. Simultaneous measurement Reynolds's pressure and turbulent kinetic energy in a two-dimensional boundary layer. Theor Bases of Eng Calculations 1975;1:312–21.

[132] Brehovskih LM. Waves in complex environments. Moscow: Science; 1973;540.

[133] Brennen C. Cavity surface wave patterns and general appearance. J Fluid Mech 1970;44(part 1):33–49.

[134] Britter RE, Hunt JGR, Mumford JC. The distortion of turbulence by a circular cylinder. J Fluid Mech 1979;92(pt. 2):269–301.

[135] Brown FNM. The physical model of boundary layer transition. In: Proceedings of the ninth midwestern mechanics conference of university of wisconsin, Madison, Wis.; August 16–18, 1965. pp. 117–129.

[136] Brown FT, Margolis DL, Shah RP. Behavior of disturbances of small amplitude, imposed on turbulent flow in hydraulic pipelines. Theor Bases of Eng Calculations 1969;4:56–68.

[137] Bushnell DM. Compliant surfaces introduction. Viscous Flow Drag Reduction Symposium Technical Paper, Dallas, Texas, 1979. New York; 1980. pp. 357–390.

[138] Bushnell DM. Turbulent Drag Reduction for External Flows. AIAA Paper 83-0227; 1983. pp. 1–20.

[139] Cantwell BJ. The organized motion in a turbulent flow. Ann Rev Fluid Mech 1981;13:457–515.

[140] Carpenter PW, Garrad AD. The hydrodynamic stability of flow over a Kramer-type. compliant surface. Part I. Tollmien-Schlichting instabilities. J Fluid Mech 1985;155:465–510.

[141] Carpenter PW, Garrad AD. The hydrodynamic stability of flow over a Kramer-type. compliant surface. Part II. Flow-induced surface instabilities. J Fluid Mech 1986;170:199–232.

[142] Cebeci T, Smith AM. Analysis of turbulent boundary layers. New York; London: Academic Press; 1974;404.

[143] Cebeci T, Bradshaw P. Momentum transfer in boundary layers. Washinton and London: McGraw-Hill; 1977;XVII, 391.

[144] Chang PK. Separation of flow. Oxford: Pergamon Press; 1970.

[145] Chang PK. Separated flow. Moscow: Mir; 1972;V.1, p. 229.

[146] Chang PK. Control of flow separation. Washington, DC: Hemisphere Publishing Corp.; 1976 (Chang P. K. (1979) A flow separation control. Moscow: Mir, p. 552)

[147] Lee C, Kim J. Control of the viscous sublayer for drag reduction. Phys Fluids 2002;14(7):2523–9.

[148] Chernishov OB, Koval AP. Peculiarities of a structure of a skin cover of the Sarda sard. Bionika 1977;11:83–86.

[149] Chernishov OB, Koval AP, Drobaha AA. Some peculiarities of morphology of the apparatus gills of fishes, connected to speed of their swimming. Bionika 1978;12:103–108.

[150] Choi KS, Yang X, Clayton BR, Glover T, Altar M, Semenov BN, et al. Turbulent drag reduction using compliant surfaces. Proc R Soc London, Ser A 1977;453:2229.

[151] Kwing-So C. Near-wall structure of turbulent boundary layer with spanwise-wall oscillation. Phys Fluids 2002;14(7):2530–42.

[152] Chu DC, Karniadakis GE. A direct numerical simulation of laminar and turbulent flow over ribet-mounted surfaces. J Fluid Mech 1993;250:1–42.

[153] Xu C-X, Huang W-X. Transient response of Reynolds stress transport to span wise-wall oscillation in a turbulent boundary layer. Phys Fluids 2004;17(1):018101–018104.

[154] Coles D. Coherent structures of turbulent boundary layers. Lehish Univ.; 1979;462.

[155] Computation of turbulent boundary layer. In: Coles PE, Hirst EA, editors. Proceedings of AFOSR-IFR Stanford Conference, vol. 2, 1969 p.

[156] Paola C, Vitori G, Blondeaux P. Coherent structures in oscillatory boundary layers. J Fluid Mech 2003;474:1–33.

[157] Daily JW, Harleman DRF. Fluid dynamics. Reading, Mass.: Addison-Wesley Publishing; 1966. p. 396.

[158] Davies JT. Turbulence phenomena. New York: Academic Press; 1972. p. 416.

[159] Davis MR. Design of flat plate leading edges to avoid flow separation. AIAA J 1980;18(5):598–600.

[160] Davis C, Carpenter PW. Instabilities in a plane channel flow between compliant walls. J Fluid Mech 1997;352:205–43.

[161] Deich ME, Filippov GA. Gas dynamics of two-phase media. Moscow: Energiya; 1968. p. 422.

[162] Denisenko OV. Flow of longitudinal grooves by of viscous gas stream. Notes of Scientists of the Central Aero-Hydrodynamic Inst 1978;9(4):19–60.

[163] Deutsch S, Fontane AA, Money MJ, Petrie HL. Drag reduction with combined micro-bubble and polymer injection. In: Proceeding of the 2nd International Symposium on Seawater Drag Reduction (ISSDR) 2005, Korea: Busan; 2005. pp. 459–468.

[164] Device for reduction of resistance of water. Inventions in the USSR and abroad. 1980; Part 43 (4), 22.

[165] Devnin SI. Aero-hydromechanics high-drag constructions. Handbook. Leningrad: Shipbuilding; 1983. p. 331.

[166] Dinkelacker A. Preliminary experiments on the influence of flexible walls on boundary layer turbulence. J Sound and Vibrat 1966;4:187–214.

[167] Dinkelacker A. Do Tornado-like vortices play a role in turbulent mixing processes. Structure of Turbulence in heat and Mass Transfer, IUTAMIICHNT Symposium. Washington, New York, London: Dubrovnik 1980, Hemisphere Publishing Co.; 1982. pp. 59–72.

[168] Dovgy SA, Moros VV, Babenko VV, Polishuk SV. Underwater apparatus. Patent of Ukraine No 65250A; 2004.

[169] Drisoll RJ, Kennedy LA. A model for the turbulent energy spectrum. Phys Fluids 1983;26(5):1228–33.

[170] Dyban E, Epik EJ. Heat-mass exchange and hydrodynamics turbulized streams. Kiev: Naukova dumka; 1985:295.

[171] Eckelman H. Pattern recognition of bounded turbulent shear flows. Advances in Fluid Mechanics: In: Proceedings of a Conference Held at Aachen, March 26–28, 1980. Lect. Notes Phys. No 148; 1981, pp. 280–290.

[172] Efimov BM, Kusnezov VB. Pulsations shearing stress on a surface of a plate. Acoustics of turbulent flows. Moscow: Science; 1983. pp. 46–53.

[173] Efimthov BM. Criteria of similarity of spectra near-wall pulsations of pressure of a turbulent boundary layer. Acoustic J 1984;30(1):58–61.

[174] Euromech Colloquium 361 Active Controls of Turbulent Shear Flows. Book of abstracts. Berlin; 1997.

[175] Euromech. 3rd European Fluid Mech. Conf. Book of abstracts. Gottingen; 1997. p. 402.

[176] Evseev AR, Maltzev LI, Malyuga AG. Experimental investigation of a water boundary layer with gas microbubbles. In: Proceedings of the 11th Drag Reduction Working Meeting, Prague; 1999, p. 13.

[177] Evtushenko AV, Pusino MG. About influence of initial turbulence of a flow on a spectrum of power near-wall pulsations of pressure. Acoustic J 1978;24(3):423–435.

[178] Experimental studying of structure near-wall pulsating fields of a turbulent boundary layer. Moscow: Central Aero-Hydrodynamic Institute (Proceedings of CAHI No 579); 1980. p. 80.

[179] Falco RE. Combined simultaneous flow visualization—hot wire anemometry for the study of turbulent flows. J Fluid Eng 1980;102:174–182.

[180] Favre-Marinet M, Binder G, Hac TV. Formation of oscillating jets. Theory Bases ASME 1981;103 (4):213–8.

[181] Favre A. Structure of spatiotemporal correlations for velocity in a boundary layer. Translation of all-Union Institute of the Scientific and Technical Information No 91011/1; 1969. p. 41.

[182] Fedjaevsky KK. Selected works. Leningrad: Shipbuilding; 1975. p. 439.

[183] Fekete GI. Coanda flow of a two-dimensional wall jet on the outside of a circular cylinder. Mech. Eng. Labs. 1963 Rept. 63-11, McGill Univ. Montreal, Quibec.

[184] Felis NA, Potter MK, Smith MK. Experimental investigation of flow of the incompressible liquid in the channel near to the transition mode. Theor Found of an Eng Calculation 1977;99(4):201−7.

[185] Fiebig M, et al. Structure of velocity and temperature fields in laminar channel flows with longitudinal vortex generators. Numerical Heat Transfer. Part A, 1989;15:281−302.

[186] Fiedler HE. Occurrence, development and influence on the subsequent flow of coherent structures in turbulent shift streams. The all-Union Center of Translations. No D -19365; 1982. p. 62.

[187] Fisher MC, Weinstein L M. et al. Compliant wall-turbulent skin-friction reduction research. AIAA Paper No 833; 1975. pp. 1−25.

[188] Fisher MK. Investigations on change of friction in a turbulent boundary layer on pliable walls. The all-Union center of translations. No A-4985; 1980. p. 48.

[189] Fletcher CAJ. Computational techniques for fluid dynamics, vols. I, II. Berlin, Heidelberg: Springer-Verlag; 1988 (translated into Russian, Moscow: Mir, 1991, v1, p. 504, v2, p. 552).

[190] Floryan JM. On the goertler instability of boundary layers. Prog Aerospace Sci 1991;28:235−71.

[191] Fontaine AA, Deutsh S, Moeny MJ, Brungart TA, Petrie HL. Synergistic drag reduction in coupled systems: micro bubbles and polymers. In: Proceedings of the 11th Drag Reduction Working Meeting, Prague; 1999. pp. 14−15.

[192] France PW, Griffits RT. The effects of a flexible surface on the boundary layer turbulence level in air. Int J Mech Eng Educ 1975;3(1):11−6.

[193] Frost U, Yu. B. Statistical concepts of the theory of turbulence. Moscow: World; 1980;67−98.

[194] Fruman DG, Galivel P. Anomalous effects connected to an ejection of polymer, reducing a resistance, in turbulent boundary layers of pure water. In: Hough, G.R. (Ed.), Proceedings of a symposium on "viscous flow drag reduction," Dallas, Tex., November 7, 8, 1979.

[195] Furuya Y, Miyata Fuzhita. Turbulent boundary layer and resistance of plates with a roughness formed by wires. Theor Bases of Eng Calculations 1976;4:146−56.

[196] Gad-el-Hak M, Blackwelder RF, Riley JF. On the interaction of compliant coatings with boundary layer flows. J Fluid Mech 1984;140:257−80.

[197] Gad-el-Hak M, Ho C-M. The pitching delta wing. AIAA J 1985;23(11):1660−5.

[198] Gad-el-Hak M. Compliant coatings research: a guide to the experimentalist. J Fluid and Structures 1987;1:55−70.

[199] Gad el Hak M. Compliant coatings for drag reduction. Prog Aerosp Sci 2002;38:77−99.

[200] Galway RD. An investigation into the possibility of laminar boundary layer stabilization using flexible surfaces. Phys Fluids 1977;20(10, pt. 2):31−8.

[201] Galish M, Gjunim R, Reutor E. Research transitive process at fluctuations of pressure in viscous-elastic pipes. Theoretical Found of Eng Calculations 1979;101(4):205−10.

[202] Ganiev RF, Ukrainian LE, Telalov AI. Experimental research of flow of a liquid in pipelines with pliable walls. Bionics 1980;14:46−50.

[203] Gilinskii MM, Stasenko AL. Supersonic gas-dispersed jets. Moscow: Mashinostroenie; 1990. p. 175.

[204] Ginevsky AS, Vlasov EV, Kolesnikov AP. Aero-acoustic interactions. Moscow: Mechanical engineering; 1978. p. 177.

[205] Gmelin C, Rist U. Active control of laminar-turbulent transition using instantaneous vorticity signals at the wall. Phys Fluids 2001;13(2):513−9.

[206] Gogish LV, Yu SG. Turbulent separation flows. Moscow: Science; 1979. p. 367.

[207] Goldshtic MA, Shtern VN. Hydrodynamic stability and turbulence. Novosibirsk: Science SD; 1977. p. 364.

[208] Goldstein ME, Sockol PM, Sanz J. The evolution of Tollmien-Schlichting waves near a leading edge. Pt.2. Numerical determination of amplitudes. J Fluid Mech 1983;129:443−53.

[209] Gorban VA, Gorban IM. Research of dynamics of vortical structures in angular area and near to a surface with deepening's. App. Hydromechanics 1999;1(1):4−11.

[210] Gorban VA, Gorban IN. Theoretical analysis of control schemes of near-wall flows with use interceptors. App. Hydromechanics 2000;2(74):43−50.

[211] Gorban VA, Gorban IM. Theoretical analysis of vortical structure of flows in cross-section grooves. Bull Herson State Tech Univ 2002;15(N2 2):146−9.

[212] Gorlin SM, Zrazhevskiy IM. Influence of external turbulence of a stream on flow in a boundary layer. Proc Acad Sci USSR Mech liquid and gas 1972;4:46−53.

[213] Grant IR, Huyer SA. Development of lagrangian vorticity methods for calculating unsteady flows. Hydrodynamics. Technical Digest. Naval Undersea Warfare Conference Division, Newport; 1985. pp. 45−58.

[214] Gray J. Animal locomotion. London: Weidenfeld and Nicolson; 1961. p. 479.

[215] Granville PS. Drag and turbulent boundary layer of flat plates at low Reynolds numbers. J Ship Res 1977;21(1):30−9.

[216] Greshilov EM, Evtushenko AV, Ljamshev LN, Shirokova NL. Некоторые особенности влияния полимерных добавок на пристеночную турбулентность. Eng Phys J 1973;25(6):999−1005.

[217] Gromeko IS. Collected works. Moscow: Science; 1952. p. 295.

[218] Grosskreutz R. Wechselwirkungen zwischen turbulenten Grenzschichten und weichen Wanden. Mitt. M. Plank Inst. Stromungsforschung und der Aerodyn. Verstichsangt. vol. 53; 1971, p. 85−93.

[219] Gul VE. Electro spending polymeric materials. Moscow: Science; 1968. p. 148.

[220] Gul VE. Structure and durability of polymers. Moscow: Science; 1978. p. 228.

[221] Gyorgyfalvy D. The possibilities of drag reduction by the use of flexible skin. (AIAA 4th Aerospace Science Meeting Los Angeles, California), AIAA Paper N66-430; 1966. pp. 1−19.

[222] Habahpasheva EM, Perepelitsa BV. On certain characteristics of boundary-layer turbulence in water streams containing some macromolecular substances. Journal of Engineering Physics 1970;18(6):758−760.

[223] Habahpasheva EM. Some data about structure of flow in a viscous sublayer. Problems of thermophysics and physical hydrodynamics. Novosibirsk: Science SD; 1974. pp. 223−235.

[224] Habahpasheva EM, Efimenko GI. Distribution of pressure tangents and velocities in near-wall areas of a turbulent boundary layer. Russia, Novosibirsk: 1−9 (Proceeding of the Thermophysics Institute of the Academy of Science of the USSR SD; 1981. pp. 67−81).

[225] Haider M, Lindsley DB. Microvibrations in man and dolphin. Science 1964;146(3648):1181−2.

[226] Hanratty TJ. Research on wall turbulence. Chem Eng Educ 1980;14(4):162−6.

[227] Hansen RJ, Hunston DL. Fluid property effects on flow-generated waves on a compliant surface. J Fluid Mech 1983;133:161−72.

[228] Hanston DL, Zakin GL. Influence of molecular parameters to an dependence of a diminution of a resistance from expenditure and similar by it of an appearance. In: Hough, GR, editor. Proceedings of a Symposium on "Viscous Drag Reduction," Dallas; Tex; 1979. New York: American Inst. of Aeronautics and Astronautics—Progress in Astronautics and Aeronautics; 1980;72:453.

[229] Harsha PT, Li SK. Connection between a turbulent stress of friction and kinetic energy of turbulent movement. AIAA J 1970;8(8):179−81.

[230] Hejns FD. Preliminary results on stability boundary layer on a flexible plate. AIAA J 1965;35(4):249−50.

[231] Henckok B. Influence of turbulence of undisturbed flow on characteristics of turbulent boundary layers. Theor Bases of Eng Calculations 1983;3:126−33.

[232] Hertel H. Structure, form and movement (Biology and Technology). Reinhold Publishing Corp.; 1966. p. 251.

[233] Hinze JO. Turbulence. Moscow: Fismatgis; 1963. p. 680.

[234] Hinze JO. Turbulence. second ed. New York: McGraw-Hill; 1975. p. 790.

[235] Howard FG, Quass BF, Weinstein LM, Bushnell DM. Longitudinal grooves for Bluff-Body Drag Reduction. AIAA. Journal 1981;19:535−537.

[236] Howard FG, Goodman WL. Axisymmetric bluff-body drag reduction through geometrical modification. J Aircraft 1985;22:516−22.

[237] Hoyt JW, Taylor JJ. Turbulence structure in a water jet discharging in air. Phys Fluids 1977;20 (10, Pt. II):253−7.

[238] Hoyt J. Reduction of hydrodynamic resistance by the fish slimes. Bio-hydrodynamics of the swimming and flight. Moscow: Mir; 1980. pp. 128–146.

[239] Hoyt JW. Polymer solution effects on turbulent friction mechanisms. In: Proceedings of International Symposium on Seawater Drag Reduction, Newport, USA; 1998. pp. 1–5.

[240] Hussain AKMF. Coherent structures—reality and myth. Phys Fluids 1983;26(10):2816–51.

[241] Hydrodynamic surfaces. Inventions in the USSR and abroad. Part 32, No 1; 1992. p. 9.

[242] Inmel PN, Bradshaw P. Length of a way of mixing in turbulent boundary layers at Reynolds low numbers. AIAA J 1981;19(7):131–3.

[243] Ionov AV. Temperature-frequency dependences of loss factor of multilayered coverings for the power equipment. New vibro-absorbing materials and coverings and their application in the industry. Leningrad: the Leningrad house of scientific and technical education; 1980. pp. 44–48.

[244] Isaac KM, Jakubowski AK. Experimental study of the interaction of multiple jets with a cross flow. AIAA J 1985;23(11):1679–83.

[245] Ito A. Breakdown. Structure of longitudinal vortices along a concave wail. J Japan Soc Aero Space Sci 1988;36:272–9.

[246] IUTAM Symposium on Flow past Highly Compliant Boundaries and in Collapsible Tubes. 26–30 March, 2001. University of Warwick, Coventry. Proc. of the IUTAM Symposium. Edited by Peter W. Carpenter and Timothy J. Pedley. Kluwer Academic publishers. Dordrecht/Boston/London, 2003.

[247] Ivanov VP, Klochkov VP, Kozlov LF. Measurement of speed structure in liquid flows of large volumes with the help of a laser Doppler measurer of speed. News AN USSR Mech Fluid and Gases 1977;5:170–3.

[248] Ivanov VP, Klochkov VP, Kozlov LF. Research of a jet flow over an ellipsoid of rotation with the help of a laser Doppler anemometer. Eng Phys J 1978;34(1):99–103.

[249] Ivanuta YuF, Chekalova LA. Experimental research of turbulent flow of dilute polymeric solutions in various diameter pipes. Inzhenerno-Physichesky J 1971;21(1):5–12.

[250] Ivanyuta YF, Pogrebnyak VG, Naumchuk NV, Frencel SY. Flow structure of polyethylene oxide solution in the input zone of a short capillary. Eng -Phys J 1985;49(4):614–21.

[251] Jang J-N, Heller RA. Random vibrations of compliant wall. Eng Mech Proc—Civil Eng ASCE 1976;46 (12):136–43.

[252] Jensen BL, Sumer BM, Fredsqe J. Turbulent oscillatory boundary layer at high Reynolds numbers. J Fluid Mech 1989;206:265–97.

[253] Kader BA, Jaglom AM. Laws of similarity for near-wall turbulent flows. In: Results of Science and Techniques. Series on Mech Liquid and Gas. Moscow, 1980;13:81–156.

[254] Kaind RDj, Guden Ê, Dvorak FA. Experimental research of currents with a tangential blow in and comparison of experimental data with results of settlement methods. Rocket Eng Astronautics 1979;17:74–80.

[255] Kalugin VN. 1973. Results of numerical calculations of nonlinear stability of Puaseil flow with elastic borders. In: Proc. of 4th All-Union Seminar on Numerical Methods of Mechanics of a Viscous Liquid. Science SD, Novosibirsk, pp. 219–233.

[256] Kanarsky MV, Teslo AP. About a turbulent flow of a plate with an elastic surface. Hydromechanics 1977;35:51–3.

[257] Kanarsky MV, Babenko VV, Kozlov LF. Experimental investigation of turbulent boundary layer on an elastic surface. Stratificated and turbulent flows. Kiev: Naukova dumka; 1979. pp. 59–67.

[258] Kanarsky MV. Experimental research of the dynamic module of elasticity of an elastic plate. Bionics 1981;15:98–101.

[259] Kanarsky MV, Babenko VV, Voropaev GA. Measuring the kinematics characteristics of turbulent boundary layer with computer support. Hydromechanika 1982;45:30–6.

[260] Karino ER, Brodki RS. Visual research near-wall area in turbulent flow. Mechanics 1971;125(I):56–82.

[261] Kaplan RE. On the wall structure of the turbulent boundary layers. J Fluid Mech 1976;76:89–112.

[262] Kawamata S, Kato T, et al. Experimental research on the possibility of reducing the drag acting on a flexible plate. Theor and App. Mech 1973;21:507−18.

[263] Kayan VP, Pjatetsky VE. Bio-hydrodynamic installation of the closed typ. for research of hydrodynamics of navigation of sea animals. Bionika 1971;5:121−5.

[264] Kayan VP. On hydrodynamic characteristic of fin-typ. propulsor of dolphin. Bionics 1979;13:9−15.

[265] Keller JP, South JC, Jr. RAXBOD- A FORTRAN Program for Inviscid Transonic Flow Over Axisymmetric Bodies. NASA TM X-72831; 1976.

[266] Kendall JM. The turbulent boundary layer over a wall with progressive surface waves. J Fluid Mech 1940;41(pt. 2):259−81.

[267] Kidun SM. Investigation of propagation velocity of fluctuations on a dolphin's cover. Bionics 1979;13:52−8.

[268] Kiknadze GI, Krasnov YK. Evolution tornado-like flows of a viscous liquid II Reports of the Academy of Sciences of the USSR. 290, N2 6; 1986. pp. 1315−1319.

[269] Kim S, Tagori T. Drag Measurements on flat plates with uniform injection of polymer solutions and their direct application to the wall. Proc. Annual Meeting, ASME.; 1969.

[270] Kim J. On the structure of wall-bounded turbulent flows. Phys Fluids 1983;26(8):2088. p. 97.

[271] Kapinos VM, Tarasov FI. Definition a stress of friction on a wall on the measured velocity profile in an external part of a boundary layer. Eng -Phys J 1981;40(5):787−92.

[272] Kistler AL, Tan FC. Some properties of turbulent separated flow. Boundary layers and turbulence. Phys of Fluids Suppl 1967; S165−Si74.

[273] Klebanoff PS. Characteristics of turbulence in a boundary layer with zero pressure gradients. NACA Rep 1955;1247:1−19.

[274] Klebanoff PS, Nidstrom KD, Sargent LM. The three-dimensional nature of boundary layer instability. J Fluid Mech 1962;12:1−34.

[275] Klein A. Development of turbulent current in a pipe: (Review). Theor Bases Eng Calculation 1981;103 (2):180−8.

[276] Kley DP, Baker KB, Eizenhoot DD. Action on a boundary layer of an axisymmetric skew field with the help of diluted polymers in water. In: Hough, GR, editor. Proc. of a symp. on "viscous flow drag reduction," Dallas; 1979.

[277] Kline SJ, Reynolds U, Schraub F, et al. Structure of turbulent boundary layers. Mechanics 1969;4:41−8.

[278] Kline SJ, Reynolds WC, Schraub FA, Rundstadler PW. The structure of turbulent boundary layers. J Fluid Mech 1967;30(4):741−73.

[279] Klinzing GE, Kubovichek RJ, Marmo JF. Frictional losses in foam-damped flexible tubes. Ind Eng Chem Process Des Dev 1969;8(1):112−7.

[280] Knapp CF, Roach PI. A combined visual and hot-wire anemometer investigation of boundary-layer transition. AIAA J 1968;6(1):29−36.

[281] Knisely C, Rockwell D. Self-sustained low-frequency components in an impinging shear layer. J Fluid Mech 1982;116:157−86.

[282] Kochin NE. Vector calculation and the beginnings tensor analysis. Moscow: Publishing house of the Academy of sciences of the USSR; 1961. p. 426.

[283] Kohama Y. Three-dimensional boundary layer transition on a concave−convex curved wall. In: Difference-Modular Theory of Elasticity. Physics and Mathmatics State Publishing House, Moscow, pp. 215−226.

[284] Kont-Bello Zh. Turbulent flow in the channel with parallel walls. Moscow: World; 1968. p. 176.

[285] Kornev AL, Larin SL. A numerical research of a turbulent boundary layer on a slice with an injection of a polymeric solution. Physical Hydrodynamics. Donetsk: Visha Shkola, Kiev; 1990. pp. 47−53.

[286] Korobov VI. Experimental research of influence of pliable surfaces on integrated characteristics of a boundary layer. Eng-phys J 1979;37(3):518−23.

[287] Korobov VI. Experimental research integral characteristics of the boundary layer on pliable plates. Bionics 1980;14:53−7.

[288] Korobov VI, Babenko VV, Kozlov LF. Integral characteristics of the boundary layer on elastic plates. Eng-Phys J 1981;XLI(2):351–2.

[289] Korobov VI, Babenko VV. On the method of measuring of boundary layer integral characteristics. Hydromechanika 1983;48:57–63.

[290] Korobov VI, Babenko VV. About one mechanism of the interaction between elastic wall and flow. Eng-Phys J 1983;44(5):730–3.

[291] Korobov VI, Babenko VV, Kozlov LF. Interaction of turbulent boundary layer with elastic plate. Eng-Phys J 1989;56(2):220–5.

[292] Korobov VI, Babenko VV, Belinsky VG. Fin propulsive device. Patent of Ukraine No 1671515 Al; 1991.

[293] Korobov VI. Joint influence of a compliant surface and polymeric additives on a boundary layer. Bionika 1993;26:27–31.

[294] Kostritsky SN, Zirkin MZ, Ekelchin VS, et al. Research of visco-elastic properties of composite polymeric materials by a resonant method. Res Mech Composite Mater Des 1981;Part 344:4–12.

[295] Koval AP. Structures of crypts of skin and of gills covers, forming slime of the swordfish. Bionika 1978;12:108–11.

[296] Koval AP. Histological structures of a skin cover of the sailfish and swordfish. Bionika 1982;16:21–7.

[297] Koval AP. Roughness and some features of a structure of a swordfish skin. Bionika 1987;21:73–7.

[298] Koval AP, Zajats VA, Kaljuzhnaja TA. Morphological peculiarities of a skin cover structure of fishes of various high-speed groups. Bionika 1987;21:77–84.

[299] Koval AP, Butusov SV. Some features of a microrelief of a surface tail fin blue marlin and ox-eyed tuna. Bionika 1991;24:88–91.

[300] Koval AP, Koshovskij AA. Structure heat exchanger of a tunas of various high-speed groups. Bionika 1992;25:94–8.

[301] Kovasznay LSG, Kibens V, Blackwelder RF. Large-scale motion in the intermittent region of a turbulent boundary layer. J Fluid Mech 1970;41(pt.2):432–56.

[302] Kozlov LF, Leonenko IV. Influence xiphoid tip on drag reduction of a sphere. Rep Acad Sci Ukraine SSR 1971;6:622–4.

[303] Kozlov LF, Leonenko IV. Research of influence xiphoid tip on resistance of the extended body of rotation. Bionika 1973;7:8–14.

[304] Kozlov LF, Shakalo VV. Some results of measurements of pulsations of speed in a boundary layer of dolphins. Bionics 1973;7:50–2.

[305] Kozlov LF, Shakalo VM, Burjanova LD, Vorobev NN. About influence non-stationary on a condition of flow in a boundary layer of the Black Sea bottle-nosed dolphin. Bionica 1974;8:13–6.

[306] Kozlov LF, Babenko VV. Experimental investigations of the boundary layer. Kiev: Naukova dumka; 1978. p. 184.

[307] Kozlov LF. Hydrodynamics of the water animals with a half-lunar tail fin. Bionika 1979;13:3–9.

[308] Kozlov LF, Shakalo VM. About a condition of flow in a quasi-stationary boundary layer some cetacean. Bionica 1980;14:74–81.

[309] Kozlov LF. Theoretical bio-hydrodynamics. Kiev: Visha shkola; 1983. p. 240.

[310] Kozlov LF, Pershin SV. Complex researches of active regulation by a skin of the dolphins of drag reduction. Bionika 1983;17:3–12.

[311] Kozlov LF, Thiganjuk AI, Babenko VV, Voropaev GA, Nikishova OD. Forming of turbulence in shear flows. Kiev: Naukova dumka; 1985. p. 283.

[312] Kramer MO. Boundary layer stabilization by distributed damping. J Am Soc Naval Eng 1960;72 (1):25–33.

[313] Kramer MO. Improvements in method of making drag-reducing covering. Patent UK 855.224; 1960.

[314] Kramer MO. Improvements in means for reducing frictional drag in fluids. Patent UK 864.593; 1961.

[315] Kramer MO. Device for stabilizing laminar boundary layer flow. Patent UK 881.570; 1961.

[316] Kramer MO. Boundary layer stabilization by distributed damping. Naval Eng J 1962;72(1):25–73.

[317] Kramer MO. Means and method for stabilizing laminar boundary layer flow. Patent USA 3.161.385; 1964.

[318] Kramer MO. Hydrodynamics of the dolphin. Advances in hydroscience, vol. 2. New York, London: Academic Press; 1965.

[319] Kramer MO. Means and method for stabilizing laminar boundary layer flow. Patent USA 3.585.953; 1971.

[320] Krasilshikova BA. Velocity field excited by a spectrum of small disturbances on an elastic plate. In: Proc. of a symposium on mechanics of the continuous environment and related problems of the analysis. Tbilisi: Science. vol. 2; 1974. pp. 147−158.

[321] Per-Age K, Anatoly K. Some effects of localized injection on the turbulence structure in a boundary layer. Phys Fluids 2000;12(11):2990−9.

[322] Kruka V, Eskinezi S. The wall jet in a moving stream. J Fluid Mech 1964;20(Part 4):555.

[323] Kubota H, Stollery JL. An experimental study of the interaction between a glancing shock wave and a turbulent boundary layer. J Fluid Mech 1982;128:431−58.

[324] Kulik VM, Poguda IS, Semenov BN. Experimental investigation of one-layer visco-elastic coating acting on turbulent friction and wall pressure pulsations. Eng-Phys J 1984;47(2):189−96.

[325] Kulik V, Rodyakin S, Lee I, Chun HH. The response of Compliant Coating to Unstationary Disturbances. In: Proc. of the 2nd intern. symp. on seawater drag reduction (ISSDR), May 23−26, 2005. Korea: Busan; 2005. pp. 427−438.

[326] Kutateladze SS. Near-wall turbulence. Novosibirsk: Science SD; 1973. p. 227.

[327] Kutateladze SS, Mironov VP, Nakorjakov VE, et al. Experimental investigations near-wall turbulent flows. Novosibirsk: Science SD; 1975. p. 166.

[328] Kutateladze SS, Habahpasheva EM, Orlov VV, et al. Experimental investigation of structure near-wall turbulence and a viscous sublayer. Turbulent shift flows. Moscow: Mechanical engineering. vol. 1; 1982. pp. 92−108.

[329] Kutateladze S. Selected works. Novosibirsk: Science SD; 1989. p. 427.

[330] Laghthill MJ. About the wave theory of turbulent flows with cross-section shift. All-Union institute of the scientific and technical information No 94745/1; 1971. p. 56.

[331] Lamley J. Spectral distribution of energy in near-wall turbulence. Translation of all-Union institute of the scientific and technical information No 90980/1; 1966. 27c.

[332] La Monte FR. Scales of the Atlantic Species of *Makaira*. Bull Am Nat/Hist 1958;114:381−95.

[333] Landahl MT. Influence of the additives to dynamics of turbulent ejections. In: Hough, GR, editor. Proc. of a symp. on "viscous flow drag reduction." Dallas; 1979.

[334] Landahl MT, Kaplan RE. The effects of compliant walls on boundary layer stability and transition. Naples: AGARD Boundary layer Technology Meeting; 1965. p. 67.

[335] Lang TG. Hydrodynamic analysis of dolphin fin profiles. Nature 1971;104(9):1110−1.

[336] Latorre R, Babenko VV. Role of bubble injection technique drag reduction. In: Proc. Int. Symp. on Seawater Drag Reduction. Newport, RI; 1998. pp. 319−326.

[337] Lebedev L. Machines and devices for tests of polymers. Moscow: Mechanical engineering; 1967. p. 212.

[338] Lecture Notes in Engineering. In: M. Gad-el-Hak. Editor. Berlin, Heidelberg, New York, London: Springer-Verlad; 1985. p. 532.

[339] Leonenko IV. About hydrodynamic meaning of a rostrum a swordfish. Bionika 1974;8:21−3.

[340] Levina MO. About influence of elastic pliable border on a spectrum of pressure pulsations of a turbulent flow. News all-union research inst.. Hydraulic Eng 1974;106:34−50.

[341] Lighthill J. Mathematical biofluid dynamics. Philadelphia, Pa: SIAM; 1983. p. 281.

[342] Lil LG, Fuller GG, Olbriht UL. A research of expansion, induced by a stream, macromolecules in the diluted solutions. In: Hough GR, editor. Proc. of a symp. "viscous flow drag reduction," Dallas; 1979.

[343] Lin JK, Howard FG, Selby GV. Application of longitudinal grooves for stream separation control, which is flowing round a ledge. Space 1990;12:10−2.

[344] Lipatov YS, Kercha YY, Sergeeva LM. Structure and properties polyurethanes. Kiev: Naukova dumka; 1970. p. 279.

[345] Lissamen PV, Harris GL. Turbulent superficial friction on deformable surfaces. AIAA J 1966;7 (8):243–4.

[346] Logvinovich GV. Hydrodynamics of current with free borders. Kiev: Naukova dumka; 1969. p. 215.

[347] Logvinovich GV. Hydrodynamics of fine flexible body. Bionica 1970;4:3–5.

[348] Logvinovich GV. Hydrodynamics of fish's navigation. Bionika 1973;7:3–8.

[349] Lojtsansky LG. Mechanics of a liquid and gas. Moscow: Science; 1973;848стр.

[350] Mager, Shauer, Justis. Development of flow and of friction coefficient in semi-closed flat turbulent jet. Technical Mechanics, vol.1. Moscow: World; 1963. p. 58.

[351] Martinov AK. Applied aerodynamics. Moscow: Mechanical Engineering; 1972. p. 447.

[352] Matvievsky SA. Some empiric dependencies of turbulent flow of diluted polymeric solution. Hydromechanics 1984;43:162–70.

[353] Matvievsky SA. A turbulent boundary layer on a slice in a stream of a homogeneous solution of polymer. Hydromechanics 1984;43:171–9.

[354] May C, Voropaev GA. Design of viscoelastic coatings to reduce turbulent friction drag. US Patent No 6.516.652;2003.

[355] McCarthy JH. Some fundamental problems of ship resistance and flow; new methods to reduce frictional drag. Scientific and methodological seminar on ship hydrodynamics. Bulgarian ship hydrodynamics center; 1983. pp. 37–39.

[356] McCoermick ME, Johnson B, Lee WM. Effects of a flexible surface on surface shear stress fluctuations beneath a turbulent boundary layer. J of Hydraulics 1972;6(1):62–4.

[357] McMichael JM, Klebanoff PS, Mease NE. Experimental investigation of drag on a compliant surface. Viscous Flow Drag Reduct. Techn. Pap. Symp. Dallas, Texas, 1979. New York; 1980. pp. 410–438.

[358] Mechanics of non-uniform and turbulent flows. In: Struminsky V. V. Editor Science: Moscow. 1989. p. 248.

[359] Meng JCS. Engineering insight of near-wall micro turbulence for drag reduction and derivation of a design map for seawater electromagnetic turbulence control. In: Proc. of the intern. symp. on seawater drag reduction, Newport, Rhode Island, USA; 1998. pp. 359–367.

[360] Meng JCS. Wall layer microturbulence phenomenological model and a semi-marcov probability predictive model for active control of turbulent boundary layers. In: Panton. RL, editor. Self-sustaining mechanism of wall turbulence. Southampton, UK and Boston, USA: Computational Mechanics Publications; 1997. pp. 201–52.

[361] Merkulov VI. Control of flow movement. Novosibirsk: Science SD; 1981. p. 173.

[362] Min T, Yoo JY, Choi H, Joseph DD. Drag reduction by polymer additives in a turbulent channel flows. J Fluid Mech 2003;486:213–38.

[363] Monin AS, Jaglom AM. Statistical hydromechanics: in 2 Parts. Moscow: Science; 1965. p. 630.

[364] Monson J, Hichuchi H. Measurement of superficial friction by means of two-beam laser interferometer. AIAA J 1981;19(8):80–8.

[365] Moore KJ, Rajan T, Gorban VA, Babenko VV. Method and apparatus for increasing the effectiveness and efficiency of multiple boundary-layer control techniques. United States Patent No US 6,357,374 B1; 2002.

[366] Moore KJ. Engineering an efficient shipboard friction drag reduction system. In: Proceedings of the 2nd Int. Symp. on Seawater Drag Reduction. ISSDR 2005, Busan, Korea, 2005. pp. 345–358.

[367] Morkovin MV, Reshotko E. Dialogue on progress and issues in stability and transition research. Opening Invited Lecture. Third IUTAM symp. on laminar turbulent transition, Toulouse, France, Sept.; 1989. pp. I-24.

[368] Morrison WR, Bullock KJ, Kronauer RE. Experimental evidence of waves in the sublayer. J Fluid Mech 1971;47(4):639–56.

[369] Mostafa AA, Mondzhia AA, McDonell VG, Samuelsen GS. Propagation of dust-laden jets. Theoretical and experimental study. Aerokosmicheskaya Tekhnika 1990;3:65–81.

[370] Movchan VT, Shkvar EA.Modeling of dynamics and heat transfer processes in turbulent near-wall shear flows. In: Proc. 2nd Europ. thermal-sciences and 14th UIT national heat transfer conf., Rome, Italy. Vol. 1, Edizioni ETS, Pisa; 1996. pp. 535–540.

[371] Movchan VT, Shkvar EA. Modeling of turbulent near-wall shear flows properties. In: Proc. NATO/AGARD Fluid Dynamics Panel Workshop on High Speed Body Motion in Water. Kiev, Ukraine. AGARD. Rep. 827; 1997. pp. 10.1–10.7.

[372] Mudrov OA, Savchenko IM, Shitov VS. Handbook on elastomeric coatings and heretics in shipbuilding. Leningrad: Shipbuilding; 1982. p. 184.

[373] Mueller TI, et al. Smoke visualization of boundary-layer transition on a spinning axisymmetric body. AIAA J 1981;12:1607–8.

[374] Mulhearn PJ. Turbulent flow over a periodic rough surface. Phys Fluids 1978;21(7):1113–5.

[375] Musker AJ. Explicit expression for the smooth wall velocity distribution in turbulent boundary layer. AIAA J 1979;17(6):655–7.

[376] Nakagushi H. Jet along a curved wall. Res Mat. 1961;4.

[377] Nakayama H, Nezu I. Structure of space-time correlations of bursting phenomena in an open-channel flow. J Fluid Mech 1981;1–4:1–43.

[378] Nakorjakov VE, Burdukov AP, Kashinsky OK, et al. Electro-diffusion method of investigation of local structure of turbulent flows. Novosibirsk: Thermo physics Inst. of the academy of science of the USSR SD; 1986. p. 247.

[379] Nedobezhkin AE. Ship with system of injection of air on its bottom. Patent of the USSR No 1273292; 1986.

[380] Newman BG. The deflection of plane jets by adjacent boundaries, coanda effect. In: Lachmen GV, editor. Boundary layer and flow control. Pergamon Press; 1964. p. 232.

[381] Nickel K, Schonauer W. Eine einfache exeprimentelle methode zur Sichtbarmachung von Tollmien — Wellen und Goertler wirbein. ZAMM 1961;41:145–7.

[382] Nikitin IK. Complex turbulent flows and processes warmly—mass transfer. Kiev: Naukova dumka; 1980. p. 238.

[383] Nikiforov AS. Vibroabsorption on ships. Leningrad: Shipbuilding; 1979. p. 184.

[384] Nikulin VA. A model of near-wall turbulence in weak solutions of polymers. Kiev, Donetsk: Vysha Shkola; 1990. pp. 34–46.

[385] Nikishova OD, Babenko VV. Flow of elastic body by fluid. Bionika 1975;9:55–60.

[386] Oldaker DK, Tiederman WG. Spatial structure of the viscous sub layer in drag-reducing channel flows. Phys Fluids 1977;20(10, pt. 2):133–44.

[387] Oliger AA. Wave guides for superficial acoustic waves. Review TIIER (Technical Investigations Institute of Electronics and Radio) 1976;64(5):SL1–65.

[388] Oran ES, Boris JP. Numerical simulation of reactive flow. New York: Elsevier; 1987 (translated into Russian, Moscow, Mir, 1990. p. 661.)

[389] Organized structures and turbulence in fluid mechanics. Book of abstracts in Conf. Grenobe, France; 1989. p. 234.

[390] Ostisty VV, Babenko VV, Koslov LF, Holjavchuk SD. Fin propulsive device. Patent of Ukraine No 1297377; 1987.

[391] Ostisty VV, Babenko VV, Koslov LF, Ribko VD. Apparatus for damping the energy of two-phase flows and the separation. Patent of Ukraine No 1325243; 1987.

[392] Ovcharov OP. About hydrodynamic functions of the apparatus gills of the swordfish. Bionika 1982;16:11–5.

[393] Ovchinnikov VV. Swordfish and a sailfish. Kaliningrad: AtlantNIRO; 1970. p. 269.

[394] Paskonov VM, Polezhaev VI, Chudov LA. Numerical simulation of the processes of heat- and mass-transfer. Moscow: Nauka; 1984. p. 285.

[395] Pathak SK. Characteristics and potentials of flows with flexible boundaries. J of Inst Eng Civil Engin Div (India) 1978;58(15):231–5.

[396] Patterson RW, Abernathy FH. Turbulent flow drag reduction and degradation with dilute-phase polymer solution. J Fluid Mech 1970;43(4):689–710.

[397] Pershin SV. Biohydrodynamics laws of the water animal's navigation as a principle of optimization in a nature of shipped body's movement. Questions in Bionics. Moscow: Science; 1967. pp. 555–560.

[398] Pershin SV, Chernishov OB, Kozlov LF, Koval AP, Zajats VA. Laws in coverings of high-speed fishes. Bionika 1976;10:3–21.

[399] Pershin SV. Biohydrodynamical phenomenon a swordfish as a limiting case high-speed hydrobionts. Bionika 1978;12:40–8.

[400] Pershin SV. Fundamentals of a hydrobionics. Leningrad: Shipbuilding; 1988. p. 263.

[401] Petrie HL, Fontane AA, Money MJ, Deutsch S. Experimental study of slot injected polymer drag reduction. In: Proc of the 2nd int. symp. on seawater drag reduction. ISSDR 2005, Busan, Korea; 2005. pp. 605–619.

[402] Petrovscy VS. Hydrodynamic problems of turbulent noise. Leningrad: Shipbuilding; 1966. p. 314.

[403] Pfenninger W, Viken J, Vemuru CS, Volpe G. All laminar supercritical lfc airfoils with natural laminar flow in the region of the main wing structure. In: Liepmann NW, Narasimka R, editors. Turbulence management and relaminarisation. Berlin, Heidelberg, New York, London: Springer-Verlag; 1988. pp. 349–405.

[404] Philip. R, Castano J, Stace J. Combined polymer and micro bubble drag reduction. In: Proc. Int. Symp. on Seawater Drag Reduction. Newport, RI; 1998, pp. 319–326.

[405] Pilipenko VN. Modeling of turbulent flows in fluid with the additives of polymers. Turbulent flows. Moscow: Nauka; 1977. pp. 145–150.

[406] Pilipenko VN. Influence of the additives on near-wall turbulent flows. The totals of a science and engineering. Mech of Fluid and Gas 1980;15:156–257.

[407] Pisarenko GS, Jakovlev AP, Matveev VI. Vibroabsorption properties of constructional materials. Hand-book. Kiev: Naukova dumka; 1971. p. 375.

[408] Piva R, Orlandi P. Numerical solution for flows in atmospheric boundary layer over the street canyons, Proc. Of the 4th Int. Conf. on Numerical Methods in Fluid Dynamics, June, 24–28, 1974, Univ. Colorado, Springer-Verlag, Berlin (translated into Russian, In book: Numerical solving of fluid-dynamical problems, Moscow, Mir, 1977; 1975. pp. 127–134).

[409] Pjadishus AA, Janushas VI, Zigmantas GP. Deformation of structure of a turbulent boundary layer at the increased turbulence of an external flow. Proc Academy of Sciences of Lithuanian SSR Energy 1990;1:30–44.

[410] Pjatezky VE, Shakalo VM, Tsiganjuk AI, Sisov II. Investigations of a flow regime of water animals. Bionica 1982;16:31–6.

[411] Pletcher RH. Computation of incompressible turbulent separated flow. Teoreticheskie Osnovy Inzhenernyh Rasschetov 1978;100(4):167–73 (in Russian, translated from Trans. ASME, ser. D).

[412] Pogrebnyak VG. Polymer macromolecules as a tool for studying wall-adjacent turbulent flow. In: Proc of the 2nd Int. Symp. on Seawater Drag Reduction. ISSDR 2005, Busan, Korea; 2005. pp. 79–90.

[413] Polishuk SV, Babenko VV. Fin propulsive device. Patent of Ukraine No 1754578 Al; 1992.

[414] Polishuk SV, Babenko VV. Fin propulsive device. Patent of the Russian Federation No 2013305 Cl; 1994.

[415] Polishuk SV, Babenko VV. Fin propulsive device. Patent of the Russian Federation No 2033938 Cl; 1995.

[416] Polishuk SV, Babenko VV, Korobov VI. Fin propulsive device of the apparatus. Patent of Ukraine No 25355 A; 1998.

[417] Polishuk SV, Babenko VV, Korobov VI. Colapsible fin propulsive device. Patent of Ukraine No 25356 A; 1998.

[418] Polishuk SV, Babenko VV, Korobov VI. Fin propulsive device. Patent of Ukraine No 25621 A; 1998.

[419] Polishuk SV, Babenko VV, Korobov VI. Apparatus with fin propulsive device. Patent of Ukraine No 25646 A; 1998.

[420] Polishuk SV, Babenko VV. Apparatus with fin propulsive device. Patent of Ukraine No 25799 A; 1998.

[421] Polishuk SV, Babenko VV, Korobov VI. Underwater glider. Patent of Ukraine, N 28282 A; 2000.

[422] Polishuk SV, Babenko VV, Majster VI. Underwater apparatus with fin propulsive device. Patent of Ukraine No 41724; 2001.

[423] Polishuk SV, Babenko VV. Body of a apparatus. Patent of Ukraine, N 46638 A; 2002.

[424] Poturaev VN, Dirda VN, Krush MI. Applied mechanics of rubber. Kiev: Naukova dumka; 1975. p. 215.

[425] Povh IL. Technical hydromechanics. Leningrad: Mechanic engineering; 1976. p. 502.

[426] Povh IL, Nikulin VA. Turbulent flow of viscous liquid and slight polymers solutions in pipes. Physic hydromechanics. Donetsk: Higher School; 1977. pp. 25−34.

[427] Povh IL, Stupin AB, Aslanov PV. Peculiarities of turbulent structure of flows with admixtures of surface-active materials and polymers. Problems of turbulent. Moscow: Nauka; 1987. pp. 152−162.

[428] Prandtl L. Hydro-air-mechanics. Moscow: Foreign Lit; 1949. p. 520.

[429] Poc. Colloquium on Goertler Vortex Flows. Euromech 261. Nantes, France; 1990. p. 101.

[430] Proc. of the 11th European Drag Reduction Workin Meeting. Prague. Czech Republic; 1999. p. 86.

[431] Proc. of the Symp. on Seawater Drag Reduction. Newport, Rhode Island; 1998. p. 494.

[432] Protasov VR, Staroselskaya AG. Hydrodynamic peculiarities of fishes. Atlas. Moscow: Nauka; 1978. p. 103.

[433] Rabinovich MI, Sushik MM, Shagalov SD. Influence of an acoustic field on coherent structures in turbulent flows. Acoustics of turbulent streams. Moscow: Science; 1983. pp. 93−128.

[434] Reynolds AJ. Turbulent flows in engineering appendices. Moscow: Energy; 1979. p. 408.

[435] Resnikovsky MM, Lukomskaja AI. Mechanical tests of caoutchouc and rubbers. Moscow: Chemistry; 1964. p. 184.

[436] Repik EU, Sosedko YuP, Tronina NS. Research of flow structure in near-wall areas of a turbulent boundary layer. Near-wall turbulent flows, Part 2. Novosibirsk: Science SD; 1975. pp. 186−202.

[437] Repik EU, Sosedko JP. Spectral research quasi-ordered structures of flow in a turbulent boundary layer. Proc of the Acad Sci USSR Mech Liquid and Gas 1982;3:10−7.

[438] Rockwell D. Oscillations of impinging shear layers. AJAA J 1983;5:645−64.

[439] Romanenko EV. Foundations statistical bio-hydrodynamics. Moscow: Science; 1976. p. 167.

[440] Romanenko EV. Theories of swimming of fishes and dolphins. Moscow: Science; 1986. p. 152.

[441] Roshko A, Phishdon U. About role of transition in a near trace. Mechanica 1969;6:50−8.

[442] Rotta IK. Turbulent boundary layer in an incompressible liquid. Leningrad: Shipbuilding; 1967. p. 232.

[443] Saric WS, Reed HL, Kerschen EJ. Boundary-layer receptivity to free stream disturbances. Ann Rev Fluid Mech 2002;34:291−319.

[444] Scen SF. Calculated amplified oscillation in the plane Poiseulle and Blasius flows. J Aeron Sci 1954;21.

[445] Schilz W. Untersuchungen uber den Einfluss biegeformiger Wandschwingungen auf die Entwicklung der Stromungsgrenzschicht. Acustica 1965;15(21):27−36.

[446] Schlichting H, Gersten K. Boundary-layer theory. Eighth Revised Enlarger ed. Berlin Heidelberg New York: Springer-Verlag; 2000. p. 801.

[447] Schlichting H. Entstehung der turbulenz. Handbuch der physik, Bd. VIII/1: Stromungmechanik1. Berlin: Springer-Verlag; 1959. pp. 351−450.

[448] Schubauer GB, Skramstad HK. Laminar boundary layer oscillation and stability of laminar flow. J Aeronaut Sci 1947;14:69−81.

[449] Scott J. Physical tests of caoutchouc and rubbers. Moscow: Chemistry; 1963. p. 315.

[450] Sedov LI, Ioselevich VA, Pilipenko VN. Mechanism of lowering of friction by polymeric additives. Problems of turbulent flows. Moscow: Nauka; 1987. pp. 9−14.

[451] Sedov LI, Iosilevitch VA, Pilipenko VN. Friction and heat exchange in near-wall turbulent streams of fluids with the polymeric components. Turbulent flows. Moscow: Nauka; 1977. pp. 7−19.

[452] Sedov LI. Methods of similarity and dimensions in the mechanics. Moscow: Science; 1977. p. 438.

[453] Self-Sustaining Mechanism of Wall Turbulence. In: Panton, RL, editor. Southampton, UK and Boston, USA: Computational Mechanics Publications; 1997. p. 422.

[454] Semenov BN. Interaction of elastic border with a viscous sublayer of a turbulent boundary layer. App Mech Tech Phy 1971;12(3):58−62.

[455] Semenov BN. Analysis of deformation characteristics of visco-elastic coatings. Hydrodynamics and acoustics near-wall and free flows. Novosibirsk: Science SD; 1981. pp. 57−76.

[456] Semenov BN, Kulik VM, Lopirev VA, et al. About joint influence of small polymeric additives in a flow and a pliability surfaces on turbulent friction. Proc Acad Sciences of USSR, SD. A Series of Eng Science 1984;1(4):89−94.

[457] Semenov BN. Analysis of four types of viscoelastic coating for turbulent drag reduction. Emerging techniques in drag reduction. London: Mechanical Engineering Publications; 1996. pp. 187−206.

[458] Semenov BN. The combination of polymer, compliant wall and micro bubble drag reduction schemes. In: Proc. Int. Symp. on Seawater Drag Reduction, Newport, RI; 1998. pp. 319−326.

[459] Semenov BN. The combination of polymer compliant wall and microbubble drag reduction schemes. Proc. of the Intern. Symp. on Seawater Drag Reduction, Newport, Rhode Island, USA; 1998. pp. 269−275.

[460] Sendethky J. Elastic properties of composites. Mechanics of composite materials, Vol. 2. Moscow: World; 1978. pp. 61−101.

[461] Shankar V, Kumaran V. Stability of wall modes in fluid flow past a flexible surface. Phys Fluids 2002;14 (7):2324−40.

[462] Shen SF. Some considerations in the laminar stability of time dependent basic flow. J Aeronaut Sci 1961;15(5):394−404.

[463] Sherstjuk AN. Turbulent boundary layer. Moscow: Energy; 1974. p. 272.

[464] Shpet NG. Singularities of the form of a trunk and tail fin of whales. Bionica 1975;9:36−41.

[465] Shlanchauskas AA, Vegite NI. Model of a turbulent boundary layer made on large-scale moving. Parietal turbulent flows. Novosibirsk, vol. 2; 1975. pp. 203−208.

[466] Sickmann J. Investigation into the movement of swimming animals. ZWDI 1962;104(10):433−9.

[467] Sill BL. New flat plate turbulent velocity profiles. J Hydraulics Div 1982;108(41):1−15.

[468] Smith CR, Metzler SP. The characteristics of low-speed streaks in the near-wall region of a turbulent boundary layer. J Fluid Mech 1983;129:27−54.

[469] Smoljakov AV, Tkachenko VM. Measurement of turbulent pulsations. Leningrad: Energy; 1980. p. 164.

[470] Sodha MC, Ghatak AK. Heterogeneous optical wave guides. Moscow: Connection; 1980. p. 216.

[471] Sokolov VE. Structures of a skin cover some cetacean. Bull of Moskow Society of Nature Investigators, Dep of Biology 1955;6(60): 45−60.

[472] Stache M, Bannasch R. Bionische trageflugelender zur minimierung des induzierten widerstandes. Biona − Report 12 1988;4(2):11−224.

[473] Strickland JH, Simpson RL. The spike's frequencies get from shear layer stress fluctuations near the wall in the turbulent boundary layer. Phys Fluids 1975;18(3):306−8.

[474] Structure of a turbulent stream and the mechanism of heat exchange in channels. Moscow: Atompublishers; 1973. p. 296.

[475] Struminsky VV. About one direction of development of a problem of turbulence. Turbulent flows. Moscow: Science; 1977. pp. 20−24.

[476] Stuart JT. Hydrodynamic stability. App. Mech Rev 1965;18(7):223−31.

[477] Surkina RM. On structure and function of skin musculature of dolphins. Bionika 1971;5:81−7.

[478] Surkina RM. Location of dermal rolls on the body of common dolphin. Bionika 1971;5:87−94.

[479] Suzuki Y, Kasagi N. Turbulent drag reduction mechanism above a riblet surface. AIAA J 1994;32 (29):1781−90.

[480] Swimming and Flying in Nature. v.1. New York, London: Plenum Press; 1975. p. 421.

[481] Taneda S. Studies on wake vortices (I, II, III). Report of Res Inst for App. Mech (1954, 1955) 1956; IV:99−105.

[482] Taneda S, Honji H. The skin friction drag on flat-plates coated with flexible material. Rep on Res Ins for App. Mech 1967;15(49):121−37.

[483] Tani I. Boundary layer transition. Annu Rev Fluid Mech 1969;12(1):169–97.

[484] Tannehill JC, Anderson DA, Pletcher RH. Computation fluid mechanics and heat transfer. second ed. Taylor & Francis; 1997. p. 792.

[485] Teslo AP, Zhoga VA. Some output of the experimental investigations of turbulent flow in flexible pipes; 1973.

[486] Teslo AP, Ya FV. Influence of a pliable surface on characteristics of the turbulent stream. Hydromechanics 1973;24:45–9.

[487] Toruanik AI. Molecular aspects of a reduction of a hydrodynamic resistance by the additives of polymers. Physical Hydrodynamics. Kiev, Donetsk: Vysha Shkola; 1990. pp. 26–32.

[488] Trifonov GF. About two approaches to statement of task Prandtl for viscoelastic border Turbulent shift flows non-newton liquids. In: Mironov B. P. (ed.), Novosibirsk; 1981. pp. 82–89.

[489] Tsahalis DT. On the theory of skin friction reduction by compliant walls. AIAA Paper, No 686; 1977. p. 11.

[490] Turbulence Management and Relaminarisation. In: Liepmann NW, Narasimka R. et al., editors. Berlin, Heidelberg, New York, London: Springer-Verlag; 1988. p. 524.

[491] Turbulence: Principles and applications. In: Frost U, Moulden T, editors. World: Moscow. 1980. p. 527.

[492] Turbulent Flow. Proc of the 2nd Int. Symp. on Seawater Drag Reduction. ISSDR 2005, Korea: Busan; pp. 79–90.

[493] Turbulent shift flows. Ginevscky AS, editor. Parts 1–2; 1982–1989.

[494] Turbulent streams in the boundary layer. Moscow: Central Aero-Hydrodyn. Inst.; 1979. p. 136 (Proc. CAHI No 553).

[495] Tetjanko VA. Experimental investigation of statistical characteristics pulsating velocity at transition of the laminar boundary layer in turbulent. Novosibirsk: 43 p. (Institute of thermo-physics Academy of sciences of the USSR; 1981. pp. 70–81).

[496] Ultrasound: Small encyclopedia. Goljamin I. P. (Ed.), Moscow: Soviet Encyclopedia; 1979. p. 400.

[497] Van Dyke M. An album of liquid and gas flows. Moscow: World; 1986. p. 181.

[498] Van Heijst GIF. Spin-up phenomena in non-axisymmetric containers. J Fluid Mech 1989;206:171–91.

[499] Vanshtejn LM, Fisher MK. Experimental acknowledgement of reduction of resistance by a turbulent surface of friction on pliable walls. AIAA J 1975;13(7):144–6.

[500] Virk PS. An independent fragment model for drag reduction by dilute polymer solutions suffering degradation in recirculating and single-pass pip. flows. In: Proc. of the Int. Symp. On Seawater Drag Reduction, Newport, USA; 1998. pp. 177–186.

[501] Vogel HU. Boundary conditions for flow fields at a plane, thick, elastic wall. Fluid Dynamics Trans. In: Proceedings of the 10th Symposium on Advanced Problems and Methods in Fluid Mechanics, Rynia, Poland, September, 6–11, 1971.V, 16, pt. 2: 31–37. Wydawnictwo Naukowe PWN: Warsaw; 1971.

[502] Voytkunsky JI, Faddeev YuI, Fedyaevsky KK. Hydromechanics. Leningrad: Shipbuilding; 1982. p. 456.

[503] Voropaev GA, Babenko VV. Absorption of pulsation's energy by damping coating. Bionika 1975;9:60–9.

[504] Voropaev GA, Babenko VV. Turbulent boundary layer on an elastic surface. Hydromechanika 1978;38:71–7.

[505] Voropaev GA, Svirskaja EA. About boundary conditions for a turbulent flow on an visco-elastic border. Bionics 1982;16:47–53.

[506] Voropaev GA, Kozlov LF, Leonenko IV. Potential flow axe symmetric bodies of the complex form. Bionika 1982;16:41–4.

[507] Voropaev GA, Babenko VV. Turbulent gradient flows on compliant surface. In: Proc. USSR Workshop on hydrodynamic stability and turbulence, Springer Verlag, Novosibirsk; 1989. pp. 186–190.

[508] Voropaev GA, Rosumnyuk NV. Numerical modelling of turbulent boundary layer on deformable surface. J App. Hydromechanics 2000;2(74):23–34.

[509] Voropaev GA, Zagumenniy Y. Turbulent boundary layer over compliant and deformable surface. In: Proc. of the 2nd Intern. Symp. on Seawater Drag reduction. ISSDR 2005. Korea: Busan; 2005. pp. 447–458.

[510] Vovk VN, Kalion VA, Tsciganuk AI. Calculation of a resistance of a boundary layer in a stream of polymer of a constant concentration without pressure gradient. Hydromechanics 1984;43:44−7.

[511] Walters V. The trachipterod integument and a hypothesis on it hydrodynamic function. Copeia 1963;260−70.

[512] Wehrmann OH. Tollmien-Schlichting waves under the influence of a flexible wall. The Phys Fluids 1965;8(7):1389−90.

[513] Weinstein LM, Balasubramaian R. An electrostatically driven surface for flexible wall drag reduction studies. Drag Reduction Paper. In: Proceedings of the 2nd International Conference on Drag Reduction. Cambridge, Sept. Cambridge, 1977. pp. 67−75.

[514] Werle Fl. Hydrodynamic visualization of organized structures and turbulences in boundary layers, wakes, jets or propeller flows. In: Proc. Int. Conf. Organized structures and turbulence in fluid mechanics, Grenoble. France; 1989, pp. 9−11.

[515] White BR. Low Reynolds Number Turbulent Boundary Layer. Trans of the ASME. 1981;103:624−30.

[516] Willmarth WW, Jang CS. Wall pressure fluctuation beneath turbulent boundary layers on a flat plate and a cylinder. J Fluid Mech 1970;41(1):47−80.

[517] Willmarth WW, Sharma KK. Study of turbulent structure with hot wires smaller than the viscous length. J Fluid Mech 1984;142:121−49.

[518] Wilson D. Dj, Goldshteyn RDj. A turbulent attached to wall jet on a cylindrical surface. Theor Bases of Eng Accounts Ser D 1976;98(3):328−36.

[519] Wilson DDj. An experimental investigation of the mean velocity, temperature and turbulence fields in plane and curved now-dimensional wall jets: Coanda Effect. PhD Thesis Univ. of Minessota; 1970.

[520] Wortmann FX. Eine methode zur Beobachtung und Messung von Wasserstromung mit Tellur. Z. fur angew. Physic 1953;5(6):200−6.

[521] Wortmann FX. Visualization of transition. J Fluid Mech 1969;38(3):473−80.

[522] Wu TY. Hydromechanics of swimming of fishes and cetaceans. Adv App. Mech 1971;11:1−63.

[523] Zigmantas GP. Heat irradiation of the plate in the presence of flow disturbances. Physical hydrodynamics and thermal processes. Novosihirsk; 1980. pp. 69−75.

[524] Zedan MF, Dalton C. The Inverse Method Applied to a Body of Revolution with an Extended Favorable Pressure Gradient Forebode. Comm App. Numerical Method 1986;2:113.

[525] Zhigulev VN, Tumin AM. Occurrence of turbulence. Novosibirsk: Science SD; 1987. p. 282.

[526] Zolotov SS, Hodorkovsky YS. Peculiarities of friction resistance of a body a sword-fish. Bionika 1973;7:14−8.

[527] Yanovsky YuG, Vinogradov GV, Ivanova LI. Temperature-time visco-elastic characteristics of the polyuretans with a various microstructure of their mixes. Eng-Phys J 1984;46(6):974−81.

[528] Yampolsky DD. Contactless micrometer for measurement of micro displacement of an elastic surface. Proc Central Res Inst 1965;Part 219:97−103.

[529] Toms BA. Some observations of the flow of a linear polymer solution through straight tubes at large Reynolds number. In: Proc. 1st Intern. Congress in Rheology, vol. 2. Amsterdam: Elsevier/Noord-Holland Publishing Co.; 1948. pp. 131−145.

Index

Note: Page numbers followed by "*f*" and "*t*" refer to figures and tables, respectively.

Printed and bound by CPI Group (UK) Ltd, Croydon, CR0 4YY

14/05/2025

01871089-0002